MW00651834

World Soils Book Series

Series Editor

Alfred E. Hartemink
Department of Soil Science, FD Hole Soils Laboratory
University of Wisconsin–Madison
Madison, WI
USA

The World Soils Book Series publishes peer-reviewed books on the soils of a particular country. They include sections on soil research history, climate, geology, geomorphology, major soil types, soil maps, soil properties, soil classification, soil fertility, land use and vegetation, soil management, soils and humans, soils and industry, future soil issues. The books summarize what is known about the soils in a particular country in a concise and highly reader-friendly way. The series contains both single and multi-authored books as well as edited volumes. There is additional scope for regional studies within the series, particularly when covering large land masses (for example, The Soils of Texas, The Soils of California), however, these will be assessed on an individual basis.

More information about this series at http://www.springer.com/series/8915

Paul W. Blackburn • John B. Fisher •
William E. Dollarhide • Douglas J. Merkler •
Joseph V. Chiaretti • James G. Bockheim

The Soils of Nevada

Paul W. Blackburn
USDA Natural Resources
Conservation Service (retired)
Elko, NV, USA

John B. Fisher
USDA Natural Resources
Conservation Service (retired)
Reno, NV, USA

William E. Dollarhide
USDA Natural Resources
Conservation Service (retired)
Reno, NV, USA

Douglas J. Merkler
USDA Natural Resources
Conservation Service (retired)
Boulder City, NV, USA

Joseph V. Chiaretti
USDA Natural Resources
Conservation Service (retired)
Reno, NV, USA

James G. Bockheim
Department of Soil Science,
University of Wisconsin (retired)
Madison, WI, USA

ISSN 2211-1255 ISSN 2211-1263 (electronic)
World Soils Book Series
ISBN 978-3-030-53156-0 ISBN 978-3-030-53157-7 (eBook)
https://doi.org/10.1007/978-3-030-53157-7

This Springer imprint is published by the registered company Springer Nature Switzerland AG
The registered company address is: Gewerbestrasse 11, 6330 Cham, Switzerland

This book is dedicated to the professional Soil Scientists of the US Department of Agriculture; Natural Resources Conservation Service and US Forest Service and US Department of Interior; Bureau of Land Management that mapped soils in Nevada. We would like to recognize the Range Conservationists from these agencies that worked with the Soil Scientists during the mapping of most of Nevada. We would like to acknowledge the support of management from these agencies and other land management agencies that contributed to soil surveys. We would also like to acknowledge the support from the University of Nevada-Reno and the University of Nevada-Las Vegas personal. This book could not have been written without the support of NRCS database managers and data manipulators.

The NRCS served as the lead agency in mapping the soils in Nevada. This organization began in 1899 as the Division of Soils, became the Bureau of Soils in 1901, the Soil Conservation Service in 1935, and the Natural Resources Conservation Service in 1994.

The report that follows draws primarily on information gathered from the 16 counties of Nevada that were mapped primarily by NRCS personnel, significant area was mapped by BLM Soil Scientists, and smaller areas by the USFS and private contractors.

We thank James Komer, Nevada state Soil Scientist, and his staff for their support of this project. With assistance from Erin Hourihan, Matt Cole produced the general soil map of Nevada.

Preface

In that 84% of the land in Nevada is federally owned, this book is intended for use by employees of the Bureaus of Indian Affairs, Land Management, and Reclamation; the Departments of Defense and Energy; and the US Forest, Natural Resources Conservation, Fish and Wildlife, and National Parks Services. The book will also serve state agencies in Nevada, including the Department of Agriculture; Nevada Wildlife Service; Commissions on Rangeland Resources, Economic Development, and Mineral Resources; Department of Tourism and Cultural Affairs; and Department of Conservation and Natural Resources. The book could be used in natural resource courses at the Desert Research Institute, College of Southern Nevada, the University of Nevada-Las Vegas, the University of Nevada-Reno, the Great Basin College, and Truckee Meadows Community College, as well as universities and colleges in the adjacent states of Oregon, California, Idaho, Utah, and Arizona. The book may also be of interest to persons interested in the geography of soils, particularly in the Western Range and Irrigated land Resource Region.

Elko, USA	Paul W. Blackburn
Reno, USA	John B. Fisher
Reno, USA	William E. Dollarhide
Boulder City, USA	Douglas J. Merkler
Reno, USA	Joseph V. Chiaretti
Madison, USA	James G. Bockheim

Contents

1 **Overview** 1
  1.1 Definition of Soil 1
  1.2 Nevada History 1
  1.3 Major Soil Regions of Nevada 2
  1.4 Classification of Nevada Soils 2
  1.5 Conclusions 11
  References 12

2 **History of Soil Studies in Nevada** 13
  2.1 Introduction 13
  2.2 Soil Surveys 13
  2.3 Soil Research 13
  2.4 The State Soil 15
  2.5 Summary 15
  References 16

3 **Soil-Forming Factors** 17
  3.1 Introduction 17
  3.2 Climate 17
    3.2.1 Current Climate 17
    3.2.2 Past Climates 20
  3.3 Vegetation 20
  3.4 Relief 22
  3.5 Geologic Structure 22
  3.6 Surficial Geology 25
  3.7 Time 26
  3.8 Humans 29
  3.9 Summary 32
  References 35

4 **General Soil Regions of Nevada** 37
  4.1 Introduction 37
  4.2 Soils of the Sierra Nevada Mountains (MLRA 22A) 37
  4.3 Soils of the Malheur High Plateau (MLRA 23) 37
  4.4 Soils of the Humboldt Area (MLRA 24) 38
  4.5 Soils of the Owyhee High Plateau (MLRA 25) 38
  4.6 Soils of the Carson Basin and Mountains (MLRA 26) 39
  4.7 Soils of the Fallon-Lovelock Area (MLRA 27) 42
  4.8 Soils of the Great Salt Lake Area (MLRA 28A) 42
  4.9 Soils of the Central Nevada Basin and Range (MLRA 28B) 43
  4.10 Soils of the Southern Nevada Basin and Range (MLRA 29) 45

4.11 Soils of the Mojave Desert (MLRA 30) . . . 46
4.12 Conclusions . . . 46
Reference . . . 47

5 **Soil Geomorphology of Nevada** . . . 49
5.1 Introduction . . . 49
5.2 Fan and Remnant Terminology . . . 49
5.3 Soil Associations . . . 49
5.4 MLRA 22A—Sierra Nevada Mountains . . . 50
5.5 MLRA 23—Malheur High Plateau . . . 51
5.6 MLRA 24—Humboldt Area . . . 51
5.7 MLRA 25—Owyhee High Plateau . . . 54
5.8 MLRA 26—Carson Basin and Mountains . . . 54
5.9 MLRA 27—Fallon-Lovelock Area . . . 54
5.10 MLRA 28A—Great Salt Lake Area . . . 56
5.11 MLRA 28B—Central Nevada Basin and Range . . . 56
5.12 MLRA 29—Southern Nevada Basin and Range . . . 56
5.13 MLRA 30—Mojave Desert . . . 57
5.14 Conclusions . . . 57
References . . . 58

6 **Diagnostic Horizons and Taxonomic Structure of Nevada Soils** . . . 59
6.1 Introduction . . . 59
6.2 Diagnostic Horizons . . . 59
6.3 Orders . . . 60
6.4 Suborders . . . 60
6.5 Great Groups . . . 60
6.6 Subgroups . . . 61
6.7 Families . . . 61
6.8 Soil Series . . . 64
6.9 Depth and Drainage Classes . . . 64
6.10 Comparison of Nevada Soil Taxonomic Structure with Other States . . . 67
6.11 Key to Classifying Nevada Soils to the Great-Group Level . . . 70
6.12 Conclusions . . . 70
References . . . 74

7 **Taxonomic Soil Regions of Nevada** . . . 75
7.1 Introduction . . . 75
7.2 Torriorthents (Great Group Associations 2, 5, 6, 7, 8, 9, 14, and 15) . . . 75
7.3 Haplargids (Great Group Associations 3, 5, 6, 7, 10, 11, 13, and 18) . . . 78
7.4 Argixerolls (Great Group Associations 3, 6, 7, 9, 11, 12, 13, 16, 17, and 18) . . . 78
7.5 Haplocalcids (Great Group Associations 8, 9, and 15) . . . 81
7.6 Haplocambids (Great Group Associations 10 and 14) . . . 82
7.7 Haplodurids (Great Group Associations 8, 10, and 17) . . . 83
7.8 Argidurids (Great Group Associations 3, 17, and 18) . . . 84
7.9 Natrargids . . . 85
7.10 Petrocalcids (Great Group Association 5) . . . 86
7.11 Torripsamments . . . 87
7.12 Haploxerolls (Great Group Association 13) . . . 88
7.13 Durixerolls . . . 88
7.14 Halaquepts (Great Group Associations 2 and 14) . . . 89
7.15 Endoaquolls . . . 90
7.16 Calcixerolls . . . 90

7.17 Natridurids 91
7.18 Torrifluvents 92
7.19 Haplocryolls (Great Group Associations 4 and 12) 92
7.20 Argicryolls (Great Group Association 12) 94
7.21 Argiustolls (Great Group Association 11) 95
7.22 Calciargids 95
7.23 Calcicryolls (Great Group Association 4) 95
7.24 Cryrendolls (Great Group Association 4) 95
7.25 Haploxeralfs (Great Group Association 16) 96
7.26 Xeropsamments (Great Group Association 16) 96
7.27 Other Great Groups 96
7.28 Conclusions 96
References 96

8 Aridisols 97
8.1 Distribution 97
8.2 Properties and Processes 97
8.3 Use and Management 98
8.4 Conclusions 103
Reference 103

9 Mollisols 105
9.1 Distribution 105
9.2 Properties and Processes 105
9.3 Use and Management 109
9.4 Conclusions 109

10 Entisols 111
10.1 Distribution 111
10.2 Properties and Processes 113
10.3 Use and Management 113
10.4 Conclusions 117

11 Inceptisols 119
11.1 Distribution 119
11.2 Properties and Processes 120
11.3 Use and Management 120
11.4 Conclusions 125

12 Alfisols, Vertisols, and Andisols 127
12.1 Distribution 127
12.1.1 Alfisols 127
12.1.2 Vertisols 127
12.1.3 Andisols 127
12.2 Properties and Processes 128
12.2.1 Alfisols 128
12.2.2 Vertisols 128
12.2.3 Andisols 128
12.3 Use and Management 131
12.4 Conclusions 131

13 Soil-Forming Processes in Nevada 133
13.1 Introduction 133
13.2 Argilluviation 133
13.3 Melanization 133

13.4 Silicification 133
13.5 Calcification 134
13.6 Gleization 134
13.7 Cambisolization 134
13.8 Vertization 134
13.9 Solonization 135
13.10 Salinization 135
13.11 Gypsification 135
13.12 Andisolization 135
13.13 Paludization 135
13.14 Conclusions 135
References 135

14 Benchmark, Endemic, Rare, and Endangered Soils in Nevada 137
14.1 Introduction 137
14.2 Benchmark Soils 137
14.3 Endemic Soils 137
14.4 Rare Soils 137
14.5 Endangered Soils 138
14.6 Shallow Soils 138
14.7 Conclusions 138
References 138

15 Land Use in Nevada 139
15.1 Introduction 139
15.2 Rangeland 139
15.3 Pasture 139
15.4 Agricultural Crops 142
15.5 Forest Products 142
15.6 Wildlife Habitat 144
15.7 Development 145
15.8 Conclusions 146
References 147

16 Conclusions 149
References 150

Appendix A: Taxonomy of Nevada Soils 151

Appendix B: Thickness of Diagnostic Horizons of Nevada Soil Series with Areas in Excess of 78 km^2 211

Appendix C: Soil-Forming Factors for Soil Series with an Area Greater than 78 km^2 229

Appendix D: List of Benchmark Soils in Nevada 257

Appendix E: Soil Series in Nevada that are Endemic, Rare, and Endangered 261

Bibliography 297

Index 299

Authors' Note

With a land area of 286,380 km^2, Nevada is the seventh largest state in the US. Because it has a population of less than 3 million people, Nevada has one of the lowest state population densities in the US. With an average mean annual precipitation of 175 mm (7 in), Nevada is the driest state in the US. More than three-quarters (89%) of the state has been mapped, with the first soil survey being completed in 1909. Dr. C. F. Marbut, a historical figure in the history of soil science in the USA, played a prominent role in delineating Nevada's soils.

Nevada is divided into 10 Major Land Resource Areas and features two major deserts—the Great Basin Desert and the Mojave Desert—and over 100 north-south-trending enclosed basins separated by mountain ranges (Basin and Range Province), several of which have peaks exceeding 3,400 m (11,000 ft).

The soils of Nevada represent 7 of the 12 orders recognized globally, 29 suborders, 69 great groups, and over 1,800 soil series. Some of the classic research on the origin of duripans and petrocalcic horizons has been conducted in Nevada.

This study is the first report of the soils of Nevada and provides the first soil map of Nevada utilizing *Soil Taxonomy*.

About the Authors

Paul W. Blackburn began working as a Soil Scientist for the SCS, later to become the NRCS, in June 1976. Paul retired as an MLRA Project Office Leader in January of 2018 with a total of 41 years of service. Paul is enjoying staying close to home, doing small home remodel projects, and teaching grandsons about soils.

John B. Fisher worked over 41 years as a Soil Scientist in Nevada with NRCS before retiring as a Senior Regional Soil Scientist in 2017. John lives in Reno and enjoys reading, gardening, and playing with his grandsons.

William E. Dollarhide transferred to Nevada NRCS in 1969. He served as a Soil Scientist, Project Leader, Assistant State Soil Scientist, State Soil Scientist, and Major Land Resource leader before retiring in 2010, after 41 years of service. Bill lives in Reno with his family and enjoys gardening and playing senior softball.

Douglas J. Merkler began working as a Soil Scientist for the SCS, later to become the NRCS, in September 1978. Douglas retired as a Resource Soil Scientist for Nevada in July of 2017, just shy of 39 years of service. Douglas remains active in the Soil Science Society of America and the International Biogeographic Society, is currently teaching at Nevada State College, and has started a resource-oriented, drone-based consulting firm with his wife in retirement.

Joseph V. Chiaretti began his career as a Soil Scientist with the BLM in south-central New Mexico in 1978 and then transferred to the SCS (now NRCS) in 1979. He mapped soils on four soil survey areas in New Mexico over 19 years, serving as lead field mapper and Project Leader. Joe, along with co-authors Paul, John, and Douglas, is recognized by the NRCS as a million-acre mapper for the National Cooperative Soil Survey (NCSS) program. In January 1998, Joe transferred to Nevada and conducted quality assurance on soil survey products in the former Great Basin MLRA region until November 2008. He then served on the Soil Survey Standards staff of the National Soil Survey Center in Lincoln, Nebraska as an instructor, the principal editor of NCSS standards documents such as the National Soil Survey Handbook and the Keys to Soil Taxonomy, and the national soil classification expert. Joe retired from federal service in early 2014 and is now enjoying his hobbies of gardening, hiking, and traveling back in his adopted State of Nevada.

James G. Bockheim was Professor of Soil Science at the University of Wisconsin from 1975 until his retirement in 2015. He has conducted soil genesis and geography in many parts of the world. His interest in Nevada stemmed from its high pedodiversity. His previous books include *Pedodiversity* (2013; with J. J. Ibáñez); *Soil Geography of the USA: a Diagnostic-Horizon Approach* (2014); *Cryopedology* (2015); *The Soils of Antarctica* (2015; editor), and *The Soils of Wisconsin* (2017; with A. E. Hartemink), and *Soils of the Laurentide Great Lakes, USA and Canada.*

1 Overview

Abstract

This chapter considers the definitions of soils and briefly reviews the history, major soil regions, and classification of soils in Nevada.

1.1 Definition of Soil

There are many definitions for soil ranging from the utilitarian to a description that focuses on material. Soil has been recognized as (i) a natural body, (ii) a medium for plant growth, (iii) an ecosystem component, (iv) a vegetated water-transmitting mantle, and (iv) an archive of past climate and processes. In this book, we follow the definition given in the *Keys to Soil Taxonomy* (Soil Survey Staff 2014, p. 1) that the soil "is a natural body comprised of solids (minerals and organic matter), liquid, and gases that occurs on the land surface, occupies space, and is characterized by one or both of the following: horizons, or layers, that are distinguishable from the initial material as a result of additions, losses, transfers, and transformations of energy and matter or the ability to support rooted plants in a natural environment".

1.2 Nevada History

The Spanish meaning of Nevada is "snow-clad," in reference to the state's more than 100 mountain ranges, some of which rise above 3,400 m (11,300 feet). The highest elevation in Nevada is 4,007 m (13,147 feet) on the summit of Boundary Peak in the northern end of the White Mountains adjacent to California. With a land area of 286,380 km^2, Nevada is the seventh-largest state in the US. Nevada has only 3 million residents and is ranked 42nd in terms of population density. About one-half (45%) of Nevada's population resides in the Greater Las Vegas area (Table 1.1). The state originally was settled by Native Americans, including the Paiute, Shoshone, and Washoe tribes, and later by the Spanish (Elliott and Rowley 1987). Trappers and experienced scouts such as Joseph R. Walker passed through the region in the 1820s, and John C. Frémont and Kit Carson explored and mapped in what is now western Nevada and eastern California during expeditions in 1843–1844 and 1845. Frémont verified that all the land centered on modern-day Nevada (between Reno and Salt Lake City) was endorheic, without any outlet rivers flowing toward the sea. He is credited with coining the term "Great Basin" to describe the internal drainage of the region he explored in the mid-1840s. In 1848 Nevada, then part of the Utah Territory, was transferred to the U.S. by Mexico following the Treaty of Guadalupe Hidalgo, which ended the Mexican–American War (1846–1848). In 1859, silver was discovered near Mount Davidson in the Virginia Range and was named the Comstock Lode after the discoverer, Henry Comstock. Subsequent minerals mined in Nevada include gold, copper, lead, zinc, mercury, barite, and tungsten. The Nevada Territory gained statehood prior to the presidential election in 1864 as a new State on the side of the Union. The silver mines declined after 1874. Nevada today is officially known as the "Silver State" because of the importance of silver to its history and economy. It is also known as the "Battle Born State", because it achieved statehood during the Civil War.

Nevada has been divided into 16 counties that range in size from 373 km^2 (144 mi^2) (Carson City County) to 47,001 km^2 (18,147 mi^2) (Nye County) (Fig. 1.1). The economy of Nevada is based on cattle ranching, entertainment, government infrastructure, and tourism. From 84 to 87% of the land in Nevada is under federal jurisdiction, including the Bureau of Indian Affairs, the Bureau of Land Management, the Bureau of Reclamation, the Department of Defense, the Department of Energy, the U.S. Fish and Wildlife Service, the US Forest Service, and the National Park Service (Figs. 1.1 and 1.2).

P. W. Blackburn et al., *The Soils of Nevada*, World Soils Book Series,
https://doi.org/10.1007/978-3-030-53157-7_1

Table 1.1 Nevada's most populated cities

City	Population	Key industries
Las Vegas*	648,000	tourism
Henderson	308,000	tourism
Reno	245,000	tourism
North Las Vegas	249,000	tourism
Sparks	98,000	warehousing
Carson City	55,000	government

1.3 Major Soil Regions of Nevada

To date, there has not been a book describing Nevada's soils, nor has there been a detailed soil map of the state. However, about 89% of the state has been mapped (Fig. 1.3), and these data are available via the Web Soil Survey. The state is divided into 10 Major Land Resource Areas that reflect differences in physiography, geology, climate, water, soils, biological resources, and land use (Fig. 1.4; Table 1.2). The state also is divided into 5 Level III ecoregions, each containing from 2 to 25 Level IV ecoregions (Fig. 1.5; Table 1.3). The ecoregions approach is comparable to the MLRA approach and is based on differences in geology, physiography, vegetation, climate, soils, land use, wildlife, and hydrology.

1.4 Classification of Nevada Soils

The first soil survey in Nevada was conducted in 1909 for the Fallon Area that included parts of Churchill and Lyon Counties. The soil map legend included six soil series, nine soil types, and one land unit (Fig. 1.6). The Las Vegas Area was mapped in 1926 and the Moapa Valley Area in 1928. The former survey showed the distribution of nine soil series; 34 soil types differentiated on the basis of texture, relief, soil thickness, and degree of erosion; and two land types.

From 1905 until 1955 classification of soils in Nevada was limited to soil series and parent material texture, although a national soil classification scheme had been available since 1928 (Marbut 1927; Baldwin et al. 1938). Virtually no soil mapping occurred between 1929 and 1958. The Soil Survey of the Lovelock Area in 1966 was the last to use of the 1938 soil classification system. The Soil Survey of the Las Vegas and Eldorado Valleys Area in 1957 was the first in Nevada to use the *Seventh Approximation* (Soil Survey Staff 1960) for classifying soils of Nevada; this document was the precursor to the first edition of *Soil Taxonomy: A Basic System of Soil Classification for Making and Interpreting Soil Surveys* (Soil Survey Staff 1975).

All soils in Nevada are now classified using *Soil Taxonomy* (Soil Survey Staff, 1999) and is used throughout this book. The *Keys to Soil Taxonomy* (Soil Survey Staff, 2014) is an abridged companion document that incorporates all the amendments that have been approved to the system since publication of the second edition of soil taxonomy in 1999, in a form that can be used easily in a field setting. *Soil Taxonomy* is a hierarchical classification system that classifies soils based on the properties as contained of diagnostic surface and subsurface horizons. For classification purposes, the upper limit of the soil is defined as the boundary between the soil (including organic horizons) and the air above it. The lower limit is arbitrarily set at 200 cm. The definition of the classes (taxa) is quantitative and uses well-described methods of analysis for the diagnostic properties. The assumed genesis of the soil is not used in the system and the soil is classified "as it is" using morphometric observations in the field coupled with laboratory analysis and other data. The nomenclature in soil taxonomy is mostly derived from Greek and Latin sources, as is done for the classification of plants and animals.

Soil taxonomy classifies soils, from broadest to narrowest levels, into orders, suborders, great groups, subgroups, and families. Families occur in one or more soil series. Soil associations (composed of soil series and miscellaneous areas) and consociations (composed of a single major component) constitute the primary soil map units.

There are eight diagnostic surface horizons (epipedons) defined in soil taxonomy and six of them occur in Nevada: anthropoic, folistic, histic, mollic, umbric, and ochric (Table 1.4). The anthropic epipedon forms in parent materials are strongly influenced by human activity, may contain artifacts, and occur in soils of urban areas and some gardens. The folistic and histic epipedons consist primarily of organic soil materials. The folistic epipedon occurs in well-drained

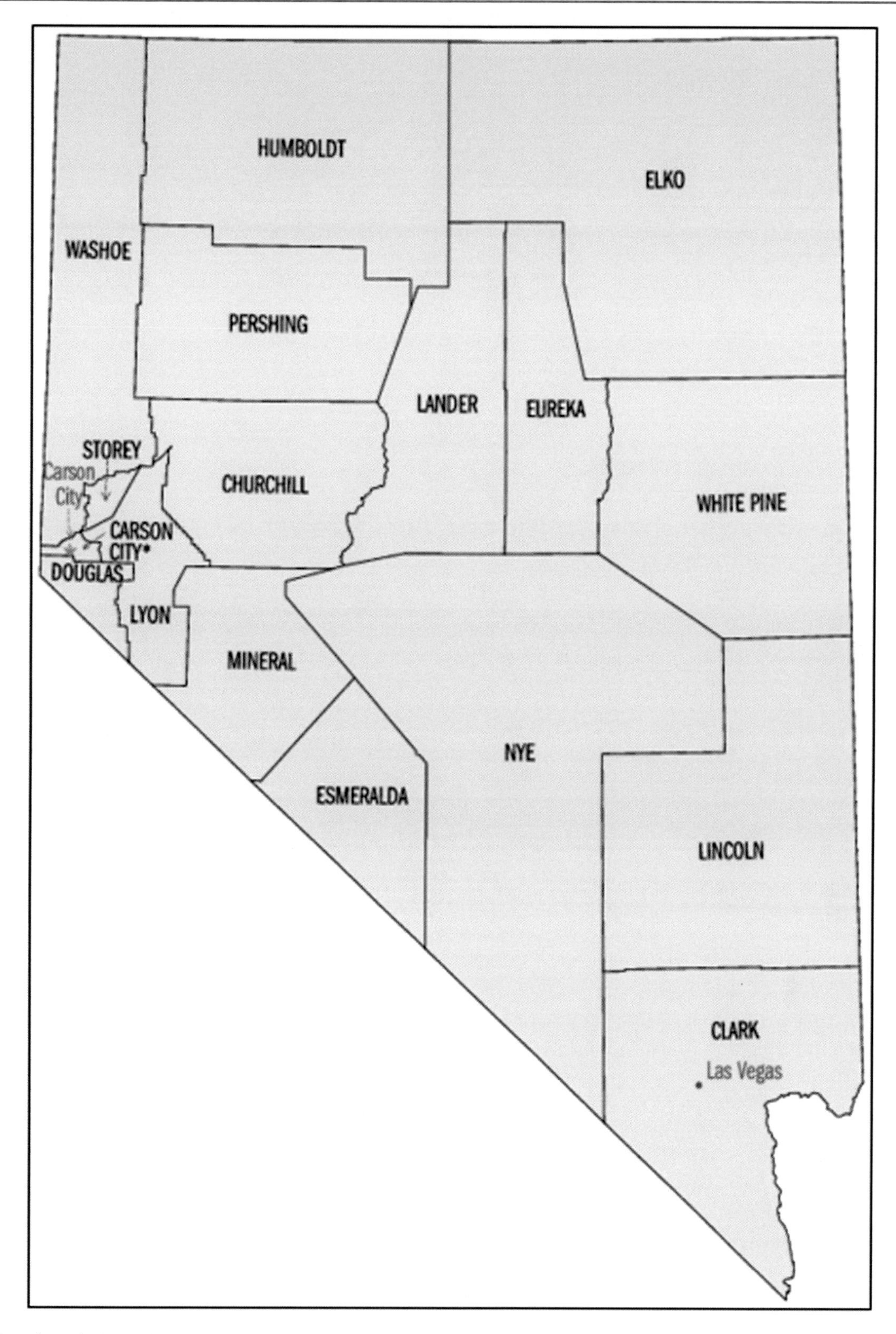

Fig. 1.1 Location of 16 counties in Nevada

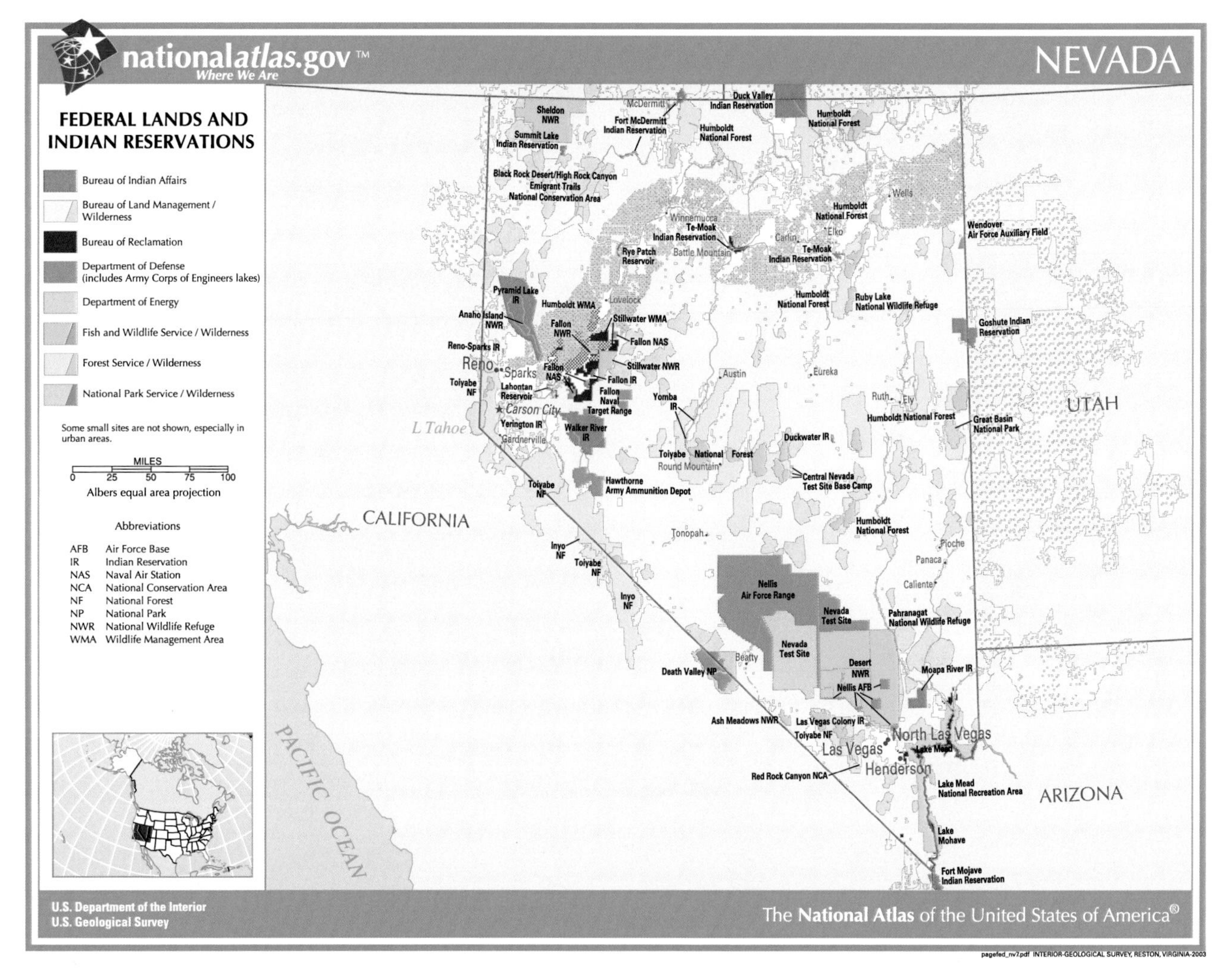

Fig. 1.2 Location of federal lands and Native American lands in Nevada. Legend: red = Bureau of Indian Affairs; yellow = Bureau of Land Management/Wilderness; purple = Bureau of Reclamation; blue = Department of Defense & Army Corps of Engineers; pink = Department of Energy; orange = US Fish & Wildlife Service; green = US Forest Service; light blue = National Park Service/ Wilderness (*Source* nationalatlas.com). (*Source* Nationalatlas.gov)

soils and is not saturated for prolonged periods during the year, while the histic epipedon contains primarily organic materials and is saturated for prolonged periods during the year. The mollic and umbric epipedons occur in mineral soils and are thick, dark-colored, and enriched in organic matter. The mollic epipedon is enriched in base cations, such as calcium, magnesium, and potassium, while the umbric epipedon contains low amounts of these cations. The ochric epipedon is thin, commonly light-colored, and often low in organic matter content.

Ten of the 20 diagnostic subsurface horizons identified in soil taxonomy are present in the soils of Nevada (Table 1.3). The albic horizon is composed of materials from which clay and/or free iron oxides have been removed by eluviation to a degree that primary sand and silt particles impart a light color to the horizon. The argillic horizon is enriched in clay that has moved down the profile from percolating water. The natric horizon is a type of argillic horizon, which shows evidence of clay illuviation that has been accelerated by the dispersive properties of sodium. The cambic horizon shows minimal development other than soil structure and color. The calcic horizon features a significant accumulation of secondary calcium carbonates; this horizon must be 15 cm or more thick, have a 5% or more $CaCO_3$ equivalent, and not be cemented. The petrocalcic horizon is at least 10 cm thick, has sufficient $CaCO_3$ that it is cemented, and roots are unable to penetrate it except along vertical fractures with a horizontal spacing of 10 cm or more. The gypsic horizon

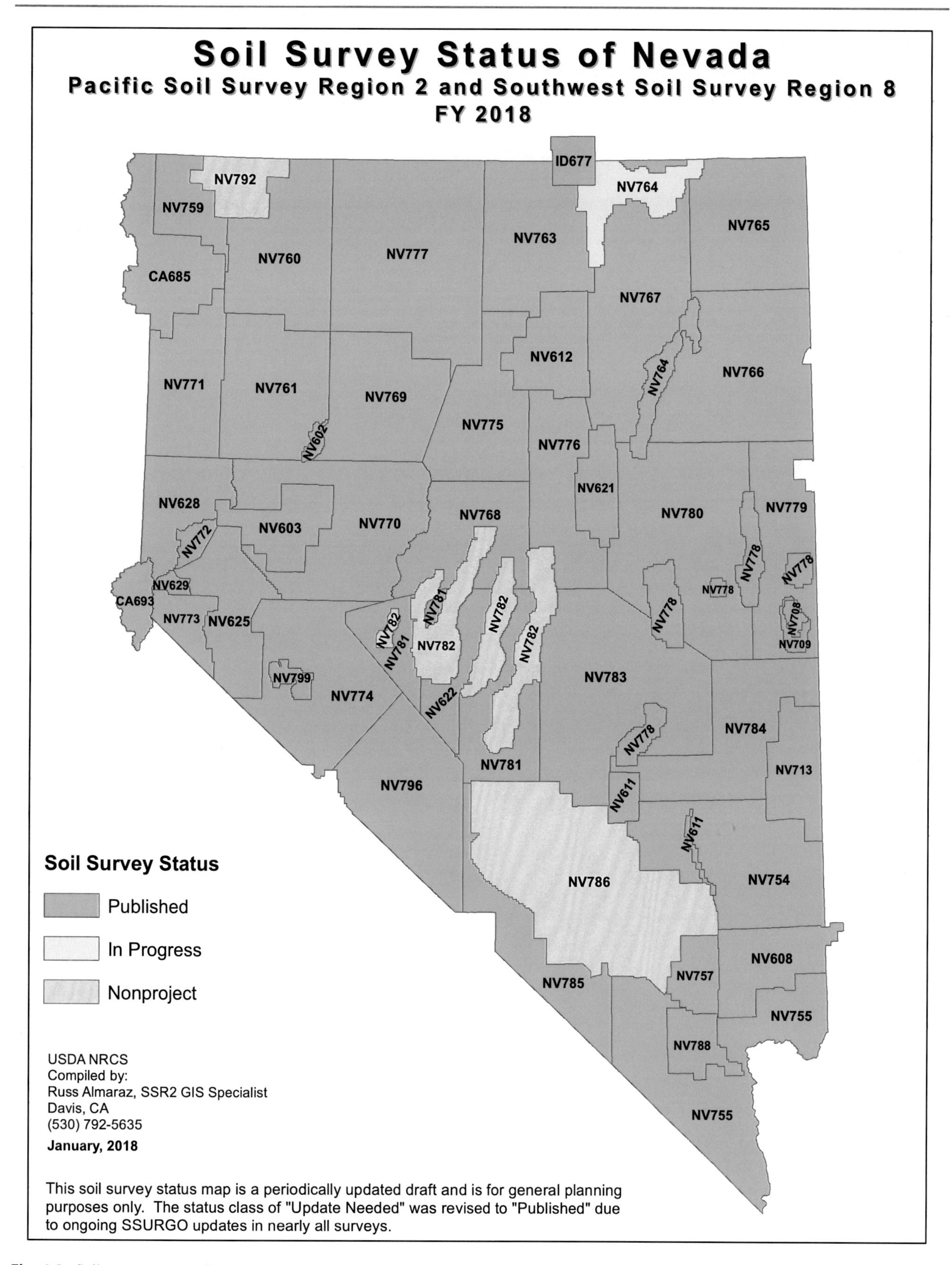

Fig. 1.3 Soil survey status of Nevada, 2018 (*Source* USDA, NRCS)

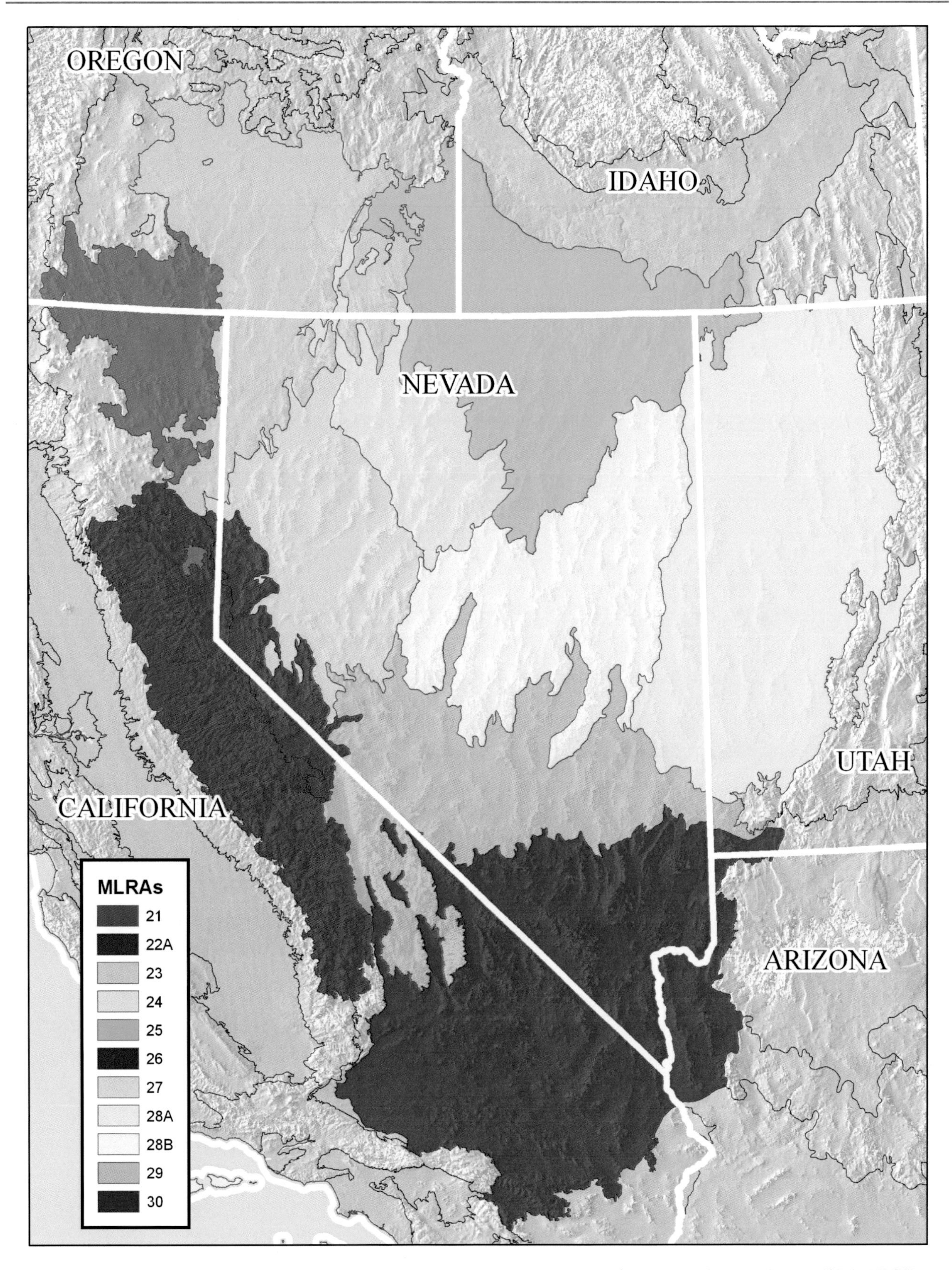

Fig. 1.4 Distribution of Major Land Resource Regions (MLRAs) in Nevada (see Table 1.2 for names and areas) (*Source* USDA, NRCS)

Table 1.2 Major Land Resource Areas represented in Nevada

MLRA	Description	Area (km^2)	%
22A	Sierra Nevada Mountains	975	0.3
23	Malheur High Plateau	14,830	5.1
24	Humboldt Area	30,884	10.7
25	Owyhee High Plateau	38,979	13.5
26	Carson Basin and Mountains	12,668	4.4
27	Fallon-Lovelock Area	32,560	11.3
28A	Great Salt Lake Area	15,248	5.3
28B	Central Nevada Basin and Range	61,035	21.1
29	Southern Nevada Basin and Range	49,742	17.2
30	Mojave Desert	31,744	11.0
		288,665	100.0

features a significant accumulation of gypsum; this horizon must be at least 15 cm thick, have 5% or more gypsum by weight, and not be cemented. The petrogypsic horizon is cemented so that roots are unable to penetrate it except in vertical fractures with a horizontal spacing of 10 cm or more.

A duripan is a silica-cemented subsurface horizon with or without auxiliary cementing agents. The duripan is cemented in more than 50% of the volume of some horizon and shows evidence of the accumulation of opal or other forms of silica, such as laminar caps, coatings, lenses, partly filled interstices, bridges between sand-sized grains, or coatings on rock fragments. Less than 50% of the volume of air-dry fragments slakes in 1 M HCl, but more than 50% slakes in concentrated KOH or NaOH. Roots can only penetrate the pan in vertical fractures with a horizontal spacing of 10 cm or more. Secondary calcium carbonate is often an accessory cementing agent in duripans. Duripans are very common in the soils of central and northern Nevada due in large part to the presence of soluble volcanic glass in soil parent materials. A salic horizon features the accumulation of salts that are more soluble than gypsum in cold water. This horizon must be 15 cm or more thick, have an electrical conductivity for 90 consecutive days or more of 30 dS/m or more in the water extracted from a saturated past; and have the product of the EC and thickness be 900 or more. Photographs of the subsurface horizons are given in chapters describing soils in each of the orders represented in Nevada. The formative elements used in constructing soil names are given in Appendix A.

Soil orders are defined primarily on the basis of diagnostic soil characteristics and diagnostic surface and subsurface horizons. Eight of the 12 orders in soil taxonomy occur in Nevada: Alfisols, Andisols, Aridisols, Entisols, Histosols, Inceptisols, Mollisols, and Vertisols (Table 1.5). Alfisols are base-enriched forest soils with an argillic horizon. Andisols are soils derived from amorphous clays. Aridisols are dry soils that feature the accumulation of clay or some salts. Entisols are very poorly developed recent soils that may have only an anthropic or ochric epipedon. Histosols are organic soils that are rare in Nevada due to the dominantly arid climate. Inceptisols are juvenile soils that contain an epipedon and either a cambic horizon, a salic horizon, or a high exchangeable sodium percentage. Mollisols are dark-colored, base-enriched grassland soils. Vertisols are derived from abundant swelling clays that lead to cracks and slickensides.

Suborders are distinguished by soil climate for five of the seven orders occurring in Nevada: the Alfisols, Andisols, Inceptisols, Mollisols, and Vertisols. Soil parent materials are used to differentiate Vertisols and among Entisols; and the amount of clay or types of salts are used to differentiate among Aridisols. There are 29 suborders of soils in Nevada. Great groups are distinguished from a variety of soil characteristics; there are 69 great groups of soils in Nevada.

On an area basis, 52% of the soil series in Nevada are Aridisols, followed by Mollisols (23%) and Entisols (22%); Inceptisols (1.9%), Alfisols (0.5%), Vertisols (0.4%), and Andisols (< 0.1%) comprise the remaining soil orders (Fig. 1.7). A list of all soil series recognized in Nevada, along with their areas and classification, is given in Appendix A. The thickness of diagnostic horizons of major Nevada soil series is given in Appendix B. Soil-forming factors of major Nevada soil series are listed in Appendix C.

Fig. 1.5 Ecoregions of Nevada (*Source* EPA, NRCS, USGS)

Table 1.3 Legend for Fig. 1.5, Ecoregions of Nevada

5	**Sierra Nevada**
5a	Mid-Elevation Sierra Nevada
5b	High Elevation Sierra Nevada
13	**Central Basin and Range**
13a	Salt Deserts
13b	Shadscale-Dominated Saline Basins
13c	Sagebrush Basins and Slopes
13d	Woodland- and Shrub-Covered Low Mountains
13e	High Elevation Carbonate Mountains
13 g	Wetlands
13 h	Lahontan and Tonopah Playas
13j	Lahontan Salt Shrub Basin
13 k	Lahontan Sagebrush Slopes
13 l	Lahontan Uplands
13 m	Upper Humboldt Plains
13n	Mid-Elevation Ruby Mountains
13o	High Elevation Ruby Mountains
13p	Carbonate Sagebrush Valleys
13q	Carbonate Woodland Zone
13r	Central Nevada High Valleys
13 s	Central Nevada Mid-Slope Woodland and Brushland
13t	Central Nevada Bald Mountains
13u	Tonopah Basin
13v	Tonopah Sagebrush Foothills
13w	Tonopah Uplands
13x	Sierra Nevada-Influenced Ranges
13y	Sierra Nevada-Influenced High Elevation Mountains
13z	Upper Lahontan Basin
13aa	Sierra Nevada-Influenced Semiarid Hills and Basins
14	**Mojave Basin and Range**
14a	Creosote Bush-Dominated Basins 14b Arid Footslopes
14c	Mojave Mountain Woodland and Shrubland
14d	Mojave High Elevation Mountains
14e	Arid Valleys and Canyonlands
14f	Mojave Playas
14 g	Amargosa Desert
22	**Arizona/New Mexico Plateau**
22d	Middle Elevation Mountains
80	**Northern Basin and Range**
80a	Dissected High Lava Plateau
80b	Semiarid Hills and Low Mountains 80d Pluvial Lake Basins
80e	High Desert Wetlands 80 g High Lava Plains
80j	Semiarid Uplands
80 k	Partly Forested Mountains
80 l	Salt Shrub Valleys

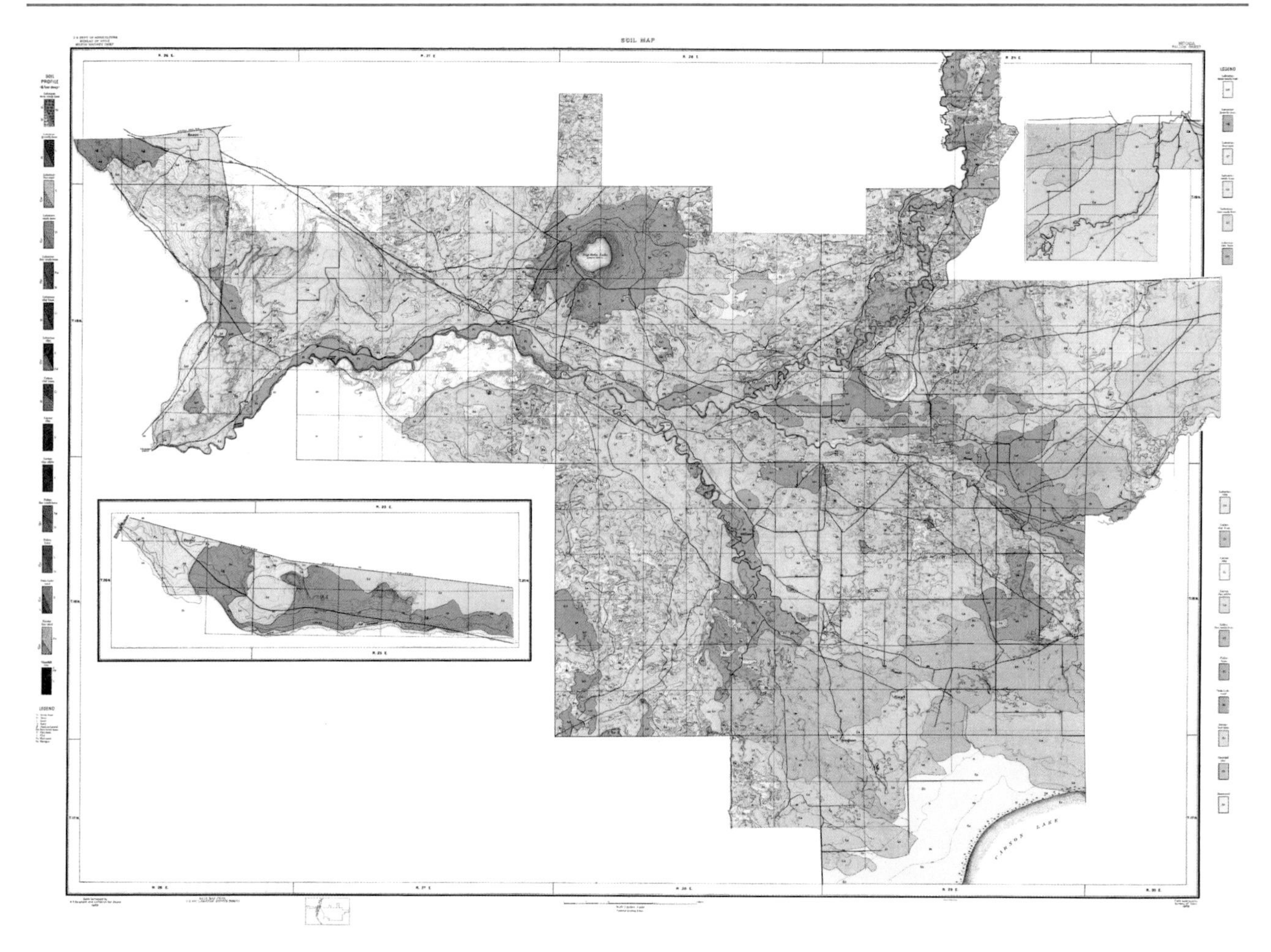

Fig. 1.6 Soil Survey of Fallon Area, Nevada in 1909. The map shows six soil series, nine soil types, and one land unit (*Source* Strahorn and Van Duyne, 1909)

Table 1.4 Definitions of diagnostic horizons present in Nevada soils[a]

Diagnostic surface horizons (epipedons)	
Anthropic	at least 25 cm thick; formed in human-altered or human-transported materials; contains artifacts, midden material, or anthraquic conditions
Folistic	at least 15 cm thick; organic matter content 16% or more; saturated for less than 30 days in normal years
Histic	greater than 20 cm thick; organic matter content 16% or more; saturated for more than 30 days in normal years unless artificially drained
Mollic	at least 18 cm thick; dark-colored; organic C 0.6% or more; base saturation 50% or more
Umbric	at least 18 cm thick; dark-colored; organic C 0.6% or more; base saturation less than 50%
Ochric	an altered horizon that fails to meet the requirements of other epipedons; lacks rock structure or finely stratified fresh sediments; includes underlying eluvial horizons such as albic
Diagnostic subsurface horizons	
Albic	a light-colored eluvial horizon 1 cm or more in thickness; composed of albic materials
Argillic	an illuvial horizon that gives evidence of translocation of clay, based on the ratio of that in the clay-enriched horizon to an overlying eluvial horizon, the presence of clay films (argillans)
Calcic	a non-cemented horizon of secondary carbonate accumulation with at least 15% calcium carbonate equivalent in a horizon, that is at least 15-cm thick, and has at least 5% more carbonate than an underlying layer
Cambic	an altered horizon that shows color and/or structure development, is at least 15-cm thick, and has a texture of very fine sand, loamy very fine sand, or finer

(continued)

Table 1.4 (continued)

Duripan	a horizon that is cemented in more than 50% of the volume by opaline silica; air-dry fragments do not slake in water of HCL but do slake in hot concentrated KOH; restricts rooting of plants except in vertical cracks that have a horizontal spacing of 10 cm or more
Gypsic	a non-cemented horizon of secondary gypsum accumulation that is 15 cm or more in thickness, contains 5% or more gypsum (1% visible), and the product of horizon thickness in cm and gypsum percentage by weight is 150 or more
Natric	meets the requirements of an argillic horizon but also has prismatic, columnar, or blocky structure, an exchangeable sodium percentage of 15 or more, or a sodium adsorption ratio of 13 or more
Petrocalcic	a carbonate-cemented horizon that is 10 cm or more thick and restricts rooting of plants except in vertical cracks that have a horizontal spacing of 10 cm or more
Petrogypsic	a gypsum-cemented horizon that is 5 mm or more thick with at least 40% gypsum by weight and restricts rooting of plants except in vertical cracks that have a horizontal spacing of 10 cm or more
Salic	a horizon of accumulation of salts more soluble than gypsum that is at least 15 cm thick; the electrical conductivity (EC) is at least 30dS/m for 90 consecutive days; the product of horizon thickness in cm and the EC is 900 or more

[a]Revised from Buol et al. (2011)

Table 1.5 Simplified key to soil orders in Nevada[a]

Histosols	Soils that do not have andic soil properties in 60% or more of the upper 60 cm and have organic soil materials in two-thirds or more of the total thickness
Andisols	Other soils with andic soil properties in 60% or more of the upper 60 cm
Vertisols	Other soils with a layer 25 cm or more thick containing either slickensides or wedge-shaped peds, have more than 30% clay in all horizons between depths of 18 and 50 cm or a root-limiting layer if shallower, and have cracks that open and close periodically
Aridisols	Other soils with either an aridic soil moisture regime and some diagnostic surface and subsurface horizons or a salic horizon accompanied by both saturation within 100 cm of the soil surface and dryness in some part of the soil moisture control section during normal years
Mollisols	Other soils with a mollic epipedon and a base saturation (by ammonium acetate at pH 7) of 50% or more in all depths above 180 cm
Alfisols	Other soils with an argillic or natric horizon
Inceptisols	Other soils with an umbric or mollic epipedon, or a cambic horizon, or a salic horizon, or a high exchangeable sodium percentage which decreases with increasing depth accompanied by groundwater within 100 cm of the soil surface
Entisols	Other soils

[a]Revised from Buol et al. (2011)

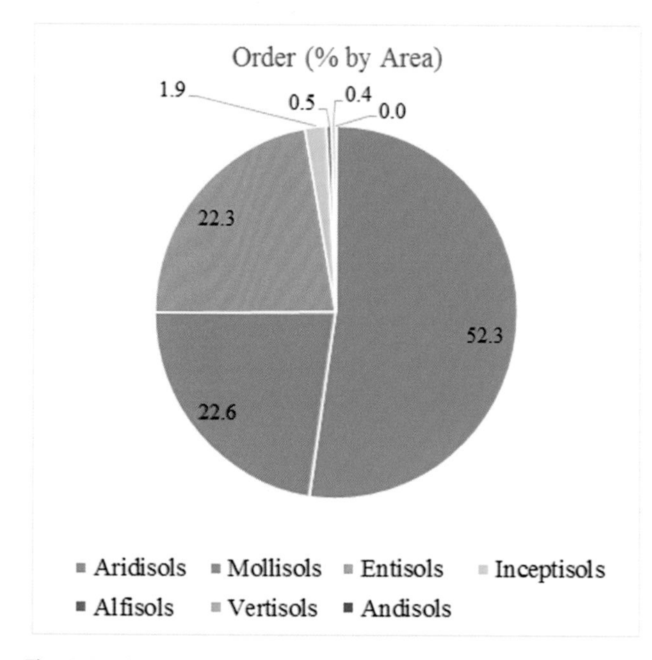

Fig. 1.7 Distribution of soil orders in Nevada (*Source* NRCS database)

1.5 Conclusions

Soil is viewed in this book as a natural body comprised of solids (minerals and organic matter), liquid, and gases that occurs on the land surface, occupies space, and is characterized by one or both of the following: horizons, or layers, that are distinguishable from the initial material as a result of additions, losses, transfers, and transformations of energy and matter *or* the ability to supported rooted plants in a natural environment (Soil Survey Staff 1999).

About 89% of Nevada has been mapped. The state is divided into 10 Major Land Resource Areas and 43 ecoregions that reflect differences in physiography, geology, climate, water, soils, biological resources, and land use. The first soil survey in Nevada was conducted in 1909 in the Fallon Area. This book uses *Soil Taxonomy* (Soil Survey Staff 1999) and the *Keys to Soil Taxonomy* (Soil Survey Staff 2014), which classifies soils into 12 orders based on the presence of diagnostic horizons and characteristics, as well as soil climate (e.g., Aridisols); suborders are delineated primarily on soil climate but also on the types of parent material and salts; other categories in this hierarchical systems are great groups, subgroups, families, and soil series. Nevada has 6 of the 8 diagnostic surface horizons identified in soil taxonomy and 10 of the 20 diagnostic subsurface horizons. Aridisols (52%) are the dominant soil order in Nevada, followed by Mollisols (23%), Entisols (22%), Inceptisols (1.9%), Alfisols (0.5%), Vertisols (0.4%), and Andisols (0.1%).

References

Baldwin M, Kellogg CE, Thorp J (1938) Soil classification. Soils and Men. U.S. Dep. Agric. Yearbook. U.S. Govt. Print., Washington, DC, pp 979–1001

Buol SW, Southard RJ, Graham RC, McDaniel PA (2011) Soil Genesis and Classification. 6th edit. Wiley & Blackwell, West Sussex, UK

Elliott RR, Rowley WD (1987) History of Nevada. 2nd edit. University of Nebraska Press, Lincoln, NE

Marbut CF (1927) In: Glinka KD (ed.) The Great Soil Groups of the World and their Development. Edwards Bros., Ann Arbor, MI

Soil Survey Staff (2014) Keys to Soil Taxonomy. 12th edit. U.S. Dept Agric Natural Resour Conserv Serv, Linoln, NE

Soil Survey Staff (1960) Soil Classification, a Comprehensive System, 7th Approximation. Soil Conserv. Serv., U.S. Dept. Agric. U.S. Govt. Printing Office, Washington, DC. 265 pp

Soil Survey Staff (1975) Soil Taxonomy, a Basic System of Soil Classification for Making and Interpreting Soil Surveys. Soil Conserv. Serv., U.S. Dept. Agric. Handbook, vol. 436. U.S. Govt. Printing Office, Washington, DC. 754 pp

Soil Survey Staff (1999) Soil Taxonomy: a Basic System of Soil Classification for Making and Interpreting Soil Surveys. 2nd edit. Agric. Handbook, vol. 436. U.S. Govt. Print. Office, Washington, DC. 869 pp

Strahorn AT, Van Duyne C (1909) Soil Survey of the Fallon Area, Nevada. In: Field Operations of the Bureau of Soils, Washington, D.C., pp. 1477–1515

History of Soil Studies in Nevada

2

Abstract

Nevada has a rich and long history of soil investigations, which began with the Soil Survey of the Fallon Area in 1909 by the Bureau of Soils and continued with the mapping of all or portions of all 16 counties in Nevada under the leadership of the Soil Conservation Service and later the Natural Resources Conservation Service. This chapter traces the history of soil surveys, briefly describes the nature of soil research, and highlights the Orovada soil series, Nevada's official state soil.

2.1 Introduction

Nevada has a rich and long history of soil investigations that began with the Soil Survey of the Fallon Area in 1909 by the Bureau of Soils (Fig. 1.6) and continued with the mapping of all or portions of all 16 counties in Nevada generally at a scale of 1:24,000 (Table 2.1). This work has been complemented by soil research by university and NRCS, BLM, and USFS personnel over the past 60 years.

2.2 Soil Surveys

Following the initial soil survey of the Fallon Area in 1909, only two other surveys, the Las Vegas Area and the Moapa Valley) were completed in Nevada prior to 1967 (Fig. 2.1). Maps completed between 1909 and 1928 showed a limited number of soil series, the primary soil map unit. In the early 1960s, great soil groups identified by Baldwin et al. (1938) were depicted on soil maps in Nevada. Beginning in the early 1960s, the *Seventh Approximation*, a precursor to *Soil Taxonomy* (Soil Survey Staff 1975, 1999) was employed, and from 1975 to the present, soil taxonomy has been used exclusively throughout the U.S. From the mid-1970s to 2008, the cumulative number of soil surveys in Nevada increased exponentially.

The only areas that have not been mapped in Nevada are the Sheldon Antelope Refuge Area that includes parts of Humboldt and Washoe Counties (NV792); the north part of the Humboldt National Forest that includes part of Elko County (NV764); the central part of the Toiyabe National Forest that includes parts of Eureka, Lander, and Nye Counties (NV782); Death Valley National Park Area that includes parts of Esmeralda and Nye Counties (CA793), and the Energy and Defense Area that includes parts of Clark, Lincoln, and Nye Counties (NV786) (Fig. 1.3).

In 1909, only four soil series had been identified in Nevada, including the Carson, Churchill, Lahontan, and Soda Lake series. From 1909 through 1965, only 126 soil series had been mapped in the state, but accelerated soil survey beginning in the late 1970s under the leadership of Ed Naphan (Blackburn 2000) led to an exponential increase in the number of established soil series (Fig. 2.2). By 2018, over 1,800 soil series had been identified in Nevada, of which 1,309 are found only in Nevada.

2.3 Soil Research

Nevada has benefited from considerable soil research by university and NRCS investigators over the past 60 years. Key soil research efforts have focused on the nature, properties, and genesis of desert soils (Harper 1957; Springer 1958; Morrison 1964; Gile et al. 1966; Gardner 1972; Chadwick 1984; Richmond 1986); the role of eolian deposition on polygenesis of soils (Marion 1989; Chadwick et al. 1995; Reheis and Kiel 1995; Blank et al. 1996; Rehei 2003; Ernst et al. 2003); the timing of fluctuations of pluvial lakes, such as Pleistocene Lake Lahontan, in response to climate change (Chadwick and Davis 1990; Adams and Wesnouskey 1999); the origin of petrocalcic horizons (Amundson et al. 1989a, b; Brock and Buck 2005, 2009; Robins et al. 2012); the origin of duripans and durinodes (Chadwick et al. 1987, 1989); the origin of argillic and natric horizons in

P. W. Blackburn et al., *The Soils of Nevada*, World Soils Book Series,
https://doi.org/10.1007/978-3-030-53157-7_2

Table 2.1 Availability of printed soil surveys by county in Nevada

County	Area (km^2)	Scale	Soil survey area
Carson City	373	1:24,000	Carson City Area (1979); Tahoe Basin Area (1974; 2007)
Churchill	12766	1:24,000	Churchill County Area (2001); Fallon-Fernley Area (1975); Lovelock Area (1965)
Clark	20489	1:24,000	Clark County Area (2006); Las Vegas Valley Area (1985); Las Vegas and Eldorado Valley Areas (1967); Virgin River Area (1980)
Douglas	1839	1:24,000	Carson Valley Area (1971); Douglas County Area (1984); Tahoe Basin Area (1974; 2007)
Elko	44501	1:24,000	Diamond Valley Area (1980); Duck Valley Indian Reservation (1986); Elko County, Central Pt. (1997); Elko County, Northeastern Pt. (1998); Elko County, Southeast Pt. (2002); Northwest Elko Area (1997); Tuscarora Mountain Area (1980)
Esmeralda	9295	1:63,360	Esmeralda County Area (1991)
Eureka Humboldt	10816 25014	1:63,360 1:24,000	Diamond Valley Area (1980); Eureka County Area (1989); Northwest Elko Area (1997); Tuscarora Mountain Area (1980); Western White Pine County Area (1998) Humboldt County, East Pt. (2002); Humboldt County, West Pt. (2003)
Lander	14229	1:24,000	Lander County, North Pt. (1992); Lander County, South Pt. (1991); Tuscarora Mountain Area (1980)
Lincoln	27545	1:24,000	Lincoln County, North Pt. (2007); Lincoln County, South Pt. (2000); Meadow Valley Area (1976); Pahranagat-Penoyer Area (1968); Virgin River Area (1980)
Lyon	5164	1:20,000	Churchill County Area (2001); Lyon County Area (1984)
Mineral	9731		Mineral County Area (1991)
Nye	47001	1:24,000	Big Smoky Valley Area (1980); Nye County, Northeast Pt. (2002); Nye County, Northwest Pt. (2002); Nye County, Southwest Pt. (2004)
Pershing	15563	1:24,000	Lovelock Area (1965); Pershing County, East Pt. (1994); Pershing County, West Pt. (1998)
Storey	684	1:24,000	Fallon-Fernley Area (1975); Storey County Area (1990)
Washoe	16426	1:24,000	Fallon-Fernley Area (1975); Surprise Valley-Home Camp Area (1974; 2006); Tahoe Basin Area (1974; 2007); Washoe County, Central Pt. (1997); Washoe County, North Pt. (1999); Washoe County, South Pt. (1983)
White Pine	22991	1:24,000	Diamond Valley Area (1980); Great Basin National Park (2009); Western White Pine County; Area (1998)

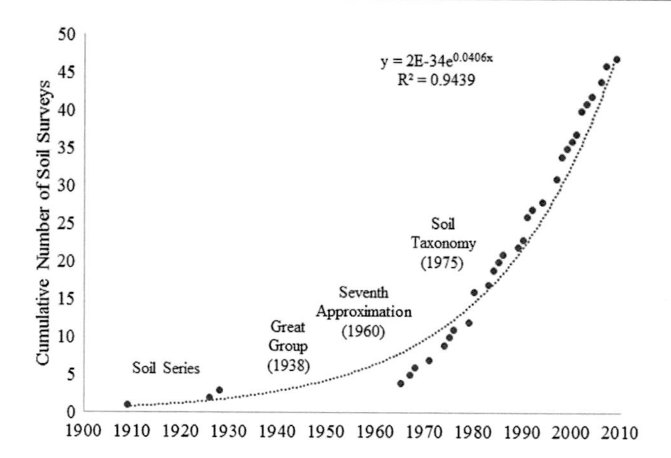

Fig. 2.1 History of soil mapping in Nevada (*Source* NRCS database)

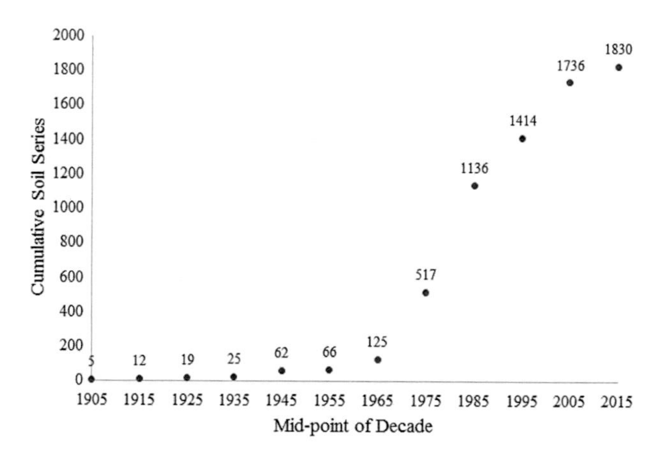

Fig. 2.2 Cumulative number of soil series in Nevada by midpoint of decade (*Source* NRCS database)

desert soils (Nettleton et al. 1975; Alexander and Nettleton 1977; Elliott and Drohan 2009); the importance of soil chronosequences in studying soil evolution (Harden et al. 1991a, b; Reheis et al. 1992); and soil-plant relationships (Nettleton et al. 1986). The Land soil series, a Typic Aquisalids, was the type locality for the salic horizon.

2.4 The State Soil

In 2001, the Orovada soil series was approved as the state soil of Nevada (Fig. 2.3). Orovada became recognized as the state soil thanks to the efforts of the Orovada Elementary School students, with soil scientist Paul Blackburn taking the proposal to the state legislators and the governor. Orovada soils are extensive in northern Nevada, comprising over 1,473 km^2. The soil is common on semiarid rangeland with sagebrush-grass plant communities. The Orovada soil is arable when irrigated and is considered prime farmland. Alfalfa for hay and seed, winter wheat, and barley, and grass for hay and pasture are the principal crops grown on these soils.

2.5 Summary

The first soil survey in Nevada was completed in 1909. Since then the entire state has been mapped except for the Sheldon Antelope Refuge Area, portions of the Humboldt and Toiyabe National Forests, Death Valley National Park Area that includes parts of Esmeralda and Nye Counties (CA793), and

Fig. 2.3 Nevada's state soil, the Orovada soil series. Photo by John Fisher

the Energy and Defense area in southern Nevada. Soil mapping increased exponentially from 1965 to 2008. Soil surveys in Nevada reflect historical changes in soil map units in the U. S., progressing from a limited number of soil series prior to 1938, mapping of zonal great soil groups until 1960, and the use of soil taxonomy thereafter. The number of soil series recognized in Nevada increased markedly from 1965 to 2008. Nearly three-quarters (72%) of the soil series recognized in Nevada occur only in the state.

References

Adams KD, Wesnousky SG (1999) The Lake Lahontan highstand: age, surficial characteristics, soil development, and regional shoreline correlation. Geomorphology 30:357–392

Alexander EB, Nettleton WD (1977) Post-Mazama Natrargids in Dixie Valley. Nevada Soil Sci Soc. Am. J. 41:1210–1212

Amundson RG, Chadwick OA, Sowers JM, Doner HE (1989a) Soil evolution along an altitudinal transect in the eastern Mojave Desert of Nevada, U.S.A. Geoderma 43:349–371

Amundson RG, Chadwick OA, Sowers JM, Doner HE (1989b) The stable isotope chemistry of pedogenic carbonates at Kyle Canyon. Nevada Soil Sci Soc Am J. 53:201–210

Baldwin M, Kellogg CE, Thorp J (1938) Soil classification. Soils and Men. U.S. Dep. Agric. Yearbook. U.S. Govt. Print, Washington, DC, pp 979–1001

Blackburn PW (2000) A history of soil survey in Nevada: a compilation of short stories commemorating the 100th anniversary of the soil survey program. http://www.nv.nrcs.usda.gov. Accessed 04 Nov 2018

Blank RR, Young JA, Lugaski T (1996) Pedogenesis on talus slopes, the Buckskin Range, Nevada, USA. Geoderma 71:121–142

Brock AL, Buck BJ (2005) A new formation process for calcic pendants from Pahranagat Valley, Nevada, USA, and implication for dating Quaternary landforms. Quat Res 63:359–367

Brock AL, Buck BJ (2009) Polygenetic development of the Mormon Mesa, NV petrocalcic horizons: geomorphic and paleoenvironmental interpretations. CATENA 77:65–75

Chadwick OA, Davis JO (1990) Soil-forming intervals caused by eolian sediment pulses in the Lahontan Basin, northwestern Nevada. Geology 18:243–246

Chadwick OA, Hendricks DM, Nettleton WD (1987) Silica in duric soils: I. A depositional model. Soil Sci Soc Am J 51:975–982

Chadwick OA, Hendricks DM, Nettleton WD (1989) Silicification of Holocene soils, in northern Monitor Valley. Nevada. Soil Sci. Soc. Am. J. 53:158–164

Chadwick OA, Nettleton WD, Staidl GJ (1995) Soil polygenesis as a function of Quaternary climate change, northern Great Basin, USA. Geoderma 68:1–26

Chadwick, O.A. 1984. A soil chronosequence at Terrace Creek: studies of late Quaternary tectonism in Dixie Valley, Nevada. US Geol. Surv. Open-File Rep. 84–90

Elliott PE, Drohan PJ (2009) Clay accumulation and argillic-horizon development as influenced by aeolian deposition vs. local parent material on quartzite and limestone-derived alluvial fans. Geoderma 151:98–108

Ernst WG, Van de Ven CM, Lyon RJP (2003) Relationships among vegetation, climatic zonation, soil, and bedrock in the central White-Inyo Range, eastern California: a ground-based and remote-sensing study. GSA Bull. 115:1583–1597

Gardner R (1972) Origin of the Mormon Mesa caliche, Clark County. Nevada. GSA Bull. 83:143–156

Gile LH, Peterson FF, Grossman RB (1966) Morphological and genetic sequences of carbonate accumulation in desert soils. Soil Sci 101:347–360

Harden JW, Taylor EM, Hill C, Mark RK, McFadden LD, Reheis MC, Sowers JM, Wells SG (1991a) Rates of soil development from four chronosequences in the southern Great Basin. Quat Res 35:383–399

Harden, J.W., Taylor, E.M., McFadden, L.D. and Reheis, M.C. 1991b. Calcic, gypsic, and siliceous soil chronosequences in arid and semi-arid environments. In: Nettleton, W.D (Ed.) Occurrence, Characteristics, and Genesis of Carbonate, Gypsum, and Silica Accumulations in Soils. SSSA Spec. Pap., No. 26, pp. I–16

Harper WG (1957) Morphology and genesis of Calcisols. Soil Sci Soc Am 21:420–424

Marion GM (1989) Correlation between long-term pedogenic $CaCO_3$ formation rate and modern precipitation in deserts of the American Southwest. Quat. Res. 32:291–295

Morrison RB (1964) Lake Lahontan: Geology of the southern Carson Desert, Nevada. US Geol Surv. Prof. Pap. 401, 156 pp

Nettleton WD, Brasher BR, Spencer EL, Langan LN, Peterson FF (1986) Differentiation of closely related Xerolls that support different sagebrush plant communities in Nevada. Soil Sci Soc Am J 50:1277–1280

Nettleton WD, Witty JE, Nelson RE, Hawley JW (1975) Genesis of argillic horizons in soils of desert areas of the southwestern United States. Soil Sci Soc Am J 39:919–926

Reheis MC, Kihl R (1995) Dust deposition in southern Nevada and California, 1984-1989: relations to climate, source area, and source lithology. J Geophys Res 100(D5):8893–8918

Reheis MC, Sowers JM, Taylor EM, McFadden LD, Harden JW (1992) Morphology and genesis of carbonate soils on the Kyle Canyon fan, Nevada, USA. Geoderma 52:303–342

Reheis, M.C. 2003. Dust deposition in Nevada, California, and Utah, 1984–2002. USGS Open-file Rep. 03-138. 11 pp

Richmond GM (1986) Stratigraphy and correlation of glacial deposits of the Rocky Mountains, the Colorado Plateau and the ranges of the Great Basin. Quat. Sci. Rev. 5:99–127

Robins CR, Brock-Hon AL, Buck BJ (2012) Conceptual mineral genesis models for calcic pendants and petrocalcic horizons. Nevada. Soil Sci. Soc. Am. J. 76:1887–1903

Soil Survey Staff (1975) Soil Taxonomy, a Basic System of Soil Classification for Making and Interpreting Soil Surveys. Soil Conserv. Serv., U.S. Dept. Agric. Handbook, vol. 436. U.S. Govt. Printing Office, Washington, DC. 754 pp

Soil Survey Staff (1999) Soil Taxonomy: a Basic System of Soil Classification for Making and Interpreting Soil Surveys. 2nd edit. Agric. Handbook, vol. 436. U.S. Govt. Print. Office, Washington, DC. 869 pp

Springer ME (1958) Desert pavement and vesicular layer of some soils of the desert of the Lahontan Basin. Nevada Soil Sci Soc Am Proc 22:63–66

Soil-Forming Factors

3

Abstract

The soils of Nevada result from the interplay of five soil-forming factors: climate, organisms, relief, parent material, and time. This chapter considers the effects of present and past climate, vegetation, relief, geologic structure, surficial geology, time, and humans on soil formation in the state.

3.1 Introduction

The expression of a soil results from five factors operating collectively: climate, organisms, relief, parent material, and time. The factors interact and cause a range of soil processes (e.g., illuviation) that result in a diversity of soil properties (e.g., high clay content in the subsoil). Human activities cause soil changes and are often considered a sixth factor. Following the "Russian school of soil science," Kellogg (1930) illustrated the importance of geology, climate, and native vegetation on the distribution of soils in the USA. The following is a review of the role of soil-forming factors in the development of Nevada soils.

3.2 Climate

3.2.1 Current Climate

Nevada lies on the eastern, lee side of the Sierra Nevada Range, a massive mountain barrier that strongly influences the climate of the state. The prevailing winds are from the west; as the warm, moist air from the Pacific Ocean ascends the western slopes of the Sierra, the air cools and condensation takes places, causing most of the moisture to fall as precipitation. As the air descends the eastern slope, it is warmed by compression, and very little precipitation occurs. The seasonal distribution of precipitation in Nevada varies by region. A winter precipitation maximum occurs in the western and south-central portions of the state; a spring maximum occurs in the central and northeastern sections; and a summer maximum occurs primarily in the eastern portion where thunderstorms are most frequent.

With a state average of 175 mm/yr (7 in/yr), Nevada is the driest state in the U.S. The mean annual precipitation ranges from 114 mm/yr (4.5 in/yr) in southern and eastern Nevada to 890 mm/yr (35 in/yr) or more in the higher mountain ranges (Fig. 3.1). The mean annual snowfall in Nevada ranges from 0 cm in Boulder City to 178 cm (70 in) or more in the mountains, especially in the White Mountains, Snake range, Toquima Range, Spring Mountains, Schell Creek Range, Toiyabe Range, and White Pine Range.

The mean annual temperature is dependent on latitude, elevation, and aspect. The mean annual temperature varies from 5.5 °C (42 °F) at Mountain City in northernmost Nevada to 24 °C (73 °F) at Laughlin in southern Nevada. The lowest temperatures occur on the Malheur and Owyhee High Plateaus, the Sierra Nevada Range, and north-south trending mountain ranges throughout the state. The warmest temperatures occur in the Mojave Desert south of latitude 37 °N. There is strong surface heating during the day and rapid nighttime cooling, largely because of the dry air. This results in wide diurnal temperature changes that often range from 16 to 20 °C (30–35 °F). Extreme temperatures in Nevada have ranged from 49 °C (120 °F) to −46 °C (−50 °F).

The soils of Nevada are classified into four soil moisture regimes and five soil temperature regimes that vary according to latitude and elevation and reflect the large climate diversity in Nevada. The soil moisture regimes (SMR) are ranked from greatest to least in terms of area: aridic (torric), xeric, aquic, and ustic (Fig. 3.2). With the *aridic* and *torric* SMRs, the soil is dry in all parts for more than half of the cumulative days per year when the soil temperature at a depth of 50 cm (20 in) is above 5 °C (41 °F) and moist in some or all parts for less than 90 consecutive days when he soil temperature at a depth of 50 cm is above 8 °C (46 °F). With the *xeric* SMR, the winters

P. W. Blackburn et al., *The Soils of Nevada*, World Soils Book Series,
https://doi.org/10.1007/978-3-030-53157-7_3

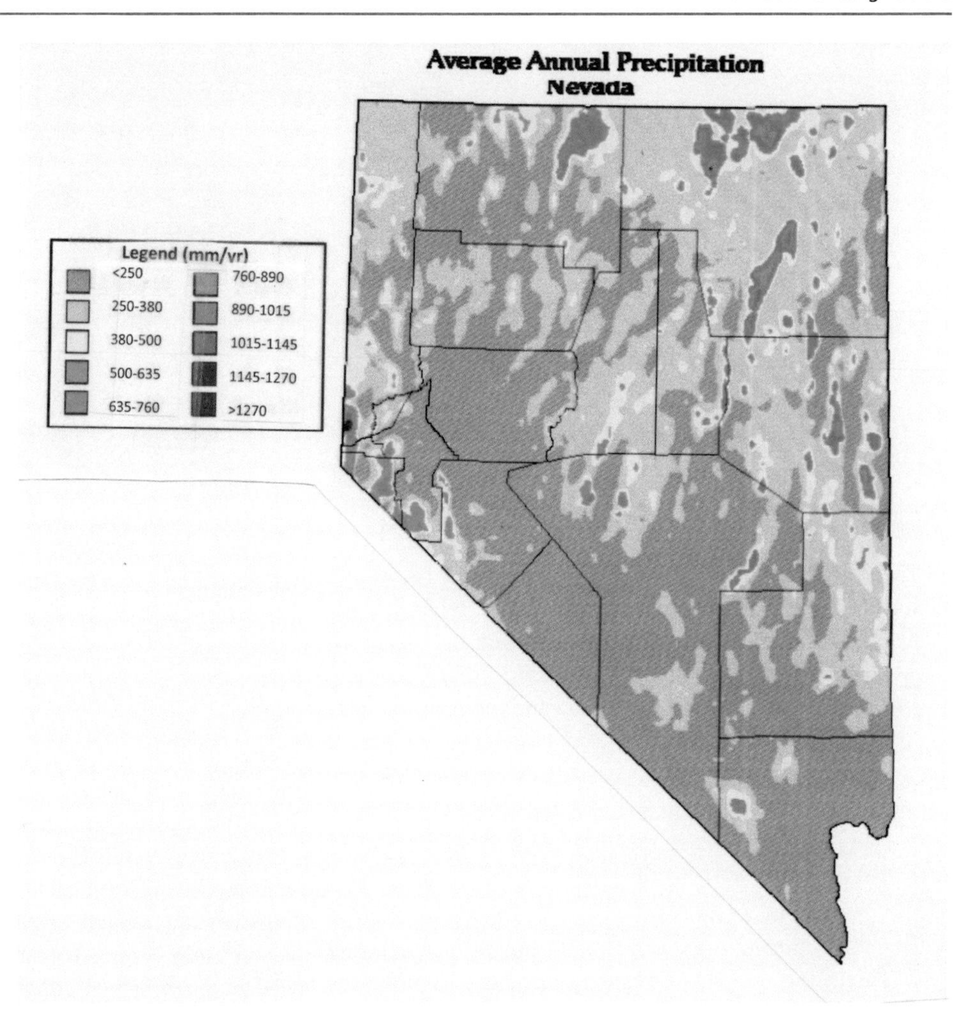

Fig. 3.1 Mean annual precipitation in Nevada (*Source* Oregon State University, Prism Group and Western Region Climate Center)

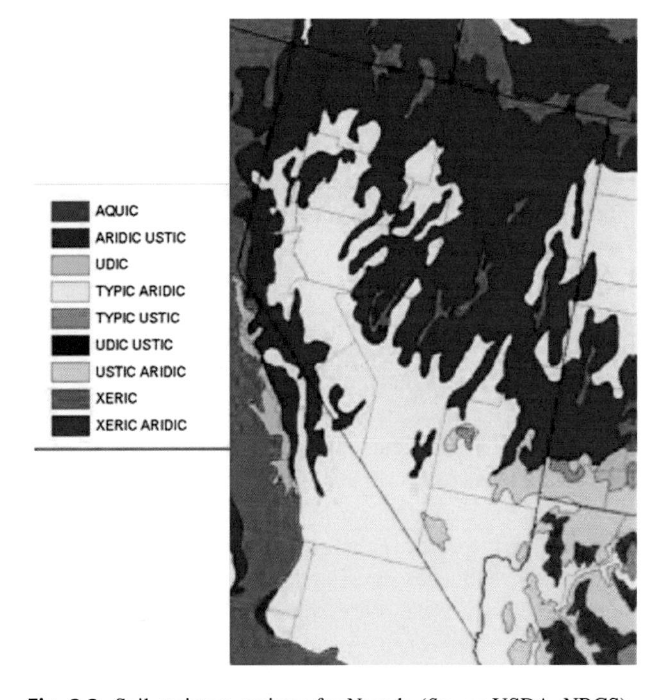

Fig. 3.2 Soil moisture regimes for Nevada (*Source* USDA, NRCS)

are moist and cool, and the summers are warm and dry; more specifically, the soil is dry in all parts for 45 or more consecutive days in the 4 months following the summer solstice and moist in all parts for 45 or more consecutive days in the 4 months following the winter solstice. The xeric SMR occurs primarily in the mountains. With the *aquic* SMR, the soil is saturated with moisture from above (episaturation) or below (endosaturation) for varying times of the year. With the *ustic* SMR, soil moisture is limited, but is generally available when conditions are suitable for plant growth; more specifically, the soil is dry in some or all parts for 90 or more cumulative days in normal years, but it is not dry in all parts for more than half of the cumulative days when the soil temperature at a depth of 50 cm is higher than 5 °C.

The soil temperature regimes (STR) are ranked in Nevada from coldest to warmest: cryic, frigid, mesic, thermic, and hyperthermic (Fig. 3.3). In the *mesic* STR, the mean annual soil temperature at a depth of 50 cm ranges between 8 and 15 °C (46 and 59 °F). In the *thermic* and *hyperthermic* STRs, the mean annual soil temperature is between 15 and 22 °C (59 and 72 °F) and >22 °C (>72 °F), respectively (Soil Survey Staff 2014). In the *frigid* STR, the mean annual soil

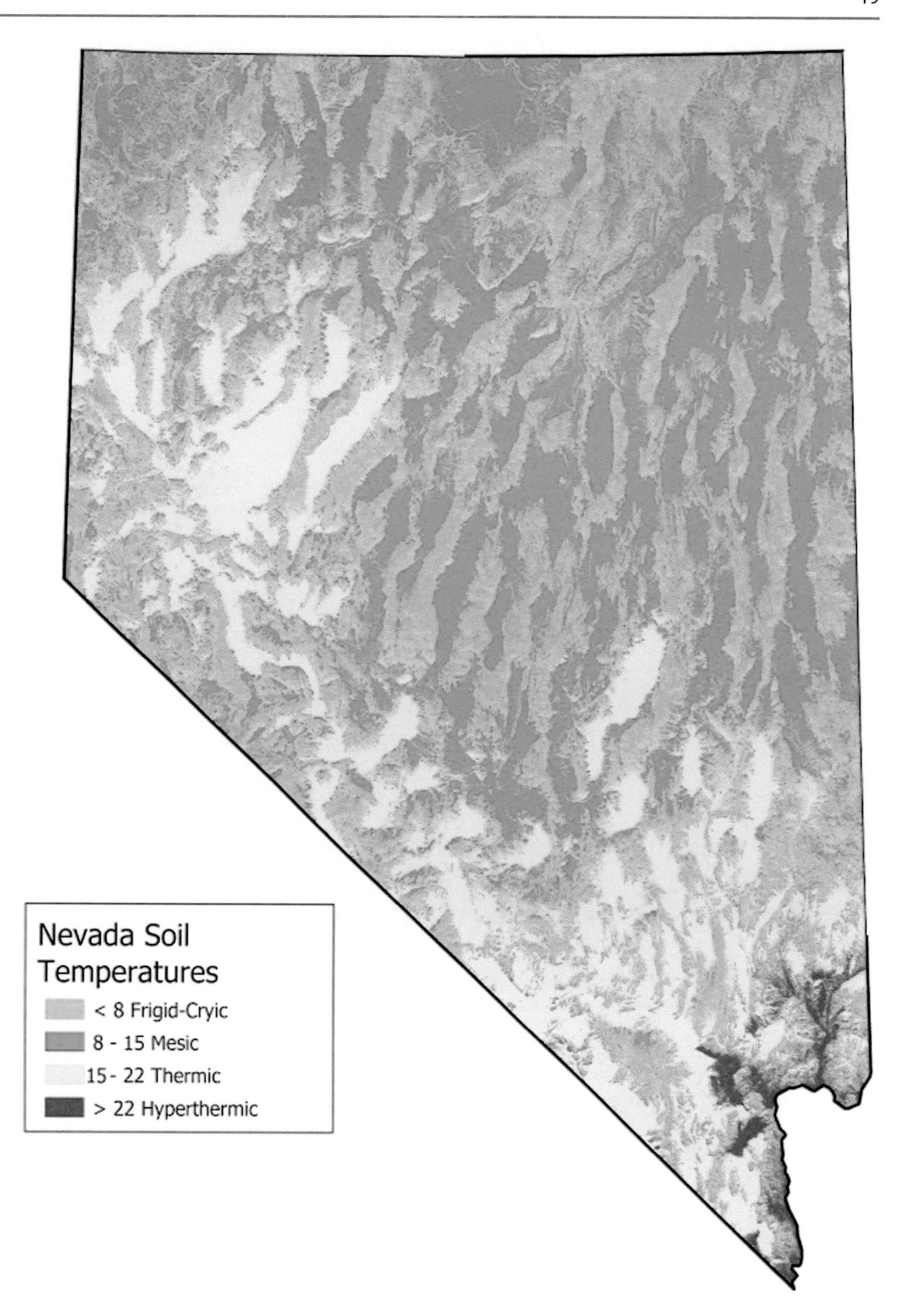

Fig. 3.3 Soil temperature regimes in Nevada. Prepared by Cathy Wilson, US Fish, and Wildlife Service

temperature is between 0 and 8 °C (32 and 46 °F) and the difference between the mean summer (June, July, and August) and mean winter (December, January, and February) soil temperatures is 6 °C (11 °F) or more at a depth of 50 cm. The cryic STR is comparable to the frigid, except that the summer-winter differences range between 0 and 6 °C (0 and 11 °F). From ranges in elevation for soil series found only in Nevada, we established approximate elevation boundaries for soil temperature regimes based on the following aspects (north- vs. south-facing slopes): hyperthermic 150–750 m, thermic 750–1,100 or 1,500 m, mesic 1,100 or 1,500–1,600 or 1,800 m, frigid 1,600 or 1,800–2,000 or 2,300 m, and cryic 2,000 or 2,300–4,000 m (Table 3.1).

According to data from the National Center for Environmental Information, the average annual temperature for Nevada has increased to 2 °C since the early 1950s (Fig. 3.4).

Table 3.1 Effects of elevation, aspect, and latitude on soil temperature regimes in Nevada

Soil-temperature regime	Aspect	Elevation (m)	
		Lower	Upper
Latitude 42°00′–37°48′N			
Cryic	N	2000	4000
Cryic	S	2300	4000
Frigid	N	1600	2000
Frigid	S	1800	2300
Mesic	N	1100	1600
Mesic	S	1500	1800
Latitude 37°45′–35°00′N			
Thermic	N	147	1100
	S	147	1500
Hyperthermic	N, S	147	750

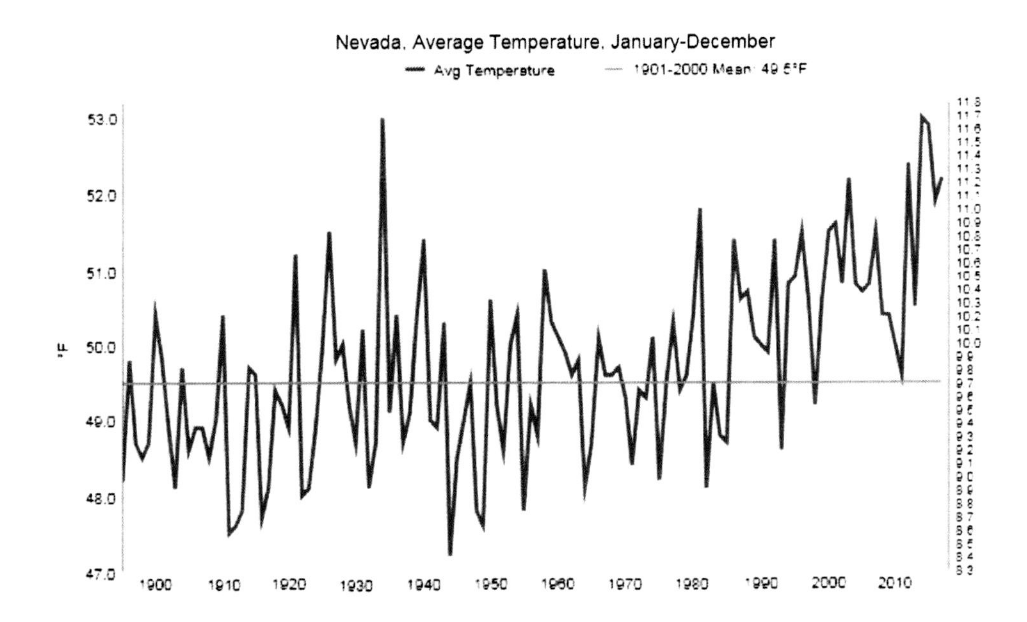

Fig. 3.4 Average annual air temperature for Nevada from 1902 to 2014 (National Centers for Environmental Information 2018)

3.2.2 Past Climates

The climate of Nevada has varied considerably over geologic time, including the past 2.5 million years. In the recent past, pluvial lakes formed in Nevada. A pluvial lake is a body of water that accumulates in a basin during periods of greater moisture availability. Pluvial lakes typically are formed in closed (endorheic) basins. In Nevada, Pleistocene pluvial lakes included ancestral Lake Lahontan in the northwest, Lake Bonneville in the east, and lakes in the Mojave Region of southern Nevada (Fig. 3.5). The lakes ranged from 12,000 to 52,000 km^2 in the area at their maxima and were up to 300 m (1,000 ft) deeper than at present, as evidenced by strandlines, or former shorelines (Table 3.2; Fig. 3.6). These lakes occupied basins in Nevada many times, beginning over 1 million years ago; by 8.7 kyr BP, the lakes became dry except for smaller terminal lakes such as Pyramid Lake and Walker Lake in Nevada and the Great Salt Lake in Utah.

3.3 Vegetation

The natural vegetation of Nevada can be divided broadly into seven biomes along an elevational gradient (Tueller 1975). From the lowest to the highest elevation, these include salt desert, sagebrush grassland, pinyon-juniper woodland, mountain shrub, montane forest, subalpine forest, and alpine tundra (Table 3.3).

Salt desert features Bailey's greasewood (*Sarcobatus baileyi*), black greasewood (*S. vermiculatus*) (Fig. 3.7 upper), fourwing saltbush (*Atriplex canescens*) (Fig. 3.7

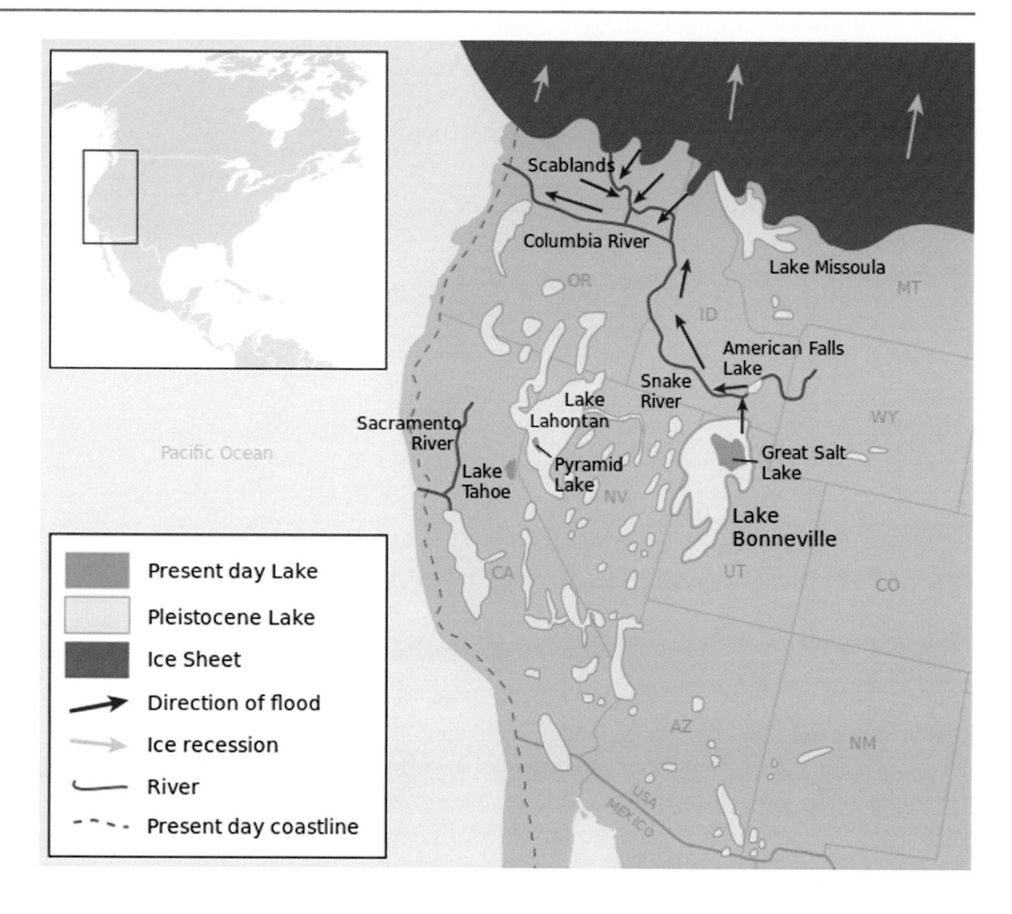

Fig. 3.5 Extent of Bonneville Flood and Pleistocene lakes in northwestern USA around 17.5 kyr BP. (*Source* Wikipedia, based on the map by DeGrey et al. Idaho State Univ.)

Table 3.2 Pluvial lakes in Nevada

Max. depth					
Pluvial lake	Current lake(s)	Time interval (kyr BP)	Area (km^2)	(m)	Key Reference
Lahontan	Pyramid, Walker	9–12.7	22,000	270	Thompson et al. (1986)
Bonneville	Great Salt, Utah, Sevier	13–30	52,000	300	Eardley et al. (1973)
Mojave	Soda, Silver	13.8–25	12,000	245	Kirby et al. (2018)

middle), shadshale saltbush (*Atriplex confertifolia*) (Fig. 3.7 lower), saltgrass (*Distichlis stricta*), alkali sacaton (*Sporobolus airoides*), and other species. Salt desert is most prevalent in low, saline valleys of the Lahontan Basin in western Nevada, the Bonneville (Great Salt Lake) Basin in eastern Nevada, and the Mojave Basin in southern Nevada.

Sagebrush-grasslands can be divided into southern and northern components (Tueller 1989). The northern component is dominated by Wyoming big sagebrush (*Artemisia tridentata* ssp. *wyomingensis*) (Fig. 3.8 upper), little sagebrush (*A. arbuscula* var. *arbuscula*), and black sagebrush (*A. nova*), which are characteristics of valley bottoms in the Great Basin of northern Nevada. The southern component of sagebrush-grassland features creosote bush (*Larrea tridentata*) (Fig. 3.8 lower), burrobush (*Ambrosia dumosa*), saltbush (*Atriplex polycarpa*), Joshua tree (*Yucca brevifolia*), spiny hopsage (*Grayia spinosa*), and blackbrush (*Coleogyne ramosissima*). Northern sagebrush-grasslands often are invaded by cheatgrass (*Bromus tectorum*), especially where overgrazing and fires have occurred. Table 3.4 is a modification of a summary of the general soil, topographic, climatic, and landscape position of major sagebrush species in Nevada modified from Schultz and McAdoo (2002). There are distinct differences in soils, climate, and landforms of the seven predominant sagebrush-grassland communities in Nevada.

Pinyon-juniper woodlands occur above the canyon floors and are dominated by single-leaf pinyon (*Pinus monophylla*) and Utah juniper (*Juniperus osteosperma*), commonly with an understory of short (bunch) grasses (Fig. 3.9 upper). At

Fig. 3.6 Pluvial Lake Lahontan shorelines near Sand Mountain, Nevada (*Source* F. Hopson)

higher elevations and on moister sites, mountain shrub develops; this vegetation type is dominated by curl-leaf mountain mahogany (*Cercocarpus ledifolius*) (Fig. 3.9 lower) and mountain big sagebrush (*Artemisia tridentata* ssp. *vaseyana*). In Nevada, the development of a substantial forest cover requires at least 300 mm (12 in) of annual precipitation. Ponderosa pine (*Pinus ponderosa*) and quaking aspen (*Populus tremuloides*) are common in the montane forest (Fig. 3.10). The subalpine forest contains Engelmann spruce (*Picea engelmannii*), and subalpine fir (*Abies lasiocarpa*) (Fig. 3.11 upper) bristlecone pine (P. longaeva) (Fig. 3.11 middle), and limber pine (P. flexilis). At elevations of 3,000–3,600 m (10,000-12,000 feet), the subalpine forest transitions to alpine tundra, which contains grasses, sedges, forbs, and cushion plants (Fig. 3.11 lower).

About one-half (51%) of the soil series in Nevada are covered with Northern Desert Shrub-Grassland, followed by Southern Desert Shrub-Grassland (23%), and Alkali Desert shrubs (14%) (Fig. 3.12). The remaining vegetation zones cover about 12% of the soil series.

The U.S. EPA has produced a 1:350,000-scale map of 43 ecoregions in Nevada (Fig. 1.5; Table 1.3) that enables an understanding of the vegetation of the state.

3.4 Relief

The relief map of Nevada is very striking. The Columbia Plateau extends into the north-central portion of the state and the Sierra Nevada Range into the west-central region, but the majority of the state is included in the Basin and Range physiographic province. The basins and ranges are oriented roughly NNE to SSW (Fig. 3.13). The highest elevations are found in five extensive and high-elevation mountain ranges, including the Schell Creek Range, the Toiyabe Range, the Ruby Mountains, the Shoshone Mountains, and the Snake Range. Elevations in these ranges exceed 3,400 m (11,300 ft). At 4,007 m (13,147 ft) in the White Mountains, Boundary Peak is Nevada's highest point. The lowest elevation is on Nevada's extreme southern border with California on the Colorado River, at 147 m (481 ft).

3.5 Geologic Structure

The oldest rocks in Nevada are metamorphic and intrusive rocks from the early and middle Protozoic (2,500–1,600 My ago; they are exposed in only a few locations in southernmost Nevada (Fig. 3.14; dark brown color). These rocks are overlain by limestone and other sedimentary rocks of the late Paleozoic (254 My ago), which are exposed in central and southern Nevada (light purple color). Volcanic assemblages intruded the sedimentary rocks from the late Paleozoic through the Mesozoic (254–72 My ago). Most of the ranges in central Nevada represent lower volcanic rocks from 17 to 42 My ago. Tuffaceous and sedimentary rocks dated at 6–16 My ago compose ranges in the northeastern part of the state. The upper volcanic rocks (6–17 My ago) are exposed in northern (MLRAs 23 and 25; Fig. 1.4) Major Land Resource Areas (MLRAs) and south-central (MLRA 29) Nevada. Volcanic rocks less than 6 My ago exist as volcanoes at several small locations in northwestern and south-central Nevada. The basins are composed of alluvial and playa deposits of late Quaternary age.

Table 3.3 Some dominant plants in Nevada

Zone	Latin name
Alpine	
Cushion plants	
Sedges, grasses	
Subalpine Spruce-Fir Forest	
Great Basin bristlecone pine	*Pinus longaeva*
Whitebark bine	*Pinus albicaulis*
Engelmann spruce	*Picea engelmannii*
Subalpine fir	*Abies lasiocarpa*
Mountain hemlock	*Tsuga mertensiana*
Upper Montane Forest	
Red fir	*Abies magnifica*
White fir	*Abies concolor*
Jeffrey pine	*Pinus jeffreyi*
Limber pine	*Pinus flexis*
Western white pine	*Pinus monticola*
Ponderosa pine	*Pinus ponderosa; Pinus ponderosa subsp. brachyptera*
Douglas fir	*Pseudotsuga menziesii var. glauca*
Trembling aspen	*Populus tremuloides*
Montane Shrubs	
Curly-leaf mountain mahogony	*Cercocarpus ledifolius*
Mountain big sagebrush	*Artemisia tridentata var. wyomingensis*
Snowberry	*Symphoricarpus oreopilus*
Antelope bitterbrush	*Purshia tridentata*
Stansbury cliffrose	*Purshia stansburiana*
Gambel oak	*Quercus gambelii*
Utah serviceberry	*Amelanchier utahensis*
Pinyon-Juniper Forest	
Utah juniper	*Juniperus osteosperma*
Rocky Mountain juniper	*Juniperus scopulorum*
Single-leaf pinyon pine	*Pinus monophylla*
Twoneedle pinyon pine	*Pinus edulis*
Montane Shrubs	
Curly-leaf mountain mahogony	*Cercocarpus ledifolius*
Mountain big sagebrush	*Artemisia tridentata var. wyomingensis*
Snowberry	*Symphoricarpus oreopilus*
Antelope bitterbrush	*Purshia tridentata*
Stansbury cliffrose	*Purshia stansburiana*
Gambel oak	*Quercus gambelii*
Utah serviceberry	*Amelanchier utahensis*
Northern Desert Shrubs	
Black sagebrush	*Artemisia nova*
Basin big sagebrush	*Artemisia tridentata*
Wyoming big sagebrush	*Artemisia tridentata var. wyomingensis*
Little sagebrush	*Artemisia arbuscula*

(continued)

Table 3.3 (continued)

Zone	Latin name
Horsebrush	*Tadedymia tetramores*
Shadscale saltbush	*Atriplex confertifolia*
Winterfat	*Krascheninnikovia lanata*
Bud-sagebrush	*Picrothamnus desertorum*
Mat saltbush	*Atriplex corrugata*
Green molly	*Bassia americana*
Indian ricegrass	*Oryzopsis hymenoides*
Bottlebush squirreltail	*Elymus elymoides*
Sandberg bluegrass	*Poa secunda*
Muttongrass	*Poa fendleriana*
Needlegrass	*Stipa spp.*
Cheatgrass	*Bromus tectorum*
Wildrye	*Elymus*
Southern Desert Shrubs	
Nevada dalea	*Psorothamnus polydenius*
Cattle saltbush	*Atriplex polycarpa*
Creosote bush	*Larrea tridentata*
Joshua tree	*Yucca brevifolia*
White bursage (burrobush)	*Ambrosia dumosa*
Anderson's wolfberry	*Lycium andersonii*
Blackbrush	*Coleogyne ramosissima*
Mormon tea	*Ephedra viridis*
Big galleta	*Pleuraphis rigida*
Sandpaper plant	*Petalonyx* spp.
Salt Desert Shrubs	
Greasewood, black	*Sarcobatus vermiculatus*
Greasewood, Bailey	*Sacrobatus baileyi*
Shadscale saltbush	*Atriplex confertifolia*
Spiny hopsage	*Grayia spinosa*
Seepweed	*Sueda spp.*
Pickleweed	*Salicornia spp.*
Desert saltgrass	*Distichlis spicata*
Alkali rabbitbrush	*Chrysothamnus albidus*
Halogeton	*Halogeton glomeratus*
Gardner's saltbush	*Atriplex gardneri*
Great Basin wildrye	
Kochia	*Kochia americana*
Alkali sacaton	*Sporobolus airoides*
Rubber rabbitbrush	*Ericameria nauseousus*
Four-winged saltbush	*Atriplex canescens*

Sources OSDs; Tueller 1989

Fig. 3.7 Common plants in salt deserts of Nevada, including black greasewood (*Sacrobatus vermiculatus*) (upper) from Sarcobatus Valley, NV (*Source* US Geological Survey), fourwing saltbush (*Atriplex canescens* var. *canescens*) (middle) from Red Rock Canyon, NV (*Source* Stan Shebs), and shadscale (*Atriplex confertifolia*) (lower) from the Great Basin (*Source* R.E. Rosiere)

Fig. 3.8 Dominant species of sagebrush-grasslands, including Wyoming big sagebrush (*Artemisia tridentata* var. *wyomingensis*) (*Source* Matt Levins, Wikimedia Commons) and creosote bush (*Larrea tridentata*) in the Mojave Desert (*Source* steemit.com)

3.6 Surficial Geology

The surficial geology of Nevada includes seven kinds of materials: (i) colluvium and residuum from limestone; (ii) colluvium and residuum from igneous and metamorphic materials; (iii) alluvium; (iv) lacustrine materials; (v) loess, (vi) eolian sand deposits; and (vii) volcanic ash deposits (Fig. 3.15; Table 3.5). Nearly half (48%) of the soil series in Nevada are derived from alluvium, followed by colluvium (32%), loess (8.7%), and residuum (6.2%) (Fig. 3.16). Many of the landforms in Nevada, both on the mountain ranges and in the basins, are covered with a mantle of volcanic ash

Table 3.4 Soil, topographic, climatic, and landscape features of predominant sagebrush communities in Nevada (modified from Schultz and McAdoo (2002)

Sagebrush species	Dominant diagnostic subsurface horizons	General soil chemistry	Particle-size classes	Elevation (m)	Precipitation (mm/yr)	MAAT (°C)	Max. slope (%)	Landform
Basin big sagebrush	Argillic, none	Non-alkaline, nonsaline, noncalcareous	Fine-loamy, coarse-loamy, loamy-skeletal	1475–2100	277 ± 118	7.4 ± 1.7	34 ± 25	Valley side slopes, ephemeral stream channels, degraded meadows
Wyoming big sagebrush	Argillic, duripan, none, cambic	Can be mildly alkaline	Loamy, loamy-skeletal, coarse-loamy, fine	1350–2050	242 ± 26	8.7 ± 1.2	31 ± 23	Mid-elevation valley bottoms, valley slopes, lower foothills
Mountain big sagebrush	Argillic	Nonsaline, non-alkaline, noncalcareous	Loamy-skeletal	1700–2500	366 ± 80	6.4 ± 1.4	62 ± 13	Mountain sideslopes, summits, & ridges; high-elevation plateaus & valley bottoms
Black sagebrush	Argillic, duripan, none, calcic	Can be calcareous; nonsaline	Loamy-skeletal	1500–2400	258 ± 50	8.8 ± 1.7	47 ± 21	Valley slopes, mountain slopes
Low sagebrush	Argillic	Nonsaline, non-alkaline, usually noncalcarous	Clayey-skeletal, clayey	1575–2350	307 ± 65	7.2 ± 2.2	54 ± 20	Valley slopes, mountain slopes, mesas
Lahontan sagebrush	Argillic none	Nonsaline, non-alkaline	Clayey-skeletal	1350–2150	230 ± 17	9.6 ± 1.1	67 ± 12	Lower foothills, mountain slopes valley bottoms
Bud-sagebrush	Argillic, duripan, cambic, natric	Noncalcareous sometimes alkaline and saline tolerant	Clayey loamy-skeletal, sandy-skelet	1300–1900	161 ± 20	9.9 ± 1.5	25 ± 22	Lower foothills, mountain slopes, mesas, plains

and/or loess that ranges from 13 to 97 cm and averages 35 cm in thickness. This material has an ashy silt loam or fine sandy loam texture and is most common in north-central Nevada and likely originates from westerly winds eroding sediments from pluvial LakeLahontan (Table 3.6). The volcanic ash in northern Nevada originates partly from the Mount Mazama eruption 6.9 ky ago. However, volcanic ash throughout the state is present in Miocene-aged volcanic rocks and is released by physical weathering. The eolian deposits of northern Nevada are derived from the former Lake Lahontan and Lake Mojave basins. In soils in west-central Nevada, much of the eolian volcanic ash originates from the Long Valley caldera and Mono Craters volcanic fields of eastern California. Because alluvium is reworked in Nevada's bolsons and semi-bolsons, it is often enriched in loess and volcanic ash.

Glacial deposits from alpine glaciations exist to a limited extent in the highest mountain ranges of Nevada (Blackwelder 1931; Sharp 1938; Richmond 1986), including Boundary Mountain in the Sierra Nevada Mountains of west-central Nevada, Wheeler Peak in the Snake Range of Great Basin National Park in east-central Nevada, the Spring Mountains in southern Nevada (Orndorff et al. 2003); the Ruby Mountains and East Humboldt Range in northeast Nevada (Wayne 1984), the Pine Forest Range in northwestern Nevada, and possibly elsewhere. A small glacier in a cirque on Wheeler Peak is reputedly the southernmost glacier in the conterminous USA.

In soil survey reports of Nevada, map unit components of the highest mountain ranges often feature the miscellaneous area named Rubble land (Fig. 3.17 upper); these stony deposits may be derived from till and/or colluvium. Other landforms in Nevada include sand dunes (Fig. 3.17 lower), playas (Fig. 3.18 upper), and sand sheets (Fig. 3.18 lower).

3.7 Time

There are two schools of thought regarding the role of time in pedogenesis: some pedologists favor the soil evolution, suggesting that soils develop progressively over time and that an array of soils influenced primarily by time can be

Fig. 3.9 Mountain forest-shrub communities in Nevada, including single-leaf pinyon pine-Utah juniper (*Pinus monophylla-Juniperus osteosperma*) in White Pine County, NV (upper) (Source: Creative Commons; Famartin) and mountain mahogany (*Cercocarpus intricatus*) in the Spring Mountains, NV (lower) (*Source* Stan Shebs)

Fig. 3.10 Montane forests in the Great Basin National Park, including ponderosa pine (*Pinus ponderosa*) and trembling aspen (*Populus tremuloides*). *Source* Soil Survey of the Great Basin National Park, Nevada

studied as a soil chronosequence or mathematically as a soil chronofunction. Other soil scientists suggest that most soils, especially those in Nevada, are polygenetic, i.e., they develop cyclically and have been influenced by intermittent changes in climate, soil erosion, and sedimentary deposition.

Soils in Nevada range from late Holocene in age, in the case of some Entisols and Aquolls, to $\sim$5 My, in the case of soils containing petrocalcic horizons (Figs. 3.19 and 3.20). Adams and Wesnousky (1999) observed soils developed on materials from the pluvial LakeLahontan highstand that were $\sim$120–280 ky in age. Natrargids derived from post-Mazama basin fill in the Dixie Valley are less than 6.9 ky in age (Alexander and Nettleton 1977). Blank et al. (1996) conjectured that talus influenced by eolian dust in the Buckskin Range of west-central was likely 100 ky in age. Chadwick and Davis (1990) identified several soil-forming intervals in northwestern Nevada related to fluctuations of pluvial LakeLahontan. Many of the soils of the Great Basin show evidence of polygenesis due to climate change and varying rates of eolian activity (Chadwick et al. 1995). Notable silicification leading to the formation of durinodes was noted in Holocene soils of the northern Monitor Valley, Nevada (Chadwick et al. 1989). Harden et al. (1991a, b) compared two Holocene chronosequence in southern Nevada at Fortymile Wash and Kyle Canyon with two chronosequences in southeast California. Pedogenesis was from four to ten times less in Nevada because of lower rates

Fig. 3.11 Examples of vegetation in subalpine and alpine areas of Nevada. The upper photograph shows subalpine fir (*Abies lasiocarpa*) and whitebark pine (*Pinus albicaulis*) in the Copper Mountains, NV (Source: Creative Commons). The middle photograph shows bristlecone pine (*Pinus longaeva*), with specimens in some areas more than 4,000 years old, in Great Basin National Park (J. Bockheim photo). The lower photograph is from the alpine zone on Wheeler Peak in the Great Basin National Park (*Source* Soil Survey of Great Basin National Park, Nevada)

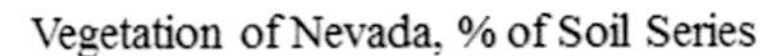

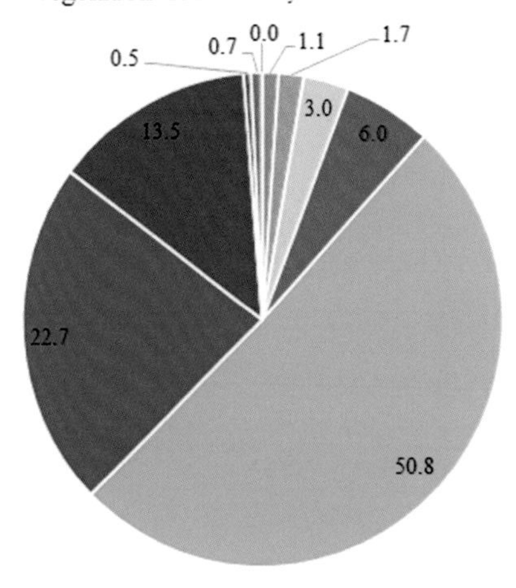

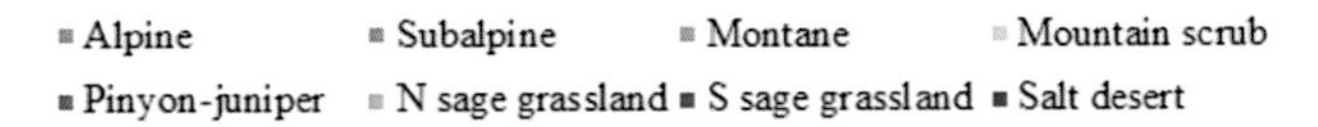

Fig. 3.12 The distribution of vegetation types in Nevada, based on the percent of soil series *Source* NRCS database

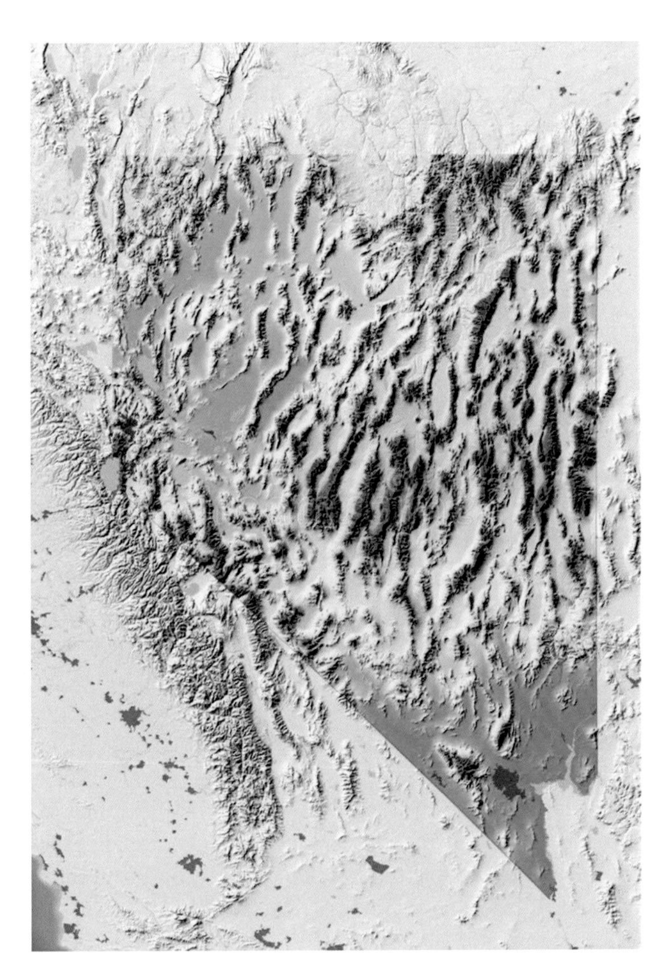

Fig. 3.13 Shaded relief map of Nevada (*Source* Nevada Bureau of Mines and Geology)

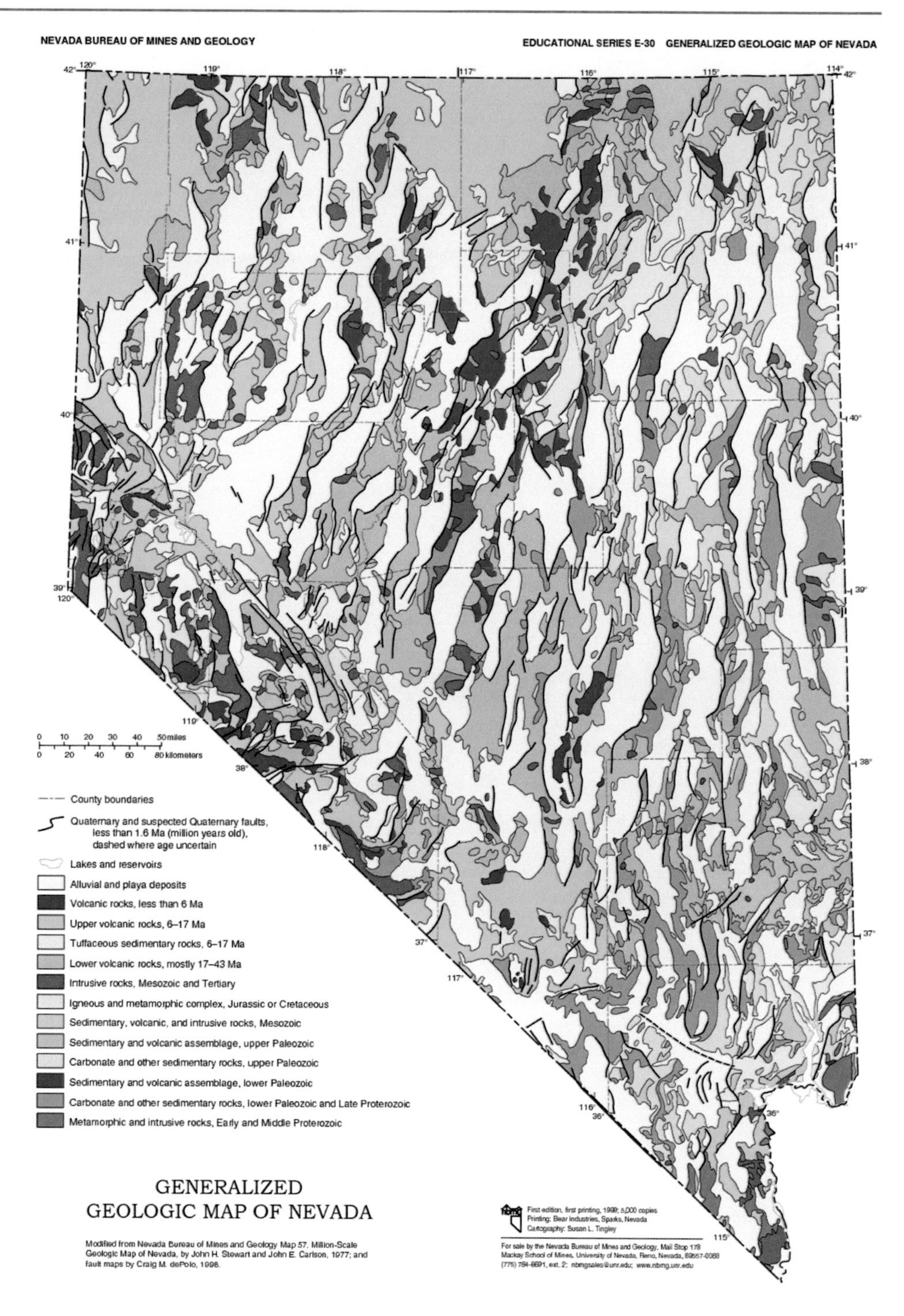

Fig. 3.14 Bedrock geology map of Nevada (*Source* Nevada Bureau of Mines and Geology)

of dust influx. Nettleton et al. (1975) reported Torripsamments on surface younger than 12 ky and Haplargids and Paleargids on older surfaces in the Lahontan Basin of Nevada. Reheis et al. (1992) identified Haplocalcids on 15 ky surfaces, Haplargids on 130 ky surfaces, and Petrocalcic Paleargids on 800 ky surfaces in Kyle Canyon, Nevada.

3.8 Humans

The earliest humans, the Pueblo Grande de Nevada, lived in limestone caves in Nevada as long as 17–12 ky ago. In modern times, four principal Native American groups have inhabited

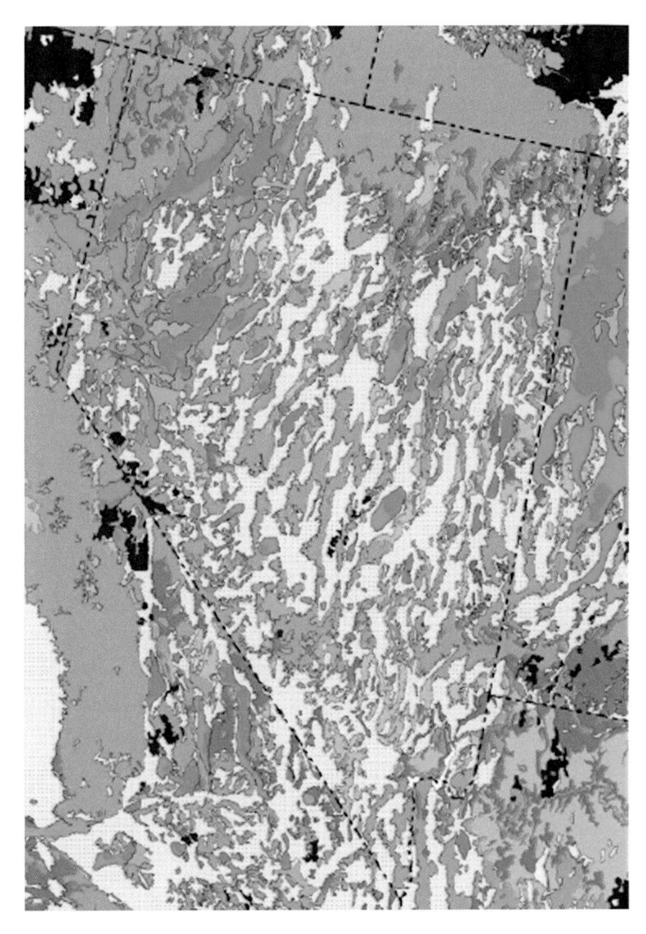

Fig. 3.15 Quaternary geology map of Nevada (see Table 3.5 for legend; from Soller and Reheis 2004)

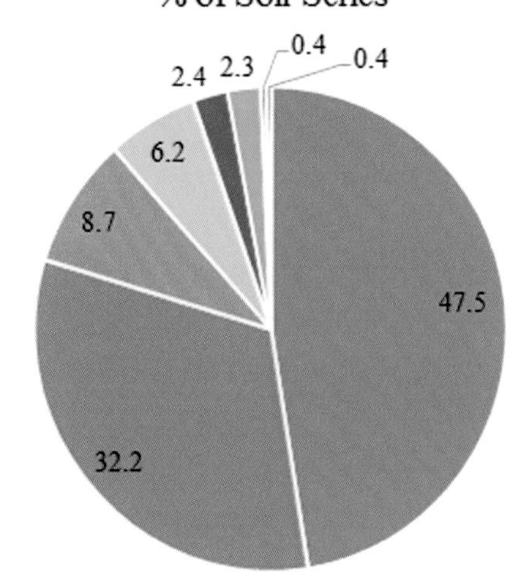

Fig. 3.16 The distribution of parent materials in Nevada, based on the percent of soil series

Table 3.5 Legend for Fig. 3.15, Surficial materials in Nevada

Color	Parent material
Yellow	Alluvium
Blue	Lacustrine; pluvial lakes
Pink	Residuum—igneous and metamorphic
Gray	Residuum—limestone
Brown	Eolian
Purple	Volcanic ash

Table 3.6 Thickness and texture of loess cap in Nevada soil series > 100 km^2

Soil series	Thick. (cm)	Volcanic ash component		Location
Chiara	33	ashy sil	X	NE
Boton	30	sil		NW
Broyles	33	ashy sil	X	NC
Mormon Mesa	41	fsl		SE
Beoska	61	sil		C
Weso	66	vfsl	X	NW
Peeko	13	sil		NE, C
Tenabo	18	vfsl	X	C
Cherry Spring	38	sil	X	NC
Pumper	30	l	X	NW
Dun Glen	25	vfsl, sil		NC
Valmy	25	fsl	X	N
Connel	51	vfsl, sil	X	N
Bioya	97	vfsl, sil		NE
Whirlo	33	fsl		C

(continued)

Table 3.6 (continued)

Soil series	Thick. (cm)	Volcanic ash component		Location
Bucan	18	l, cl	X	NC
Bartome	28	ashy vfsl	X	NC
Blackhawk	76	sil		NC
Golconda	18	sil	X	NC
Clurde	13	sil		NC
Adelaide	20	sil	X	NC
Igdell	13	sil		NE
Flue	33	ashy vfsl	X	NC
Cortez	25	ashy sil		NC
	35			
Avg.Stdev	21			

Fig. 3.17 Common landforms in Nevada, including rubble on Wheeler Peak (upper) (*Source* Soil Survey of the Great Basin National Park, Nevada) and sand dunes at Sand Mountain (lower) (J. Bockheim photo)

Fig. 3.18 Additional land types in Nevada, including a playa east of Fallon (J. Bockheim photo) and a sand sheet in the Ivanpah Valley (Nevada Bureau of Mines & Geology photo)

Nevada: Southern Paiute, Northern Paiute, Shoshoni, and Washo. The first white explorer to enter the state likely was the Spanish priest, Francisco Garces, in 1776. Nevada's first permanent white settlement was Mormon Station (later Genoa) in 1850 in western Nevada. Today the population of Nevada is around 3 million, but is concentrated mainly in the Las Vegas in the far south, Reno-Carson City-Sparks in the far west. Nevada is ranked 42nd in terms of population density in the USA. Because of the dry climate and declining water reserves in the Colorado River, only 8.5% of the state is cultivated and irrigated. The imprint of modern humans on Nevada's soils is mainly from former mining activities, deforestation in the mountains, overgrazing in some areas, and localized urbanization. These will be discussed in more detail in Chap. 16.

3.9 Summary

The soils of Nevada result from a complex interplay of five soil-forming factors: climate, organisms, relief, parent material, and time. With a state average of 175 mm/yr (7 in/yr), Nevada is the driest state in the U.S. The dry climate accounts for the abundance of Aridisols (42% of area) in Nevada. The climate was much moister during glacial periods as evidenced by pluvial lakes, such as Lake Lahontan in the northwest and Lake Bonneville in the east. These lakes were up to 300 m deeper than today. The natural vegetation of Nevada can be divided broadly into six biomes along an elevational gradient. From the lowest to the highest

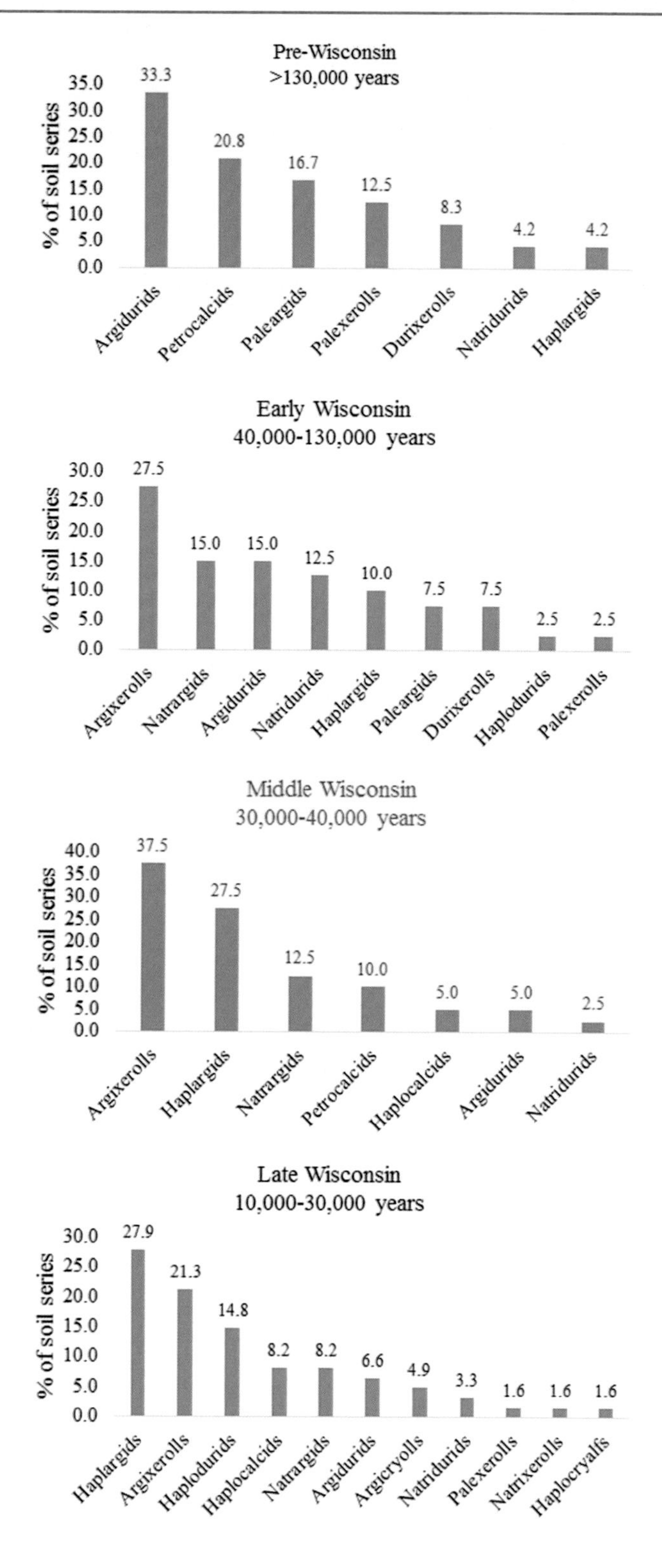

Fig. 3.19 Distribution of Pre-Wisconsin, Early Wisconsin, Middle Wisconsin, and Late Wisconsin soils in Nevada by great group *Source* NRCS database

elevations, these include salt desert, sagebrush grassland, pinyon-juniper woodland, mountain shrub, subalpine forest, and alpine tundra.

The relief map of Nevada is very striking. The Columbia Plateau extends into the north-central portion of the state and the Sierra Nevada Range into the west-central region, but the

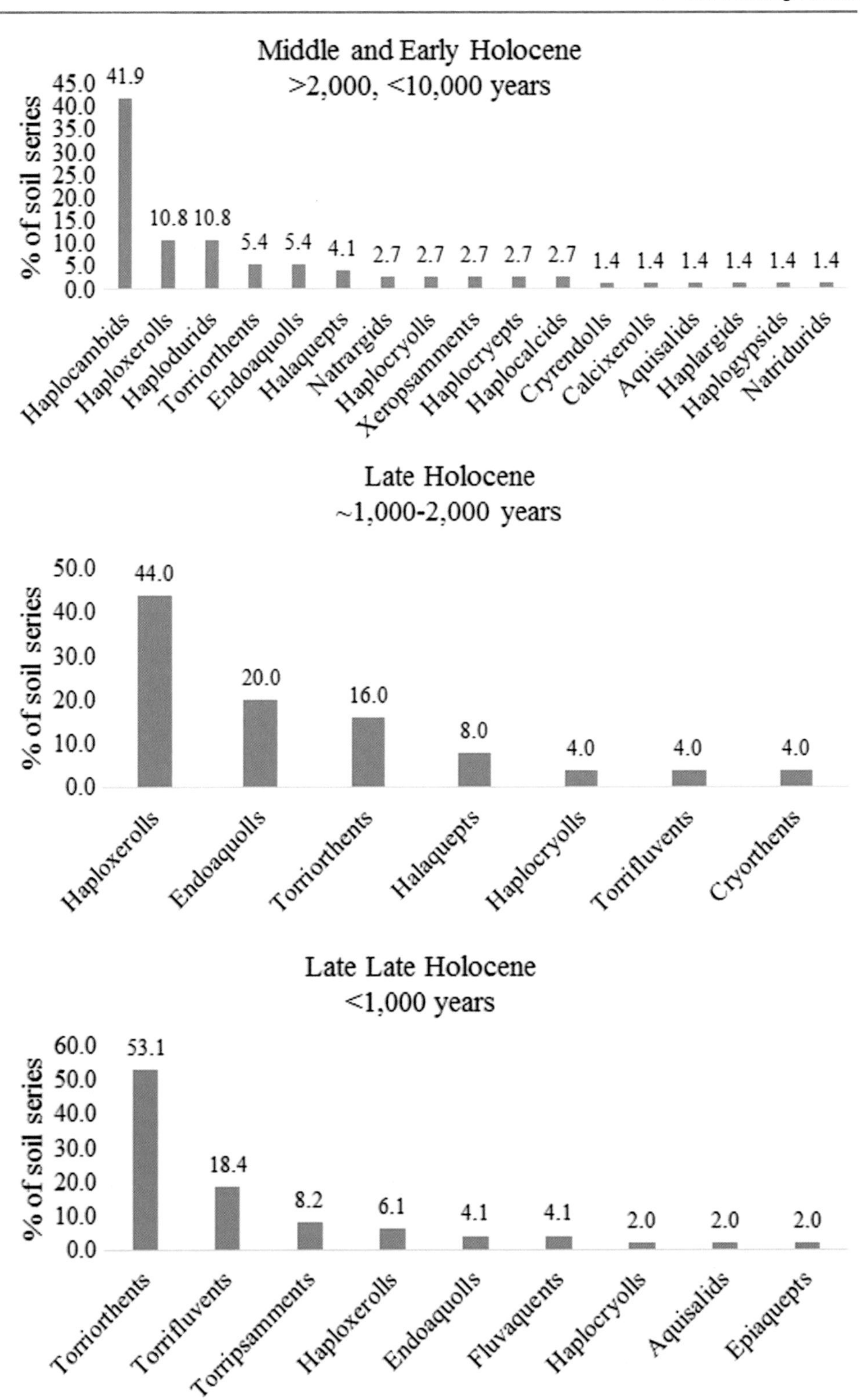

Fig. 3.20 Distribution of Holocene soils in Nevada by great group *Source* NRCS database

majority of the state is included in the Basin and Range physiographic province, with up to 100 basins and ranges oriented roughly NNE to SSW. Nevada features a variety of bedrock types reflecting ancient intrusive rocks overlain by sedimentary rocks and volcanic rocks subject to extensive faulting. Most of the rocks exposed at the surface are of volcanic origin, especially the plateaus and ranges. The basins are composed of alluvial and playa deposits of Quaternary age. The surficial geology of Nevada includes five kinds of materials: (i) colluvium and residuum from limestone; (ii) colluvium and residuum from igneous-metamorphic materials; (iii) alluvium in basins; (iv) lacustrine materials in basins, (v) eolian sand deposits, and (vi) volcanic ash deposits. Soils in Nevada range from late Holocene in age, in the case of some Entisols and Aquolls, to ~5 My, in the case of soils containing petrocalcic horizons, such as the Mormon Mesa soil series.

The imprint of modern humans on Nevada's soils is mainly from former mining activities, deforestation in the mountains, overgrazing in some areas, and localized urbanization.

References

Adams KD, Wesnousky SG (1999) The Lake Lahontan highstand: age, surficial characteristics, soil development, and regional shoreline correlation. Geomorphology 30:357–392

Alexander EB, Nettleton WD (1977) Post-Mazama Natrargids in Dixie Valley. Nevada. Soil Sci. Soc. Am. J. 41:1210–1212

Blackwelder EB (1931) Pleistocene glaciation in the Sierra Nevada and basin ranges. Geol Soc Am Bull 42:865–922

Blank RR, Young JA, Lugaski T (1996) Pedogenesis on talus slopes, the Buckskin Range, Nevada, USA. Geoderma 71:121–142

Chadwick OA, Davis JO (1990) Soil-forming intervals caused by eolian sediment pulses in the Lahontan Basin, northwestern Nevada. Geology 18:243–246

Chadwick OA, Hendricks DM, Nettleton WD (1989) Silicification of Holocene soils, in northern Monitor Valley. Nevada. Soil Sci. Soc. Am. J. 53:158–164

Chadwick OA, Nettleton WD, Staidl GJ (1995) Soil polygenesis as a function of Quaternary climate change, northern Great Basin, USA. Geoderma 68:1–26

Harden JW, Taylor EM, Hill C, Mark RK, McFadden LD, Reheis MC, Sowers JM, Wells SG (1991a) Rates of soil development from four chronosequences in the southern Great Basin. Quat. Res. 35:383–399

Harden, J.W., Taylor, E.M., McFadden, L.D. and Reheis, M.C. 1991b. Calcic, gypsic, and siliceous soil chronosequences in arid and semi-arid environments. In: Nettleton, W.D (Ed.) Occurrence, Characteristics, and Genesis of Carbonate, Gypsum, and Silica Accumulations in Soils. SSSA Spec. Pap., No. 26, pp. I-16

Kellogg CE (1930) Preliminary study of the profiles of the principal soil types of Wisconsin. Wisc Geol Natural Hist Surv Bull. No. 77A. 113 pp

Nettleton WD, Witty JE, Nelson RE, Hawley JW (1975) Genesis of argillic horizons in soils of desert areas of the southwestern United States. Soil Sci Soc Am J 39:919–926

Orndorff RL, Van Hoesen JG, Saines M (2003) Implications of new evidence for late Quaternary glaciation in the Spring Mountains, southern Nevada. J Ariz-Nev Acad Sci 36:37–45

Reheis MC, Sowers JM, Taylor EM, McFadden LD, Harden JW (1992) Morphology and genesis of carbonate soils on the Kyle Canyon fan, Nevada, U.S.A. Geoderma 52:303–342

Richmond GM (1986) Stratigraphy and correlation of glacial deposits of the Rocky Mountains, the Colorado Plateau and the ranges of the Great Basin. Quat Sci Rev 5:99–127

Schultz, B., McAdoo, K. 2002. Common sagebrush in Nevada. Univ. of Nevada Coop. Ext. Spec. Publ. SP-02–02. 9 pp

Sharp RP (1938) Pleistocene glaciation in the Ruby-East Humboldt Range, northeastern Nevada. J Geomorph 1:296–323

Soil Survey Staff (2014) Keys to Soil Taxonomy. 12th edit. U.S. Dept. Agric., Natural Resour. Conserv. Serv., Linoln, NE

Tueller PT (1975) The natural vegetation of Nevada. Mentzelia 1:3–28

Tueller PT (1989) Vegetation and land use in Nevada. Rangelands 11 (5):204–210

Wayne WJ (1984) Glacial chronology of the Ruby Mountains-East Humbolt Range. Nevada. Quat. Res. 21:286–303

General Soil Regions of Nevada

4

Abstract

This chapter discusses the soil taxonomy of each of the 10 Major Land Resource Areas in Nevada, including the Sierra Nevada Mountains, the Malheur High Plateau, the Humboldt Area, the Owyhee High Plateau, the Carson Basin and Mountains, the Fallon-Lovelock Area, the Great Salt Lake Area, the Central Nevada Basin and Range, the Southern Nevada Basin and Range, and the Mojave Desert.

4.1 Introduction

To date there has been no state soil map for Nevada. However, the state is contained within 10 Major Land Resource Areas (MLRAs) (USDA, NRCS, 2006) (see Fig. 1.4 and Table 1.2) for which soils information has been compiled. The Basin and Range Province covers 81% of the state, including the Malheur High Plateau (MLRA 23) in the northwestern corner, the Humboldt Area (MLRA 24) in the north-central part, the Owyhee High Plateau (MLRA 25) in the northeastern corner, the Carson Basin and Mountains (26) and the Fallon-Lovelock Area (27) in the west-central part, the Great Salt Lake Area (MLRA 28A) in the east-central part, the Central Nevada Basin and Range (28B) in the central part, and the Southern Nevada Basin and Range (MLRA 29) and the Mojave Desert (MLRA 30) in the southern part. Nevada contains a small segment (0.3% of state area) along its far western border of MLRA 22A, the Sierra Nevada Mountains.

4.2 Soils of the Sierra Nevada Mountains (MLRA 22A)

The Nevada portion of the Sierra Nevada is small (2% of MLRA 22A, 975 km^2) compared to that of California. The unincorporated communities of Incline Village and Stateline, Nevada, are in this MLRA. Figure 4.1 shows part of the Sierra Nevada from the Spooner Lake area in Lake Tahoe Nevada State Park. This part of the Sierra Nevada has bedrock geology that is dominated by a quartz monzonite and granodiorite batholith. Volcanic activity during the Miocene Epoch produced lava flows, ash flows, and lahar deposits, which mantled the older plutonic rocks of the Sierra batholith. The alluvium eroded from the Sierra Nevada has yielded placer deposits of gold that were mined during the California Gold Rush that began in 1848. Glacial deposits mapped in Nevada occur in moraines and cirques of the Sierra Nevada.

This area receives the greatest amount of annual precipitation recorded in Nevada, up to 1,300 mm/yr (49 in/yr). The area supports montane coniferous forest vegetation, including Douglas-fir, white fir, California red fir, ponderosa pine, Jeffrey pine, lodgepole pine, and bristlecone pine. The soil parent materials are primarily colluvium and residuum.

MLRA 22A contains primarily Haploxerolls (Glean, Calpine, and Franktown soil series), Haploxeralfs (Inville and Jorge soil series), Dystroxerepts (Tallac and Umpa soil series), and Xeropsamments (Cagwin and Toiyabe soil series) (Table 4.1). Most of the Alfisols and all of the Andisols identified in Nevada occur in MLRA 22A.

4.3 Soils of the Malheur High Plateau (MLRA 23)

Most of the Malheur High Plateau is in southeastern Oregon; Nevada has 14,830 km^2 in MLRA 23. This MLRA has a few small towns but no incorporated cities. Figure 4.2 shows The Granite Range south of Wagon Wheel Pass. Elevation ranges from 1,190 to 2,105 m (3,900 to 6,900 ft) in most of the area, but it exceeds 2,745 m (9,000 ft) on some mountains. Most of the surficial geology of the Malheur High Plateau consists of volcanic rocks–andesite and basalt–that range from 6 to 17 Ma in age. Plateaus and mesas (locally called "tables") occurring on flat-lying formations of volcanic rock are common landforms in MLRA 23. The few north-south trending mountain ranges represent uplifted fault blocks that are more common in the Great Basin Section of

P. W. Blackburn et al., *The Soils of Nevada*, World Soils Book Series,
https://doi.org/10.1007/978-3-030-53157-7_4

Fig. 4.1 Spooner Lake area in Lake Tahoe Nevada State Park (MLRA 22A). Photo by NRCS

the Basin and Range Province to the south. Soils on the plateaus and mountain ranges are derived from colluvium and residuum; soils in the valleys are composed of alluvium and lacustrine material. The soil parent materials are enriched with volcanic ash.

The mean annual precipitation in MLRA 23 ranges from 150 to 305 mm (6–12 in). At higher elevations, the MAP may be as high as 1,450 mm (57 in). Snowfall is common in the mountains during the winter months. The average annual temperature is 4–11° (39–52 °F).

MLRA 23 contains primarily Argixerolls (Devada, Cleavage, Ninemile, Sumine, Reluctan, Chen, Bucklake, and Softscrabble soil series), followed by Haplargids (Old Camp, Anawalt, Roca, and Vanwyper soil series), and Argidurids (Hunnton, Fulstone, Roval, Ratto, and Indian Creek soil series) (Table 4.1). Most of the Vertisols in Nevada occur in MLRA 23 with the Haploxererts (Brubeck, Manogue, and Tunnison soil series) and Epiaquerts (Boulder Lake and Weimer soil series) the dominant great groups.

4.4 Soils of the Humboldt Area (MLRA 24)

Nevada contains 94% of the Humboldt Area, including an area of 30,884 km^2 (11,925 mi^2). The city of Winnemucca and the town of Battle Mountain, Nevada, are along Interstate 80, which crosses this MLRA. Figure 4.3 is of the Water Canyon in the Sonora Range southeast of Winnemucca. The Humboldt Area is part of the Basin and Range Province, with elevations ranging from 1,205 to 1,800 m (3,950–5,900 ft). The roughly north-south-trending, fault-block mountain ranges are separated by wide valleys filled with alluvium and lacustrine materials.

In MLRA 24, the mean annual precipitation ranges from 150 to 305 mm (6–12 in), with greater amounts in the ranges. Much of the rainfall occurs as high-intensity, convective thunderstorms in spring and autumn. Snowfall occurs at the higher elevations. The average annual temperature ranges from 3 to 12 °C (38–53 °F).

MLRA 24 contains primarily Haplargids (Old Camp, Wieland, Soughe, Roca, and Atlow soil series), Argixerolls (Cleavage, Sumine, Reluctan, Chen, and Softscrabble soil series), Torriorthents (Mazuma, Boton, Puett, Tulase, and Bubus soil series), and Haplocambids (Orovada, Enko, Broyles, and McConnel soil series) (Table 4.1). Other dominant great groups include the Natrargids (Oxcorel, and Beoska soil series), Argidurids (Dacker and Dewar soil series), Torripsamments (Isolde and Hawsley soil series), Halaquepts (Ocala and Wendane soil series), and Haplodurids (Chiara soil series). Argixerolls and Haplargids are dominant in the mountain ranges, and Torriorthents, Haplocambids, Natrargids, Halaquepts, and Torripsamments are common in the valleys. Alluvium and lacustrine deposits occur in closed basins at the lowest elevations.

4.5 Soils of the Owyhee High Plateau (MLRA 25)

The Owyhee High Plateau occurs in northeastern Nevada; the Nevada portion is 52% (38,979 km^2; 15,050 mi^2). The city of Elko, Nevada, which is along Interstate 80, is in this MLRA. Figure 4.4 shows the Harrison Pass summit in the Ruby Mountains on the Owyhee High Plateau. The plateau is composed of isolated, uplifted fault-block mountain ranges separated by narrow, aggraded desert plains. Elevations range from 1,600 to 2,300 m (5,200–7,500 ft). Some of the

Table 4.1 Dominant suborders and great groups by Major Land Resource Area in Nevada

MLRA		Total	% in	Area NV	No. soil	
No.	MLRA Description	area (km^2)	NV	(km^2)	series	Dominant Great Groups
22A	Sierra Nevada Mountains	48745	0.02	975	70	Haploxerolls; Haploxeralfs, Dystroxerepts; Xeropsamments
23	Malheur High Plateau	59320	0.25	14830	256	Argixerolls; Haplargids; Argidurids
24	Humboldt Area	32855	0.94	30884	281	Haplargids; Argixerolls; Torriorthents; Haplocambids
						Natrargids; Argidurids; Halaquepts; Haplodurids
25	Owyhee High Plateau	74960	0.52	38979	369	Argixerolls; Haplargids; Haplocambids; Argidurids
						Haplodurids; Durixerolls
26	Carson Basin and Mountains	16890	0.75	12668	282	Argixerolls; Haplargids; Haplocambids
27	Fallon-Lovelock Area	32560	0.98	31909	296	Torriorthents; Haplargids; Natrargids; Argixerolls
						Haplocambids
28A	Great Salt Lake Area	95300	0.16	15248	254	Torriorthents; Haplocalcids; Haplodurids; Haploxerolls
						Argidurids; Haplocambids
28B	Central Nevada Basin and Range	61035	0.99	60425	429	Torriorthents; Argixerolls; Haplocalcids; Haplocambids
						Haplargids; Haplodurids; Argidurids
29	Southern Nevada Basin and Range	68140	0.73	49742	293	Torriorthents; Haplargids; Argidurids; Haplocambids
						Haplodurids; Argixerolls
30	Mojave Desert	113370	0.28 Total	31744 287403 286382	263	Torriorthents; Haplocalcids; Haplodurids

mountains exceed 3,000 m (9,840 ft). Soil parent materials are typically colluvium and residuum in the mountains and plateaus and alluvium in the valleys. A surficial mantle of loess high in volcanic ash occurs in many of the soils on plateaus and piedmont slopes.

MLRA 25 receives 180–405 mm (7–16 in) of annual precipitation, but it can exceed 1,270 mm (50 in) in the mountains. The precipitation is somewhat evenly distributed in the fall, winter, and spring, and it is lowest from mid-summer to early autumn. The average annual temperature ranges from 2 to 12 °C (35–53 °F).

The mountain ranges in MLRA 25 contain mainly Argixerolls (Devada, Ninemile, Cleavage, Sumine, Reluctan, Chen, McIvey, and Vitale soil series), Haplocryolls (Hapgood and Hackwood soil series), and Haplargids (Anawalt, Old Camp, Wieland, Roca, Burrita, and Soughe soil series). Soils at the lower elevations include Haplocambids (Orovada, Enko, McConnel, and Kelk soil series), and Argidurids (Hunnton, Dewar, Snowmore, and Dacker soil series) (Table 4.1). Other dominant great groups include the Haplodurids (Chiara, Shabliss, and Jericho soil series), and the Durixerolls (Stampede and Donna soil series).

4.6 Soils of the Carson Basin and Mountains (MLRA 26)

Part of the Basin and Range Province, 75% of the Carson Basin and Mountains (MLRA 26) is in Nevada, accounting for 12,668 km^2 (4,890 mi^2). The cities of Carson City, Reno, and Sparks, Nevada, are in this MLRA Fig. 4.5 is of the Pinenut Mountains in Mill Canyon, Douglas County. This land resource region contains isolated north-south-trending, fault-block ranges such as the Carson Range that are separated by aggraded desert plains such as the Carson Valley. The ranges are composed of granite intrusives, andesite, and basalt, and the valleys are filled with alluvium. Elevations range from 1,190 to 1,995 m (3,900–6,550 ft), but Boundary Peak reaches 3,995 m (13,100 ft), the highest in the state (Fig. 4.6).

Fig. 4.2 The Malheur High Plateau (MLRA 23) as seen from the Granite Range south of Wagon Wheel Pass. Photo by Patti Novak, NRCS

The average annual precipitation ranges from 125 to 915 mm (5-36 in), increasing with elevation. Most of the rainfall occurs as high-intensity, convective storms in spring and autumn. Precipitation is mostly snow in winter. Summers are dry. The mean annual temperature ranges from 3 to 12 °C (37–54 °F).

MLRA 26 contains primarily Argixerolls (Devada, Sumine, Chen, Deven, Softscrabble soil series), Haplargids (Old Camp and Theon, soil series), and Haplocambids (Davey and Rebel soil series). Argixerolls and Haplargids predominate in the mountain ranges and Haplocambids dominate in the basins. The flood plains of Carson Valley contain mainly Endoaquolls (Kimmerling, Bishop, and Cradlebaugh soil series) and Haploxerolls (Heidtman soil series).

Fig. 4.3 The Humboldt Valley (MLRA 24) in Water Canyon in the Sonora Range, with Winnemucca in the background. Photo by Patti Novak, NRCS

Fig. 4.4 Harrison Pass summit in the Ruby Mountains on the Owyhee High Plateau (MLRA 25). Photo by NRCS

Fig. 4.5 The Carson Basin and Mountains (MLRA 26) at a soil-temperature monitoring site in the Pinenut Mountains in Mill Canyon. Photo by Ed Blake, NRCS

Fig. 4.6 At 3,995 m (13,100 feet), Boundary Peak is located in MLRA 26 and is the highest mountain in the state (*Source* public domain)

4.7 Soils of the Fallon-Lovelock Area (MLRA 27)

Located almost entirely in Nevada, the Fallon-Lovelock Area (MLRA 27) comprises 32,560 km^2 (12,570 mi^2) in Nevada. The cities of Fallon, Lovelock, and Yerington and the town of Hawthorne, Nevada, are in this MLRA. Figure 4.7 is of the Buena Vista Valley In east-central Pershing County south of Unionville. Part of the Basin and Range Province, the land resource area is composed of isolated mountain ranges trending north-to-south separated by broad, aggraded desert plains and valleys. Nearly half of the area features valley-fill sediments of alluvium and lacustrine deposits, with the remaining area containing colluvium and residuum derived from andesitic and basaltic rock. Elevations range from 1,000 to 1,800 m (3,300–5,900 ft) with some peaks being more than 2,400 m (7,870 ft).

This land resource area receives from 125–255 mm (5–10 in) of precipitation per annum, but may reach 485 mm (19 in) on higher slopes. Most of the rainfall occurs as high-intensity, convective thunderstorms during the growing season. The amount of precipitation is very low from summer to mid-autumn. The precipitation in winter occurs mainly as gentle rain and snow. The mean annual temperature ranges from 6 to 12 °C (43–54 °F).

MLRA 27 contains largely Torriorthents (Mazuma, Bluewing, Boton, Singatse, Ragtown, Sondoa, Hopeka, Benin, Sojur, Roic, Swingler, Kram, and Rustigate soil series) and Haplargids (Old Camp, Theon, Atlow, Shawave, Pineval, Ceejay, Granshaw, Grumblen, Jung, Rednik, Bombadil, Colbar, and Patna soil series). Other common great groups are Natrargids (Dorper, Jerval, Ricert, Appian, Genegraf, Biddleman, Biga, Pokergap, and Buckaroo soil series), Argixerolls (Ninemile, Cleavage, Sumine, Pickup, Acrelane, and Itca soil series), and Haplocambids (McConnel, Davey, Whirlo, Wholan, and Dun Glen soil series). The Halaquepts (Wendane, Umberland, and Nuyobe) and Endoaquolls (Welch, Humboldt, and Kolda soil series) are also well represented in MLRA 27. Whereas the ranges contain primarily Torriorthents, Haplargids, and Argixerolls, the valleys contain Torriorthents, Haplargids, Natrargids, Haplocambids, and Endoaquolls. Alluvium and lacustrine deposits occur in closed basins at the lowest elevations.

4.8 Soils of the Great Salt Lake Area (MLRA 28A)

Only 16% (15,248 km^2; 5,887 mi^2) of MLRA 28A, the Great Salt Lake Area, occurs in Nevada. The town of Pioche and the small cities of West Wendover and Caliente,

Fig. 4.7 The Buena Vista Valley in the Fallon-Lovelock Area (MLRA 27). Photo by NRCS

Nevada, are in this MLRA. Figure 4.8 is of the Cave Valley in southeastern Lincoln County. Part of the Basin and Range Province, MLRA 28A is an area of nearly level basins between widely separated ranges trending north-to-south. The Great Salt Lake, a large terminal saline lake and surrounding salt-desert playas (formerly pluvial Lake Bonneville), composes much of MLRA 28A. Elevations range from 1,800 to 2,900 m (5,900-9,500 ft) in the basins and 2,000 to 3,400 m (6,560–11,150 ft) in the mountains.

The mean annual precipitation ranges from 125 to 300 mm (5–12 in) but may reach 1,245 mm (49 in) in the mountains. Most of the rainfall occurs as high-intensity, convective thunderstorms during the growing season. The driest period is from mid-summer to early autumn. Precipitation in winter typically occurs as snow. The mean annual temperature ranges from 4 to 12°C (39–53°F).

The predominant great groups represented in MLRA 28A include the Torriorthents (Cliffdown, Tooele, Linoyer, Izo, Puett, Timpie, Medburn, Izamatch, Eaglepass, Ragtown, Leo, Geer, Sheffit, Penoyer, Sondoa, Hopeka, Logring, Izar, Theriot, Hundraw, Swingler, Solak, Wala, and Wrango soil series) and the Haplocalcids (Amtoft, Hiko Peak, Armespan, Tecomar, Kunzler, Escalante, Pyrat, Tarnach, Sycomat, Tosser, Gravier, Summermute, and Hardhat soil series). Other abundant great groups are the Haplodurids (Ursine, Shabliss, Jericho, Lien, and Eastmore soil series), Haploxerolls (Hymas, Hyzen, Agassiz, and Sevenmile), Argidurids (Chuska, Acana, Chuckridge, Chuckmill, and Deerlodge), and Haplocambids (Koyen, Heist, Keefa, Haybourne, and Chuffa soil series). The mountains feature mainly Haploxerolls; the basins support Torriorthents, Haplocalcids, Haplodurids, Argidurids, and Haplocambids.

4.9 Soils of the Central Nevada Basin and Range (MLRA 28B)

Located entirely in Nevada, the Central Nevada Basin and Range is the largest MLRA in Nevada, accounting for 21% of the state area (61,035 km^2; 23,565 mi^2). The towns of Austin, Eureka, and Ely, Nevada, are in MLRA 28B. Figure 4.9 is of Independence Valley and shows the basin and range relief of central Nevada. Part of the Basin and Range Province, MLRA 28B contains nearly level, aggraded desert basins and valleys between a series of north-south-trending mountain ranges. Elevations range from 1,500 to 2,000 m (4,900–6,550 ft) in the valleys and basins and from 2,000 to 3,600 m (6,550–11,900 ft) in the mountains. The mountains in the north part of the MLRA are composed of carbonate rocks (i.e., limestone), and the mountains in the south are dominated by volcanic rocks, including andesite and basalt. The valleys are composed of alluvium, with lacustrine deposits in closed basins at the lowest elevations.

The mean annual precipitation ranges from 100 to 300 mm (4–12 in) along the valley floors and 200 to

Fig. 4.8 Southeast Lincoln County, near Cave Valley, part of the Great Salt Lake Area (MLRA 28A). Photo by NRCS

900 mm (8–36 in) in the mountains. Most of the rainfall occurs as high-intensity, convective thunderstorms during the growing season. The driest period is from mid-summer to mid-autumn. The mean annual temperature ranges from 1 to 11°C (34–52°F).

The most represented great groups in MLRA 28B are the Torriorthents (Mazuma, Kyler, Cliffdown, Tooele, Bluewing, Wardenot, Katelana, Linoyer, Puett, Timpie, Tulase, Medburn, Izamatch, Bubus, Ragtown, Sheffit, Sondoa, Batan, Hopeka, Izar, Blimo, Benin, Swingler, Kram, Okan,

Fig. 4.9 Independence Valley in Elko County is part of the Central Nevada Basin and Range (MLRA 28B). Photo by Joseph V. Chiaretti, NRCS

Rustigate, Valmy, and Solak soil series), Argixerolls (Ninemile, Cleavage, Sumine, Chen, Softscrabble, Mclvey, Quarz, Itca, Cropper, Cotant, Segura, Duco, Hutchley, Graley, Walti, Douhide, Upatad, Clanalpine, Loomer, and Suak soil series), Haplocalcids (Amtoft, Pookaloo, Tecomar, Kunzler, Wintermute, Tarnach, Pyrat, Duffer, Automal, Sycomat, Tosser, Gravier, Loray, and Biken soil series), and the Haplocambids (Orovada, Enko, McConnel, Broyles, Heist, Kelk, Silverado, Wholan, Rebel, Pumper, Dun Glen, and Kobeh soil series). Other highly represented great groups are the Haplargids (Wieland, Roca, Robson, Pineval, Jung, Bucan, Colbar, Gance, Abgese, and Hooplite soil series), Haplodurids (Palinor, Chiara, Shabliss, Jericho, Peeko, Parisa, Lien, Rubyhill, and Umil soil series), and Argidurids (Hunnton, Dewar, Yody, Grassval, Ratto, Acana, Broland, and Minu soil series). The mountains contain primarily Argixerolls and some Haplocalcids; the valleys and basins contain Torriorthents, Haplocalcids, Haplocambids, Haplargids, Haplodurids, Natrargids, and Argidurids.

4.10 Soils of the Southern Nevada Basin and Range (MLRA 29)

Nearly three-quarters (73%) of MLRA 29, the South Nevada Basin and Range, is contained in Nevada, occupying an area of 49,742 km^2 (19,205 mi^2) in the state. The town of Tonopah, Nevada, is in this MLRA. Figure 4.10 shows the southern portion of the basin and range in Nevada. Part of the Basin and Range Province, MLRA 29 is an area of broad, nearly level, aggraded desert basins and valleys between a series of north-south-trending mountain ranges. The mountains are uplifted fault blocks with steep side slopes; the basins are bordered by sloping fans and terraces. The mountains are composed of carbonate rocks (i.e., limestone) and volcanic rocks of ash-flow tuff and rhyolite lava which were erupted from a set of large, overlapping calderas. The valleys contain alluvium and, in closed basins at the lowest elevations, lacustrine deposits. Elevations range from 600 to 1,700 m (1,950–5,600 ft) in valleys and up to 4,007 m (13,147 ft) on Boundary Peak, the highest in Nevada.

The mean annual precipitation ranges from 75 to 300 mm (3–12 in) in valleys and 300 to 735 mm (12–29 in) in the mountains. Most of the rainfall occurs as high-intensity, convective thunderstorms during the growing season. Summers are dry, but sporadic storms are common in July and August. The mean annual temperature ranges from -2 to 22°C (28-72°F).

The most extensive great groups in MLRA 29 are the Torriorthents (Kyler, Cliffdown, Wardenot, Boxspring, Izo, Blacktop, Pintwater, Gynelle, Eaglepass, Leo, Geer, Penoyer, Logring, Theriot, Beelem, Luning, Roic, Rustigate, and Potosi soil series), followed by Haplargids (Stewval, Unsel, Downeyville, Gabbvally, Chubard, Watoopah, Richinde, Zaqua, Wyva, Winklo, and Lathrop soil series), the Argidurids (Handpah, Zadvar, Lojet, Delamar, Belted, Acana, Lyda, Littleailie, Chuckridge, Tokoper, and Minu soil series), and Haplocambids (Koyen, Veet, Heist, Yomba, Whirlo, Keefa, Annaw, Lyx, Hollywell, and Eastgate soil

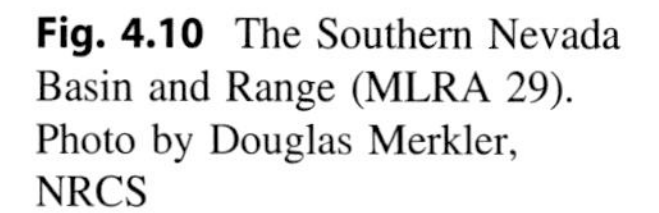

Fig. 4.10 The Southern Nevada Basin and Range (MLRA 29). Photo by Douglas Merkler, NRCS

series). Other great groups well represented in MLRA 29 include the Haplodurids (Ursine, Peeko, Greyeagle, Tybo, Lien, Garhill, Umil, and Osobb soil series) and Argixerolls (Bellehelen, Itca, Cropper, Brier, Squawtip, and Gochea soil series). The dominant great groups in the mountains are Torriorthents, Haplargids, and Argixerolls; in the valleys and basins, they are Torriorthents, some Haplargids, Argidurids, Haplocambids, and Haplodurids.

4.11 Soils of the Mojave Desert (MLRA 30)

About 28% of the Mojave Desert (MLRA 30) occurs in Nevada, accounting for 31,744 km^2 (12,256 mi^2). The cities of Las Vegas, Henderson, and Mesquite and the towns of Beatty, Pahrump, Searchlight, and Laughlin, Nevada are in this MLRA. Figure 4.11 is from near Bitter Ridge. Within the Basin and Range Province, the Mojave Desert contains Quaternary alluvial deposits on alluvial fans and valley floors. Playas occur at the lowest elevations in closed basins that are fringed by eolian deposits. The mountain ranges are isolated and are composed of metamorphic and igneous rocks (plutonic and volcanic subtypes) and some carbonate rocks (i.e., limestone).

The mean annual precipitation ranges from 50 to 200 mm (2–8 in), but may exceed 950 mm (27 in) at the higher elevations. Most of the rainfall occurs in the winter months as low-intensity precipitation from Pacific storms that are frontal in nature. High-intensity, convective thunderstorms can occur during the summer, but they contribute little to soil moisture. The average annual temperature ranges from 6°C (43°F) in the highest mountains to 25°C (76°F) along the Colorado River.

The dominant great groups in MLRA 30 are Torriorthents (Arizo, Carrizo, Yermo, St. Thomas, Sunrock, Boxspring, Goldroad, Haleburu, Theriot, Orwash, Potosi, Seanna, Threelakes, Nipton, Birdspring, Hypoint, Canoto, Nolena, Sanwell, and Upspring soil series) and Haplocalcids (Nickel, Weiser, Huevi, Zeheme, Tonopah, Commski, Riverbend, Meadview, Arada, Kurstan, Joemay, and Corbilt soil series). Haplodurids are also common in MLRA 30, including the Greyeagle, Crosgrain, Skelon, and Longjim soil series. Nevada's Gypsids occur mainly in this MLRA, around the Lake Mead area. Torriorthents occur in the mountains; basins contain Torriorthents, Haplocalcids, and Haplodurids.

4.12 Conclusions

A provisional analysis of the soils of Nevada by Major Land Resource Area suggests

- Torriorthents, Haplargids, Argixerolls, Argidurids, Haplocambids, and Haplodurids are the most conspicuous great groups in Nevada;

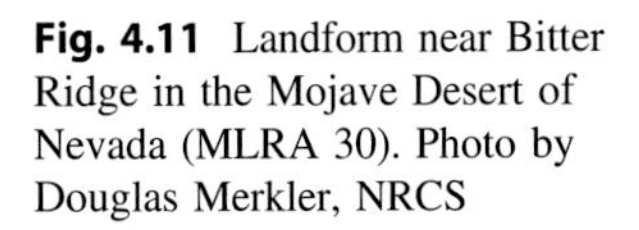

Fig. 4.11 Landform near Bitter Ridge in the Mojave Desert of Nevada (MLRA 30). Photo by Douglas Merkler, NRCS

- Mountain ranges contain Argixerolls, some Haplargids, and Torriorthents;
- Piedmont slopes contain Haplocambids, Natrargids, Torriorthents, Haplocalcids, Haplodurids, Argidurids, and some Haplargids; and
- Halaquepts and Aquisalids occupy moist areas of basin floors, and Torripsamments occur on dunes and sand sheets.

Reference

U.S. Dept. of Agric., Natural Resour. Conserv. Service (2006) Land Resource Regions and Major Land Resource Areas of the United States, the Caribbean, and the Pacific Basin. U.S. Dept. Agric, Handbook, p 296

Soil Geomorphology of Nevada

5

Abstract

This chapter focuses on soil geomorphology of each of the 10 Major Land Resource Areas in Nevada. Block diagrams are used to show the relations among soils, landforms, and geologic materials.

5.1 Introduction

When soil mapping began in earnest in Nevada in the early 1970s, it became apparent that the intermontane-basin topography of the Great Basin was unique and that a consistent system of geomorphic description in soil surveys was required. Dr. Frederick F. Peterson, Professor of Soil Science at the University of Nevada-Reno, who played a major role in the Desert Soil-Geomorphology Project (Gile and Grossman, 1979; Monger et al. 2009), was called upon to prepare his now classic, "Landforms of the Basin and Range Province: Defined for Soil Survey".

Peterson (1981) separated intermontane basins into bolsons (Fig. 5.1), which have internal drainage, and semi-bolsons (Fig. 5.2), which have external drainage. The basins and their flanking ranges were divided secondarily into landscapes (i.e., small-scale assemblages of spatially associated features) defined as mountains, piedmont slopes, and basin floors. The mountain ranges flanking the basins were defined on both a landscape and a landform scale and described appropriately. Major landforms (i.e., discrete, survey scale earth-surface features) within these three distinct landscapes include alluvial fans, rock pediments, inset fans, fan remnants, mountain-valley fans, fan piedmonts, fan skirts, alluvial flats, and playa or axial-stream floodplains (Fig. 5.3). This terminology was used extensively in the 57 soil surveys that cover Nevada's 16 counties.

5.2 Fan and Remnant Terminology

A fan is a generic term for constructional landforms that are built of more-or-less stratified alluvium and that occur on the piedmont slope, downslope from their source of alluvium. Types of fan landforms include (i) alluvial fans, which are cone- or fan-shaped deposits of alluvium that have their apex at a point-source of alluvium; (ii) fan aprons, which are landforms comprised of a sheet-like mantle of relatively young alluvium covering part of an older fan piedmont; (iii) fan piedmonts, which are an extensive, major landform formed by the lateral coalescing of mountain-front alluvial fans downslope into one generally smooth slope; (iv) fan skirts, which are landforms composed of laterally coalescing, small alluvial fans that issue from gullies cut into fan piedmonts; (v) inset fans, which are composed of deposits of stream alluvium on a narrow fan located between flanking fan remnants (defined below); (vi) mountain-valley fans, which result from alluvial filling of a mountain valley by coalescent valley fans whose toeslopes meet from either side of the valley along an axial drainage way; most mountain-valley fans have been dissected.

A remnant is defined as a remaining part of some landform which has been destroyed by erosion and/or buried under sediment. Types of remnant landforms include (i) basin-floor remnants, which are flattish-topped, erosional remnants of any landform on the basin floor resulting from stream dissection; and (ii) fan remnants, which are flattish-topped landforms which represent the remaining parts of older fan landforms that have been dissected or partially buried.

Other terms that are used in describing intermontane landforms include (i) ballenas, a form of fan remnant which are composed of round-topped ridgeline remnants of fan alluvium; (ii) lake plains (also called lake terraces), which are landforms on some bolson floors that are built of nearly level, fine-textured, stratified, bottom sediments of a Pleistocene lake; and (iii) playas, which are ephemerally flooded, barren areas on basin floors that are covered with silty or fine-textured sediment.

5.3 Soil Associations

Soil associations consist of two or more dissimilar major components occurring in a regular and repeating pattern on the landscape (Soil Science Division Staff 2017). Soil

P. W. Blackburn et al., *The Soils of Nevada*, World Soils Book Series,
https://doi.org/10.1007/978-3-030-53157-7_5

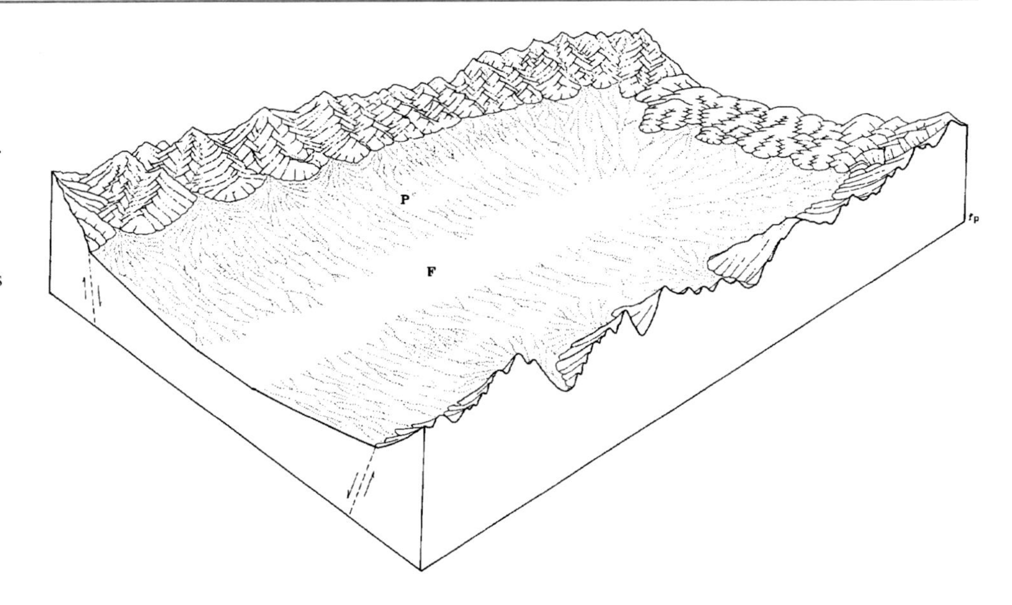

Fig. 5.1 The major physiographic parts of an internally drained intermontane basin, or bolson: the piedmont slope (P), and the basin floor (F). The bolson is bounded by mountain ranges on the left and right and by hills in the back. Drainageways are shown by dotted lines and suggest positions of major landforms. The drainageways and playas of the basin floor are not shown. (Source: Soil Survey of Eureka County, Nevada)

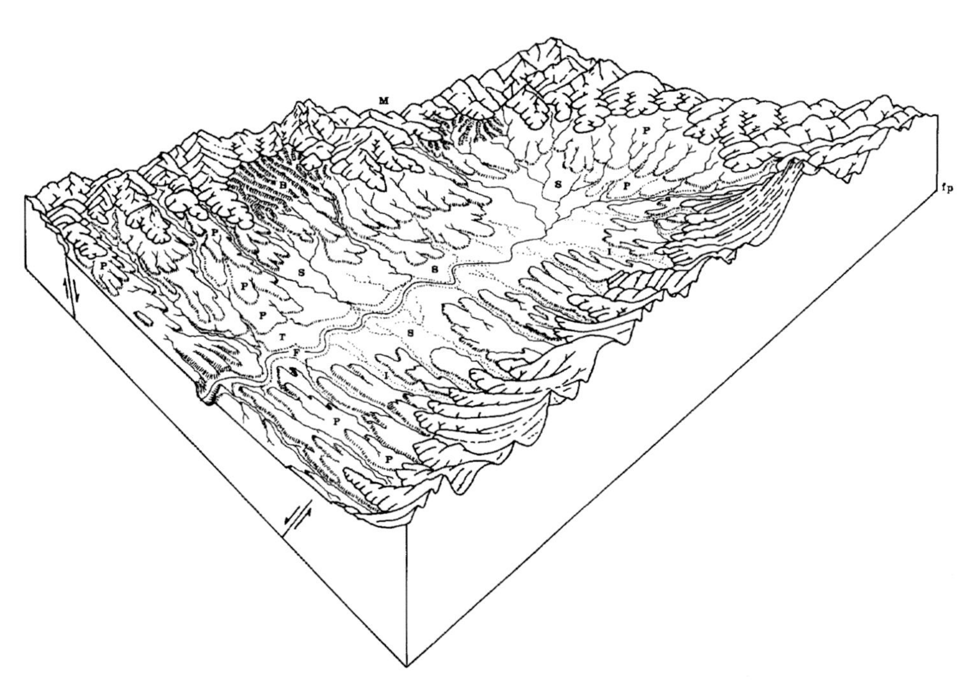

Fig. 5.2 A semi-bolson shown here displays the effects of several cycles of dissection and deposition. The major landforms are: ballenas (B), fan piedmonts composed of several levels or ages of fan remnants (P), fan skirts (S), an axial stream terrace (T), and an axial stream flood plain (F). All alluvial fans are not distinguished from fan piedmonts. Component landforms of inset fans (I) are between fan remnants. The basin is bounded on two sides by mountains (M). Source: Soil Survey of Eureka County, Nevada

associations generally are mapped at a scale of 1:24,000 and can be illustrated in block diagrams. Approximately 27 block diagrams have appeared in Nevada soil surveys at nine locations. This chapter employs block diagrams of multiple soil associations from published soil surveys of different MLRAs to illustrate soil-geomorphic relations in Nevada.

5.4 MLRA 22A—Sierra Nevada Mountains

The Sierra Nevada Mountains just reach into the state in southern Washoe, Carson City, and Douglas Counties (Fig. 1.4). The mountains are composed of plutonic rocks, dominantly quartz monzonite, and granodiorite, as a batholith. Although it is located just across the border in Alpine County, CA, the Meiss Meadows Area is representative of MLRA 22A; seven of the ten soil series shown are mapped in Nevada (Fig. 5.4). A part of the north-facing slope is mantled with till; colluvium mantles most of the area. A creek exits to the west. The Callat soil series (Humicryepts) occurs on the higher slopes and is derived from colluvium over till. The Meiss and Cagwin soil series are formed in residuum, with the Meiss soil (Humicryepts) on the uppermost cooler slopes and the Cagwin soil (Xeropsamments) on the lower, warmer slopes. The Dagget, Temo, and Toem soil series are formed in colluvium over residuum.

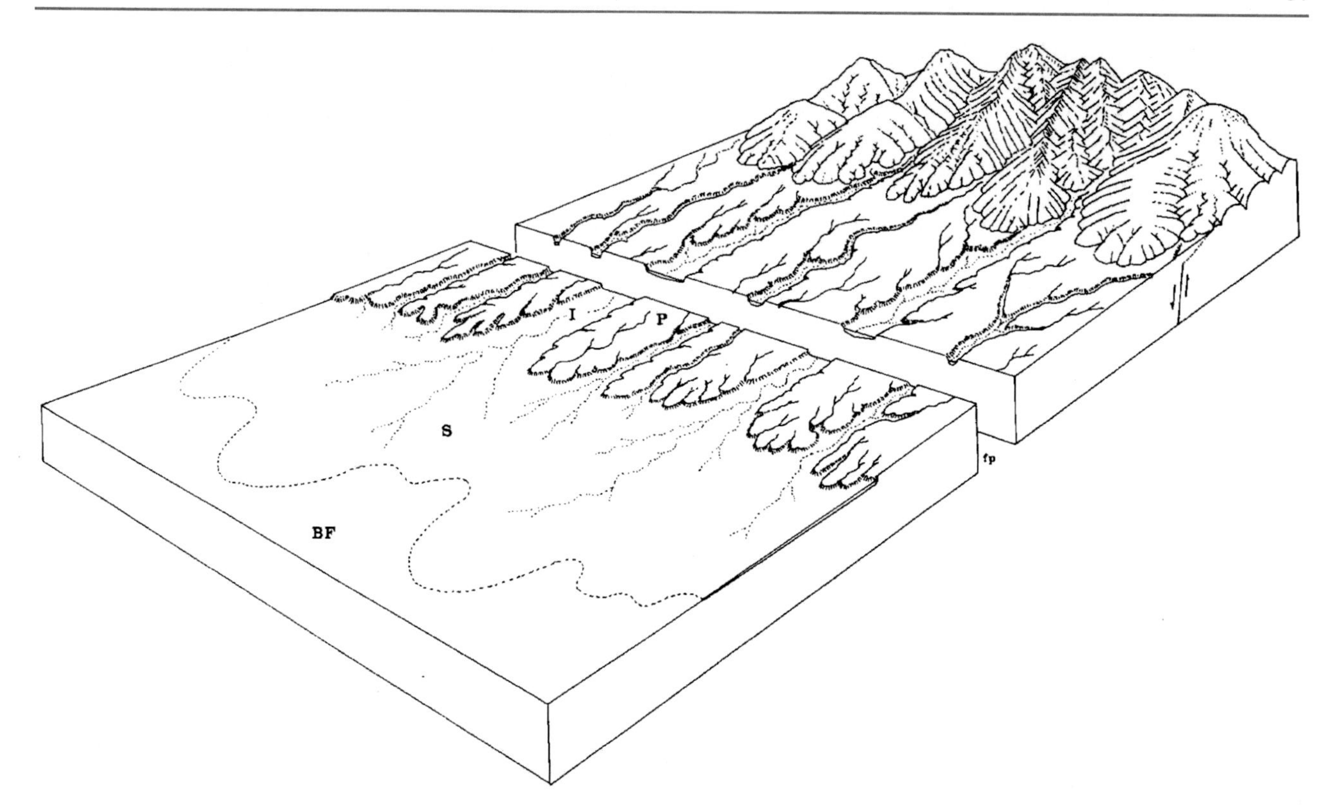

Fig. 5.3 A fan skirt (S) that merges along its lower boundary with a basin floor (BF) was formed by coalescing alluvial fans originating at gullies cut in a dissected fan piedmont (P) and by debouching inset fans (l) of the fan piedmont. The erosional fan piedmont remnants and mouths of the inset fans form the upper boundary of the fan skirt. It is the same age surface as the inset fans, but is younger than the relict summits of the fan remnants. The fan skirt may be the same age or younger than the basin floor surface, but in this diagram it is younger, because its alluvium overlaps the basin floor surface. Source: NRCS

Whereas the Dagget (Cryorthents) and Temo soils (Cryopsamments) are on steep (up to 75%) upper slopes, the Toem soil (Xeropsamments) is on steep lower slopes. The Bidart soil series (Cryaquepts) is derived from alluvium in the flood plain of the creek.

5.5 MLRA 23—Malheur High Plateau

The Malheur High Plateau (MLRA 23) is an intermontane plateau that is part of the Basin and Range Province and is partially located in northwestern Nevada. The area features uplifted fault blocks that are flat-lying or gently sloping and composed of volcanic rocks, mainly basalt and tuff. No block diagrams are available for the Nevada portion of the Malheur High Plateau. However, the southern part of Lake County, OR, contains soils similar to those in northwestern Nevada (Fig. 5.5). The Harcany soil series (Haplocryolls) is located on steep (up to 75%) upper side slopes and is derived from loess and ash over colluvium. The Newlands soil series (Argicryolls) is formed on less steep slopes (up to 50%) in colluvium over residuum. The Ratto soil series (Argidurids) is formed from alluvium and colluvium at lower elevations on fan remnants with slopes of 2–15%. The Brace and Floke soil series are also Argidurids but do not occur in Nevada.

5.6 MLRA 24—Humboldt Area

The Humboldt Area (MLRA 24) is in the Great Basin Section of the Basin and Range Province in north-central Nevada. The area consists of wide valleys filled with alluvium and isolated, dissected mountain ranges extending roughly north to south. Two block diagrams from northern Lander County will be used to illustrate soil-landform relationships, one of a bolson with internal drainage and one of a semi-bolson with external drainage. The bolson area (Fig. 5.6) features, along an elevational gradient, three soil associations in the mountains (3152, 3121, and 3846) that are composed of Argixerolls and Haplargids derived from loess and colluvium over residuum from rhyolite; a soil association on a fan piedmont (1281) composed of

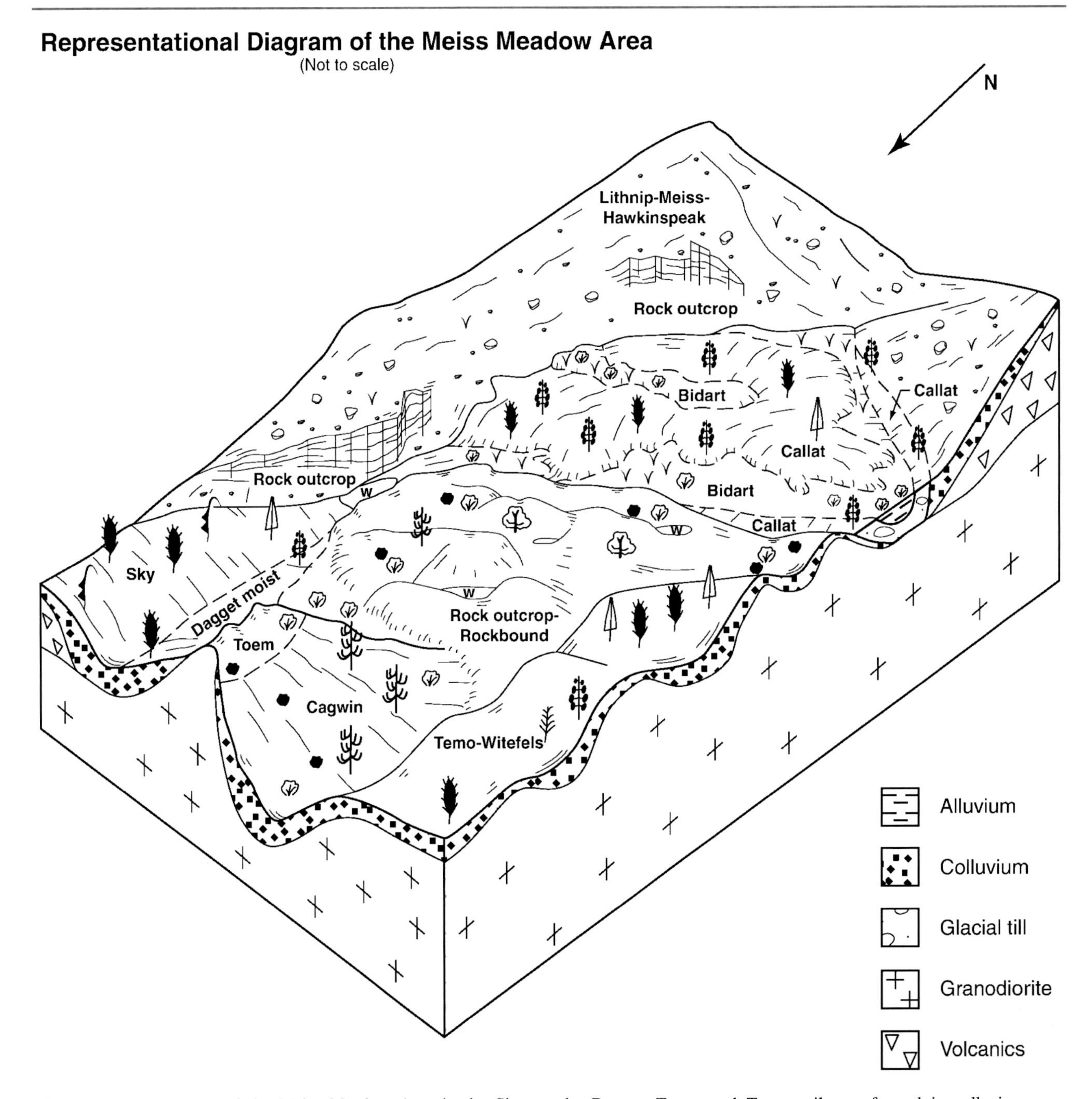

Fig. 5.4 Block diagram of the Meiss Meadow Area in the Sierra Nevada Mountains (MLRA 22A) along the border of California and Nevada. The Meiss soil is formed in residuum derived from volcanic rocks; the Cagwin soil is formed in residuum derived from granodiorite; the Dagget, Temo, and Toem soils are formed in colluvium over residuum derived from granodiorite; and the Bidart soil is derived from alluvium. Source: NRCS

Natrargids, Haplocambids, and Haplargids; a fan skirt (701) with Haplocambids; and alluvial flats (1143) with Halaquepts. Figure 5.7 of a semi-bolson shows three soil associations in the mountains (16, 19, and 21) that are composed of Argixerolls and some Haplargids on colluvium, often with a loess mantle over residuum. Association 16 is at the break with a fan remnant. The mid-slope association contains a fan piedmont (12) with Haplodurids and Natridurids and two fan remnants (7 and 11) with Argidurids, Haplargids, Haplodurids, and Haplocambids). The basin contains an alluvial flat (3) with Fluvaquents, Haploxerolls, and Endoaquolls.

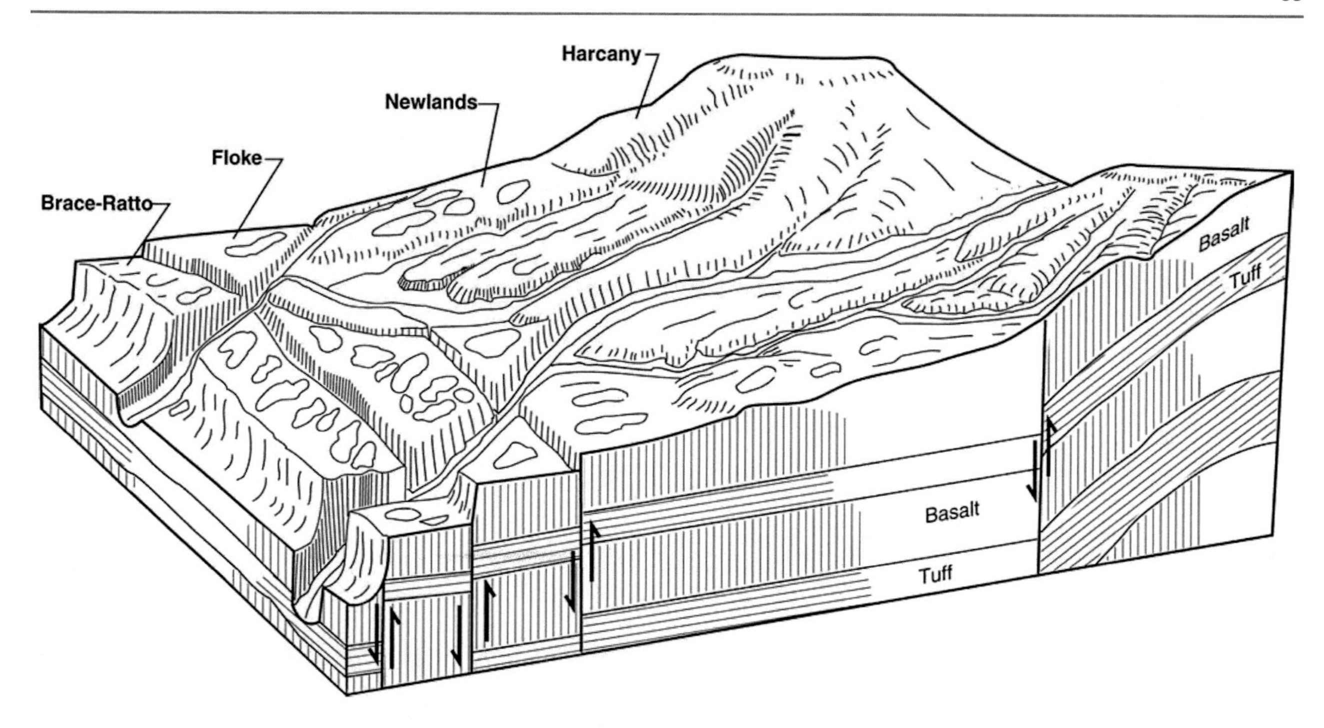

Fig. 5.5 The Ratto soil association is from southern Lake County, Oregon, just across the Nevada border. It is representative of the Malheur High Plateau (MLRA 23). The Harcany soil (Haplocryolls) is derived from loess and ash over colluvium on side slopes; the Newlands soil (Argicryolls) is formed in colluvium over residuum derived from basalt on side slopes; the Ratto soil (Argidurids) is formed in alluvium and colluvium on a fan remnant. The Brace and Floke soils do not occur in Nevada. Source: NRCS

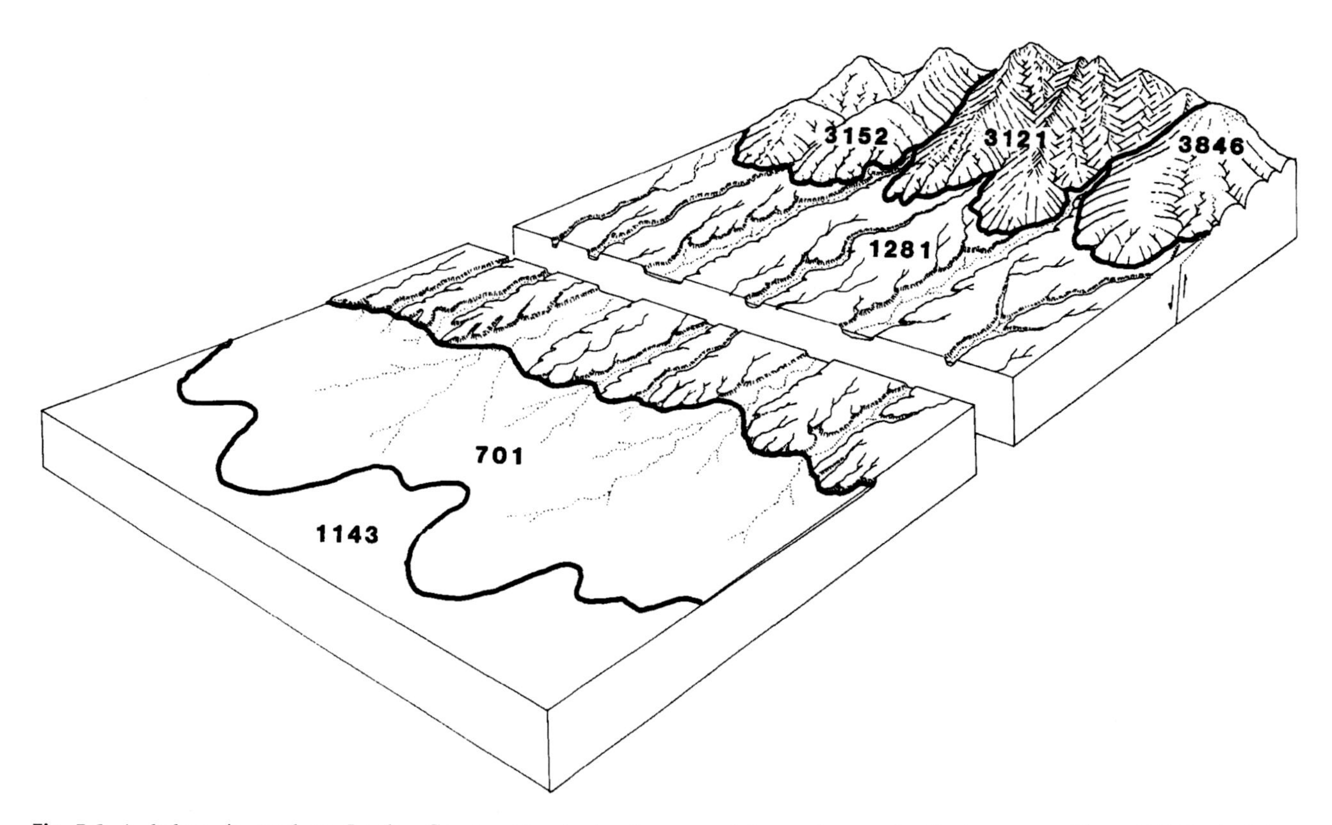

Fig. 5.6 A bolson in northern Lander County, Nevada in the Humboldt Area (MLRA 24). The mountains contain the Robson (Haplargids)-Reluctan (Argixerolls) association on colluvium over residuum from rhyolite (3152) and the Walti-Softscrabble-Bucan (Argixerolls) association (3121) and Jung-Wiskan (Haplargids) association on loess and colluvium over residuum (3846). The Ricert (Natrargids)-Whirlo (Haplocambids)-Pineval (Haplargids) association (1281) occurs on a fan piedmont; the Orovada (Haplocambids) association (701) occurs on a fan skirt; and the Wendane (Halaquepts) association (1143) occurs on alluvial flats. Source: NRCS

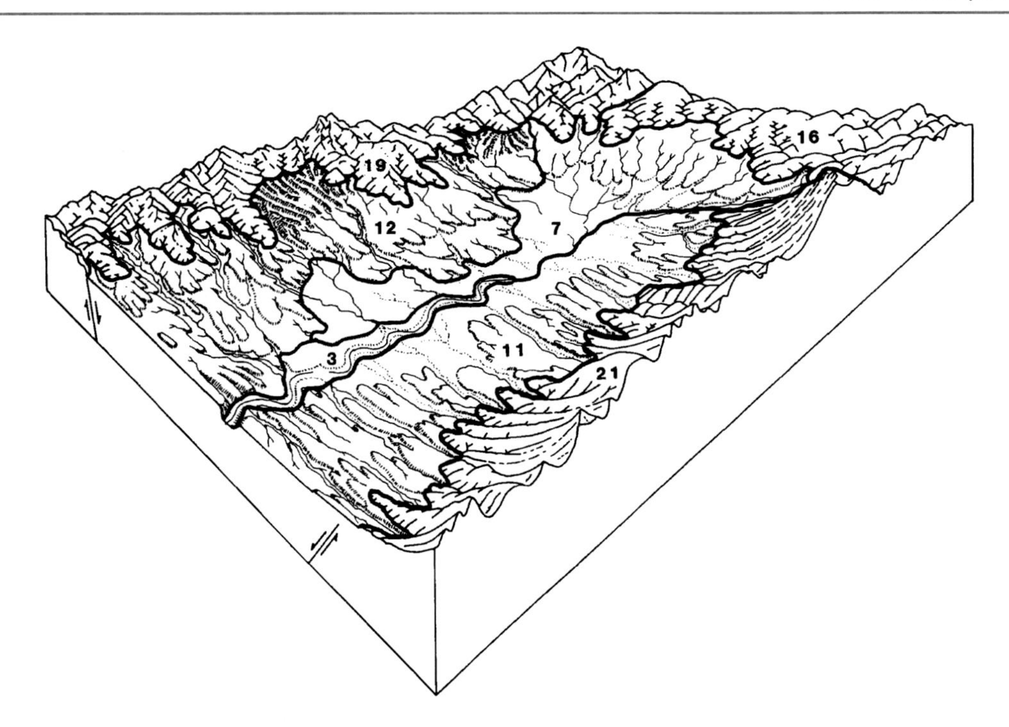

Fig. 5.7 This block diagram is on a semi-bolson in the Humboldt Area (MLRA 24) in northern Lander County, Nevada. The mountains contain three associations, including the Quarz (Argixerolls)-Walti (Argixerolls)-Glean (Haploxerolls) association (21) and the Sumine-Chen (Argixerolls) association (19) on colluvium over residuum. The Robson (Haplargids)-Akerue (Argidurids)-Buffaran (Argidurids) association (16) is on hills and is formed in alluvium and residuum. The fan remnant contains three associations including the Orovada (Haplocambids)-Broyles (Haplocambids)-Shabliss (Haplodurids) association (7), the Buffaran (Argidurids)-Allor (Haplargids)-Chiara (Haplodurids) association (11), and the Broyles-Chiara-Cortez (Natridurids) association (12). The Sonoma (Fluvaquents)-Rixie (Haploxerolls)-Paranat (Endoaquolls) association (3) occupy alluvial flats, terraces, and lake plains on the basin floor. Source: NRCS

5.7 MLRA 25—Owyhee High Plateau

The Owyhee High Plateau (MLRA 25) is an intermontane plateau within the Basin and Range Province that is located in northeastern Nevada. The dominant rock types are volcanic. Figure 5.8 shows a semi-bolson in central Elko County. The mountains features three soil associations (261, 576, and 241) composed mainly of Haplargids and Argixerolls formed in colluvium over residuum derived from rhyolite, with a discontinuous mantle of loess and volcanic ash. Flanking the mountains are a fan piedmont (511) composed of Argidurids, Haplargids, and Haplocambids; a fan skirt (110) with Torriorthents, Halaquepts, and Haplocambids); and the basin floor (440) with Endoaquolls.

5.8 MLRA 26—Carson Basin and Mountains

The Carson Basin and Mountains (MLRA 26) are in the Great Basin Section of the Basin and Range Province and contain isolated north-south-trending mountain ranges separated by aggraded semi-desert valleys (i.e., semi-bolsons). No block diagrams exist to illustrate the common soil associations in the area. However, a sequence of soils in a typical portion of the MLRA includes flanking mountains of igneous origin (volcanic and plutonic rock) and semi-bolsons with dissected piedmont slopes. The mountains contain the Oppio and Xman soil series (Haplargids) derived from residuum, while the Terca soil series (Argixerolls) occurs in areas with colluvium and residuum derived from volcanic rocks. The semi-bolsons contain the Fulstone soil series (Argidurids), which occurs on fan remnants and formed in alluvium derived from mixed igneous rocks.

5.9 MLRA 27—Fallon-Lovelock Area

The Fallon-Lovelock Area (MLRA 27) is part of the Great Basin Section of the Basin and Range Province and contains isolated mountain ranges trending north-to-south that are separated by broad, aggraded desert valleys (i.e., bolsons). A block diagram from the eastern part of Pershing County (Fig. 5.9) shows a broad valley with piedmont slopes and a basin floor flanked by mountains and hills. The mountains (11) contain Argixerolls, Haplargids, and Haploxerolls formed in colluvium and residuum derived from mixed sources. Hills (13) contain Torriorthents, Haplocalcids, and Calcixerolls formed in colluvium and residuum derived from limestone. The fan remnants and fan skirts (6) contain

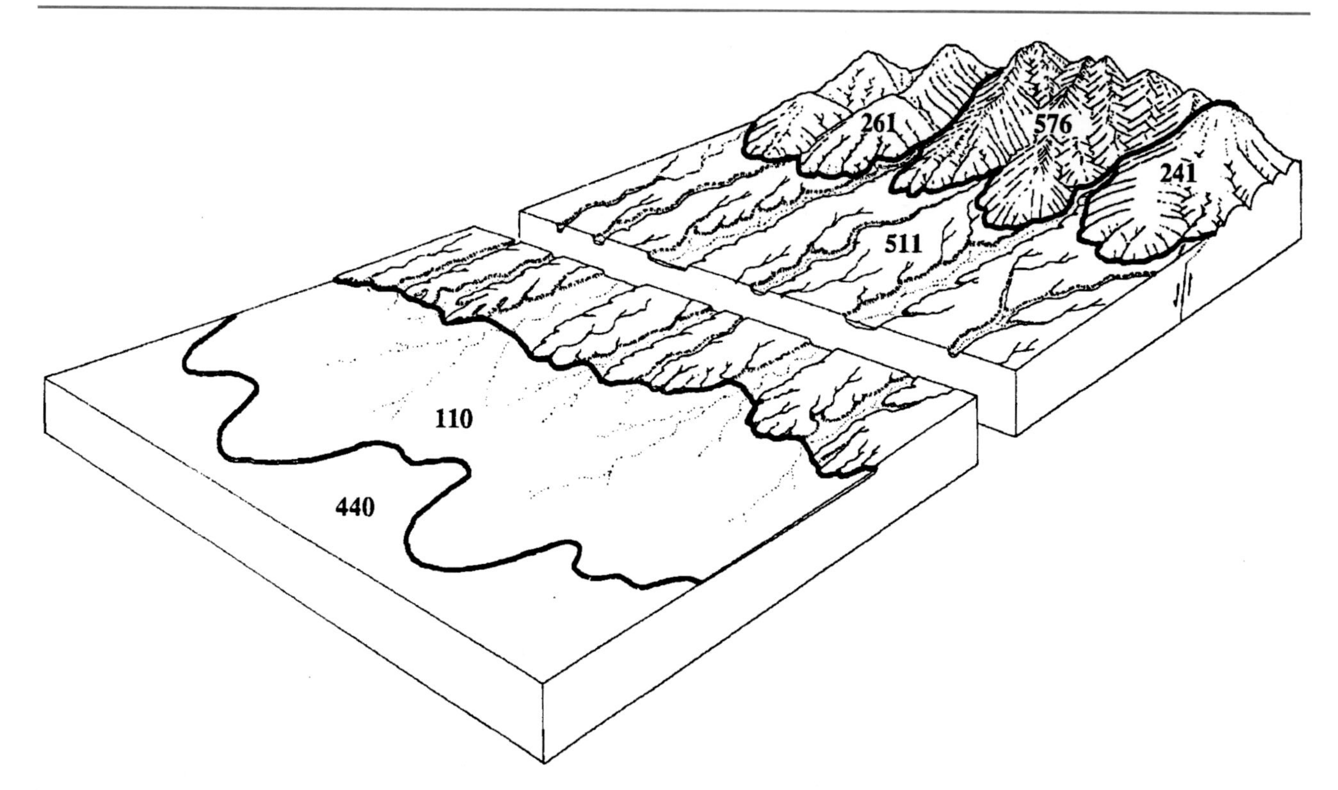

Fig. 5.8 This block diagram is of a semi-bolson on the Owyhee High Plateau (MLRA 25) in central Elko County, Nevada. The mountains flanking the basin contain three associations, including the Sumine (Argixerolls)-Cleavage (Argixerolls)-Hapgood (Haplocryolls) association (576), the Cleavage-Loncan (Haploxerolls) association (241), and the Linkup-Roca-Vanwyper (all Haplargids) association (261) on colluvium and residuum derived from rhyolite and other materials. The Hapgood soil series has a mantle of loess and volcanic ash. The dissected fan piedmont (511) has the Dacker (Argidurids)-Gance (Haplargids)-Kelk (Haplocambids) association; the fan skirt (110) has the Moranch (Torriorthents)-Ocala (Haplaquepts)-Orovada (Haplocambids) association; and the basin floor (440) has the Devilsgait-Woofus (both Endoquolls) association. Source: NRCS

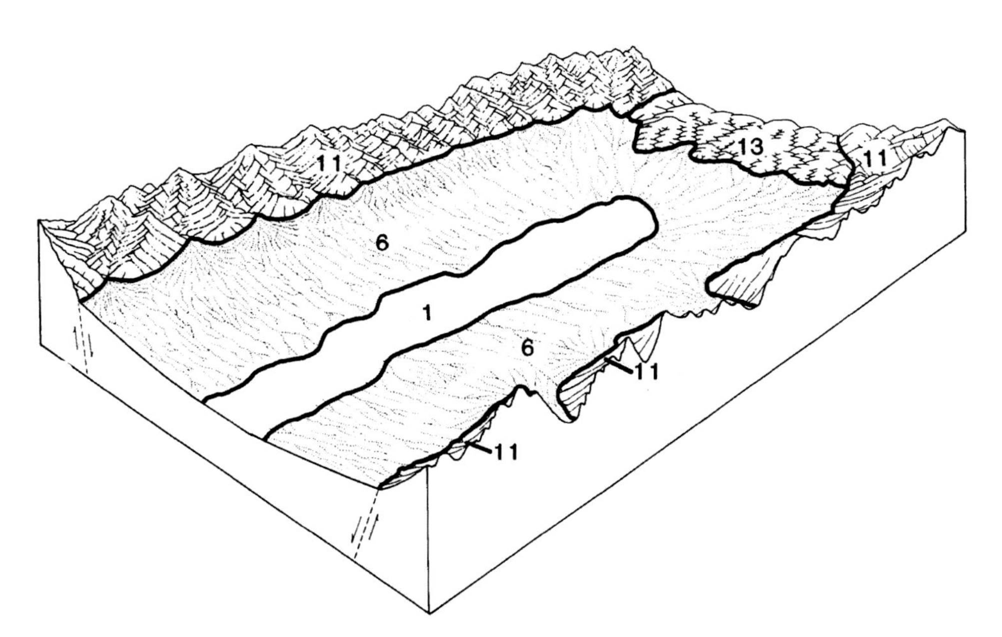

Fig. 5.9 Block diagram in the Fallon-Lovelock Area (MLRA 27) of eastern Pershing County, Nevada. Two associations include the Puffer (Torriorthents)-Mulhop (Haplocalcids)-Xine (Calcixeroll) association (13) on hills and the Reluctan (Argixerolls)-Roca (Haplargids)-Iver (Haploxerolls) association (11) on mountains, and formed in colluvium and residuum derived from limestone and volcanic rocks, respectively. The fan remnants and fan skirts (6) contain the Jerval (Natrargids)-Dun Glen (Haplocambids)-Tenabo (Natridurids) association, with a barren playa (1) situated on the basin floor. Source: NRCS

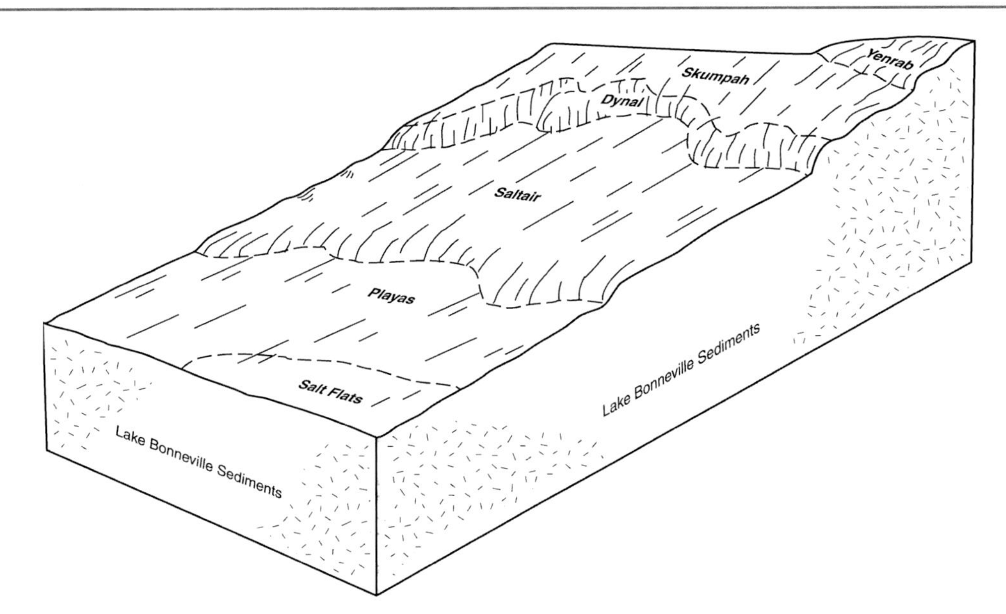

Fig. 5.10 This block diagram of the Bonneville Basin in northwestern Utah is representative of the Great Salt Lake Area (MLRA 28A). The diagram shows lake terraces with the Yenrab soil series (Torripsamments), alluvial flats with the Skumpah soil series (Natrargids), a beach cliff with the Dynal soil series (Torripsamments). The Saltair soil series (Aquisalids) is on a portion of the basin floor adjacent to playas and salt flats. Of the four soil series, only the Saltair soil occurs in Nevada. Source: NRCS

Natrargids, Haplocambids, and Natridurids formed in alluvium derived from mixed sources. The playa on the basin floor is a miscellaneous area with no discernible soils.

5.10 MLRA 28A—Great Salt Lake Area

The Great Salt Lake Area (MLRA 28A) is an area of nearly level basins between widely separated mountain ranges trending north to south and is part of the Great Basin Section of the Basin and Range Province. The block diagram is a bolson from the Tooele Area in Western Utah that includes parts of White Pine and Elko Counties in Nevada (Fig. 5.10). The parent materials are pluvial Lake Bonneville sediments. The Skumpah soil series (Natrargids) and Yenrab soil series (Torripsamments) are mapped right up to the Nevada border and occur on alluvial flats and lake terraces, respectively. The Saltair soil series (Aquisalids) occurs in Nevada and formed in lacustrine deposits situated on basin floors. Playas and salt flats are miscellaneous areas with no discernible soils.

5.11 MLRA 28B—Central Nevada Basin and Range

The Central Nevada Basin and Range (MLRA 28B) is part of the Great Basin Section of the Basin and Range Province, is located entirely in Nevada, and contains aggraded desert basins between a series of mountain ranges trending north to south. Figure 5.11 is a block diagram from western White Pine County. The mountains are composed of three soil associations (436, 1430, and 113) that vary in elevation. The highest mountains (1430) contain soils with a cryic soil temperature regime, including Cryrendolls, Haplocryalfs, and Calcicryolls, which formed in colluvium and residuum derived from limestone. The lower mountain slopes contain calcareous soils derived from limestone, including Haplocalcids, Calcixerolls, and associated Haploxerolls and Torriorthents. The fan remnant (290) contains Haplodurids and Torriorthents; the fan skirt (351) has Haplocambids and Torriorthents; and the lake plain (250) has Torriorthents.

5.12 MLRA 29—Southern Nevada Basin and Range

The Southern Nevada Basin and Range (MLRA 29) is part of the Great Basin Section of the Basin and Range Province. Figure 5.12 is a block diagram of a bolson and flanking mountain range in Mineral County, Nevada. The mountain range contains two soil associations (4170 and 1241) with Haplargids and Torriorthents formed in colluvium and residuum derived from basic igneous rocks; the dissected fan piedmont (5100) contains Calciargids and Torriorthents; the fan skirt (1155) has Torriorthents; and the basin floor (1441) has Torrifluvents.

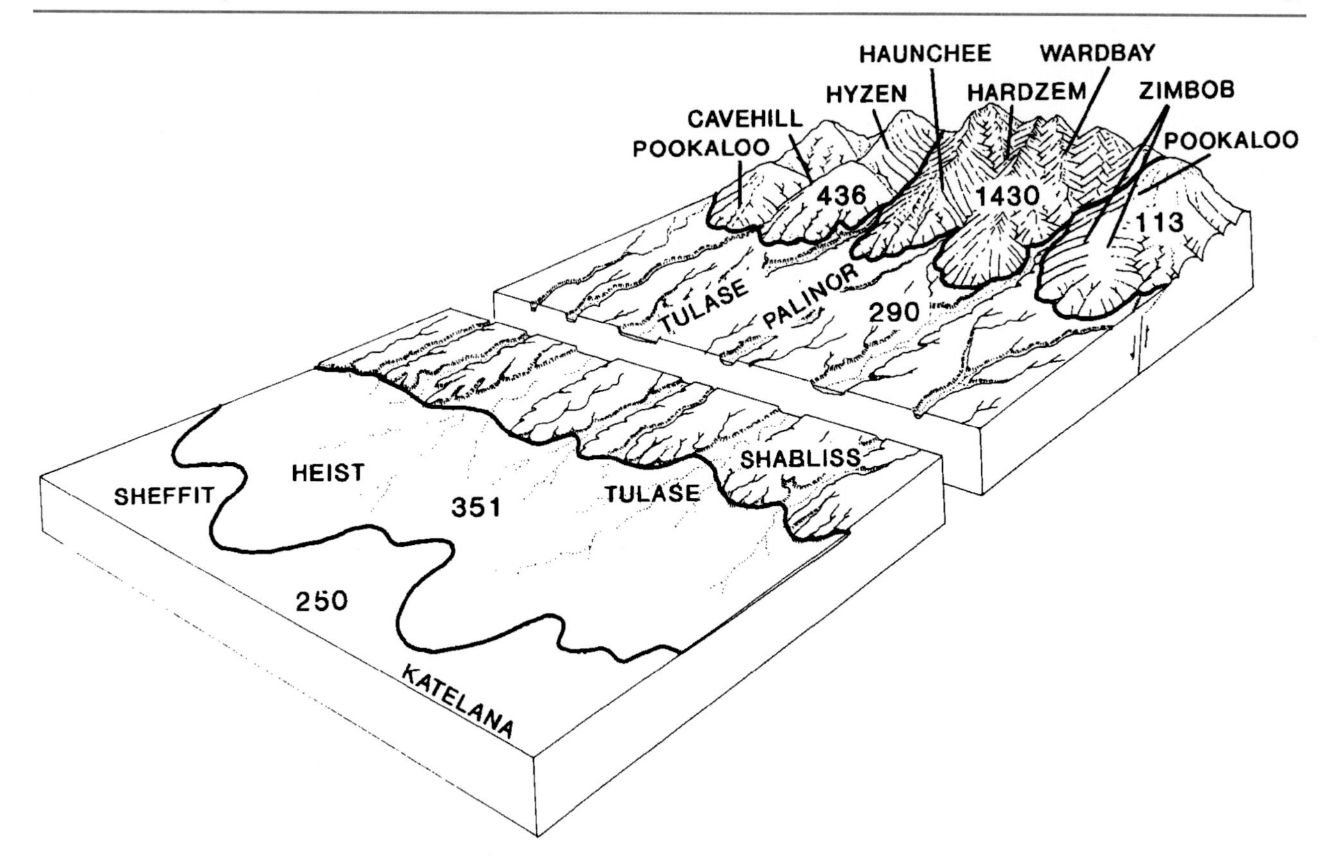

Fig. 5.11 This block diagram occurs in western White Pine County in the Central Nevada Basin and Range (MLRA 28B). Three associations in the mountains have soils formed in colluvium and residuum derived from limestone, including the Pookaloo (Haplocalcids)-Cavehill (Calcixerolls)-Hyzem (Haploxerolls) association (436), the Haunchee (Cryrendolls)-Hardzem (Haplocryalfs)-Wardbay (Calcicryolls) association (1430), and the Zimbob (Torriorthents)-Pookaloo association (113). The fan remnant (290) contains mainly the Shabliss and Palinor soil series (Haplodurids); the fan skirt (351) contains the Tulase soil series (Torriorthents) and Heist soil series (Haplocambids); and the lake plain (250) has the Sheffit and Katelana (both Torriorthents) soil series. Source: NRCS

5.13 MLRA 30—Mojave Desert

The Mojave Desert (MLRA 30) is in an externally drained portion of the Basin and Range Province. Most of the area is underlain by Quaternary alluvial and lacustrine deposits on piedmont slopes and basin floors. There is no block diagram shown for MLRA 30. A typical association of soils is described that extends from mountains, which are composed mainly of limestone, to basin floors. The mountains contain the Zeheme (Haplocalcids) and Birdspring soil series (Torriorthents) that are formed in colluvium and residuum derived from limestone. At the base of the mountain slopes, a fan piedmont contains the Mormount and Wechech soil series (Petrocalcids). Fan remnants contain the Irongold and Lastchance soil series (Petrocalcids) and the Joemay soil series (Haplocalcids). Rounded fan remnants (i.e., ballenas) contain the Purob and Ferrogold soil series (Petrocalcids) and the basin floor contains the Las Vegas soil series (Petrocalcids).

5.14 Conclusions

Block diagrams showing multiple soil associations exist for most of the ten Major Land Resource Areas (MLRA) in Nevada. Examples are reproduced here to show the relations between geomorphology (i.e., landscapes, landforms) and the soil series and some miscellaneous areas in the state. Since most of the state is located in the Basin and Range

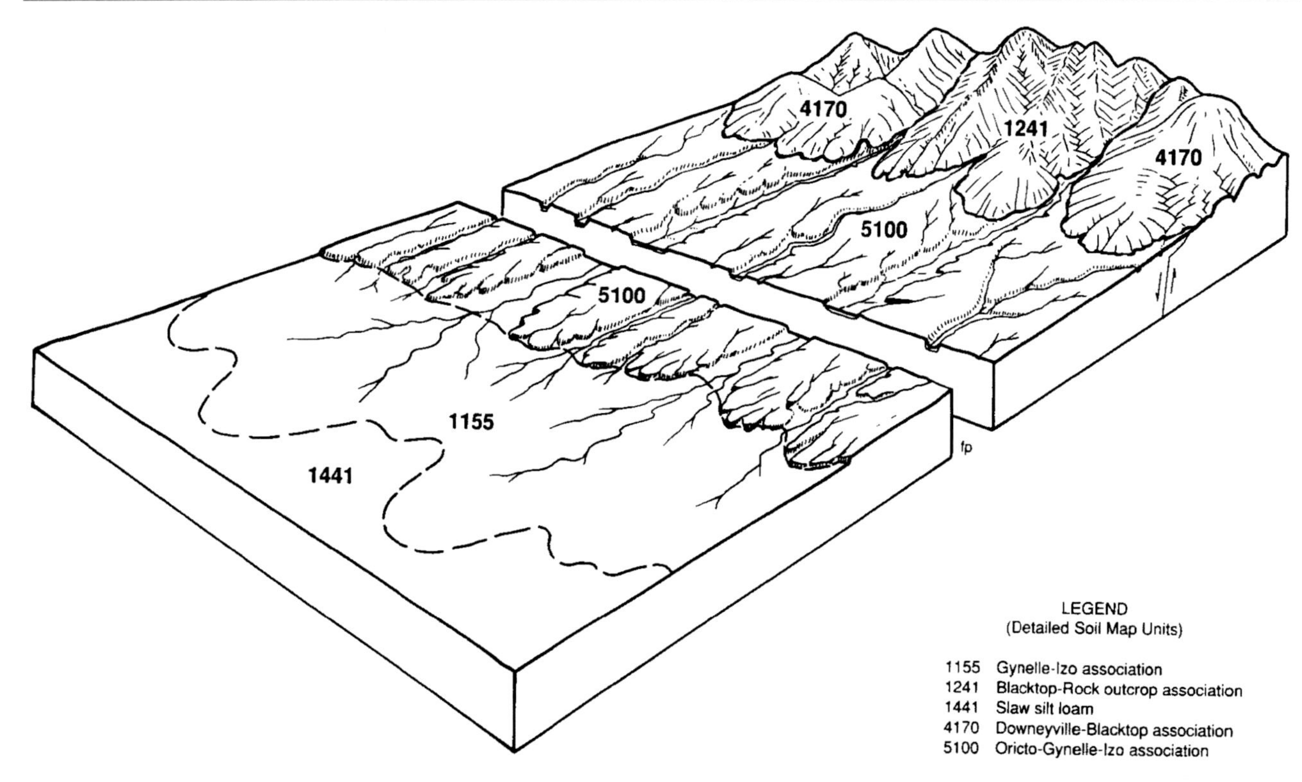

Fig. 5.12 A block diagram from Mineral County, Nevada, illustrates part of the Southern Nevada Basin and Range (MLRA 29). The mountains are composed of the Downeyville (Haplargids)-Blacktop (Torriorthents) association (4170) and the Blacktop-Rock outcrop association (1241) on colluvium and residuum derived from volcanic rocks. The heavily dissected fan piedmont contains the Oricto (Calciargids)-Gynelle (Torriorthents)-Izo (Torriorthents) association (5100). The fan skirt (1155) has the Gynelle-Izo association; and the basin floor has the Slaw soil series (Torriorthents). Source: NRCS

Province, the diagrams show mountain ranges with their associated soils. The basins are either bolsons (internal drainage) or semi-bolsons (external drainage). These basins include the piedmont slopes and their component landforms of alluvial fans, fan aprons, fan piedmonts, inset fans and the basin floors that may include component landforms of alluvial flats, axial-stream flood plains, and playas. Whereas Haplargids and Argixerolls may dominate in the mountains, the piedmont slopes contain soils with duripans as well as cambic, argillic, natric, calcic, and petrocalcic horizons. The basin floor landforms contain Endoaquolls, Halaquepts, and Aquisalids as well as miscellaneous areas such as playas and salt flats with lack soils.

References

Gile LH, Grossman RB (1979) The Desert Project soil monograph: soils and landscapes of a desert region astride the Rio Grande Valley near Las Cruces, New Mexico. Lincoln, NE: US Dept. of Agriculture, Soil Conservation Service

Monger HC, Gile LH, Hawley JW, Grossman RB (2009) The desert project—an analysis of aridland soil-geomorphic processes. NM State Univ Agr Exp Stn Bull, p 798

Peterson FF (1981) Landforms of the Basin & Range Province defined for soil survey. Univ. Nevada-Reno, College of Agriculture, Tech Bull 28, 52 pp http://www.nv.nrcs.usda.gov. Accessed 04 Nov 2018

Soil Science Division Staff (2017) Soil Survey Manual. 2nd edit. U.S. Dept. Agric., Handb. No. 18. 603 pp

6 Diagnostic Horizons and Taxonomic Structure of Nevada Soils

Abstract

The soils of Nevada contains 6 of 8 diagnostic surface horizons and 10 of the 20 diagnostic subsurface horizons recognized in Soil Taxonomy. In addition, the state contains soils representative of 7 of the 12 orders, 27 of the 68 suborders, 69 of the 337 great groups, 300 subgroups, 1,012 families, and over 1,800 soil series. Compared to other states, Nevada has a complex soil taxonomic structure and a high pedodiversity. A key is provided that enables classifying Nevada soils to the great-group level.

6.1 Introduction

Six of the 8 epipedons (diagnostic surface horizons) recognized in *Soil Taxonomy* (1999, 2014) and 10 of the 20 diagnostic subsurface horizons occur in the soils of Nevada. The State of Nevada contains soils representative of 7 of the 12 orders, 27 of the 68 suborders, 69 of the 337 great groups, 300 subgroups, 1,012 families, and over 1,800 soil series.

6.2 Diagnostic Horizons

Recognizing the diagnostic surface horizons (epipedons) and diagnostic subsurface horizons that occur in the soils of Nevada is fundamental in classifying them using soil taxonomy. Based on occurrence in soil series, diagnostic surface horizons can be ranked: ochric (74%), mollic (24%), umbric (1.6%), histic (0.4%), and folistic (0.1%) (Table 6.1). The aerial extent of soils having the recently redefined anthropic epipedon (Soil Survey Staff 2014) is unknown at this time but is probably less than 0.1% in Nevada. The ranking of occurrence of epipedons in Nevada soils is comparable to that of the USA (Bockheim 2014), except that the mollic epipedon is slightly underrepresented in Nevada.

Thicknesses of epipedons for Nevada soil series can be ranked: umbric 47 × 25 cm, mollic 38 × 23 cm, histic 23 × 6 cm, and ochric 15 × 8 cm. Soils with mollic or umbric epipedons greater than 40 cm in thickness are typically classified in Pachic and Cumulic subgroups. There are 67 soil series in Nevada that are classified in Pachic subgroups and 22 in Cumulic subgroups; and all except one are Mollisols. Soils in Pachic (Gr. *pachys*, thick) subgroups have a progressive accumulation of soil organic C, which results in an over thickened epipedon with no evidence of new parent material at the surface. In contrast, soils classified in Cumulic (L. *cumulus*, heap) subgroups also have a over thickened epipedon, but occur on slope gradients of less than 25%, have received new parent materials at the soil surface, and have an irregular distribution of soil organic C with depth, which is evidence of recent alluvial stratification. In Nevada, soils in Pachic and Cumulic subgroups support either big sagebrush or a pinyon-juniper canopy over short (bunch) grasses. 11 soil series with an umbric epipedon supports subalpine conifers or sedges and rushes. Only three soil series in Nevada have a histic epipedon. These soils occupy less than 10 km^2 and occur in Humaquepts or Cryaquepts great groups.

Diagnostic subsoil horizons can be ranked by occurrence in soil series: argillic (41%), duripan (17%), calcic (14%), cambic (11%), natric (4.7%), and petrocalcic (2.2%) (Table 6.1). Albic, gypsic, petrogypsic, and salic horizons occur only minimally in soils of Nevada but are locally important. The thickness of diagnostic subsurface horizons in Nevada soil series (more than three occurrences) can be ranked: salic 93 ± 84 cm, gypsic 79 ± 42 cm, petrocalcic 78 ± 44 cm, calcic 60 ± 41 cm, duripan 44 ± 33 cm, argillic 40 ± 28 cm, natric 32 ± 18 cm, and cambic 30 ± 18 cm. One-third (33%) of the soil series containing a duripan in the U.S. occur in Nevada.

Argillic horizons that are 50 cm or greater in thickness occur in Argixerolls, Palexerolls, Haplargids, and Paleargids that are formed in colluvium and alluvium that have received

P. W. Blackburn et al., *The Soils of Nevada*, World Soils Book Series,
https://doi.org/10.1007/978-3-030-53157-7_6

Table 6.1 Diagnostic horizon thicknesses for Nevada soil series

Horizon	Mean	Std Dev	% of soil series
Umbric	47	25	1.3
Mollic	39	24	27.5
Histic	23	6	0.3
Folistic	18	–	0.1
Ochric	15	7	70.8
Salic	111	67	0.3
Gypsic	84	39	0.7
Petrocalcic	79	42	0.3
Petrogypsic	61	75	0.3
Calcic	60	41	15.2
Duripan	43	33	16.2
Argillic	40	28	41.7
Natric	32	18	4.3
Cambic	30	18	11.2
Albic	17	19	0.7

loess and volcanic ash additions and support sagebrush and pinyon-juniper vegetation. Calcic horizons that are 60 cm or greater in thickness occur mainly in Haplocalcids formed in alluvium and eolian material and support shadscale, creosote bush, or sagebrush. Duripans that are 60 cm or greater in thickness are mainly in Argidurids and Haplodurids that are formed exclusively in alluvium; more than two-thirds also have an argillic horizon; and sagebrush is the dominant vegetation. Natric horizons occur in Natrargids and Natridurids that are formed in alluvium but have received loess and volcanic ash additions and support shadscale or greasewood vegetation. Petrocalcic horizons occur mainly in Petrocalcids formed in alluvium on fan remnants and ballenas; 77% also have a calcic horizon; and 69% support creosote bush or Black Bush vegetation.

Five soil series in Nevada contain three diagnostic subsurface horizons. Whereas the Ursine soil has calcic and petrocalcic horizons and a duripan; the Acana, Ackett, Lojet, and Troughsprings soils have argillic and calcic horizons and a duripan. In desert soils, the argillic horizon normally forms close to the surface. Clay particles subject to translocation are often in the medium (0.2–0.08 μm) and fine (<0.08 μm) clay fractions; and movement in the field is evidenced by clay films on faces of peds, lining pores, and clay bridging coarse silt and sand grains. The calcic horizon, where it occurs in conjunction with an argillic horizon, occurs below the argillic horizon. Dissolved $CaCO_3$ and clay-sized dust enriched in $CaCO_3$ move downward in the soil profile, coating peds, and fragments and gradually filling voids over time. In extreme cases and over long periods of soil development, all of the pores become filled and cemented, resulting in a petrocalcic horizon (colloquially called calcrete or caliche). When a duripan occurs in soils that also have an argillic and/or calcic horizon, it forms the lowermost in the profile. The duripan is formed by finely divided opaline silica (Chadwick et al. 1987) and likely forms early in the desert weathering process. Duripans can occur in conjunction with petrocalcic horizons, but take precedence for taxonomic classification as indicated by the key to suborders for Aridisols, where Durids precede Calcids in the keying sequence.

The relative positions of argillic and calcic horizons and duripans is supported by analytical data for the Acana soil series. Based on the distribution of clay, the argillic horizon occurs from 8 to 25 cm; based on the depth-distribution of $CaCO_3$, the calcic horizon occurs from 25 to 61 cm; and based on the ratio of 1500 kPa water/clay ratios, a relative measure of "microagglomeration" of silt and clay particles (Chadwick et al. 1989), the duripan occurs from 61 to 81 cm in depth (Fig. 6.1).

6.3 Orders

The Aridisols, Mollisols, and Entisols compose 97% of the soil area in Nevada (Table 6.2; Fig. 6.2). The remaining soil area is composed of Inceptisols (1.9%), Alfisols (0.5%), and Vertisols (0.4%), with a small area containing Andisols.

6.4 Suborders

More than three-quarters (88%) of the soil area in Nevada are in six suborders: the Argids (20%), Orthents (19%), Xerolls (18%), Durids (14%), Calcids (10%), and Cambids (7.1%) (Fig. 6.2).

6.5 Great Groups

Three-quarters (75%) of the soil area in Nevada are in seven great groups, including the Torriorthents (19%), Haplargids (16%), Argixerolls (13%), Haplocalcids (7.8%), Haplocambids (7.1%), Haplodurids (6.7%), and Argidurids (6.4%) (Fig. 6.2).

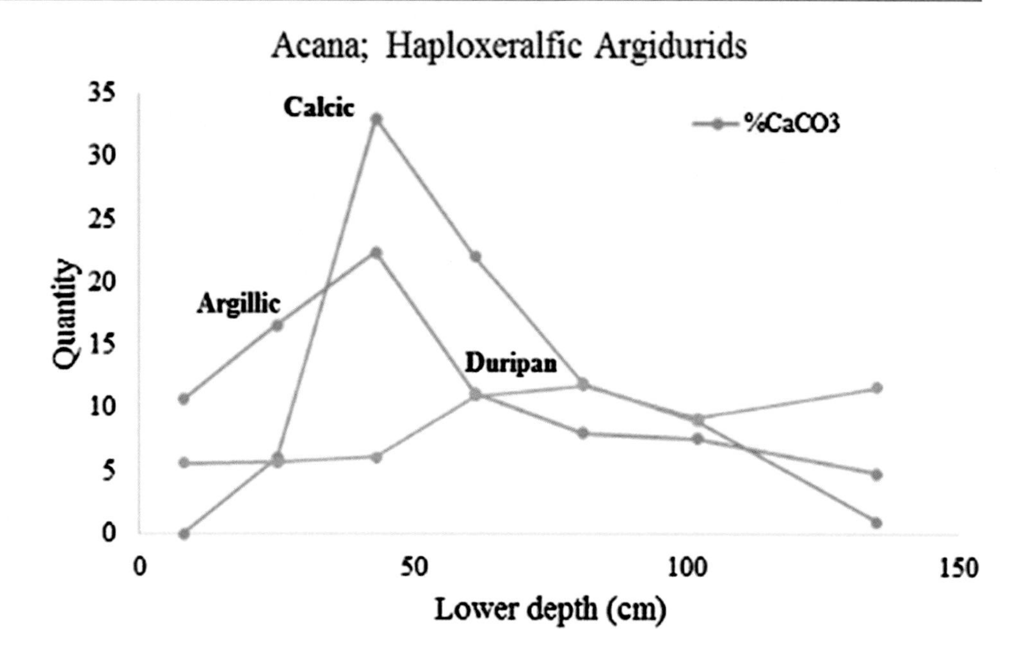

Fig. 6.1 Position of the argillic and calcic horizons and the duripan in the Acana soil series (Haploxeralfic Argidurids). *Source* NRCS database

6.6 Subgroups

About 19% of Nevada's soil series are classified in Lithic (Gr. lithos, stone) subgroups, which represent soils with a lithic contact (i.e., hard, slightly fractured bedrock) that occurs within 50 cm of the surface. These soils occur primarily on the hills, mountain ranges, and high plateaus of the Great Basin. About 15% of the Nevada soil series are classified in Xeric subgroups, primarily reflecting the presence of Aridisols which border the xeric soil moisture regime. About 10% of the Nevada soil series are classified in Aridic subgroups; these are primarily Xerolls which have an aridic soil moisture regime that borders on xeric. About 9% of the soil series of Nevada are classified in Durinodic and related ("dur") subgroups of Aridisols, Entisols, and Mollisols, reflecting either the presence of durinodes or a firm, brittle horizon matrix within 100 cm, or a duripan which occurs between 100 and 150 cm from the soil surface. About 4.5% of the soil series are classified in Pachic and Cumulic subgroups.

More than three-quarters (68%) of the soil series in Nevada are in extragrade subgroups that identifies soil individuals with properties that do not clearly intergrade toward specifically defined categories but have one or more properties common to soils in several categories or to non-soil materials (e.g., Lithic and Calcic). Nearly one-quarter (21%) are in intragrade soil subgroups that have specific properties differing from the Typic or Haplic subgroup and one or more characteristics similar to another order, suborder, or great group (e.g., Haploxeralfic and Vitritorrandic). Finally, 11% of the soil subgroups follow the central concept of the great group are named Typic (typical) or Haplic (simple).

6.7 Families

Nearly, two-thirds (57%) of Nevada's soil series have a typic aridic soil moisture regime and 30% have a xeric aridic soil moisture regime (Fig. 6.3 upper). Only 5% of Nevada's soil series have an aquic soil moisture regime that is present for a least a few days in normal years. More than half (57%) of the soil series in Nevada have a mesic soil temperature regime; 32% are cryic or frigid; 9.3% are thermic; and 1.7% are hyperthermic (Fig. 6.3 lower).

More than two-thirds (68%) of Nevada's soil series have the mixed mineralogy class, implying that no single mineral is dominant (Fig. 6.4 upper). Nearly, one-quarter (18%) of the soils are smectitic, indicating the dominance of swelling (i.e., smectite family) clays in soils with high clay content; 8% are carbonatic (high $CaCO_3$ content); and 5% are glassy (high volcanic glass content). Nearly, two-thirds (63%) of the soils series in Nevada are in loamy particle-size classes and half (46%) are in skeletal particle-size classes (Fig. 6.4 middle). More than a third (34%) of the soil series are in the loamy-skeletal particle-size class. In view of the abundance of both 2:1 clay minerals and high cation-exchange capacity, perhaps it is not surprising that 59% of Nevada's loamy soils are in the superactive cation-exchange activity class (Fig. 6.4 lower).

Table 6.2 Taxonomic structure of soil series from Nevada

Order	Suborder	Great group	Subgroups	No.	%	Area (km^2)	%
Alfisols	Cryalfs	Haplocryalfs	Andic (2), Lamellic (1), Vitrandic (1), Xeric (2)	6		277.2	
	Ustalfs	Haplustalfs	Oxyaquic (1), Vitrandic (1)	2		49.1	
	Xeralfs	Durixeralfs	Abruptic (2), Typic (1)	3		195.4	
	Xeralfs	Haploxeralfs	Andic (2), Fragiaqic (1), Lithic (2), Mollic (3), Ultic (6)	14		515.9	
	Xeralfs	Palexeralfs	Ultic (3), Vitrandic (1)	4		48.3	
			Total	29	1.6	1085.9	0.5
Andisols	Cryands	Vitricryands	Lithic (1), Xeric (2)				
	Xerands	Vitrixerands	Humic (1)	3		0	
			Total	1		0.5	
Aridisols	Argids	Calciargids	Durinodic (2), Durinodic Xeric (2), Lithic (1), Petro-	4	0.2	0.5	0.0
			nodic (1), Typic (6), Xeric (3)				
	Argids	Haplargids	Durinodic (10), Durinodic Xeric (24), Lithic (12), Lithic-	15		1579.9	
			Ruptic (1), Lithic Ustic (2), Lithic Xeric (34), Typic (27),				
			Ustic (8), Vitrixerandic (5), Xeric (76)				
	Argids	Natrargids	Aquic (2), Durinodic (12), Durinodic Xeric (7), Haplic				
			(2), Lithic (2), Typic (20), Vertic (1), Vitrixerandic (1),	199		35735	
			Xerertic (1), Xeric (7)				
	Argids	Paleargids	Calcic (3), Typic (1), Vertic (7), Xeric (5)				
	Argids	Petroargids	Duric (4), Typic (1)	55		7212	
	Calcids	Haplocalcids	Aquic (5), Duric (2), Durinodic	16		487	
			(16), Durinodic Xeric (12), Lithic (8), Lithic Ustic (1),	5		220.8	
			Lithic Xeric (6), Petronodic (8), Petronodic Xeric (1),				
			Sodic (7), Typic (27), Ustic (1), Vertic (2), Vitrixerandic				
			(1), Xeric (12)				
	Calcids	Petrocalcids	Aquic (1), Argic (5), Calcic (20), Typic (10), Ustalfic (3),				
			Xeric (2)	109		17373	
	Cambids	Aquicambids	Fluventic (1), Sodic (1)				
	Cambids	Haplocambids	Durinodic (11), Durinodic Xeric (19), Lithic (2), Sodic	41		5427	
			(7), Typic (20), Ustic (2), Ustifluventic (1), Vertic (1),	2		11.7	
			Vitrixerandic (4), Xeric (25), Xerofluventic (1)				
	Cambids	Petrocambids	Xeric (1)				
	Durids	Argidurids	Abruptic (4), Abruptic Xeric (9), Argidic (8), Haplo-	93		15946	
			xeralfic (18), Typic (19), Ustic (2), Vertic (2), Vitrixer-	1		217	
			andic (2), Xeric (38)				
	Durids	Haplodurids	Aquicambidic (3), Cambidic (18), Typic (27), Vitrixer-				
			andic (3), Xereptic (15), Xeric (18)	102		14383	
	Durids	Natridurids	Aquic (1), Natrargidic (6), Natrixeralfic (3), Typic (6),				
			Vitrixerandic (2), Xeric (3)	84		15078	
	Gypsids	Calcigypsids	Typic (4)				
	Gypsids	Haplogypsids	Leptic (7), Sodic (1), Typic (1)	21		2276	
	Gypsids	Petrogypsids	Typic (2)	4		111.3	
	Salids	Aquisalids	Calcic (1), Typic (4)	9		480	
	Salids	Haplosalids	Typic (2)	2		29.8	

(continued)

Table 6.2 (continued)

Order	Suborder	Great group	Subgroups	No.	%	Area (km^2)	%
			Total	5		459.8	
Entisols	Aquents	Fluvaquents	Aeric (3), Typic (2)	2		6.5	
	Fluvents	Torrifluvents	Aquic (2), Duric (2), Duric Xeric (3), Oxyaquic (7), Typic	765	42.2	117033.8	52.3
			(18), Vitrandic (4), Xeric (4)				
	Fluvents	Xerofluvents	Aquic (2), Mollic (1), Oxyaquic (1)	5		497.9	
	Orthents	Cryorthents	Lithic (3), Typic (6)				
	Orthents	Torriorthents	Aquic (3), Duric (19), Lithic (20), Lithic Ustic (4), Lithic	40		2061	
			Xeric (26), Oxyaquic (7), Typic (82), Vertic (3), Vitrandic	4		17.9	
			(4), Xerertic (2), Xeric (42)	9		134.1	
	Orthents	Ustorthents	Lithic (1)				
	Orthents	Xerorthents	Dystric (1), Lithic (1), Typic (2)				
	Psamments	Cryopsamments	Typic (2)	209		41922	
	Psamments	Torripsamments	Haploduridic (3), Oxyaquic (1), Typic (12),	1		0	
			Xeric (6)	4		368.6	
	Psamments	Xeropsamments	Dystric (4), Typic (2)	2		23	
			Total				
Inceptisols	Aquepts	Cryaquepts	Fluvaquentic (1), Histic (1)	22		4843	
	Aquepts	Endoaquepts	Aeric (1), Aquandic (1), Fluvaquentic (1), Typic (1)	6		108.3	
	Aquepts	Epiaquepts	Vertic (1)	302	16.7	49975.8	22.3
	Aquepts	Halaquepts	Aeric (12), Aquandic (1), Duric (6), Typic (4)				
	Aquepts	Humaquepts	Cumulic (1), Histic (2)	2		4.9	
	Cryepts	Calcicryepts	Oxyaquic (1), Typic (1), Xeric (4)	4		35.5	
	Cryepts	Dystrocryepts	Lamellic (1), Xeric (1)	1		17.2	
	Cryepts	Haplocryepts	Calcic (1), Lamellic (2), Oxyaquic (1), Xeric (3)	23		3335.4	
	Cryepts	Humicryepts	Lithic (1), Xeric (5)	3		1.9	
	Ustepts	Calciustepts	Aridic (1)	6		345.6	
	Ustepts	Haplustepts	Calcic (1)	2		40.7	
	Xerepts	Dystroxerepts	Andic (1), Aquic (1), Humic (4)	7		213.7	
	Xerepts	Haploxerepts	Vitrandic (2)	6		140.5	
	Xerepts	Humixerepts	Pachic (1)	1		5.3	
			Total	1		40.5	
Mollisols	Aquolls	Calciaquolls	Typic (2)	6		23	
	Aquolls	Cryaquolls	Typic (1)	2		59.2	
	Aquolls	Endoaquolls	Aquandic (3), Cumulic (11), Cumulic Vertic (3), Duric (2),	1		19.7	
			Fluvaquentic (21), Typic (4), Vertic (1)	65	3.6	4283.1	1.9
	Cryolls	Argicryolls	Aquic (1), Calcic (1), Lithic (8), Oxyaquic (2), Pachic (8),				
			Typic (1), Vertic (2), Vitrandic (12), Xeric (13)	2		5.3	
	Cryolls	Calcicryolls	Lithic (1), Pachic (4), Xeric (1)	1		3.9	
	Cryolls	Haplocryolls	Aquic (1), Calcic (1), Lithic (4), Pachic (21), Vitrandic (4),				
			Xeric (15)	45		2959.3	
	Rendolls	Cryrendolls	Lithic (3), Typic (2)				
	Ustolls	Argiustolls	Aridic (5), Aridic Lithic (10), Calcidic (2), Vitritorrandic (1)	48		1712.2	

(continued)

Table 6.2 (continued)

Order	Suborder	Great group	Subgroups	No.	%	Area (km^2)	%
	Ustolls	Calciustolls	Aridic (1), Pachic (5), Petrocalcic (3)	6		992.8	
	Ustolls	Haplustolls	Aridic (1), Aridic Lithic (4), Fluventic (1), Torriorthentic (1),				
			Vitrandic (1)	46		2013.6	
	Ustolls	Paleustolls	Petrocalcic (1)	5		848.9	
	Xerolls	Argixerolls	Alfic (1), Aquic (1), Argiduridic (7), Aridic (75), Aridic Lithic	18		1686	
			(38), Calciargidic (11), Calcic (5), Lithic (21), Pachic (12),	9		271.9	
			Pachic Ultic (2), Typic (26), Ultic (4), Vitrandic (13), Vitri-				
			torrandic (16)	8		472.2	
	Xerolls	Calcixerolls	Aridic (12), Aridic Lithic (2), Lithic (4), Pachic 1), Typic (6),	1		4.7	
			Vertic (1)				
	Xerolls	Durixerolls	Abruptic (4), Argidic (5), Argiduridic (14), Cambidic (1),				
			Haplic (1), Haploduridic (4), Paleargidic (2), Typic (8),				
			Vertic (3), Vitrandic (2), Vitritorrandic (3)	232		28288.5	
	Xerolls	Haploxerolls	Aquic Cumulic (1), Aquic Duric (1), Aquic (3), Aridic (13),				
			Aridic Lithic (5), Calcidic (6), Cumulic (7), Duridic (7), Entic	26		2461	
			(2), Entic Ultic (1), Fluvaquentic (2), Lithic (6), Lithic Ultic				
			(1), Oxyaquic (7), Pachic (13), Psammentic (1), Torrertic				
			(1), Torrifluentic (4), Torriorthentic (9), Torripsammentic (4),	47		3546.1	
			Typic (4), Vertic (1), Vitrandic (3), Vitritorrandic (14)				
	Xerolls	Natrixerolls	Aridic (1)				
	Xerolls	Palexerolls	Aridic (5), Duric (1), Petrocalcic (1), Petrocalcidic (2), Typic (7),				
			Ultic (1), Vertic (4)				
			Total				
Vertisols	Aquerts	Epiaquerts	Xeric (3)	116		4454.2	
	Torrerts	Haplotorrerts	Sodic (1)	1		8.7	
	Xererts	Haploxererts	Aquic (1), Aridic (6), Halic (1), Leptic (2), Lithic (1)				
			Total	21		876	
			Grand total	632	34.9	50605.3	22.6

6.8 Soil Series

Nevada contains more than 1,800 soil series (Table 6.1), of which 72% occur in Nevada only. For soil series that only occur in Nevada, there is a highly significant correlation between the number of series within a great group and the area of that great group (Fig. 6.5). For soil series that occur in Nevada only, 70% occupy 100 km^2 or less (Fig. 6.6).

6.9 Depth and Drainage Classes

Nearly one-half (43%) of Nevada soils are very deep (>1.5 m); these soils mainly are derived from alluvium and lacustrine sediments (Fig. 6.7 upper). However, 42% of the soils are shallow (0.25 to 0.50 m) or very shallow (<0.25 m); these soils tend to originate from colluvium and residuum. About 93% of the soil series in Nevada are excessively, somewhat excessively, or well drained (Fig. 6.7 lower).

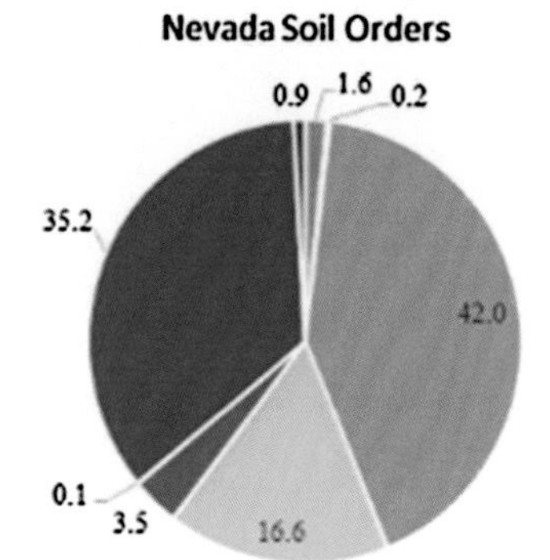

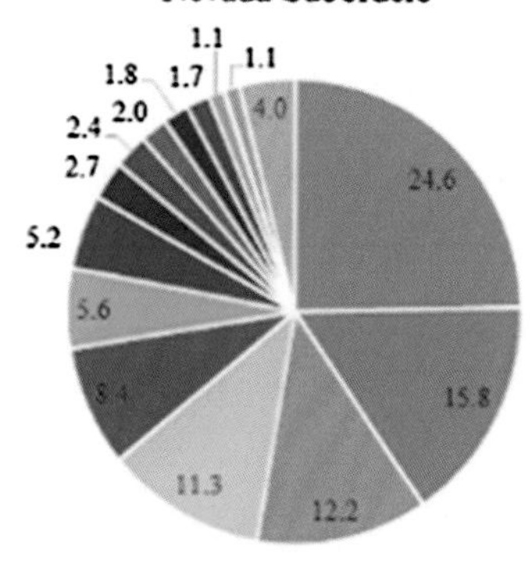

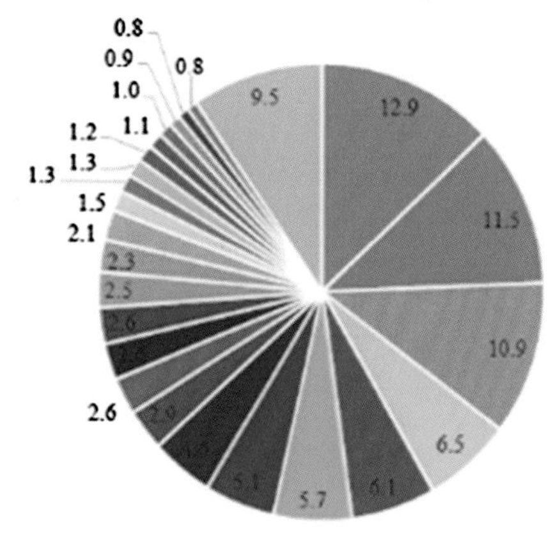

Fig. 6.2 Proportion of soil series in Nevada by order, suborder, and great group. *Source* NRCS database

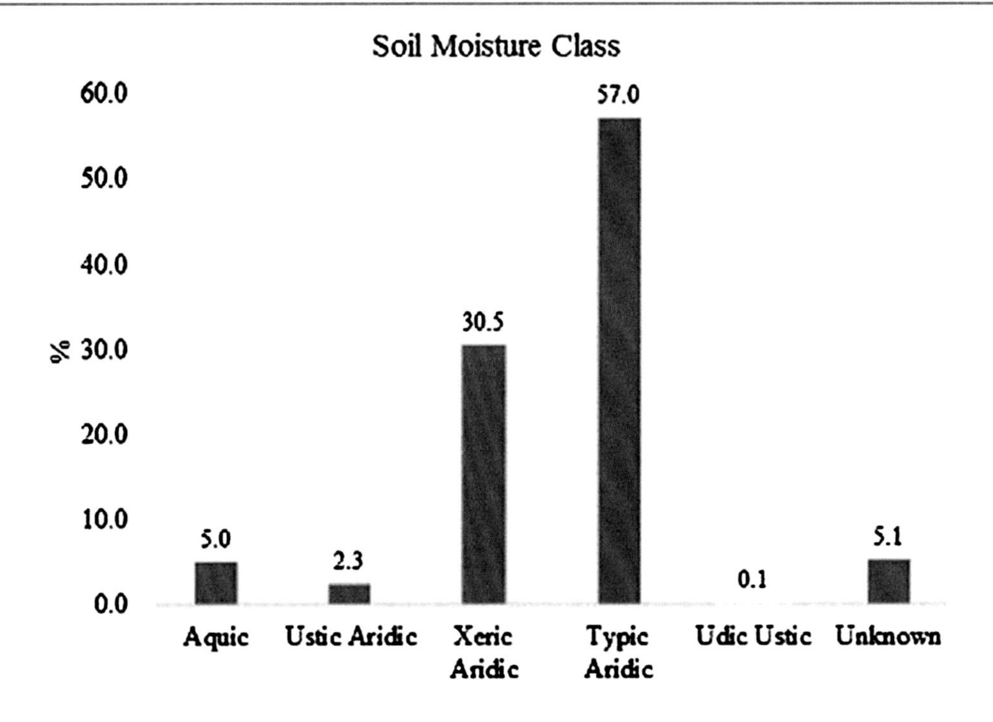

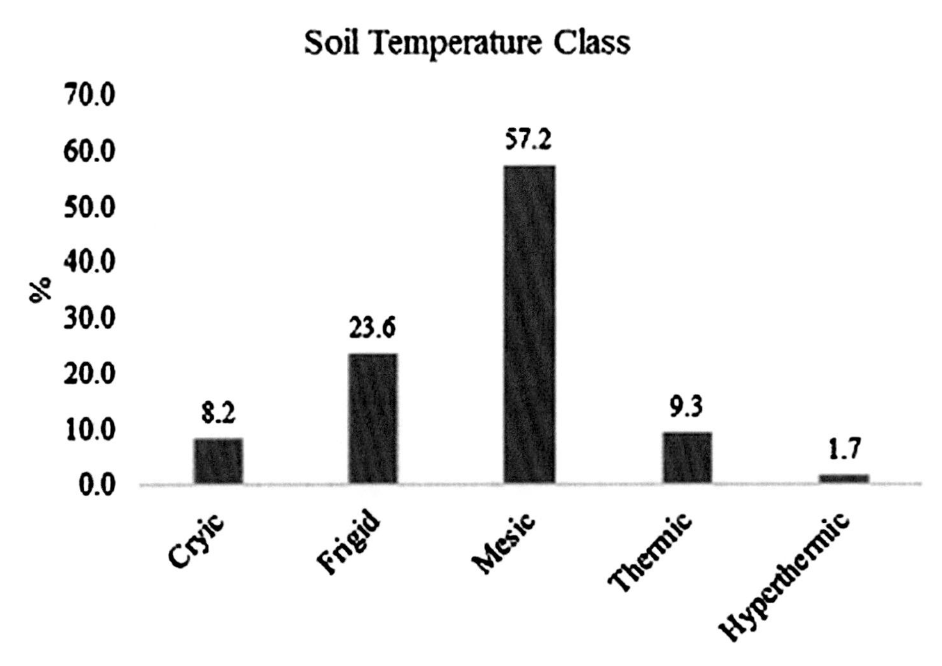

Fig. 6.3 Proportion of soil series in Nevada by soil moisture and soil temperature regimes. *Source* NRCS database

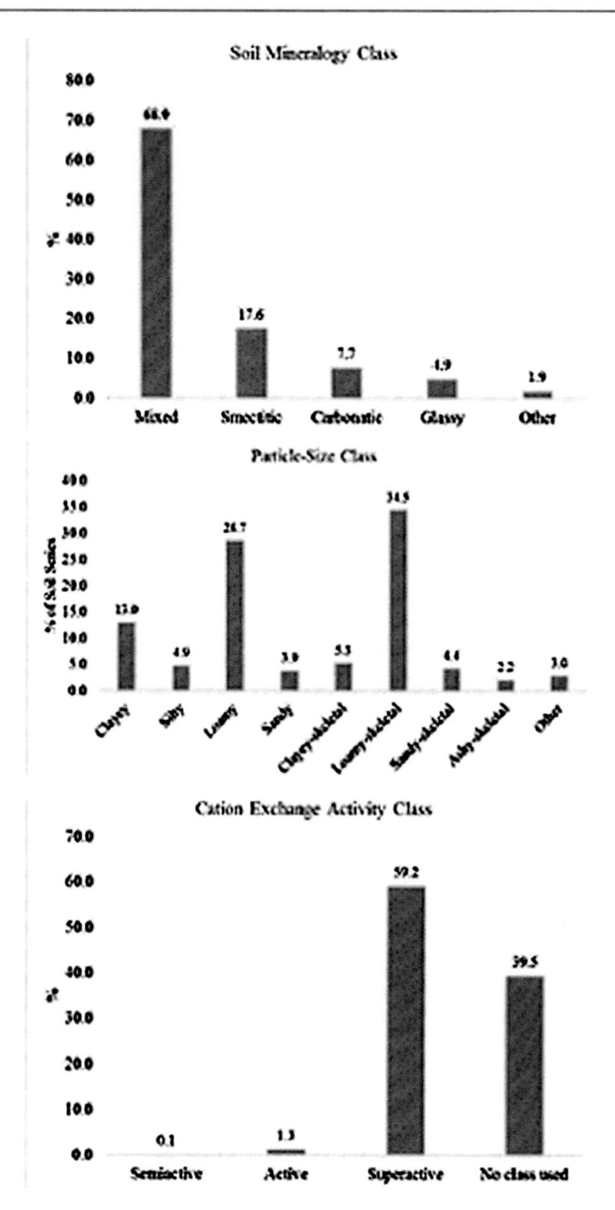

Fig. 6.4 Proportion of soil series in Nevada by soil mineralogy class, particle-size class, and cation-exchange activity class. For the particle-size classes, the label "clayey" represents the clayey, fine, and very-fine particle-size classes; the label "silty" represents the coarse-silty and fine-silty particle-size classes; and the term "loamy" represents the loamy, coarse-loamy, and fine-loamy particle-size classes. *Source* NRCS database

6.10 Comparison of Nevada Soil Taxonomic Structure with Other States

Compared to other states, Nevada has a complex soil taxonomic structure, which contributes to its comparably high pedodiversity (Chap. 14). The state has seven soil orders, which ranks it tied for 16th place nationally (Table 6.3). Nevada is ranked 8th in the number of suborders and great groups, sixth in the number of subgroups and families, second (to California) in the number of soil series, third in the number of soil series occurring only in one state, and third in the number of diagnostic subsurface horizons represented in the state.

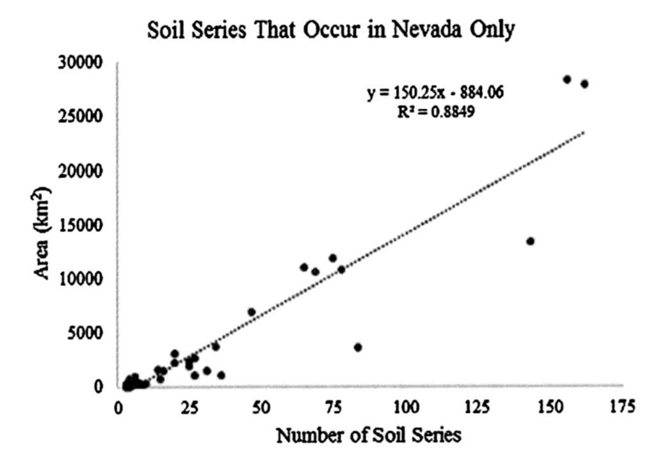

Fig. 6.5 Relation between number of soil series per great group and area of that great group for soil series occurring in Nevada only. *Source* NRCS database

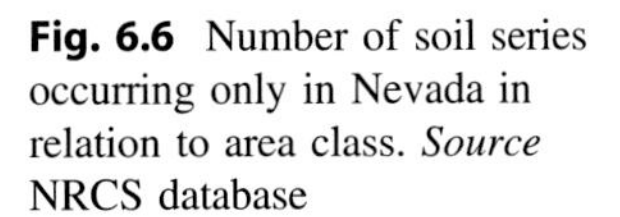

Fig. 6.6 Number of soil series occurring only in Nevada in relation to area class. *Source* NRCS database

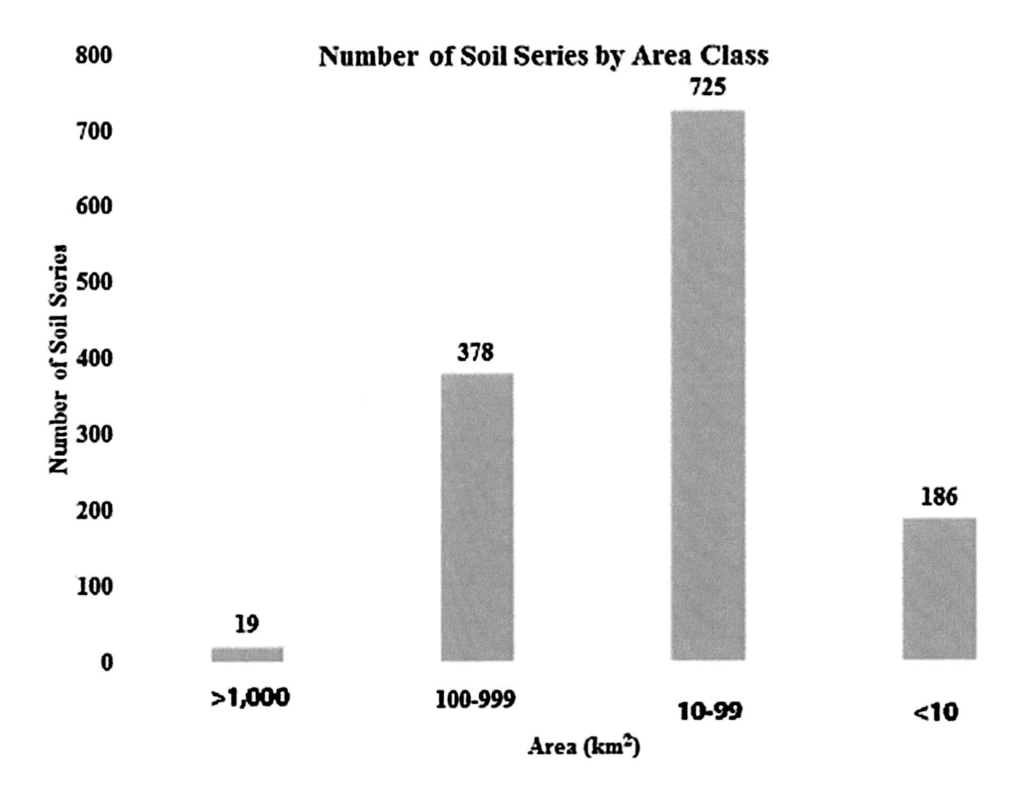

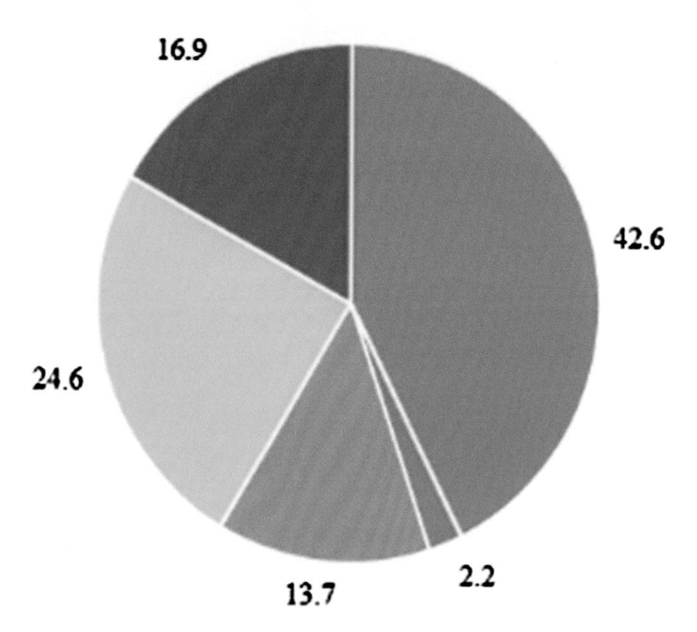

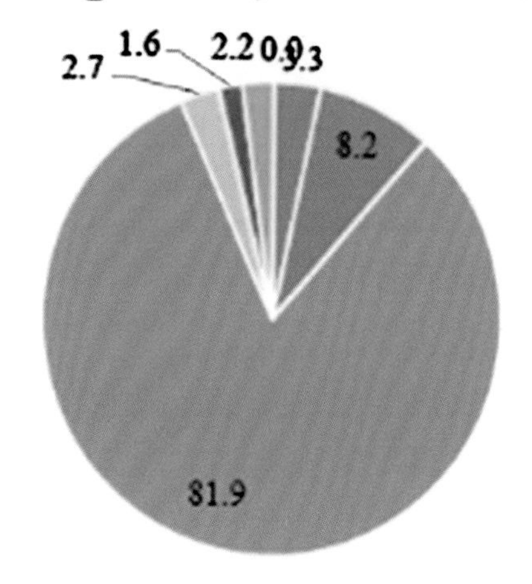

Fig. 6.7 Distribution of soil series in Nevada by depth class and drainage class. *Source* NRCS database

Table 6.3 Top US states by numbers of soil taxa

State	Land area (km^2)	Orders	Sub orders	Great groups	Sub groups	Families	Series	No. series only in state	% series only in state	No. diagnostic sub-surface horizons
CA	403466	10	42	114	434	1575	2465	2103	85.3	10
CO	268431	9	28	66	239	713	1218	680	53.2	7
HI	16635	10	27	55	112	204	250	248	99.2	9
ID	214045	9	35	93	351	1043	1668	1137	68.2	7
MT	376962	9	36	90	297	769	1291	856	66.3	9
NV	**284332**	**7**	**29**	**69**	**300**	**1012**	**1814**	**1317**	**72.6**	**10**
OR	248608	10	41	110	373	1038	1569	1210	77.1	10
PR	9104	10	32	61	132	198	212	210	99.1	7
TX	676587	9	30	83	360	1061	1481	1022	69	12
UT	212818	7	28	71	289	859	1313	699	53.2	10
WA	172119	11	37	103	353	1017	1647	1424	86.5	11
WY	251470	7	27	66	219	592	954	322	33.8	8
Rank	7	16(T)	8	8	6	6	2	3	7	3(T)

6.11 Key to Classifying Nevada Soils to the Great-Group Level

Table 6.4 is an approximate key for classifying Nevada soils to the great group level.

6.12 Conclusions

Soils of Nevada contain primarily ochric and mollic epipedons (98% of soil series) overlying argillic, calcic, cambic, natric, petrocalcic horizons or duripans (90% of soil series).

Table 6.4 Classification of Nevada soils to the great group level of *Soil Taxonomy*

Order	Suborder	Great Group
Andisols	**Cryands**	**Vitricryands**
Have > 60% andic properties in upper 60 cm or to top of densic layer	Have a cryic soil temperature regime	Have abundant volcanic ash or pumice and low water retention at 1500 kPa
	Xerands	**Vitrixerands**
	Have a xeric soil moisture regime	Have abundant volcanic ash or pumice and low water retention at 1500 kPa
Vertisols	**Aquerts**	**Epiaquerts**
Have a layer ≥ 25 cm within upper 100 cm with slickensides or wedge-shaped peds and ≥ 30% clay	Have aquic conditions	Have episaturation
	Xererts	**Haploxererts**
	cracks remain closed for < 60 consecutive days during warm periods	Other Xererts
	Torrerts	**Haplotorrerts**
	cracks open following summer solstice and close following the winter solstice	Other Torrerts

(continued)

Table 6.4 (continued)

Order	Suborder	Great Group
Aridisols	**Salids**	**Aquisalids**
Have aridic soil moisture regime, an ochric epipedon, and a diagnostic subsurface horizon, such as argillic, within 100 cm of the soil surface	Have salic horizon within 100 cm	Have saturation in 100 cm for ≥ 1 month/year
		Haplosalids
		Other Salids
	Durids	**Natridurids**
	Have duripan in 100 cm	Have natric above duripan
		Argidurids
		Have argillic above duripan
		Haplodurids
		Other Durids
	Gypsids	**Petrogypsids**
	Have gypsic or petrogypsic within 100 cm but no overlying petrocalcic	Have petrogypsic or petrocalcic within 100 cm
		Calcigypsids
		Have calcic horizon within 100 cm
		Haplogypsids
		Other Gypsids
	Argids	**Petroargids**
	Have argillic but no petrocalcic horizonwithin 100 cm	Have duripan, petrocalcic, or petrogypsic in 150 cm
		Natrargids
		Have natric horizon
		Paleargids
		Have abrupt clay increase or argillic extends to 150 cm
		Calciargids
		Have calcic horizon within 150 cm
		Haplargids
		No calcic horizon between 100 and 150 cm
	Calcids	**Petrocalcids**
	Have calcic or petrocalcic horizon in 100 cm	Have petrocalcic horizon in 100 cm
		Haplocalcids
		Other Calcids
	Cambids	**Aquicambids**
	Have cambic horizon (Other Aridisols)	Have saturation in 100 cm for ≥ 1 month/year
		Petrocambids
		Have duripan, petrocalcic, or petrogypsic within 150 cm
		Haplocambids
		Other Cambids

(continued)

Table 6.4 (continued)

Order	Suborder	Great Group
Mollisols	**Aquolls**	**Cryaquolls**
Have mollic horizon and a base saturation of > 50%	Have aquic conditions	Have cryic soil temperature regime
		Calciaquolls
		Have calcic or gypsic horizon within 40 cm but no argillic
		Endoaquolls
		Have endosaturation
	Rendolls	**Cryrendolls**
	Have $CaCO_3$ equivalent > 40% but no calci or argillic	Have cryic soil temperature regime
	Cryolls	**Argicryolls**
	Have cryic soil temperature regime	Have argillic horizon
		Calcicryolls
		Have calcic or petrocalcic horizon within 100 cm
		Haplocryolls
		Other Cryolls
	Xerolls	**Durixerolls**
	Have xeric soil moisture regime	Have duripan within 100 cm
		Natrixerolls
		Have natric horizon
		Palexerolls
		Have abrupt increase in clay from eluvial to argillic or < 20% relative decrease in clay in upper 150 cm
		Calcixerolls
		Have calcic or gypsic within 150 cm
		Argixerolls
		Have argillic horizon
		Haploxerolls
		Other Xerolls
	Ustolls	**Calciustolls**
	Have an ustic soil moisture regime or an aridic soil moisture regime that borders on xeric	Have calcic or gypsic horizon within 100 cm but no argillic above
		Paleustolls
		Have abrupt increase in clay from eluvial to argillic or < 20% relative decrease in clay in upper 150 cm
		Argiustolls
		Have argillic horizon
		Haplustolls

(continued)

Table 6.4 (continued)

Order	Suborder	Great Group
		Other Ustolls
Alfisols	**Cryalfs**	**Haplocryalfs**
Have argillic or natric horizon and does not meet requirements of preceding orders	Have cryic soil temperature regime	Other Cryalfs
	Ustalfs	**Haplustalfs**
	Have ustic soil moisture regime	Other Ustalfs
	Xeralfs	**Palexeralfs**
	Have xeric soil moisture regime	Have abrupt increase in clay from eluvial to argillic or < 20% relative decrease in clay in upper 150 cm
		Haploxeralfs
		Other Xeralfs
Inceptisols	**Aquepts**	**Halaquepts**
Have a cambic horizon and/or a folistic, histic, mollic, or umbric epipedon; or a salic horizon; or a high exchangeable sodium percentage which decreases with increasing depth accompanied by ground water within 100 cm of soil surface	Have aquic conditions	Have a salic horizon or a high exchangeable sodium percentage which decreases with increasing depth accompanied by ground water within 100 cm of soil surface
		Cryaquepts
		Have cryic soil temperature regime
		Humaquepts
		Have histic, melanic, mollic, or umbric horizon
		Epiaquepts
		Have episaturation
		Endoaquepts
		Have endosaturation
	Cryepts	**Humicryepts**
	Have cryic soil temperature regime	Have umbric or mollic horizon
		Calcicryepts
		Have calcic or petrocalcic horizon within 100 cm
		Dystrocryepts
		Have base saturation < 50%
		Haplocryepts
		Other Cryepts
	Ustepts	**Calciustepts**
	Have ustic soil moisture regime	Have calcic within 100 cm or petrocalcic within
		150 cm
		Haplustepts
		Other Ustepts
	Xerepts	**Humixerepts**
	Have xeric soil moisture regime	Have umbric or mollic horizon
		Dystroxerepts
		Have base saturation < 60%
		Haploxerepts

(continued)

Table 6.4 (continued)

Order	Suborder	Great Group
		Other Xerepts
Entisols	**Aquents**	**Fluvaquents**
Other soils	Have aquic conditions	Have slope of less than 25% and either a high organic C content at 125 cm depth or irregular decrease in organic C content with depth due to alluvial stratification
	Fluvents	**Xerofluvents**
	Have slope of less than 25% and either a high organic C content at 125 cm depth or irregular decrease in organic C content with depth due to alluvial stratification	Have xeric soil moisture regime
		Torrifluvents
		Have torric (aridic) soil moisture regime
	Orthents	**Cryorthents**
	Other Entisols	Have cryic soil temperature regime
		Torriorthents
		Have torric (aridic) soil moisture regime
	Psamments	**Cryopsamments**
	Have <35% rock fragments and texture class of loamy fine sand or coarser in all layers	Have cryic soil temperature regime
		Torripsamments
		Have torric (aridic) soil moisture regime
		Xeropsamments
		Have xeric soil moisture regime

On an area basis, the dominant orders represented in Nevada are Aridisols, Mollisols, and Entisols (97%).

The dominant suborders are Argids, Orthents, Xerolls, Durids, Calcids, and Cambids (88% of soil area); the dominant great groups are Torriorthents, Haplargids, Argixerolls, Haplocalcids, Haplocambids, Haplodurids, and Argidurids, and (77% of land area). Eleven percent of the soil series are classified in Typic subgroups; 16% are in Xeric and related ("xer") subgroups that represent Aridisols and aridic (torric) Entisols that border on the xeric soil moisture regime, 19% are in Lithic subgroups indicating that a lithic contact is within 50 cm of the surface; 10% are in Aridic and related ("id") subgroups representing the six great groups of Xerolls, and which by definition, have an aridic soil moisture regime that borders on xeric, and 9% are in Durinodic and related ("dur") subgroups that reflect the presence of durinodes, a firm, brittle horizon matrix, or duripans between 100 and 150 cm of the soil surface. There are more than 1,800 soil series in Nevada, of which 72% occur only in the state.

References

Bockheim JG (2014) Soil Geography of the USA: a Diagnostic-Horizon Approach. Springer, NY, p 320

Chadwick OA, Hendricks DM, Nettleton WD (1987) Silica in duric soils: I. A depositional model. Soil Sci Soc Am J 51:975–982

Chadwick OA, Hendricks DM, Nettleton WD (1989) Silicification of Holocene soils, in northern Monitor Valley. Nevada. Soil Sci. Soc. Am. J. 53:158–164

Soil Survey Staff. 2014. Keys to Soil Taxonomy. 12th edit. U.S. Dept. Agric., Natural Resour. Conserv. Serv., Linoln, NE

7 Taxonomic Soil Regions of Nevada

Abstract

Nevada is divided here into 17 soil regions based on great-group associations, which include 25 dominant great groups. The text is accompanied by a new General Soil Map of Nevada, which shows great-group associations at a scale of 1:2 million.

7.1 Introduction

In Fig. 7.1, Nevada is divided into 17 soil regions based on great-group associations. Map unit descriptions identify the predominant and accessory great groups and key soil-forming factors (Table 7.1). Even at a scale of 1:2 million, it is difficult to show more detailed soil-map units, because of recurring soil patterns in the basin and range complex. Each of the great groups contained in the general soil map is discussed as follows, based on their abundance. Table 7.2 summarizes key information on each great group.

7.2 Torriorthents (Great Group Associations 2, 5, 6, 7, 8, 9, 14, and 15)

Torriorthents are the dry, weakly developed soils. They are derived from alluvium on alluvial fans and from colluvium and residuum on plateaus and mountains. These soils include 209 series that cover about 42,100 km^2, 19% of the state's land area. Nevada ranks first in the area of Torriorthents in the USA. Although Torriorthents occur in all Major Land Resource Areas, they are most prevalent in MLRA 30, the Mojave Desert.

Torriorthents most commonly occur at elevations ranging from 1,200 to 2,100 m, with slopes averaging 41 ± 31% (Table 7.1). The mean annual air temperature averages 12 ± 3.8 °C, and the mean annual precipitation is 194 ± 59 mm. The vegetation is most often sagebrush (22%), shadscale saltbush (21%), greasewood (16%), and creosote bush-white bursage (10%). Parent materials are dominantly alluvium on alluvial fans, lacustrine deposits on old lake plains, and residuum on mountain slopes. Most Torriorthents are of Holocene age.

Although many Torriorthents have A-C profiles, some have tick Bk, Bq, and Bkq horizons. However, if they fail to meet the requirements of a calcic or duripan horizon, the development is not considered sufficient to be classified in other orders. More specifically, these horizons fail to meet the requirements of a cambic horizon, which would enable them to be classified as Inceptisols, because they do not have a stronger color, higher clay content, or evidence of removal of carbonates and gypsum.

Torriorthents tend to have a mixed mineralogy, a superactive cation-exchange class, a loamy-skeletal particle-size class, a calcareous reaction class, and a mesic soil-temperature regime. Nearly one-half (45%) of the Torriorthents soil series have a lithic or paralithic contact within 100 cm of the surface. The Kyler, Trocken, Boton, Singatse, Wardenot, Mazuma, and Arizo soil series each occupy 1,000 km^2 or more.

Torriorthents contain only an ochric horizon that averages 17 ± 3.7 cm in thickness. Figure 7.2 shows the Trocken soil series, a Typic Torriorthents, which contains a desert pavement, a pale brown ochric epipedon to 8 cm, and Bk horizons to the base of the pit. Analytical data for the Trocken soil series are given in Table 10.1. The pedon contains a Bw and a series of Bk horizons that fail to meet the requirements of a cambic horizon or an Inceptisol.

Torriorthents are the dominant great group of soils used in Nevada for irrigated agricultural crops, including alfalfa, small grains, corn, pasture, potatoes, and sugar beets. They also are used for woodlands (pinyon-juniper) and urban-suburban development. However, the primary use of Torriorthents, especially in the mountains, is for wildlife habitat.

P. W. Blackburn et al., *The Soils of Nevada*, World Soils Book Series,
https://doi.org/10.1007/978-3-030-53157-7_7

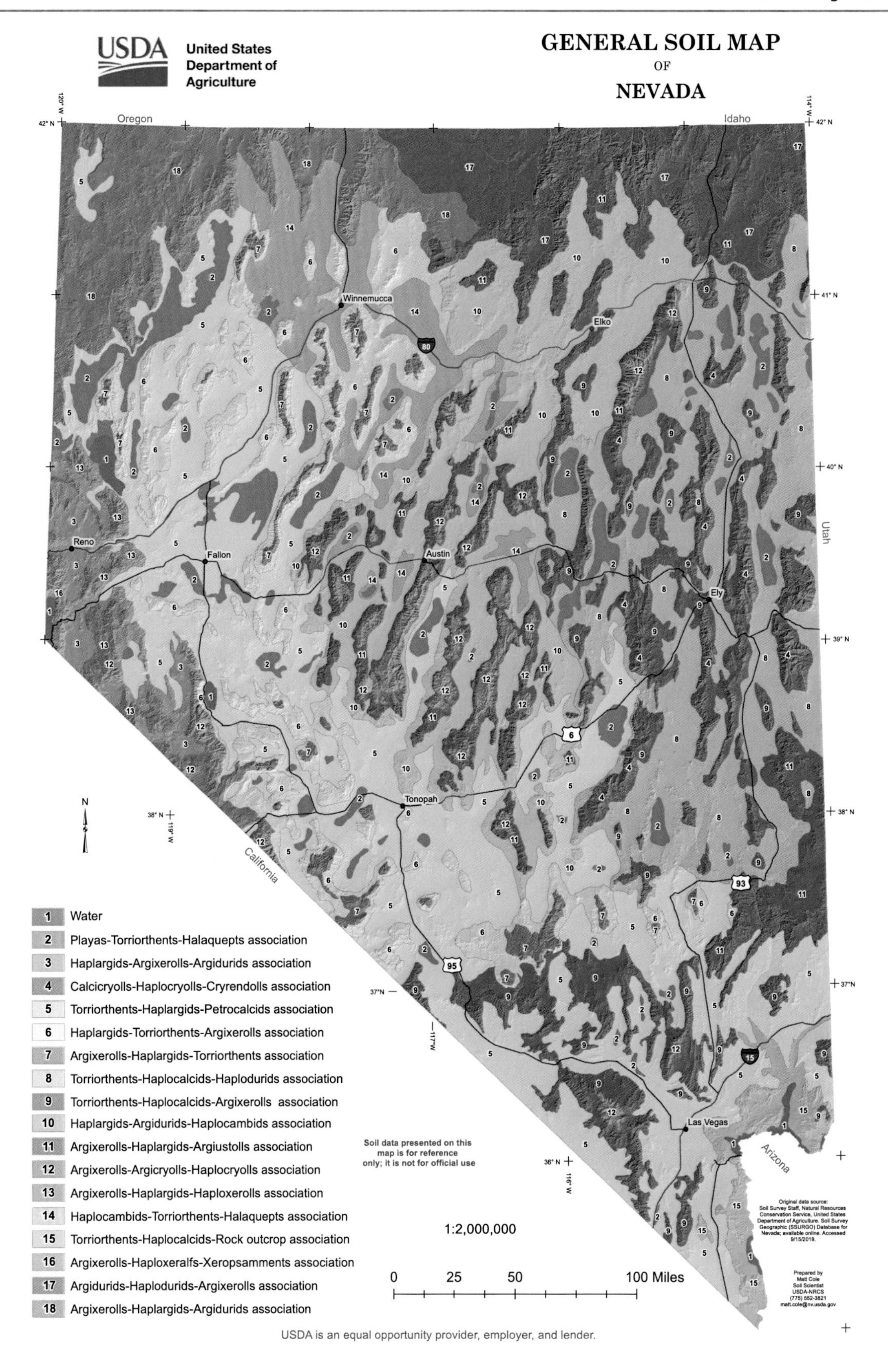

Fig. 7.1 Nevada general soil map at a scale of 1:2 million (developed by Matt Cole, NRCS)

Table 7.1 Map Unit Descriptions for the General Soil Map of Nevada

1—**Water** (100%). Includes part of Lake Tahoe, Pyramid Lake, Walker Lake, part of Lake Mead, and part of the Colorado River.

2—**Playas** (39%)-**Torriorthents** (22%)-Halaquepts (9%) association is on basin floors. These soils have formed in lacustrine deposits and alluvium derived from mixed rocks. Dominant vegetation includes black greasewood, shadscale quallbush, inland saltgrass, and Baltic rush, while playas are barren of vegetation. Some of the other great groups in this map unit include Torripsamments (7%), Torrifluvents (4%), Haplocalcids (3%), Natrargids (3%), and Haplocambids (3%).

3—**Haplargids** (29%)-**Argixerolls**(18%)-**Argidurids**(15%) association is on hills and piedmont slopes. This association has formed in residuum, colluvium, and alluvium derived from granitic and volcanic rocks. The dominant vegetation includes Wyoming big sagebrush, Lahontan sagebrush, shadscale, low sagebrush, Indian ricegrass, Thurber's needlegrass, and desert needlegrass. Some of the other great groups in this map unit are Haploxerolls (8%), Torriorthents (6%), Haplocambids (4%), Torripsamments (2%), Endoaquolls (2%), Paleargids (2%), and Rock outcrop (4%).

4—**Calcicryolls** (15%)-**Haplocryolls** (13%)-**Cryrendolls** (12%) association is on summits of mountains. These soils have formed in residuum and colluvium derived from sedimentary rocks such as limestone and dolomite. Dominant vegetation includes mountain big sagebrush, black sagebrush, white fir, limber pine, curly-leaf mountain mahogany, snowberry, bluebunch wheatgrass, and Idaho fescue. Some of the other great groups in this map unit include Argixerolls (9%), Haplocryalfs (9%), Haplocryepts (8%), Argicryolls (7%), Calcixerolls (4%), Calcicryepts (4%), Cryorthents (4%), Rendolls (3%), and Rock outcrop (6%).

5—**Torriorthents** (29%)-**Haplargids** (14%)-Petrocalcids (7%) association is on piedmont slopes and basins. This association has formed in alluvium from mixed rock sources. The dominant vegetation includes shadscale, Bailey greasewood, bud sagebrush, Indian ricegrass, bottlebrush squirreltail, galleta and in the southern portion of this map unit creosote bush, and white bursage are common. Some of the other great groups in this map unit include Haplocalcids (6%), Haplocambids (6%), Natrargids (5%), Torripsamments (5%), Argidurids (3%), Haplodurids (3%), and Torrifluvents (2%).

6—**Haplargids** (34%)-**Torriorthents** (20%)-**Argixerolls** (10%) association is on hills. These soils have formed in residuum and colluvium derived mainly from volcanic rocks. Dominant vegetation includes black sagebrush, shadscale, Wyoming big sagebrush, Indian ricegrass, bottlebrush squirreltail, and galleta. Some of the other great groups in this map unit include Natrargids (6%), Haplocambids (8%), Haplodurids (3%), Natridurids (2%), and Rock outcrop (5%).

7—**Argixerolls** (38%)-**Haplargids** (17%)-**Torriorthents** (15%) association is on mountain sideslopes. These soils have formed in residuum and colluvium derived from volcanic rocks. Dominant vegetation includes mountain big sagebrush, Wyoming big sagebrush, low sagebrush, single-leaf pinyon, Utah juniper, bottlebrush squirreltail, bluebunch wheatgrass, bluegrass, and Thurber's needlegrass. Some of the other great groups in this map unit include Haploxerolls (7%), Argicryolls (3%), Haplocryolls (3%), Haplocalcids (2%), and Rock outcrop (6%).

8–**Torriorthents** (22%)-**Haplocalcids** (19%)-**Haplodurids** (17%) association is on piedmont slopes. These soils have formed in mixed alluvium derived primarily from sedimentary rocks. Dominant vegetation includes black sagebrush, Wyoming big sagebrush, winterfat, bottlebrush squirreltail, and bluebunch wheatgrass, and galleta. Some of the other great groups in this map unit include Haplargids (9%), Argidurids (7%), Haplocambids (7%), Caciorthids (3%) Argixerolls (3%), and Durixerolls (2%).

9–**Torriorthents** (17%)-**Haplocalcids** (14%)-**Argixerolls** (13%) association is on mountain sideslopes and hills. These soils have formed in residuum and colluvium derived from sedimentary rocks mainly limestone and dolomite. Dominant vegetation includes single-leaf pinyon, Utah juniper, black sagebrush, mountain big sagebrush, curly-leaf mountain mahogany, bottlebrush squirreltail, and bluebunch wheatgrass. Some of the other great groups in this map unit include Calcixerolls (9%), Haplargids (6%), Haplodurids (4%), Haploxerolls (4%), Petrocalcids (3%), Durixerolls (3%), and Rock outcrop (8%).

10—**Haplargids** (25%)-**Argidurids** (15%)-**Haplocambids** (13%) association is on piedmont slopes. These soils have formed in mixed alluvium derived primarily from volcanic rocks. Dominant vegetation includes black sagebrush, Wyoming big sagebrush, winterfat, bottlebrush squirreltail, and bluebunch wheatgrass. Some of the other great groups in this map unit include Argixerolls (9%), Torriorthents (8%), Natrargids (3%. Haplodurids (3%), and Durixerolls (3%).

11—**Argixerolls** (29%)-**Haplargids** (26%)-**Argiustolls** (5%) association is on mountain sideslopes and hills. These soils have formed in residuum and colluvium derived from volcanic rocks. Dominant vegetation includes black sagebrush, mountain big sagebrush, low sagebrush, curly-leaf mountain mahogany, oak, bottlebrush squirreltail, blue grama, Idaho fescue, and bluebunch wheatgrass. Some of the other great groups in this map unit include Torriorthents (4%), Haploxerolls (3%), Durixerolls (2%), Argidurids (2%), Haplocalcids (2%), and Rock outcrop (4%).

12–**Argixerolls**(22%)-**Argicryolls** (14%)-**Haplocryolls** (10%) association is on mountain summits. These soils have formed in residuum and colluvium derived from volcanic rocks. Dominant vegetation includes low sagebrush, black sagebrush, mountain big sagebrush, curly-leaf mountain mahogany, Idaho fescue, and bluebunch wheatgrass. Some of the other great groups in this map unit include Haploxerolls (6%), Calciustolls (6%), Haplustolls (4%) and Rock outcrop (10%).

13—**Argixerolls** (39%)-**Haplargids** (16%)-**Haploxerolls** (8%) association are on mountains and hills. These soils have formed in residuum and colluvium derived from granitic and volcanic rocks. Dominant vegetation includes single-leaf pinyon, Utah juniper, mountain big sagebrush, Wyoming big sagebrush, curly-leaf mountain mahogany, antelope bitterbrush, and bottlebrush squirreltail. Some of the other great groups in this map unit include Torriorthents (8%), Palexeralfs (5%), Durixeralfs (4%), Argidurids (3%). Durixerolls (2%), and Rock outcrop (5%).

14— **Haplocambids** (29%)-**Torriorthents** (27%)-Halaquepts (9%) association is on piedmont slopes and basins. These soils have formed in alluvium from mixed rock sources. Dominant vegetation includes shadscale, Wyoming big sagebrush, black greasewood, bud sagebrush, Indian ricegrass, bottlebrush squirreltail, and inland saltgrass. Some of the other great groups in this map unit include Natrargids (6%), Haplargids (5%), Torripsamments (5%), Haplodurids (4%), and Endoaquolls (4%).

15—**Torriorthents**(35%)-**Haplocalcids** (25%)-Rock outcrop (9%) association is on mountain side slopes, intermountain basins, and Lake Mead. These soils have formed in sandy and gravelly residuum and colluvium derived primarily from sedimentary, metamorphic and volcanic rocks. Dominant vegetation includes creosote bush, white bursage, and white brittle brush spiny menodora, and blackbrush. Some of the other great groups in this map unit include Haplodurids (4%), Haplargids (4%), Petrocalcids (3%), Haplogypsids (3%), Torripsamments (2%), Badland (4%), and Water (2%).

(continued)

Table 7.1 (continued)

16—**Argixerolls** (22%)-**Haploxeralfs** (16%)-Xeropsamments (12%) association is in the mountains. These soils have formed in residuum and colluvium derived from granitic rocks. The dominant vegetation includes mountain big sagebrush, low sagebrush, manzanita, curly-leaf mountain mahogany, antelope bitterbrush, Jeffrey pine, lodgepole pine, red fir, bluegrass, and Idaho fescue. Some of the other great groups in this map unit include Cryopsamments (8%), Haploxerolls (6%), Cryorthents (4%), Dystroxerepts (4%), Humicryepts (4%), Haplargids (4%), and Rock outcrop (8%).
17—**Argidurids** (19%)-**Haplodurids** (17%)-**Argixerolls** (14%) association is on plateaus and hills. These soils have formed in residuum and colluvium derived from basalt rocks. Dominant vegetation includes low sagebrush mountain big sagebrush, black sagebrush, Wyoming big sagebrush, curly-leaf mountain mahogany, antelope bitterbrush, bottlebrush squirreltail, bluegrass, and bluebunch wheatgrass. Some of the other great groups in this map unit include Haplargids (13%), Durixerolls (9%), Torriorthents (7%), Haplocambids (5%), Endoaquolls (3%), and Haplocalcids (2%).
18—**Argixerolls** (39%)-**Haplargids** (18%)-**Argidurids** (8%) association is on plateaus and hills. These soils have formed in residuum and colluvium derived from basalt rocks. Dominant vegetation includes low sagebrush, Mountain big sagebrush, Wyoming big sagebrush, curly-leaf mountain mahogany, antelope bitterbrush, bottlebrush squirreltail, Idaho fescue, and bluebunch wheatgrass. Some of the other great groups in this map unit include Argicryolls (3%), Haploxerolls (3%), Haplocambids (3%), Torriorthents (2%), and Rock outcrop (2%).

7.3 Haplargids (Great Group Associations 3, 5, 6, 7, 10, 11, 13, and 18)

Haplargids are the dry, clay-enriched soils. They are derived from residuum, colluvium, and alluvium that occur in mountain ranges, especially those originating from volcanic materials, such as andesite, rhyolite, or tuff and on fan remnants. They lack a salt-enriched zone that qualifies as a calcic, petrocalcic, duripan, gypsic, petrogypsic, or salic horizon.

Haplargids range from very shallow to deep and are well drained. They most commonly occur at elevations ranging from 1,300 to 2,200 m (Table 7.1). The mean annual air temperature for Haplargids is 9.1 ± 1.8 °C, and the mean annual precipitation is 228 ± 50 mm. The vegetation on Haplargids is dominantly sagebrush, shadscale, greasewood, and Nevada ephedra, with an understory of Indian ricegrass. Slopes average 57 ± 23%. Haplargids in Nevada are derived from colluvium and alluvium on mountain slopes and alluvium on alluvial fan remnants. Haplargids range from Early Holocene (ca., 10 ka) to Pre-Wisconsin (ca, > 150 ka) in age.

There are 199 soil series in this great group, which cover about 36,000 km^2, 16% of the state's land area (Table 7.2). Nevada ranks first in the nation in the area of Haplargids. This great group is common in all MLRAs, except 22A, and 28A. The Stewval, Unsel, Theon, Downeyville, Gabbvally, Wieland, and Old Camp soil series are Haplargids that each occupies 1,000 km^2 or more. The Soughe, Roca, Chubard, Vanwyper, Watoopah, and Burrita soil series occupy from 500 to 1,000 km^2 in Nevada.

Haplargids tend to have a loamy-skeletal particle-size class, a mixed or smectitic mineralogy, a superactive cation-exchange class, and a mesic soil-temperature regime. About 70% of the Haplargids soil series have a lithic or paralithic contact within 100 cm of the surface.

Haplargids contain an ochric epipedon averaging 13 ± 7.6 cm over an argillic horizon that averages 36 ± 27 cm in thickness. Figure 7.3 shows the Jaybee soil series, a Lithic Xeric Haplargids, which contains an 8-cm-thick ochric epipedon over an argillic horizon that is in contact with unweathered basalt bedrock. Analytical data for the Unsel soil series are given in Table 8.1. The pedon has an ochric epipedon to 8 cm, an argillic horizon to 30 cm, and is underlain by Bk horizons and a Bkq horizon enriched in durinodes.

Haplargids are used typically for rangeland and wildlife habitat, because they tend to be concentrated in the mountains, hills, and fan piedmonts.

7.4 Argixerolls (Great Group Associations 3, 6, 7, 9, 11, 12, 13, 16, 17, and 18)

Argixerolls are clay-enriched soils with a mollic epipedon. They are derived from colluvium and residuum in the mountains, especially those originating from volcanic materials. There are 232 soil series in the Argixerolls, the most of any great group. Argixerolls in Nevada commonly occur at elevations of 1,400–2,600 m. The MAAT is 7.1 ± 1.2 °C, and the MAP is 335 ± 50 mm. Slopes are steep, averaging 66 ± 12%. Argixerolls support sagebrush-dominated communities. These soils most commonly are derived from colluvium and residuum on mountain slopes and less commonly from alluvium on alluvial fans and alluvial-fan remnants. Argixerolls usually are of Late Wisconsin (ca., 10–30 ka) to Early Wisconsin (ca., 40–130 ka) in age.

Argixerolls compose about 29,000 km^2, 13% of the state land area. Nevada ranks first in the nation in the area of Argixerolls. These soils are most prevalent in MLRAs 23, 25, and 26 in the northern ranges, are less common in MLRAs 28B (central ranges) and 29 (southern ranges), and are generally lacking in MLRA 30, the Mojave Desert, where precipitation is limited and salts tend to accumulate.

Argixerolls tend to have a loamy-skeletal particle-size class, a mixed or smectitic mineral class, a superactive cation-exchange class, and a frigid or less commonly mesic soil-temperature class. About 88% of the Argixerolls in Nevada have a lithic or paralithic contact within 100 cm of

Table 7.2 Relation of soil great groups in Nevada to soil-forming factors

Soil great group	Great group associations	Area (km^2)	Dominant MLRAs	MAP (mm/yr)	MAAT (°C)	Soil-moisture regime	Soil temperature regime	Vegetation type	Elev. (m)	Slope class (%)	Parent materials	Landforms	Landform age[1]
Torriorthents	13a, 13j, 13 k, 13 l, 13p, 13q, 13z, 14e	43,000	27, 30, 29, 24, 28B	100–200	10–18	torric	mesic, thermic	greasewood, creosotebush, sagebrush, shadscale	400–2300	0–30	residuum-colluvium; alluvium	mtns; alluvial fans; fan remnants	H, IQ
Haplargids	13aa, 13j, 13 k, 13 l, 13r, 13 s, 13x, 80 g	38,000	29, 26, 30, 24, 25, 27, 28B, 28A	150–250	8–12	aridic	mesic, frigid, thermic	sagebrush, shadscale, greasewood, NV ephedra, Indian ricegrass	1200–2400	3–75	residuum-colluvium; alluvium	mtns.; fan remnants	IQ
Argixerolls	13aa, 13 k, 13 l, 13q, 13 s, 13t, 13x, 5, 80a, 80 g	34,000	23, 25, 26, 28A, 28B	300–360	6–9	xeric	frigid, mesic	sagebrush, pinyon-juniper, bitter-brush, rabbitbrush, wheatgrass	1600–2400	0–75	residuum-colluvium; alluvium	mtns.; fan remnants	IQ
Haplocalcids	13p, 13q, 14e	20,000	30, 28B, 28A	250–330	8–18	aridic	mesic, thermic	sagebrush, creosotebush, shadscale, rabbitbrush	800–2400	0–75	alluvium	fan remnants; alluvial fans	IQ
Haplocambids	13r, 13z	17,000	24, 25, 27, 29, 28A, 28B	150–230	9–12	aridic	mesic, frigid	sagebrush, greasewood, shadscale	1200–2200	0–30	alluvium (often with loess, volcanic ash)	alluvial fans; inset fans; fan remnants	IQ
Haplodurids	13p, 13r, 80a	15,000	30, 29, 25, 28B	230–280	3–10	aridic	mesic, thermic, frigid	sagebrush, shadscale, creosotebush	1100–2300	0–50	alluvium (often with loess, volcanic ash)	fan remnants; fan piedmont remnants	IQ
Argidurids	13aa, 80a, 80 g	17,500	29, 26, 28B	200–250	8–12	aridic	mesic	sagebrush, Indian ricegrass, shadscale	1100–2300	0–40	alluvium; residuum-colluvium	fan remnants, mtns., fan piedmont remnants	IQ-mQ
Natrargids	[none]	7,200	27, 24, 25	125–180	9–12	aridic	mesic	sagebrush, greasewood, shadscale	1000–2000	0–35	alluvium (often with loess, volcanic ash)	fan remnants; fan piedmont remnants	IQ-mQ
Petrocalcids	13j	4,900	30	125–200	12–18	aridic	thermic, mesic	blackbrush, creosotebush	330–2000	0–40	alluvium	fan remnants; ballenas	IQ-mQ,
Torripsamments	[none]	5,530	30, 24	100–150	11–20	torric	mesic, thermic	creosotebush, greasewood, horsebrush	250–1900	0–40	eolian	sand dunes, sheets	H
Haploxerolls	13x	5,300	25, 23, 28A, 28B	250–360	5–10	xeric	frigid, mesic	sagebrush, pinyon-juniper, ID fescue	1100–2400	0–70	alluvium; residuum-	alluvial terraces; mtns.	H, IQ
Durixerolls	[none]	4,200	25, 26, 28B, 28A	280–300	6–8	xeric	frigid, mesic	sagebrush, rabbitbrush, NV ephedra	1200–2400	0–40	alluvium	fan remnants; fan	IQ-mQ
Halaquepts	13a, 13z	3,360	27, 24, 28B	150–180	9–10	aquic	mesic	greasewood, sacaton, saltgrass	1100–2000	0–4	lacustrine; alluvium	alluvial flats; lake plains	H
Endoaquolls	[none]	3,000	27, 25	180–360	6–10	aquic	mesic, frigid	wildrye, sedges, saltgrass	1200–2400	0–10	alluvium	floodplains	H, IQ
Calcixerolls	[none]	2,400	25, 28A, 28B	360–400	6–8	xeric	frigid, mesic	pinyon-juniper/sagebrush	1900–2700	4–75	residuum-colluvium	mtns.–limestone	IQ
Natridurids	[none]	2,200	24	150–230	8–11	aridic	frigid	greasewood, shadscale, sagebrush	1100–2000	0–40	alluvium (often with	fan remnants; fan pied-	IQ

(continued)

Table 7.2 (continued)

Soil great group	Great group associations	Area (km^2)	Dominant MLRAs	MAP (mm/yr)	MAAT (°C)	Soil-moisture regime	Soil temperature regime	Vegetation type	Elev. (m)	Slope class (%)	Parent materials	Landforms	Landform age[1]
Torrifluvents	[none]	2,200	27, 28A, 28B	100–175	7–12	torric	mesic	greasewood, shadscale	1100–2200	0–6	alluvium	alluvial flats; flood-	H
Haplocryolls	13e, 13f	2,100	28A, 28B, 25, 23	330–500	3–6	xeric	cryic	aspen/sagebrush	1900–3100	2–80	residuum-colluvium	mtns.	lQ
Agricryolls	13t	2,380	23, 26, 28B	360–430	6–9	xeric	cryic	sagebrush/Idaho fescue, bluegrass	1900–3100	4–75	residuum-colluvium	mtns.–volcanic	lQ
Argiustolls	13 s	1,800	30, 29	330–400	9–11	ustic	mesic	pinyon-juniper/serviceberry	1400–2400	3–65	residuum-colluvium	mtns.	lQ
Calciargids	[none]	1,580	30	150–250	7–16	aridic	mesic, thermic	shadscale	600–2100	0–30	alluvium	fan remnants	lQ–mQ
Calcicryolls	13e	990	28A, 28B, 24, 25	330–500	3–6	xeric	cryic	sagebrush, wheatgrass, mtn. mahogany	1900–3100	2–80	residuum-colluvium	mtns.–limestone	lQ
Cryrendolls	13e	840	28B, 28A	330–500	3–6	xeric	cryic	mtn. mahogany or sagebrush/ID fescue	1920–3180	15–75	residuum-colluvium	mtns.–limestone	lQ
Haploxeralfs	5	700	22A, 25	400–1000	6–7	xeric	frigid	jeffrey pine, white fir, sagebrush, Idaho fescue	1100–2500	0–50	residuum-colluvium; alluvium	plateaus; alluvial fans	lQ
Xeropsamments	5	108	22A, 25	635–1015	4–6	xeric	frigid	jeffrey pine, white fir, red fir	1500–2700	5–75	colluvium-residuum	mtn. slopes	H

Fig. 7.2 The Trocken soil series, a Typic Torriorthents. The scale is in decimeters. Photo by Burt et al. (2014)

the surface. The Cleavage, Ninemile, Devada, and Sumine soil series each occupies more than 1,000 km^2; and the Bellehelen, McIvey, Softscrabble, Wylo, Pickup, Acrelane, Quarz, Reluctan, and Chen soil series each occupies between 500 and 1,000 km^2 in Nevada.

Argixerolls have a mollic epipedon averaging 30 ± 12 cm in thickness over an argillic horizon that averages 45 ± 35 cm. Figure 7.4 is of the Millan soil series, an Aridic Argixerolls in Great Basin National Park. The pedon has a gray-brown mollic epipedon to 25 cm that is underlain by pale brown Bk horizons to 152 cm. Analytical data for the Ninemile soil series are given in Table 9.1. The pedon has a mollic epipedon to 18 cm and an argillic horizon to 31 cm directly over weathered basalt bedrock.

Argixerolls are used primarily for livestock grazing and wildlife habitat. However, some soils at the higher elevations (Bellehelen and Cropper series) are managed as pinyon-juniper or pinyon-mountain mahogany woodlands.

7.5 Haplocalcids (Great Group Associations 8, 9, and 15)

Haplocalcids are dry soils that are enriched in $CaCO_3$, i.e., have a calcic horizon, but do not contain a layer cemented by calcium carbonate (petrocalcic horizon). Haplocalcids are derived primarily from alluvium on fan remnants, ballenas, and alluvial fans. Haplocalcids commonly range from 1,200 to 1,900 m in elevation. Slopes average 41 ± 25%. They have a MAAT of 11 ± 4.4 °C and a MAP of 208 ± 62 mm. The vegetation on Haplocalcids commonly is creosote bush-white bursage or sagebrush-dominated communities. Haplocalcids generally are derived from alluvium on alluvial fans and fan remnants or less commonly from colluvium and residuum on mountain slopes. Haplocalcids usually are of Late Wisconsin (ca., 10–30 ka) to Middle Wisconsin (ca., 30–40 ka) in age.

Fig. 7.3 The Jaybee soil series, a Lithic Xeric Haplargids. The scale is in decimeters. Photo by Burt et al. (2014)

There are 109 soil series in the Haplocalcids. Haplocalcids compose about 17,500 km^2, 7.8% of the state land area. These soils are most prevalent in MLRA 30, the Mojave Desert, but are also common in MLRAs 29, 28A, and southern portions of 28B. The Pookaloo, Armespan, Tecomar, and Weiser soil series each occupy more than 1,000 km^2 in Nevada; the Zeheme, Wintermute, Candelaria, Pyrat, and Kunzler, soil series cover from 500 to 1,000 km^2.

Haplocalcids are deep soils most commonly with a loamy-skeletal or coarse-loamy particle-size class, a mixed or carbonatic mineral class, and a mesic or thermic soil-temperature regime. About one-quarter (23%) of Haplocalcids in Nevada have a lithic or paralithic contact within 100 cm of the surface.

Haplocalcids have an ochric epipedon averaging 17 ± 3.4 cm in thickness over a calcic horizon that averages 81 ± 43 cm. Figure 7.5 is of the Owyhee soil series, a Xeric Haplocalcids. Although the soil occurs in Idaho, it is on the Owyhee High Plateau (MLRA 25) and is comparable to similar subgroups in Nevada. This soil is derived from old alluvium and contains an ochric epipedon (A horizon) to 28 cm, a cambic horizon (Bw) to 50 cm, and a calcic horizon (Bk) to 70 cm. Analytical data for the Pookaloo soil series are contained in Table 8.1. This pedon has an ochric epipedon to 18 cm and a calcic horizon to 73 cm that overlies fractured limestone bedrock. Haplocalcids are used exclusively for rangeland and wildlife habitat.

7.6 Haplocambids (Great Group Associations 10 and 14)

Haplocambids are the dry soils containing a cambic horizon. They occur on alluvium—often influenced by loess and volcanic ash–on alluvial fans, fan remnants, fan skirts, and inset fans. Haplocambids commonly occur at elevations between 1,300 and 2,000 m and on slopes averaging 17 ± 13% (Table 7.1). The MAAT for Haplocambids in Nevada is 96 ± 1.6 °C, and the MAP is 197 ± 39 mm. Haplocambids support shadscale saltbush or Wyoming big sagebrush communities. They are derived from alluvium and alluvium covered with loess and/or volcanic ash on alluvial fans, alluvial-fan remnants, and old lacustrine plains. Haplocambids usually are of Early to Middle Holocene (ca., > 0.2, < 10 ka) in age.

There are 93 soil series in the Haplocambids. Haplocambids compose about 16,000 km^2, 7.1% of the state land area. These soils occur in all of the MLRAs recognized in Nevada, except 21 and 22A, and are most common in MLRAs 24 and

Fig. 7.4 The Millan soil series, an Aridic Argixerolls from the Great Basin National Park. The scale is in decimeters. Photo by USDA, NRCS (2009)

28B. The Koyen soil series has the largest area of the Haplocambids and occupies 1,215 km^2 in central Nevada. The Orovada, the state soil series for Nevada, comprises 983 km^2 in Nevada. The Broyles, Veet, Heist, Kelk, Weso, and Enko soil series each cover from 500 to 750 km^2.

Haplocambids tend to be deep soils that occur in a variety of particle-size classes, usually have a mixed mineralogy, often are superactive, and commonly have a mesic soil-temperature regime. Haplocambids have an ochric epipedon averaging 18 cm in thickness over a cambic horizon that averages 26 $\pm$ 10 cm. The Orovada soil series, Nevada's state soil, is a Durinodic Xeric Haplocambids with an ochric epipedon (A) to 5 cm, a cambic horizon (Bw) to 36 cm, and a series of horizons with secondary carbonates and durinodes to the base of the profile (Fig. 7.6). Analytical data for the Orovada soil are contained in Table 9.1. The soil has an ochric epipedon to 8 cm, a cambic horizon to 46 cm, and a duripan that extends from 140 to 196 cm that is too deep in the profile for the pedon to be recognized as a Durid or a Petrocambid.

Although Haplocambids are used primarily for livestock grazing and wildlife habitat, several widespread soils, including the McConnel, Heist, Davey, Kelk, Wholan, Orovada, Silverado, Rebel, and Creemon series, are irrigated for alfalfa and small grains.

7.7 Haplodurids (Great Group Associations 8, 10, and 17)

Haplodurids are dry soils containing a duripan. They are derived from alluvium on fan remnants, ballenas, and fan piedmonts. Haplodurids commonly occur at elevations ranging from 1,100 to 1,900 m and on slopes averaging 37 $\pm$ 15%. For Haplodurids in Nevada, the MAAT is 10 $\pm$ 3.6 °C, and the MAP is 220 $\pm$ 40 mm. The native vegetation is sagebrush, creosote bush, or shadscale saltbush. Haplodurids are derived from alluvium often containing loess and/or volcanic ash on alluvial-fan remnants and piedmonts. Haplodurids are of Late Wisconsin (ca., 10-30 ka) or older in age.

Fig. 7.5 Owyhee soil series, a Xeric Haplocalcids from southern Idaho. The scale is in decimeters. Photo by P. McDaniel

These soils include 84 series that cover about 15,100 km^2, 6.3% of Nevada's land area (Table 7.2). Haplodurids are most common in MLRAs 28B, 30, 25, and 24 and least common in MLRAs 21, 22A, 26, and 27. The Palinor soil series has the largest area of the Haplodurids at 2,510 km^2, followed by the Chiara (2,200 km^2), Ursine (1,490 km^2), and Shabliss (1,018 km^2).

Haplodurids are often shallow, occur in the loamy-skeletal or loamy particle-size class, mixed mineralogy class, superactive cation-exchange class, and mesic or thermic soil-temperature class. Although a few Haplordurids have a lithic or paralithic contact with 100 cm of the surface, the duripan often limits plant rooting to the upper 50 cm. Haplodurids have an ochric horizon averaging 18 ± 1.8 cm over a duripan that averages 42 ± 34 cm in thickness. Analytical data for the Palinor soil series are contained in Table 8.1. The pedon has an ochric epipedon to 10 cm, a calcic horizon to 41 cm, and a duripan to 83 cm. The soil is not classified as a Calcid because the duripan takes precedence in ST.

Because they often are shallow, Haplodurids are used for rangeland and wildlife habitat. The Shabliss and Rubyhill soil series are irrigated for alfalfa hay, small grains, and pasture.

7.8 Argidurids (Great Group Associations 3, 17, and 18)

Argidurids are dry soils containing an argillic horizon over a duripan. They are derived primarily from alluvium—often with a mix of loess and volcanic—on fan remnants but occasionally from residuum and colluvium derived from weathering of volcanic rocks in the mountains of Nevada. Argidurids commonly occur at elevations ranging from 1,300 to 2,000 m. Slopes average 26 ± 13%. The MAAT for Argidurids is 9.1 ± 1.4 °C, and the MAP is 215 ± 31 mm. Argidurids generally support sagebrush-dominated communities. Most Argidurids are of Early Wisconsin (40–130 ka) to Pre-Wisconsin (> 130 ka) in age.

The Argidurids contain 102 soil series in Nevada that cover about 14,500 km^2, 7.3% of the state's land area. Argidurids are most common in MLRAs 29, 24, 25, and 28B and least common in MLRAs 21 and 22A. The most extensive Argidurids are the Dewar (1,060 km^2), Hunnton (950 km^2), Zadvar (852 km^2), Handpah (657 km^2), and Dacker (509 km^2) soil series.

More than two-thirds of the Argidurids are shallow. The most common families for Argidurids are the loamy and loamy-skeletal particle-size classes, the mixed and smectitic mineralogy classes, the superactive cation-exchange class, and the mesic soil-temperature class. Only 12% of the Argidurids have a lithic or paralithic contact within 100 cm of the surface, but the presence of a duripan commonly limits rooting of plants in the upper 50 cm.

Argidurids contain an ochric epipedon that averages 12 ± 8.1 cm over an argillic horizon averaging 29 ± 14 cm and a duripan averaging 29 ± 14 cm. Some Argidurids also have a calcic horizon. Figure 7.7 is of the Colthorp soil series that occurs just across the border into Idaho on the Owyhee High Plateau (MLRA 25). The soil has an ochric epipedon (V) to 10 cm, an argillic horizon (Bt) to 30 cm, a calcic horizon (Btk) to 45 cm, and a duripan (Bkqm) to basalt bedrock at 60 cm. Analytical data for the Zadvar soil series is contained in Table 8.1. This pedon has an argillic horizon that is 28-cm

Fig. 7.6 Orovada soil series, a Durinodic Xeric Haplocambids, which is the Nevada state soil. The scale is in inches. Photo by John Fisher

thick over a duripan that extends from 43 cm to at least 152 cm.

Nearly all of the Argidurids soil series are used as rangeland and wildlife habitat. However, a small area of the Chuska soil series is irrigated for wheat, barley, and alfalfa hay production.

7.9 Natrargids

Natrargids are the dry soils containing a natric horizon. They are derived from deep alluvium on fan remnants and old lake plains on slopes averaging 27 ± 15% and at elevations commonly ranging from 1,200 to 1,800 m. The MAAT averages 10 ± 1.4 °C, and the MAP is 159 ± 33 mm. The desert shrub vegetation is dominated by greasewood and shadscale. Most Natrargids are of Late Wisconsin (10–30 ka) to Early Wisconsin (40–130 ka) in age.

The Natrargids great group contains 55 soil series in Nevada that cover about 7,200 km^2, or 3.0% of the state land area. Natrargids are most common in MLRAs 27 and 24 in central Nevada but compose an insufficient area to show on the 1:2 million-scale general soil map. The predominant soil series are Oxcorel (820 km^2), Beoska (615 km^2), Dorper (520 km^2), Jerval (487 km^2), and Ricert (464 km^2).

In Nevada, Natrargids tend to be deep, fine-textured (primarily fine and fine-loamy PSCs), in mixed or smectitic mineralogy classes, and in mesic and thermic soil-temperature classes; about half of the Natrargids are superactive.

Natrargids have an ochric epipedon that averages 14 ± 6.7 cm over a natric horizon averaging 28 ± 13 cm. Some Natrargids also have a calcic horizon. Figure 7.8 is of the Highrock soil series, a Typic Natrargids. This photo shows the pale brown ochric epipedon (A horizon) overlying the

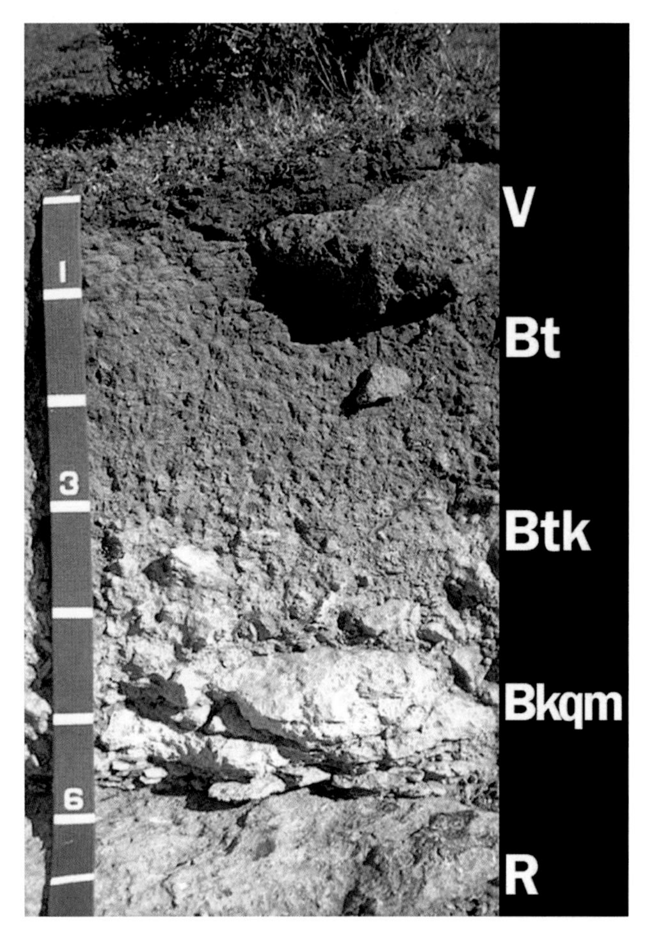

Fig. 7.7 Colthorp soil series, a Xeric Argidurids from Southern Idaho. The scale is in decimeters. Photo by P. McDaniel

brown natric horizon (Btn) at 17 cm. The Btn has a strong prismatic structure and secondary calcium carbonates. Analytical data for the Oxcorel soil series are provided in Table 8.1. The natric horizon extends from 25 to 84 cm in depth.

Although most of the Natrargids are in rangeland and wildlife habitat, parts of the Jerval and Appian soil series are irrigated for alfalfa hay and small-grain production.

7.10 Petrocalcids (Great Group Association 5)

Petrocalcids are dry soils that contain a layer that is cemented by calcium carbonate. These soils are derived from alluvium on fan remnants and often feature creosote bush, blackbrush, yellowbrush, and Mojave yucca. Petrocalcids occur at elevations between 900 and 1,400 m and occupy slopes averaging 23 $\pm$ 15%. These soils exist in areas where the MAAT averages 16 $\pm$ 2.6 °C and where the MAP is 167 $\pm$ 36 mm. Petrocalcids are of Pre-Wisconsin age (> 130 ka).

There are 41 soil series in the Petrocalcids great group in Nevada that total about 5,500 km^2, 2.2% of the state land area. Three-quarters (75%) of the Petrocalcids in Nevada are concentrated in MLRA 30. Nevada ranks third nationally in the area of Petrocalcids, exceeded only by Texas and New Mexico. The most extensive Petrocalcids in Nevada are the Irongold, Wechech, Mormon Mesa, and Bard soil series, with each occupying between 500 and 900 km^2. The most

Fig. 7.8 This photo is of the Highrock soil series, a Typic Natrargids and shows the pale brown ochric epipedon (A horizon) overlying the brown natric horizon (Btn) at 17 cm. The Btn has strong prismatic structure and secondary calcium carbonates Source: NRCS

Fig. 7.9 The Mormon Mesa soil series, a Calcic Petrocalcids. Photos by Brock and Buck (2009)

studied Petrocalcids in the USA and possibly the world is the Mormon Mesa series.

Two-thirds of the Petrocalcids are shallow because of the presence of the petrocalcic horizon. More than 80% are in the loamy-skeletal and loamy particle-size classes; they are equally divided in the carbonatic and mixed mineralogy classes; about one-third of the Petrocalcids have a superactive CEC activity class; and more than half (56%) of the soil series have a thermic soil-temperature regime.

Petrocalcids contain an ochric epipedon averaging 15 ± 5.0 cm in thickness, overlying a calcic horizon with an average thickness of 23 ± 9.1 cm and a petrocalcic horizon with an average thickness of 93 ± 55 cm. Figure 7.9 is of the Mormon Mesa soil, a Calcic Petrocalcids, near Lake Mead and the Virgin River. The photo on the left shows a mixed brecciated and eolian horizon (B), a massive horizon (M), a transitional horizon (T), and the Muddy Creek Formation (MC) on which the soil occurs. In the photo on the right, a laminar horizon (L) is outlined in black. The Mormon Mesa soil reflects soil development over the past 5 million years (Brock and Buck 2009).

The Desert Soil-Geomorphology Project (Monger et al. 2009) reported the lack of petrocalcic horizons in soils derived from limestone materials in southern New Mexico. However, in Nevada 19 of 21 soil series classified as Petrocalcids were derived from alluvium originating from limestone materials.

Analytical data for the Mormon Mesa soil are provided in Table 8.1. The pedon has an ochric epipedon to 4 cm, a calcic horizon to 49 cm, and a petrocalcic horizon to more than 6 m (20 ft)! Because they tend to be shallow, the Petrocalcids are retained almost exclusively for range and wildlife habitat.

7.11 Torripsamments

Torripsamments are the dry, poorly developed soil derived from deep eolian materials on dunes and sand sheets that contain only an ochric epipedon. The vegetation on Torripsamments is primarily saltbrush, creosote bush, and greasewood. Torripsamments occur at elevations ranging from 1,000 to 1,600 m and on slopes averaging 36 ± 19%. These soils occur in areas, where the MAAT is 12 ± 4.2 °C and the MAP is 173 ± 61 mm. Torripsamments often are of Holocene (< 10 ka) age.

There are 23 soil series in this great group in Nevada that total about 5,000 km^2, 2.3% of the state land area. Torripsamments occur in all MLRAs in Nevada except 21, 22A, and 25 but compose an insufficient area to show on the 1:2 million-scale general soil map. The most common Torripsamments in Nevada are the Hawley (1,500 km^2), Isolde (1,265 km^2), and Stumble (675 km^2) soil series.

By definition, Torripsamments have a sandy particle-size class; all of the Torripsamments are in the mixed mineralogy class. The ochric epipedon in Torripsamments of Nevada averages 18 cm in thickness. Figure 7.10 is of the Quincy soil series, a Xeric Torripsamments that has not been mapped in Nevada; however, it has been identified just across the border in Vinton, California. Analytical data for the Stumble soil series

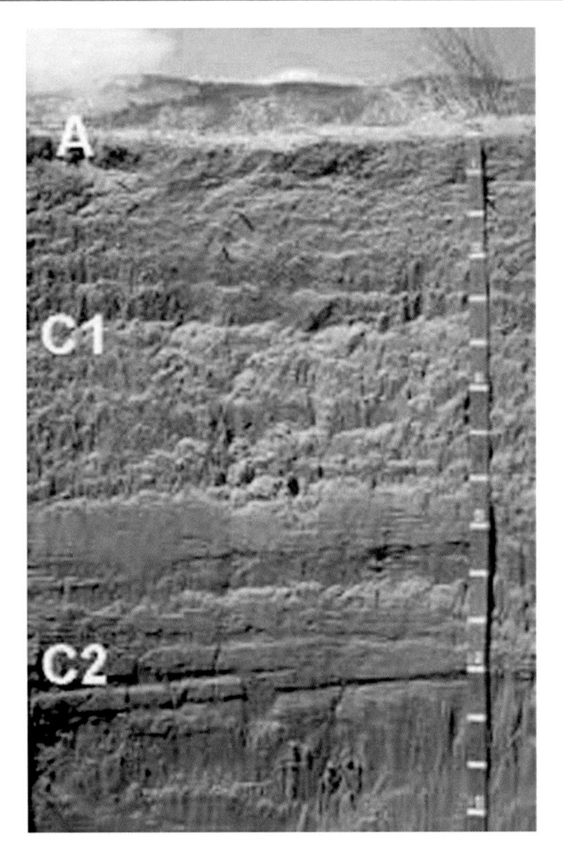

Fig. 7.10 The Quincy soil series, a Xeric Torripsamments that does not occur in Nevada but occurs just across the border near Vinton, California Source: NRCS

are provided in Table 10.1. The pedon has an ochric epipedon to 10 cm over a series of C horizons with minimal alteration.

Torripsamments are used mainly for wildlife habitat and livestock grazing, with limited use for irrigated alfalfa and pasture.

7.12 Haploxerolls (Great Group Association 13)

Haploxerolls are xeric soils that have a mollic epipedon that is not underlain by a clay-enriched layer. These soils are derived primarily from colluvium and residuum on mountain slopes, but they also form from alluvium on alluvial fans and terraces. They often support sagebrush-dominated communities with an understory of Idaho fescue, wheatgrass, needlegrass, and bluegrass. Haploxerolls occur at elevations ranging from 1,400 to 2,300 m and on slopes of 55 ± 31%. These soils exist in areas where the MAAT is 7.5 ± 2.3 °C and the MAP is 315 ± 85 mm. Haploxerolls often are of Holocene (< 10 ka) age.

There are 116 Haploxerolls soil series in Nevada that total about 4,800 km^2, 2.2% of the state land area. The Hyzen soil series (475 km^2) is the most widely distributed Haploxerolls. These soils occur in nearly all MLRAs but are most evident in MLRAs 26, 25, 23, 28A, and 28B in the northern portion of the state.

Haploxerolls tend to be in loamy PSCs and one-half are skeletal; they generally have a mixed mineralogy; two-thirds of the soil series are in the superactive CEC activity class; and the soils are equally divided between frigid and mesic STCs. Nearly three-quarters (73%) of the Haploxerolls have a lithic or paralithic contact within 100 cm of the surface. The mollic epipedon averages 47 ± 22 cm; about one-third of the Haploxerolls have a cambic horizon averaging 35 ± 6.7 cm in thickness.

Figure 7.11 is an unnamed Haploxeroll from Nevada. The soil has weakly developed mollic epipedon to 25 cm. Analytical data for the Dia soil series are given in Table 9.1. The pedon has a mollic epipedon to 25 cm over a series of minimally developed C horizons.

Most of the Haploxerolls are managed for range and wildlife habitat; however, several of the extensive Haploxerolls, including the Haypress and Atrypa soil series, produce pine and oak. The Dithod soil series is irrigated for alfalfa, small grains, corn, and legume-grass pasture. The Mottsville soil series has been developed in urban areas of Carson City.

7.13 Durixerolls

Durixerolls are semiarid soils that have a mollic epipedon over a duripan. Durixerolls are derived from alluvium on alluvial fan remnants, and plateaus that is often influenced by loess and ash deposition. They generally feature sagebrush with grasses, such as bluegrass and needlegrass, in the understory. Haploxerolls occur at elevations ranging from 1,500 to 2,000 m and on slopes of 26 ± 14%. These soils occur in areas where the MAAT is 7.8 ± 1.6 °C and the MAP is 300 ± 25 mm. Durixerolls often are of Early Wisconsin (40-130 ka) or Pre-Wisconsin (>130 ka) in age.

There are 46 soil series in the Durixerolls great group in Nevada that total about 3,800 km^2, 1.8% of the state land area. Although Durixerolls occur in most MLRAs in Nevada, they are pervasive in MLRA 25, the Owhyee High Plateau. However, the Durixerolls are too dispersed to show on the 1:2 million-scale general soil map. The Urmafot, Stampede, and Donna soil series each compose from 500 to 800 km^2 land area in Nevada.

Durixerolls tend to be fine-textured, i.e., in clayey, fine, and very-fine PSCs; nearly all of the soil series are smectitic;

Fig. 7.11 A Haploxeroll from Nevada with a weakly developed mollic epipedon to 25 cm Source: NRCS

about one-third are in the superactive CEC activity class; and three-quarters are in the mesic STC. Durixerolls contain a mollic epipedon averaging 25 ± 5.6 cm in thickness over a duripan with an average thickness of 47 ± 25 cm. Many Durixerolls also have an argillic that averages 36 ± 12 cm in thickness. Figure 7.12 is of the Stampede soil series, a Vertic Durixerolls formed in eolian material over old alluvium. The soil contains a mollic epipedon to 27 cm over an argillic horizon from 30 to 71 cm and a duripan from 71 to 84 cm. The soil is classified as a Durixerolls instead of an Argixerolls because the duripan keys out first in ST. Analytical data for the Stampede soil series are provided in Table 9.1. The pedon has a mollic epipedon to 38 cm that includes the Bt horizon and an argillic horizon from 38 to 65 cm, and a duripan from 65 to 112 cm. The duripan is extremely hard and is indurated by laminar opaline silica. Durixerolls tend to be shallow and are used for range and wildlife habitat.

7.14 Halaquepts (Great Group Associations 2 and 14)

Halaquepts are the wet, weakly developed, nonacid soils with an ochric epipedon. They are derived from deep alluvium and lacustrine materials occurring in floodplains and lake plains. The vegetation is primarily greasewood, alkali sacaton, and saltgrass. They occur at elevations ranging from 1,300 to 1,900 m and on slopes averaging 2.8 ± 1.0%. Halaquepts occur in areas where the MAAT is 9.6 ± 1.7 °C and the MAP is 172 ± 31 mm. Halaquepts often are of Holocene (<10 ka) age.

Halaquepts occur in several MLRAs but are most evident in MLRA 28B, the Central Nevada Basin and Range. Only the Wendane and Ocala soil series have areas in Nevada that are between 500 and 800 km^2. Halaquepts have a total area of about 3,335 km^2, 1.4% of the state land area.

Fig. 7.12 Stampede soil series, a Vertic Durixerolls from Harney, Oregon. The scale is in decimeters on the left and feet on the right Source: NRCS

Halaquepts commonly have a fine-silty or fine-loamy particle-size, a mixed mineralogy, a superactive CEC, and a mesic soil-temperature regime. Halaquepts in Nevada have only an ochric epipedon. They are classified as Inceptisols because they contain an exchangeable sodium percentage (ESP) of 15 or more, or have a sodium adsorption ratio (SAR) of 13 or more and a decrease in ESP or SAR values with increasing depth below 50 cm.

Halaquepts have an ochric epipedon that averages 18 cm in thickness and no diagnostic subsurface horizons. Analytical data for the Settlement soil series are given in Table 10.1. The pedon has an ochric epipedon over a saturated zone enriched in Na and Ca salts that has a prismatic structure and carbonate nodules in a material that is strongly alkaline and violently effervescent in dilute HCl. The Argenta soil series is irrigated for crops, but most of the Halaquepts are used for livestock grazing and wildlife habitat.

7.15 Endoaquolls

The Endoaquolls are the deep, poorly drained, and very poorly drained soils with a mollic epipedon. They are derived from alluvium in floodplains. They feature a variety of vegetation, but sedges and willows are common in humid environments and saltgrass in arid environments. Endoaquolls occur at elevations ranging from 1,400 to 2,100 m and on slopes averaging 5.2 +/ 4.3%. Endoaquolls in Nevada occur in areas where the MAAT is 8.1 $\pm$ 1.2 °C and the MAP is 259 $\pm$ 62 mm. Endoaquolls often are of Holocene (< 10 ka) age.

There are 45 soil series classified as Endoaquolls that compose an area of about 3,000 km^2, 1.3% of the state land area. Endoaquolls occur in a variety of MLRAs but are most common in MLRAs 25 and 27 in the northern Basin and Range. Because they are dispersed, they do not appear on the 1:2 million general soil map of Nevada. The most widespread soil series in the Endoaquolls great group in Nevada are the Welch, Crooked Creek, Humboldt, and Devilsgait, none of which comprises more than 550 km^2. The Humboldt soil series was described and sampled by C.F. Marbut in 1935.

Nevada Endoaquolls tend to be in fine and fine-loamy PSCs, the mixed or smectitic mineralogy class, the superactive or no CEC activity class, and the mesic or frigid soil-temperature class. The mollic epipedon in Endoaquolls averages 60 $\pm$ 18 cm in thickness. A few soil series have a calcic horizon. Figure 7.13 is the Kimmerling soil series, a Cumulic Endoaquoll with a mollic epipedon extending beyond 100 cm. Analytical data for the Welch soil series are included in Table 9.1. The pedon has an "over thickened" mollic epipedon (cumulic horizon) that is over 100 cm thick from regular flooding and deposition of topsoil eroded from upslope. Endoaquolls are one of the most intensively managed great group for agriculture in Nevada. Irrigated crops included alfalfa for hay, small grains, and some row crops.

7.16 Calcixerolls

The Calcixerolls are the dry Mollisols with a calcic horizon. These soils are derived from residuum and colluvium in mountain ranges underlain by limestone. The vegetation on Calcixerolls is commonly pinyon-sagebrush or mountain sagebrush with snowberry, bluegrass, and wheatgrass in the understory. These soils occur at elevations between 1,700 and 2,700 m and on slopes averaging 65 $\pm$ 20%. Calcixerolls in Nevada occur in areas, where the MAAT is 7.6 $\pm$ 1.1 °C and the MAP is 340 $\pm$ 36 cm. Calcixerolls usually are of Holocene (< 10 ka) age.

There are 26 soil series in the Calcixerolls in Nevada, which cover about 2,500 km^2, 1.0% of the land area. The major MLRAs are 25, 28A, and 28B in northeastern Nevada.

Fig. 7.13 The Kimmerling soil series, a Cumulic Endoaquolls. The mollic epipedon extends beyond 100 cm Source: NRCS

The Calcixerolls are not concentrated in one area enough to show on the general soil map of Nevada. The Cavehill series is the only widespread Calcixerolls in Nevada, occupying an area of 1,065 km^2.

Nevada Calcixerolls invariably are loamy-skeletal, have a carbonatic or mixed mineralogy, and are in the frigid soil-temperature class. More than three-quarters (78%) of the Calcixerolls have a lithic or paralithic contact within 100 cm of the surface. The mollic epipedon averages 28 $\pm$ 12 cm in thickness, and the calcic is 48 $\mp$ 44 cm. Figure 7.14 is of an Aridic Calcixerolls, the Eenreed soil series, from the Great Basin National Park. The soil has a mollic epipedon that includes the Bt horizon to 30 cm; the argillic horizon is from 13 to 30 cm and is underlain by a calcic horizon to at least 152 cm. Analytical data for the Cavehill series are given in Table 9.1. The pedon has a mollic epipedon to 28 cm and a calcic horizon to 64 cm over hard limestone bedrock.

Calcixerolls are used mainly for livestock grazing and wildlife habitat. The Cavehill soil series is managed for pinyon pine and mountain mahogany forest products.

7.17 Natridurids

Natridurids are dry soils that contain a natric horizon over a duripan. They are derived from alluvial enriched with loess and volcanic ash on alluvial-fan remnants. The dominant vegetation is shadscale, greasewood, or sagebrush along with rabbitbrush and squirreltail. Natridurids occur at elevations from 1,300 to 1,800 m and on slopes averaging 26 $\pm$ 11%. These soils occur in areas where the MAAT is 9.5 $\pm$ 1.2 °C and the MAP is 185 $\pm$ 33 mm. Natridurids are of Middle Wisconsin (30–40 ka) to Pre-Wisconsin (> 130 ka) in age.

There are 21 soil series classified as Natridurids in Nevada that cover about 2,300 km^2, 0.9% of the state land area. Natridurids are most common in MLRA 24, the Humboldt Area, and do not occur in MLRAs 22A, 28A, and 30. The Calcixerolls are not concentrated in one area enough to show on the general soil map of Nevada. The most prevalent Natridurids are the Tenabo (456 km^2) and Pirouette (428 km^2) soil series.

Fig. 7.14 Eenreed soil series, an Aridic Calcixerolls from the Great Basin National Park. The scale is in decimeters. Photo by USDA NRCS (2009)

Natridurids are in fine, clayey, loam, or fine-loamy particle-size classes, smectitic or mixed mineralogy classes; more than one-third are in the superactive CEC activity class, all of the soil series are in the mesic soil-temperature class, and about one-half of the soil series are shallow. The ochric epipedon of Natridurids in Nevada averages 15 cm and the duripan averages 25 cm in thickness. Analytical data for the Aboton soil series are given in Table 8.1. The pedon contains an ochric epipedon to 17 cm, a natric horizon to 39 cm, and a strongly cemented duripan from 48 to 56 cm in depth. Natridurids are used for livestock grazing or range and wildlife habitat. The Flue soil series is irrigated for crops.

7.18 Torrifluvents

Torrifluvents are dry, deep, poorly developed soils that are derived from alluvium on alluvial fans and floodplains. They support greasewood, shadscale, sagebrush, and other desert shrubs. Torrifluvents occur at elevations from 1,100 to 1,600 m and on slopes averaging 3.3 ± 1.5%. For Torrifluvents in Nevada, the MAAT is 12 ± 3.9 °C and the MAP is 161 ± 61 mm. Torrifluvents are of Holocene (< 10 ka) age.

There are 38 soil series in the Torrifluvents great group that total about 2,100 km^2, 0.9% of the state land area. These soils occur in many of the MLRAs in Nevada but tend to be concentrated in MLRA 27, the Fallon-Lovelock Area. However, they are not concentrated enough to appear in Fig. 7.1. The Cirac and Slaw soil series each occupy about 500 km^2 in Nevada.

Torrifluvents are deep and have PSCs that range from fine-loamy to sandy a mixed mineralogy. About three-quarters of the Torrifluvents have a mixed mineralogy, are in the calcareous reaction class, and have a mesic soil-temperature regime. Torrifluvents only have an ochric epipedon that averages 24 ± 11 cm in thickness. Figure 7.15 is a Torrifluvents from Arizona. Analytical data for the Cirac soil series are given in Table 10.1. The pedon only has a thin ochric epipedon over a series of C horizons contained in calcareous alluvium from mixed materials. Torrifluvents are used for livestock grazing and wildlife habitat. The Alhambra soil series is irrigated for alfalfa and small-grain crops.

7.19 Haplocryolls (Great Group Associations 4 and 12)

Haplocryolls are the cold mountain soils with a mollic epipedon. These soils are derived from colluvium on mountain slopes that is often influenced by loess and volcanic ash

Fig. 7.15 A Torrifluvents from Arizona, which is comparable to those in MLRA 29 in Southern Nevada Source: NRCS

Fig. 7.16 The Basinpeak soil series, a Xeric Haplocryolls from the Great Basin National Park. The scale is in decimeters. Photo by USDA NRCS (2009)

deposition. They occur at elevations of 2,000 to 3,000 m and on steep slopes averaging 76 ± 1.7%. Haplocryolls support mountain big sagebrush, with an understory of bluegrass and other grasses. For Haplocryolls in Nevada, the MAAT is 5.2 ± 1.1 °C and the MAP is 452 ± 127 mm.

Haplocryolls have a total area of about 2,000 km^2, 0.9% of the state land area. Haplocryolls occur primarily in MLRAs 23, 25, 28A, and 28B in central Nevada. The Hapgood (430 km^2) and Hackwood (345 km^2) are the major soil series in the Haplocryolls great group of Nevada.

More than half (58%) of the Haplocryolls are in the loamy-skeletal particle-size class; 93% have a mixed mineralogy; 87% are in the superactive CEC activity class; and most of the soil series have a xeric SMR. Nearly one-half (44%) of the Haplocryolls in Nevada have a lithic or paralithic contact within 100 cm of the surface. Haplocryolls have a mollic epipedon that averages 109 ± 67 cm. About 25% of the Haplocryolls are in pachic subgroups. The mollic horizon in Haplocryolls is unusually thick because of increased effective moisture as a result of aspect, elevation, and high amounts of precipitation. Figure 7.16 is of the Basinpeak soil series, a Xeric Haplocryolls from the Great Basin National Park. The soil has a mollic epipedon to 30 cm that is underlain by relatively unaltered colluvium. Analytical data for the Hapgood soil series are contained in Table 9.1. The pedon has an "over thickened" mollic epipedon (pachic horizon) that exceeds 100 cm over a C horizon. Fractured andesite bedrock occurs at a depth of 127 cm. Haplocryolls are used for livestock grazing and wildlife habitat.

7.20 Argicryolls (Great Group Association 12)

Argicryolls are soils containing a mollic epipedon and an argillic horizon that occur in a cryic region, i.e., where the mean annual soil temperature at a depth of 50 cm is between 0 and 8°C. They are formed in colluvium and residuum that is often enriched in loess and volcanic ash on mountain slopes composed of volcanic rocks. The vegetation is mountain big sagebrush and low sagebrush, along with Sandberg's bluegrass and Idaho fescue. Argicryolls in Nevada occur at elevations of 2,000 to 2,700 m and on slopes averaging 58 ± 12%. They occur in areas with a MAAT of 5.5 ± 1.2 °C and a MAP of 424 ± 85 mm. Argicryolls likely are of Late Wisconsin (10-30 ka) age.

There are 48 soil series classified as Argicryolls in Nevada which total about 1,700 km^2, 1% of the state's land area. These soils occur are most common in MLRAs 26, the Carson Basin and Mountains, and 28B, the Central Nevada Basin and Range, but they occur in all MLRAs except 30, the Mojave Desert. The most abundant Argicryolls in Nevada is the Tusel (418 km^2) soil series.

Argicryolls often are in the loamy-skeletal particle-size class, mixed or smectitic mineralogy class, the superactive CEC activity class, and have a xeric soil-moisture regime. A lithic or paralithic contact exists within 100 cm of the surface in 67% of the Argicryolls. The mollic epipedon of Argicryolls averages 38 cm and the argillic horizon averages 35 cm in thickness. Figure 7.17 is of the Lastsummer soil series, a Vitrandic Argicryolls. This pedon has a mollic epipedon (A horizon) to 18 cm and the upper part of the argillic extends to 84 cm. Stones at 84 cm mark the occurrence of a duric feature and continuation of the argillic horizon. Analytical data are provided for the Newlands soil

Fig. 7.17 The Lastsummer soil series, a Vitrandic Argicryolls. This pedon has a mollic epipedon (A horizon) to 18 cm and the upper part of the argillic extends to 84 cm. Stones at 84 cm mark the occurrence of a duric feature and continuation of the argillic horizon Source: NRCS

series in Table 9.1. The pedon has a mollic epipedon to 48 cm over an argillic horizon to 94 cm. Argicryolls are used for a range or livestock grazing and wildlife habitat.

7.21 Argiustolls (Great Group Association 11)

Argiustolls are semiarid soils with a mollic epipedon overlying an argillic horizon. They occur under pinyon-juniper forests on residuum and colluvium in mountains and mesas composed of volcanic materials. Argiustolls tend to be shallow or very shallow, i.e., the depth of rooting is less than 50 cm. They occur most commonly at elevations ranging from 1,600 to 2,300 m (Table 7.1). The MAAT is 8.9 ± 1.6 °C, and the MAP is 344 ± 38 mm. Slopes average 48 ± 17%. Argiustolls are covered most often with pinyon pine-juniper woodland. Argiustolls in Nevada commonly are of late Wisconsin age (10–30 ka) or older.

There are only 18 soil series in the Argiustolls great group that total 1,700 km^2, 0.8% of the state land area. Argiustolls occur exclusively in MLRAs 29 and 30 in southern Nevada. The Turba soil series (400 km^2) is the most widespread Argiustolls in Nevada.

More than half of the Argiustolls in Nevada have a loamy-skeletal particle-size class; the major mineralogy classes are mixed and smectitic; three-quarters of the series are superactive; and most are in the mesic soil-temperature class. The ochric epipedon averages 32 ± 12 cm, and the argillic horizon averages 30 ± 19 cm. Analytical data for the Turba soil series are contained in Table 9.1. The mollic epipedon includes the Bt1 horizon to 25 cm and the argillic extends from 13 to 56 cm and overlies weathered welded tuff. Argiustolls are used for livestock grazing and wildlife habitat. The Buckspring and Motoqua soil series are used for woodland products.

7.22 Calciargids

Calciargids are dry soils that contain an argillic horizon over a calcic horizon. They are derived from alluvium on alluvial-fan remnants, fan piedmonts, and basin floors. Calciargids occur under creosote bush, shadscale saltbush, and Wyoming big sagebrush, along with rabbitbrush, and galleta-bush muhly-Indian ricegrass. Elevations for Calciargids in Nevada range from 1,100 to 1,600 m, and slopes average 14 ± 9.5%. Calciargids occur in areas with a MAAT of 13 ± 3.3 °C and a MAP of 180 ± 67 mm.

The Calciargids contain 15 Calciargids soil series that total about 1,600 km^2, or 0.8% of Nevada's land area. These soils occur primarily in MLRA 30, the Mojave Desert, also occur in MLRAs 28A, 28B, 29, and 27, but do not occur in MLRAs 22A, 23, 24, 25, and 26. Because the Calciargids are dispersed throughout the state, they do not appear on the general soil map. The dominant Calciargids are the Oricto (409 km^2) and Cath (386 km^2) soil series.

Calciargids are in primarily loamy particle-size classes, about half of which are skeletal, the mixed mineralogy class, the superactive CEC activity class, and the mesic, thermic, and hyperthermic soil-temperature classes. The ochric horizon averages 14 cm, the argillic horizon averages 48 cm, and the calcic horizon averages 64 cm in thickness. Analytical data for the Oricto soil series are given in Table 8.1. The pedon has an ochric epipedon to 5 cm, an argillic horizon to 15 cm, and a calcic horizon to 58 cm. Calciargids are used for livestock grazing or range and wildlife habitat.

7.23 Calcicryolls (Great Group Association 4)

Calcicryolls have a mollic epipedon and a calcic horizon and occur at the higher elevations, where a cryic soil temperature regime is present. The parent materials are colluvium and residuum over limestone bedrock on mountain slopes. The vegetation is composed of mountain big sage, black sage, wheatgrass, basin rye, and occasionally mountain mahogany and scattered white fir. Calcicryolls occur at elevations between 2,300 and 2,800 m and on slopes ranging from 15 to 75%. The MAAT is between 4 and 6 °C, and the MAP is around 520 mm.

The Calcicryolls great group contains six soil series that cover 993 km^2 in Nevada. Calcicryolls occur in MLRAs 24, 25, 28A, and 28B. Major soils include the Wardbay, Hardol, and Adobe series. Calcicryolls are in the loamy-skeletal particle-size class, the carbonatic mineral class, and the xeric soil-moisture class. Table 8.1 contains analytical data for the Wardbay soil series. Calcicryolls are used for livestock grazing, rangeland, and wildlife.

7.24 Cryrendolls (Great Group Association 4)

Cryrendolls have a mollic epipedon and occur on limestone bedrock at the higher elevations. Cryrendolls are derived from colluvium and residuum, often influenced by loess, over limestone bedrock on mountain slopes. The vegetation is mountain mahogany and mountain big sagebrush, often with ponderosa fir and Douglas fir. Elevations range from 2,300 to 2,800 m, and slopes range from 8 to 75%. The MAAT is 5 to 6 °C, and the MAP is around 520 mm.

There are five soil series in the Cryrendolls great group; they cover 849 km^2 in Nevada. Cryrendolls occur in MLRAs 24, 28A, and 28B. The most extensive Cryrendolls in Nevada are the Haunchee and Halacan soil series. Cryrendolls are in loamy-skeletal particle-size class, the carbonatic mineral class, and the xeric soil-moisture class.

Table 9.1 contains analytical data for the Haunchee soil series. Cryrendolls are used for livestock grazing, rangeland, and wildlife.

7.25 Haploxeralfs (Great Group Association 16)

Haploxeralfs have an ochric epipdedon over an argillic horizon and occur in areas with a xeric soil moisture regime. Parent materials are commonly colluvium and residuum, sometimes mixed with loess and volcanic ash, over volcanic bedrock on mountain slopes. The vegetation is commonly forest, including Jeffrey pine, red fir, white fir, and lodgepole pine, or low sagebrush and Idaho fescue. Elevations range between 1,600 and 2,300 m, and slopes range from 0 to 45%. The MAAT is 6 to 7 °C, and the MAP ranges broadly between 350 and 1,000 mm.

There are 14 soil series in the Haploxeralfs great group in Nevada; they cover around 517 km^2 of the state. Haploxeralfs occur in MLRAs 22A and 25. The most extensive soil is the Bulake series.

The Haploxeralfs occur in coarse-loamy, fine-loamy, and loamy-skeletal particle-size classes, mixed, isotic, and smectitic mineral classes, the frigid soil-temperature class, and the xeric soil-moisture class. The Jorge soil series, an Andic Haploxeralfs, is shown in Fig. 12.3. The upper 126 cm is uniformly brown (7.5YR 4/4), including the argillic (Bt) horizon which occurs from 85 to 126 cm. Analytical data for the Bulake soil series are given in Table 12.1. The clay concentration doubles from the A2 to the Bt horizon. Haploxeralfs are managed as woodlands, livestock grazing, and wildlife.

7.26 Xeropsamments (Great Group Association 16)

Xeropsamments are weakly developed soils that lack a diagnostic subsurface horizon and occur in areas with a xeric soil-temperature regime. Xeropsamments are derived from colluvium and residuum over granite on mountain slopes. The vegetation is coniferous forest, including Jeffrey pine, white fir, California red fir, western white pine, and ponderosa pine. Elevations range between 1,500 and 2,700 m, and slopes range from 5 to 75%. The MAAT is 4 to 6 °C, and the MAP ranges broadly between 635 and 1,015 mm.

There are four soil series in the Xeropsamments that cover 108 km^2 in Nevada. Xeropsamments occur in MLRAs 22A and 25. The Xeropsamments occur in the mixed mineral class, the frigid soil-temperature class, and the xeric soil-moisture class.

The Christopher soil series, a Dystric Xeropsamments, is shown in Fig. 10.1. The soil contains a brown ochric epipedon (A horizon) to 20 cm and is underlain by Bw horizons that do not qualify as cambic because of their loamy coarse sand texture. Analytical data for the Cagwin soil series are provided in Table 10.1. Xeropsamments reflect multiple use, including recreation, watershed, wildlife, timber, and urban development.

7.27 Other Great Groups

Several other great groups are common (i.e., occupy from 450 to 990 km^2) in Nevada, including the Palexerolls, Haploxererts, Fluvaquents, Paleargids, Haplustolls, and Aquisalids, but their areas are insufficient to show on the general soil map of the state.

7.28 Conclusions

Nevada can be divided into 17 soil great-group associations. These contain, from greatest to least in extent, Torriorthents, Haplargids, Argixerolls, Haplocalcids, Haplocambids, Haplodurids, Argidurids, Natrargids, Petrocalcids, Torripsamments, Haploxerolls, Durixerolls, Halaquepts, Endoaquolls, Calcixerolls, Natridurids, Torrifluvents, Haplocryolls, Argicryolls, Argiustolls, and Calciargids.

References

Brock AL, Buck BJ (2009) Polygenetic development of the Mormon Mesa, NV petrocalcic horizons: geomorphic and paleoenvironmental interpretations. CATENA 77:65–75

Burt R, Chiaretti JV, Ferguson R (2014) Chemical fractionation of selected Nevada soils adjacent to the Sierra Army Depot. Soil Horizons: https://doi.org/10.2136/sh13-07-0019

Monger HC, Gile LH, Hawley JW, Grossman RB (2009) The desert project—an analysis of aridland soil-geomorphic processes. NM State Univ. Agr. Exp. Stn. Bull. 798

8 Aridisols

Abstract

Aridisols are the most abundant soil order in Nevada, in terms of numbers of soil series and land area. This chapter discusses the distribution, properties and processes, and use and management of Aridisols in Nevada.

8.1 Distribution

Aridisols are the most abundant soil order in Nevada, in terms of numbers of soil series (42%) and land area (52%) (Table 7.2). Aridisols occur in all of the 10 Major Land Resource Areas in Nevada except for MLRA 22A, the Sierra Nevada Mountains. They are most common in MLRA 30 (63% of all soil series), MLRA 29 (54%), and MLRA 24 (53%) (Table 4.1). Aridisols also occur in MLRA 27 (45%), MLRA 28B (42%), MLRA 23 (36%), MLRA 25 (36%), MLRA 28A (34%), and MLRA 26 (33%). On an area basis, 37% of the Aridisol soil series are Argids, followed by Durids (26%), Calcids (19%), Cambids (13%), with Gypsids and Salids composing less than 0.5% each (see Table 7.2). Nevada ranks first nationally in area of Durids, third for Salids, and fifth for Gypsids. Five great groups account for 84% of the Aridisols soil series, including the Haplargids (30%), Haplocalcids (15%), Haplocambids (14%), Haplodurids (13%), and Argidurids (12%) (Table 7.2).

Thirteen Aridisol soil series each cover more than 1,000 km^2, including the Stewval, Palinor, Unsel, Ursine, Pookaloo, Theon, Old Camp, Downeyville, Armespan, Gabbvally, Koyen, Weiser, Tecomar, Wieland, Dewar, and Shabliss. Photographs of major great groups of Aridisols are given in Chap. 7, including a Haplargids (Jaybee soil series; Fig. 7.3), Haplocalcids (Owyhee soil series; Fig. 7.5), Haplocambids (Orovada soil series; Fig. 7.6), Argidurids (Colthorp soil series; Fig. 7.7), Natrargids (Highrock soil series; Fig. 7.8), and Petrocalcids (Mormon Mesa soil series; Fig. 7.9). Two additional photographs are provided here of important but less common Aridisols, including a Typic Aquisalids (Fig. 8.1) and the Drygyp soil series, a Typic Petrogypsids (Fig. 8.2). The Aquisalids has white salts more soluble than gypsum on the surface and in the salic horizon (Az and Bz horizons). The Petrogypsids has gypsum from just below the surface to more than 165 cm in depth. The chunks of soil scattered around the excavation are pieces of gypsum-cemented material.

Argids occur both in basins and mountain ranges throughout the state; Durids only occur north of N37°15' and mainly in basins (they occur further south in CA and AZ); Calcids are most abundant in basins south of N37°15,' but they also occur in MLRA 28A (Great Salt Lake Area); Cambids are concentrated in basins in southernmost Nevada, but they occur elsewhere in the state; Gypsids occur in three small basins in southern Nevada; and Salids are restricted to floors of the Lahontan and Bonneville Basins. Cryids are limited in the USA to mountains in Idaho and Colorado and do not occur in Nevada.

Aridisols in Nevada commonly receive from 125 to 300 mm of water-equivalent precipitation per year. They support all of the common low- to mid-elevation vegetation types in Nevada; however, Aridisols are uncommon at the higher elevations in pinyon-juniper, mountain mahogany, and subalpine forest. Nearly three-quarters (71%) of the Aridisol soil series in Nevada are derived from alluvium, often with a thin mantle or intermixed volcanic ash and/or loess. Aridisols may form in less than 6.7 kyr (Alexander and Nettleton 1977).

8.2 Properties and Processes

By definition, the Aridisols have an aridic soil-moisture regime, which means that the soil control section is dry in all parts for more than half of the cumulative days per year when the soil temperature at a depth of 50 cm below the soil surface is above 5 °C and moist in some or all parts for less than 90 consecutive days when the soil temperature at a depth of 50 cm below the soil surface is above 8 °C.

P. W. Blackburn et al., *The Soils of Nevada*, World Soils Book Series,
https://doi.org/10.1007/978-3-030-53157-7_8

Fig. 8.1 A Typic Aquisalids from Nevada showing salts on the surface and a salic horizon (Az and Bz) (Photo by Paul McDaniel)

The key properties of Aridisols in Nevada are the presence of an ochric epipedon over a cambic, calcic, gypsic, petrocalcic, petrogypsic, duripan, argillic, or natric horizon. The most common subsurface horizons in Aridisols of Nevada are the argillic, duripan, calcic, and cambic horizons. Many Aridisols in Nevada have two or three diagnostic subsurface horizons.

Aridisols in Nevada tend to be in loamy particle-size classes (69% of soil series), have a mixed (70%) or smectitic (18%) mineralogy, and are in the superactive CEC activity class (64%). Although Aridisols commonly have a mesic soil-temperature regime (72% of soil series), they range from hyperthermic (MAST = $\geq$22 °C) to frigid (MAST = 0 to 8 °C). About one-quarter (28%) of the Aridisols are in the shallow family class.

Because of their aridity, most Aridisols in Nevada have a high base saturation, an alkaline pH, and low SOC concentrations (Table 8.1). Diagnostic horizons are designated in bold face. Argids include the Unsel soil series (Haplargids), Oxcorel (Natrargids), Phing (Paleargids), and Oricto (Calciargids). All of these soil series have at least a two-fold increase in clay concentration from the ochric epipedon to the argillic (Bt) horizon. The Phing soil series, a Paleargid, has a four-fold increase in clay from the E to the Bt1 horizon. The Oxcorel has a natric horizon (Btn), which is reflected in the high exchangeable sodium percentage (ESP) and sodium adsorption value (SAR). The Palinor, Zadvar, and Aboten soil series have a duripan, which is a horizon cemented by laminar forms of opal. The location of the duripan is reflected by the wide ratio of water retention at 1.5 mP to clay concentration.

The Mormon Mesa and Pookaloo soil series have petrocalcic and calcic horizons and high levels of $CaCO_3$. The McCarran soil series is a Haplogypsid and has a gypsic horizon. The Parran soil series is an Aquisalid and contains a salic horizon that is evidenced by exceptionally high electrical conductivities (EC), ESPs, and SARs. The Orovada and Panlee soil series lack an argillic, duripan, calcic, gypsic, and salic horizon and are classified as Cambids. Although there is a two–fold increase in clay from the A to the Bw1 horizon in the Orovada soil, no clay skins or bridging of sand grains occur, suggesting that clays have not been translocated from the surface to the subsurface. Calcium carbonate and silica have accumulated below 100 cm so that the requirements of a calcic, petrocalcic, or duripan horizon are not met.

The dominant soil–forming processes in Aridisols are argilluviation, silicification, calcificiation, cambisolization, gypsification, and salinization, which are discussed fully in Chap. 14.

8.3 Use and Management

As with all of the orders in *Soil Taxonomy*, there is considerable variation in the nature and properties of Aridisols. However, the key property that links all of the Aridisols is the aridic soil–moisture regime. For this reason, some Aridisols may be cultivated, especially the Cambids and some Argids, but irrigation is a prerequisite for most crops. Many of the Aridisols are shallow, including 15% of the Argids soil series (bedrock), 19% of the Calcids (petrocalcic horizon or bedrock), 27% of the Gypsids, and 64% of the Durids (duripan). Some Aridisols are prime agricultural soils, such as the state soil, the Orovada series (Durinodic Xeric Haplocambids). Many Aridisols are used for range and wildlife.

Fig. 8.2 NRCS personnel examining the Drygyp soil series, a Typic Petrogypsids. Most of the chunks scattered around the excavation are pieces of gypsum-cemented material. (Photo by Douglas Merkler, NRCS)

Table 8.1 Analyical properties of Aridisols found in Nevada

	Depth	Clay	Silt	Sand	SOC	CEC7	Base sat.	pH	$CaCO_3$	$CaSO_4$	EC	Ex. Na		Tot. salts	Al +1/2Fe	1.5 mPa H_2O/
Horizon[1]	(cm)	(%)	(%)	(%)	(%)	($cmol_c$/kg)	(%)	H_2O	(%)	(%)	(dS/m)	(%)	SAR	(%)	(%)	clay
McCarran; Typic Haplogypsids; Clark, NV; pedon no. 40A3174																
A	0–1				0.52			8.0	2	5	3.3					
By	1–36				0.10			8.0		37	2.9					
2By1	36–69				0.17			8.3	5	35	11.3					
2By2	69–91				0.18			8.4	26	6	20.6					
2By3	91–119				0.28			8.6	4	6	28.4					
2C	119–145				0.23			8.7	7	3	24.0					
Mormon Mesa; Calcic Petrocalcids; Clark, NV; pedon no. 89P0537																
A	0–4	5.7	19.0	75.3	0.48	6.2	100	8.3	6		0.25		tr	tr		0.72
Bk1	4–10	5.7	17.5	76.8	0.21	5.9	100	8.6	6		0.16					0.67
Bk2	10–20	7.7	13.1	79.2	0.17	5.8	100	8.4	7		0.16					0.60
Bk3	20–35	16.6	20.3	63.1	0.31	7.3	100	8.4	13		0.20	1				0.49
Bk4	35–49	21.3	24.7	54.0	0.26	5.4	100	8.3	25		0.19					0.41
2Bkm	49–51	3.6	14.8	81.4	0.09	1.7	100	9.5	76		0.12					0.97
Oricto; Typic Calciargids; Esmeralda, NV; pedon no. 73C0059																
A	0–5	7.2	28.8	64.0	0.24	12.4	100	7.2	6	tr	155.0	61	69	3.0		0.60
Bt1	5–9	18.8	41.2	40.0	0.23	17	100	7.3	8	tr	77.5	100	82	0.8		0.36
Bt2	9–15	18.8	30.8	50.4	0.30	20.9	100	7.7	7		25.1	37	35	0.6		0.43
Bk1	15–26	12.1	20.2	67.7	0.06	18.3	100	7.6	3	tr	23.0	40	34	0.6		0.56
Bk2	26–58	5.3	22.0	72.7	0.05	16.9	100	7.8	4	1	11.5	5	28	0.8		1.02
2C	58–71	5.7	12.8	81.5		27.3	100	7.8	3	tr	8.8	48	24	0.2		1.14
3C	71–78	8.8	24.9	66.3		27.9	100	8.0	9	tr	10.9	52	29	0.3		0.75
4C1	78–90	7.4	14.9	77.7		30.5	100	8.1	5	tr	9.0	51	25	0.2		0.91
4C2	90–100	6.3	15.1	78.6		23.4	100	7.9	4	tr	10.3	46	23	0.2		0.90
Orovada; Durinodic Xeric Haplocambids; Elko, NV; pedon no. 40A3179																
A	0–8	7.4	29.5	63.1	0.52	12.0	100	6.9			0.4	1	1	tr		0.60
Bw1	8–13	13.9	29.4	56.7	0.24	15.8	100	6.9			0.4	3	2	tr		0.60
Bw2	13–30	16.7	33.2	50.1	0.23	19.0	100	7.2			0.4	11	4	tr		0.67
Bw3	30–46	14.2	38.7	47.1	0.31	19.1	100	8.0			0.5	15	6	tr		0.64
Bq1	46–71	19.5	27.1	53.4	0.24	25.4	100	8.7			2.3	27	14	0.1		0.64
Bkq1	71–99	20.3	29.6	50.1	0.31	25.4	100	8.3	2		7.9	30	21	0.2		1.45
Bkq2	99–140	7.6	44.2	48.2	0.38	22.6	100	8.3	1		12.1	31	20	0.4		3.11
2Bkqm	140–173	3.5	49.4	47.1	0.37	22.3	100	8.0	1	1	13.0	25	20	0.5		1.37
2Bqm	173–196	4.9	28.5	66.6	0.09	13.1	100	8.0	1	2	11.0	22	21	0.3		1.49
2C	196–269	10.9	40.2	48.9	0.34	33.4	100	7.7	1		15.5	17	14	0.6		1.34
Oxcorel; Durinodic Natrargids; Pershing, NV; pedon no. 82P0709																
A1	0–5	8.7	54.2	37.1	0.91	15.8	100	8.2	tr		2.4	11	6	0.1		0.86
A2	5–13	10.3	52.8	36.9	0.46	17.2	100	8.7	tr		1.0	18	12	tr		0.76
A3	13–25	11.9	53.2	34.9	0.32	18.7	100	8.5			0.9	30	17	tr		0.68
Btnk1	25–38	25.8	44.1	30.1	0.50	27.9	100	8.2	1.0		3.0	56	32	0.1		0.60

(continued)

Table 8.1 (continued)

	Depth	Clay	Silt	Sand	SOC	CEC7	Base sat.	pH	$CaCO_3$	$CaSO_4$	EC	Ex. Na		Tot. salts	Al +1/2Fe	1.5 mPa H_2O/
Horizon[1]	(cm)	(%)	(%)	(%)	(%)	($cmol_c$/kg)	(%)	H_2O	(%)	(%)	(dS/m)	(%)	SAR	(%)	(%)	clay
2Btnk2	38–53	52.8	32.5	14.7	0.72	40.2	100	8.7	15.0		6.0	73	68	0.7		0.68
2Btnk3	53–66	50.7	30.1	19.2	0.37	43.4	100	8.3	4.0		22.0	68	76	1.8		0.56
2Btqkn	66–84	32.6	42.3	25.1	0.28	36.1	100	8.1	3.0		29.7	60	65	1.8		0.71
2Bqk1	84–107	18.3	49.9	31.8	0.23	30.5	100	8.0	2.0	9	33.9	60	79	1.4		0.89
2Bqk2	107–145	8.0	30.1	61.9	0.21	26.8	100	7.9	4.0		30.0	63	61	1.1		1.61
Palinor; Xeric Haplodurids; Elko, NV; pedon no. 06N0242																
A	0–10	15.1	39.7	45.2	0.70	17.6	100	8.2	7		0.9	tr	tr	tr	0.24	0.62
Bk1	10–28	17.7	36.0	46.3	0.20	17.2	100	8.5	15		0.6	5	2		0.16	0.64
Bk2	28–41	12.5	33.7	53.8		14.2	100	8.6	37		1.4	15	10	tr	0.07	1.04
Bkqm1	41–68	6.3	24.1	69.6		9.1	100	8.1	57		6.2	32	17	0.2	0.02	2.35
Bkqm2	68–83	3.9	17.3	78.8		12.4	100	8.0	48	1	12.7	40	24	0.6	0.04	4.46
Bkqy1	83–110	3.0	18.9	78.1		14.6	100	8.1	45	tr	14.4	48	31	0.6	0.09	4.97
Bkqy2	110–152	0.6	22.2	77.2		13.9	100	8.2	45		14.0	50	33	0.5	0.04	20.50
Panlee; Xeric Petrocambids; Humboldt, NV; pedon no. 88P0691																
A1	0–6	13.9	38.9	47.2		20.1	91	6.8			0.14	tr				0.63
A2	6–28	14.1	39.9	46.0		21.4	97	7.1			0.08	tr				0.66
Bw1	28–43	10.0	44.0	46.0		21.3	100	7.7			0.08	1				0.90
Bw2	43–53	13.8	38.2	48.0		26.3	100	7.7	tr		0.16	2				0.73
2Bkq	53–72	14.7	22.9	62.4		28.5	100	8.2	7		0.23	2				0.69
3Cr	72–90	12.7	18.9	68.4		29.1	100	8.4	24		0.29	7	5	tr		0.91
Parran; Typic Aquisalids; Churchill, NV; pedon no. 08N0488																
A1	0–3	41.5	39.1	19.4	0.2	24.1	100	8.7	3	tr	39.2	100	533	1.8	0.34	0.37
A2	3–8	46.5	38.0	15.5	0.1	24.2	100	8.9	3	1	93.2	100	533	4.7	0.33	0.37
Bz1	8–20	51.2	39.4	9.4	0.1	24.0	100	8.7	3	2	94.2	100	249	6.6	0.32	0.40
Bz2	20–46	46.1	45.4	8.5	tr	25.4	100	8.7	2	tr	77.9	100	237	5.8	0.29	0.47
Bz3	46–71	48.3	46.6	5.1	tr	25.6	100	8.8	1		101.1	100	240	4.0	0.29	0.47
Bz4	71–112	46.0	49.5	4.5		25.9	100	8.9	2		101.6	100	230	2.6	0.29	0.51
Bz5	112–130	48.9	46.8	4.3		28.1	100	8.9	2		116.1	100	241	1.8	0.32	0.53
C	130–166	47.1	50.3	2.6		27.9	100	9.0	1		127.7	100	303	1.5	0.29	0.68
Phing; Vertic Paleargids; Douglas, NV; pedon no. 78P0583																
A	0–10	15.2	23.1	61.7	0.57	12.2	90	7.1	tr	tr	0.02	2				0.40
E	10–23	12.7	29.4	57.9	0.32	11.6	83	7.4			0.07	5				0.53
Bt1	23–48	59.2	10.8	30.0	0.47	51.2	96	7.7			0.38	9	7	tr		0.46
Bt2	48–66	60.7	17.6	21.7	0.45	54.4	100	8.0	tr	tr	1.36	10	7	0.2		0.46
Bt3	66–79				0.34	63.2		7.9	tr	tr	1.84					
Bq1	79–119	50.7	23.9	25.4	0.14	53.7	100	7.9	2	2	1.75	10	9	0.2		0.52
Bq2	119–50	33.9	50.0	16.1	0.21	67.5	98	7.7	tr	tr	1.69	11	9	0.3		0.90

(continued)

Table 8.1 (continued)

	Depth	Clay	Silt	Sand	SOC	CEC7	Base sat.	pH	$CaCO_3$	$CaSO_4$	EC	Ex. Na		Tot. salts	Al +1/2Fe	1.5 mPa H_2O/
Horizon[1]	(cm)	(%)	(%)	(%)	(%)	(cmol$_c$/kg)	(%)	H_2O	(%)	(%)	(dS/m)	(%)	SAR	(%)	(%)	clay
Pookaloo; Lithic Xeric Haplocalcids; White Pine, NV; pedon no. 11N0082																
A	8–18	22.4	42.8	34.8	5.0	35.1	100	7.8	16							0.83
Bk1	18–33	22.7	44.2	33.1	4.4	30.4	100	8.0	25							0.96
Bk2	33–48	15.2	44.6	40.2	2.8	20.1	100	8.0	31							1.31
Bk3	48–73	15.3	46.8	38.1	2.9	23.3	100	8.1	28			tr				1.10
R	73															
Unsel; Durinodic Haplargids; Esmeralda, NV; pedon no. 73C0066																
A1	0–3	4.8	24.9	70.2	0.25	9.3	100	9.2	5			22				0.73
A2	3–8	8.0	39.4	52.6	0.07	13.6	100	9.3	5			36				0.6
Bt1	8–15	17.9	23.8	58.3	0.05	18.2	100	9.4	7		1.2	34	21	tr		0.41
Bt2	15–30	19.7	18.3	62.0		22.4	100	9.3	11		0.94	46	14	tr		
2Bk	30–51	14.2	27.6	58.2		15.6	100	9.2	9		5.7	60	48	0.1		0.54
2Bkq	51–81	12.2	17.2	70.6		14.8	100	9.0	5		13.2	76	74	0.3		0.57
3Bk	81–110	5.6	9.2	85.2		9.9	100	8.8	2		13.8	68	53	0.3		0.77
Zadvar; Haploxeralfic Argidurids; Nye, NV; pedon no. 73C0048																
A1	0–4	3.6	9.7	86.7	0.70	6.7	100	6.9				1				1.19
A2	4–10	7.1	37.1	55.8	0.27	7.9	100	7.7				1				0.54
Bt1	10–19	27.1	21.0	51.9	0.42	14.6	100	7.4				3				0.29
Bt2	19–38	49.1	6.7	44.2	0.36	28.8	100	7.5			0.54	5	7	tr		0.32
B2	38–43	34.9	13.5	51.6	0.60	26.5	100	8.4	5.0		0.72	6	4	tr		0.45
Bqm2	43–63	7.0	13.9	79.1	0.23	18.2	100	8.9	9		0.74	9	5	tr		1.47
Aboten; Natrargid Natridurids; Churchill, NV; pedon no. 72C0019																
A11	0–2	4.6	33.0	62.4	0.51	13.4	100	8.5			1.11	13	6	tr		1.22
A12	2–9	4.6	32.3	63.1	0.21	13.7	100	8.8				13				1.11
A2	9–17	6.8	28.0	65.2	0.12	16.3	100	8.8			0.62	14	4	tr		0.84
Bt1	17–28	20.9	37.4	41.7	0.21	27.3	100	9.0	1		1.26	16	12	tr		0.56
Bt2	28–39	20.2	35.0	44.8	0.38	30.5	100	8.4	1		4.87	23	16	0.2		0.66
Bkq	39–48	10.5	30.4	59.1	0.33	29.1	100	7.7	3		13.7	25	17	0.5		1.29
Bkqm	48–56	6.5	11.4	82.1	0.21	20.3	100	7.5	6	tr	21.8	35	15	0.6		1.52
C	56–83	4.6	7.6	87.8		19.6	100	7.5	3	tr	21.7	40	19	0.6		1.54

[1]Horizons in bold–face are diagnostic: Bz = salic; Bkm = petrocalcic; Bk = calcic; Bt = argillic; Bw = cambic; Btn = natric; Bqm = duripan; By = gypsum

8.4 Conclusions

Aridisols are the most common soil order in Nevada, accounting for 42% of the soil series and 52% of the land area. They occur in all MLRAs, except 22A, the Sierra Nevada Mountains. On an area basis, Aridisols can be ranked in Nevada: Argids > Durids > Calcids > Cambids >> Salids, Gypsids. Aridisols in Nevada commonly receive from 125 to 300 mm of water–equivalent precipitation per year, occur at low– to mid–elevations under desert shrubs, are usually derived from alluvium, and may form in less than 6.7 kyr. The key properties of Aridisols in Nevada are the presence of an ochric epipedon over a single or combination of argillic, duripan, calcic, and cambic horizons.

Reference

Alexander EB, Nettleton WD (1977) Post–Mazama Natrargids in Dixie Valley. Nevada. Soil Sci. Soc. Am. J. 41:1210–1212

9 Mollisols

Abstract

Mollisols are the second most abundant soil order in Nevada, in terms of numbers of soil series and land area. This chapter discusses the distribution, properties and processes, and use and management of Mollisols in Nevada.

9.1 Distribution

Mollisols are the second most abundant soil order in Nevada, in terms of numbers of soil series (35%) and land area (23%) (Table 7.2). Mollisols occur in all 10 Major Land Resource Areas in Nevada, but are most common in MLRA 21 (88% of all soil series), MLRA 26 (55%), and MLRA 25 (51%) (Table 4.1). Mollisols also occur in MLRA 28A (40%), MLRA 28B (36%), MLRA 23 (29%), MLRA 22A (28%), MLRA 24 (26%), MLRA 27 (19%), MLRA 29 (19%), and MLRA 30 (11%). Xerolls compose 78% of the Mollisol suborder area, followed by Cryolls (9.4%), Aquolls (5.9%), Ustolls (4.7%), and Rendolls (1.6%) (Table 6.2). More than one-half (56%) of the Mollisol soil area is Argixerolls (Table 7.2).

Mollisols exceeding 1,000 km^2 include the Cleavage, Ninemile, Sumine, Devada, and Cavehill soil series. Mollisols covering between 500 and 1,000 km^2 each in Nevada include the Chen, Bellehelen, Urmafot, McIvey, Softscrabble, Reluctan, Wylo, Stampede, Pickup, Acrelane, Welch, Haunchee, Quarz, and Donna soil series. Photographs of major great groups of Mollisols are given in Chap. 6, including an Argixerolls (Milan soil series; Fig. 7.4), Haploxerolls (Fig. 7.11), Durixerolls (Stampede soil series; Fig. 7.12), Endoaquolls (Kimmerling soil series; Fig. 7.13), Calcixerolls (Eenreed soil series; Fig. 7.14), Haplocryolls (Basinpeak soil series; Fig. 7.16), and Argicryolls (Lastsummer soil series; Fig. 7.17).

Xerolls occur on plateaus in northwestern and northeastern Nevada and in mountain ranges throughout the state; Cryolls occur in the highest ranges, especially in the northwest and in the Great Basin National Park in east-central Nevada; Aquolls occur in scattered river valleys and basins, such as the Carson Sink; Ustolls from colluvium and limestone and volcanic residuum in mountains; and Rendolls are limited to limestone areas, especially in MLRA 28B, the Central Nevada Basin and Range.

The mean annual precipitation for Mollisols in Nevada is commonly between 300 and 400 mm/yr. Mollisols in Nevada form under cool temperatures; 64% of the Mollisol soil series have a cryic or frigid soil-temperature regime. The dominant vegetation on Mollisols is sagebrush (*Artemisia* spp.) but a variety of short (bunch) grasses is usually present in the understory. More than two-thirds (69%) of the Mollisol soil series in Nevada occur on colluvium and residuum on mountain ranges. Argixerolls and Haploxerolls occur almost exclusively on colluvium and residuum; the other semi-dominant great groups (Haplocryolls, Durixerolls, and Endoaquolls) are derived mainly from alluvium on mountain sideslopes, fan remnants, and floodplains.

9.2 Properties and Processes

The key properties of Mollisols in Nevada are the presence of a mollic epipedon over an argillic (47% of Mollisols), calcic, or cambic horizon. The mean thickness of the mollic epipedon in Nevada soil series is 37 +/– 20 cm. Mollisols in Nevada tend to be in loamy (63% of soil series) and skeletal (41%) particle-size classes; they have a mixed (62%) or smectitic (22%) mineralogy; and they often occur in the superactive CEC activity class (57% of soil series). More than three-quarters (85%) of Mollisol soil series are in the xeric soil-moisture class and 63% are in the cryic and frigid soil-temperature classes. In Nevada, Mollisols form under some of the coolest and moistest conditions present in the

P. W. Blackburn et al., *The Soils of Nevada*, World Soils Book Series,
https://doi.org/10.1007/978-3-030-53157-7_9

Table 9.1 Analyical properties of Mollisols found in Nevada

Horizon	Depth (cm)	Clay (%)	Silt (%)	Sand (%)	SOC (%)	CEC7 ($cmol_c/kg$)	Base sat. (%)	pH H_2O	$CaCO_3$ (%)	$CaSO_4$ (%)	EC (dS/m)	Ex. Na (%)	SAR	Tot. salts (%)	Al+1/2Fe (%)	1.5 mPa H_2O/clay
Cavehill; Typic Calcixerolls; White Pine, NV; pedon no. 11N0080																
A1	7–15	2.3	26.9	70.8	7.7	24.9	100	7.8								7.22
A2	15–28	3.0	25.6	71.4	3.6	13.9	100	8.1								3.37
Bk1	28–41	3.4	25.6	71.0	4.5	9.7	100	8.2								3.26
Bk2	41–64	16.9	54.4	28.7	4.7	6.5	100									0.62
R	64															
Dia; Oxyaquic Haploxerolls; Churchill, NV; pedon no. 00P0240																
Ap1	0–8	23.4	30.3	45.9	1.0	22.2	100	7.1			1.24	2	1	tr		0.45
Ap2	8–25	21.8	34.3	43.9	tr	23.6	100	8.0			0.70	3	3	tr		0.56
C1	25–46	24.5	36.8	38.7	tr	28.2	100	7.9				5				0.56
C2	46–69	23.2	38.0	38.8	tr	27.2	100	7.8				5				0.58
C3	69–91	18.2	38.7	43.1	tr	23.4	98	7.8				5				0.59
2C	91–152	4.7	4.3	91.0		7.5	85	7.8				5				0.72
Hapgood; Pachic Haplocryolls; Lander, NV; pedon no. 82P0322																
A1	0–8	14.2	41.7	44.1	4.88	26.1	80	6.1			0.13	tr				1.09
A2	8–25	14.6	44.5	40.9	3.40	21.1	81	6.2			0.09	tr				0.84
A3	25–41	14.3	41.7	44.0	2.27	18.6	85	6.5			0.09	2				0.60
A4	41–56	14.7	41.2	44.1	1.92	21.6	71	6.6			0.08	1				0.67
A5	56–84	14.7	41.2	44.1	1.87	18.3	87	6.6			0.08	1				0.65
2A	84–107	15.4	41.2	43.4	1.59	16.1	83	6.7			0.09	1				0.59
2C	107–152	10.6	36.7	52.7	0.63	8.4	88	6.8			0.06	2				0.51
Haunchee; Lithic Cryrendolls; White Pine, NV; pedon no. 81P0667																
A1	0–5	23.2	46.7	30.1	13.4	65.5	100	7.1	2		1.18	tr	tr	0.1		1.61
A2	5–25	29.2	45.9	24.9	8.5	52.4	100	7.5	3		0.68	tr	tr	tr		0.82
R	25															
Newlands; Xeric Argicryolls; Washoe, NV; pedon no. 88P0683																
A1	0–5	15.3	37.0	47.7	3.60	32.3	89	7.2				tr			0.54	1.11
A2	5–18	16.4	35.0	48.6	2.03	28.5	89	6.9				tr			0.52	0.99
A3	18–44	16.0	34.5	49.5	1.38	26.6	90	6.7				1			0.51	1.00
Bt	44–59	16,6	37.0	46.4	0.89	24.1	91	7.0				1			0.51	0.96
2Bt1	59–69	22.3	34.5	43.2	0.79	24.9	92	6.9				1			0.49	0.80
2Bt2	69–94	23.0	28.1	48.9	0.62	23.7	92	6.9				2				0.75

(continued)

Table 9.1 (continued)

Horizon	Depth (cm)	Clay (%)	Silt (%)	Sand (%)	SOC (%)	CEC7 ($cmol_c/kg$)	Base sat. (%)	pH H_2O	$CaCO_3$ (%)	$CaSO_4$ (%)	EC (dS/m)	Ex. Na (%)	SAR	Tot. salts (%)	Al+1/2Fe (%)	1.5 mPa H_2O/clay
3C1	94–115	13.8	30.4	55.8	0.57	27.7	90	7.4				1			0.28	1.07
Ninemile; Aridic Lithic Argixerolls; Washoe, NV; pedon no. 88P0684																
A1	0–8	17.6	51.6	30.8	2.80	28.0	83	6.0			0.55	1	1	tr	0.42	0.88
A2	8–18	35.4	38.1	26.5	2.05	30.0	82	6.1			0.25	1	1	tr	0.44	0.49
Bt1	18–29	40.4	31.6	28.0	1.21	30.1	92	6.5			0.10	2			0.48	0.51
Bt2	29–49	48.9	19.4	31.7	0.38	63.4	100	6.9			0.17	2			0.39	0.75
R	49															
Stampede; Vertic Durixerolls; Elko, NV; pedon no. 89P0566																
A1	0–7	15.1	41.2	43.7	2.77	22.5	100	7.3			0.49	tr	tr	tr		0.68
A2	7–13	14.8	41.6	43.6	1.47	20.7	100	8.0	tr		0.26	tr	tr	tr		0.86
Bt	13–38	9.9	44.0	46.1	1.16	20.1	100	8.0	tr		0.21	tr				0.99
Btk	38–65	13.6	37.0	49.4	1.04	15.3	100	8.6	14		0.20	1				0.26
Bkqm1	65–86	8.5	33.1	58.4	0.61	11.3	100	8.6	27		0.16	4	2.0	tr		0.25
Bkqm2	86–112	6.9	31.6	61.5	0.32	10.9	100	8.2	27		0.34	7				1.09
Bkq	112–135	3.3	34.7	62.0	0.05	12.5	100	8.9			0.28	3				2.79
2Ck	135–155	2.0	40.8	57.2	0.08	11.4	100	8.8			1.35	3				6.35
Thesisters; Aridic Lithic Haplustolls; Clark, NV; pedon no. 05N0129																
A	0–3	16.4	51.8	31.8	1.1	16.0	100	8.1	32		0.23					0.54
ABk	3–15	20.6	56.2	23.2	1.2	21.9	100	8.2	15		0.23					0.55
R	15															
Tinpan; Vertic Paleoxerolls; Washoe, NV; pedon no. 40A3173																
A1	0–8	24.9	48.6	26.5	5.64	35.6	80	7.0			0.73			tr		0.74
A2	8–13	19.6	57.5	22.9	2.33	25.8	77	7.2			0.58			tr		0.63
Btss1	13–18	58.3	30.7	11.0	1.20	54.1	91	7.9			0.82			tr		0.57
Btss2	18–28	70.6	21.9	7.5	1.13	64.0	96	8.1			1.21			0.1		0.56
Btss3	28–41	70.9	21.9	7.2	0.92	74.6	99	8.9			2.4			0.2		0.56
Btkss1	41–48	70.9	21.9	7.2	0.67	81.5	100	8.9	2		4.21			0.4		0.52
Btkss2	48–66	67.8	24.2	8.0	0.41	80.6	100	8.7	1		6.6			0.6		0.49
R	66–91															
Turba; Vitritorrandic Argiustolls; Lincoln, NV; pedon no. 86P0885																
A	0–13	11.6	28.2	60.2	3.41	18.4	100	6.1			0.48	1	tr	tr		0.85
Bt1	13–25	25.1	26.5	48.4	1.40	21.3	100	6.9			0.14	tr				0.47

(continued)

Table 9.1 (continued)

Horizon	Depth (cm)	Clay (%)	Silt (%)	Sand (%)	SOC (%)	CEC7 ($cmol_c$/kg)	Base sat. (%)	pH H_2O	$CaCO_3$ (%)	$CaSO_4$ (%)	EC (dS/m)	Ex. Na (%)	SAR	Tot. salts (%)	Al+1/2Fe (%)	1.5 mPa H_2O/clay
Bt2	25–46	29.7	19.8	50.5	1.16	24.0	100	6.7			0.1	tr				0.48
Bt3	46–56	27.4	13.6	59.0	0.67	23.2	100	6.6			0.08	1				0.49
R	56															
Wardbay; Pachic Calcicryolls; Elko, NV; pedon no. 89P0571																
A1	0–8	16.1	44.0	39.9	6.57	28.3	100	7.5	2		1.12	tr	tr	0.1		0.85
A2	8–23	17.8	55.9	26.3	3.43	25.9	100	7.9	1		0.55	tr	tr	tr		0.66
A3	23–51	17.1	47.8	35.1	2.96	23.2	100	8.0	3		0.32	tr	tr	tr		0.65
Bk1	51–72	17.7	47.1	35.2	2.29	18.7	100	8.4	12		0.24	1				0.63
Bk2	72–117	7.7	31.4	60.9	0.24	5.5	100	8.4	29		0.12	2				0.81
Bk3	117–158	10.6	36.6	52.8	0.50	5.6	100	8.5	34		0.18	2				0.69
Welch; Cumulic Endoaquolls; Elko, NV; pedon no. 81P0759																
A1	0–5	16.7	52.6	30.7	2.39	27.2	100	8.4	tr		0.94	6	6	0.1	0.21	0.78
A2	5–24	20.5	61.8	17.7	2.16	31.6	100	8.2	tr		0.67	4	3	tr	0.24	0.71
A3	25–53	26.4	59.4	14.2	1.96	31.2	100	7.1			0.23	2			0.23	0.64
A4	53–81	26.9	49.0	24.1	1.14	25.5	99	6.9			0.07	2			0.17	0.60
A5	81–92	13.9	19.6	66.5	0.41	11.5	97	6.8			0.05	2			0.10	0.58
Cg	92–118	19.0	35.1	45.9	0.40	16.2	97	6.9			0.04	2			0.14	0.57

Bold-face text identifies mollic epipedon (A horizon & sometimes upper B); calcic horizon (Bk), argillic horizon (Bt), duripan (Bkqm)

state. Only 12% of the Mollisols are in the shallow family class; however, many of the Mollisols in the ranges have bedrock within 1 m of the surface.

All of the Mollisols shown in Table 9.1 have a mollic epipedon that exceeds 18 cm. In some cases, the upper Bt horizon is sufficiently dark and enriched in SOC to be included in the mollic. A second feature of Mollisols is the high base saturation. The Welch soil series is an example of an Aquolls. The soil is poorly drained and has a thick (96 cm) mollic epipedon that qualifies as a cumulic horizon. Mollisols at higher elevations in the mountain ranges of Nevada are delineated as Cryolls, because they have a cryic (0–7°C) soil-temperature regime. Accumulations of clay, calcium carbonate, or salts are recognized at the great-group levl, i.e., Argicryolls (Newlands soil series), Calcicryolls (Wardbay), and Haplocryolls (Hapgood). Ustolls in Nevada may have an argillic horizon (Turba soil series; Argiustolls) or lack an accumulation of clay or salts (Thesisters; Haplustolls).

Examples of Xerolls in Table 9.1 include the Ninemile series (Argixerolls), Cavehill (Calcixerolls), Stampede (Durixerolls), and Dia (Haploxerolls). The Stampede soil series has abundant calcium carbonate as well as opaline silica. The Haunchee soil series is derived from thin residuum over limestone bedrock.

The dominant soil-forming processes in Mollisols are melanization and biological enrichment of bases; ancillary processes are argilluviation, calcificiation, cambisolization, and gleization. These are discussed more fully in Chap. 13. Fire probably plays an important role in the occurrence of Mollisols on rangelands.

9.3 Use and Management

Mollisols often are highly fertile soils and when irrigated are used in Nevada for crop production and livestock grazing. Nearly three-quarters (71%) of the Aquolls are irrigated for pasture or cropland. Livestock is grazed on 71% of the Ustolls, 60% of the Cryolls, and 34% of the Xerolls. Xerolls are used predominantly as range. Woodlands are managed on 16% of the Xerolls soil series.

9.4 Conclusions

Mollisols are the second most common soil order in Nevada, accounting for 35% of the soil series and 23% of the land area. They are most common in MLRA 21 (88% of all soil series), MLRA 26 (55%), and MLRA 25 (51%). On an area basis, Mollisols can be ranked: Xerolls (78%), Cryolls (9.4%), Aquolls (5.9%), Ustolls (4.7%), and Rendolls (1.6%). The mean annual precipitation for Mollisols in Nevada is commonly between 300 and 400 mm/yr. Mollisols in Nevada form under cool temperatures, with 64% of the Mollisol soil series having a cryic or frigid soil-temperature regime. The dominant vegetation on Mollisols is sagebrush (*Artemisia* spp.), but a variety of short (bunch) grasses are usually present in the understory. More than two-thirds (69%) of the Mollisol soil series in Nevada occur on colluvium and residuum on mountain ranges. The key properties of Mollisols in Nevada are the presence of a mollic epipedon over an argillic (47% of Mollisols), calcic, or cambic horizon. Mollisols are used extensively for range and wildlife protection.

Entisols

10

Abstract

Entisols are the third most abundant soil order in Nevada, in terms of numbers of soil series and land area. This chapter discusses the distribution, properties and processes, and use and management of Entisols in Nevada.

10.1 Distribution

Entisols are the third most abundant soil order in Nevada, in terms of numbers of soil series (17%) and land area (22%) (Table 7.2). Entisols occur in all 10 Major Land Resource Areas in Nevada, but are most common in MLRA 30 (33% of all soil series), MLRA 27 (32%), and MLRA 29 (25%) (Table 4.1). Entisols are also fairly common in MLRA 28A (21%), MLRA 22A (21%), MLRA 28B (18%), and MLRA 24 (18%). More than three-quarters (84%) of the Entisols on an area basis are in the Orthents suborder, followed by Psamments (10%), Fluvents (4.4%), and Aquents (0.9%) (Table 7.2). Orthents occur in mountain ranges and basins, Fluvents along the rivers, Psamments in the former pluvial Lake Lahontan basin, and Aquents along the Humboldt River and its tributaries. More than three-quarters (84%) of the Entisol soil series are Torriorthents (Table 7.2). The Kyler, Mazuma, Hawsley, Trocken, Boton, Singatse, Wardenot, Isolde, and Arizo soil series each covers more than 1,000 km^2 in Nevada.

Photographs of key Entisol great groups are given in Chap. 8, including Torriorthents (Trocken soil series; Fig. 7.2), Torripsamments (Quincy soil series; Fig. 7.10), and Torrifluvents (Fig. 7.15). Three additional photographs are provided here for less common Entisols in Nevada. The Christopher soil series, a Dystric Xeropsamments, is derived from glacial outwash in the Tahoe Basin (Fig. 10.1). The soil contains a brown ochric epipedon (A horizon) to 20 cm and is underlain by Bw horizons that do not qualify as cambic because of their loamy coarse sand texture. The Genoapeak soil series, a Dystric Xerorthents, has formed in colluvium over highly fractured trachyte bedrock (Fig. 10.2). This soil has a dark grayish brown ochric epipedon (A horizon) to 10 cm that is underlain by a pale brown Bw horizon that is insufficiently thick (9 cm) to be recognized as a cambic horizon. The Bricone soil series, a Lithic Cryorthents, has formed from residuum and colluvium derived from limestone and dolomite (Fig. 10.3). The knife is resting on fractured limestone bedrock.

Entisols in Nevada commonly are in loamy (54%) and sandy (25%) particle-size classes, have a mixed mineralogy (84%), a superactive CEC activity class (58%), a mesic soil-temperature class (70%), and a torric (i.e., aridic) soil-moisture regime. Only 18% of the Entisol soil series in Nevada are in the shallow family class.

Although most of the Entisols contain only an ochric epipedon, more than one-third have evidence of silica accumulation (Bq and Bkq horizons), identifiable secondary carbonates (Bk and Bkq horizons), or cambic-like horizons with development of color and structure and textures finer than loamy fine sand.

P. W. Blackburn et al., *The Soils of Nevada*, World Soils Book Series,
https://doi.org/10.1007/978-3-030-53157-7_10

Fig. 10.1 The Christopher soil series, a Dystric Xeropsamments, is derived from glacial outwash in the Tahoe Basin. The soil contains a brown ochric epipedon (A horizon) to 20 cm and is underlain by Bw horizons that do not qualify as cambic because of their loamy coarse sand texture. Source: NRCS

Fig. 10.2 The Genoapeak soil series, a Dystric Xerorthents, has formed in colluvium over highly fractured trachyte bedrock. This soil has a dark grayish brown ochric epipedon (A horizon) to 10 cm that is underlain by a pale brown Bw horizon that is insufficiently thick (9 cm) to be recognized as a cambic horizon. Source: NRCS

10.2 Properties and Processes

Entisols are the least developed soils in Nevada usually containing only an ochric epipedon. The mean thickness of the ochric epipedon in Entisol soil series is 17 +/− 4.6 cm. Table 10.1 provides data for seven Entisols in various regions of Nevada. The thick A horizons in the Cagwin, Sonoma, and Temo soil series fail to meet the color and SOC requirements for a mollic epipedon. The Bw horizons in the Trocken and Graylock soil series are either too thin or too coarse-textured to meet the requirements of a cambic horizon. The soils with a cryic STR (Temo and Graylock) have acidic pH values and low proportions of base cations, reflecting leaching of base cations. All of the soils have low levels of salts, except the Trocken soil series, which contains abundant calcium carbonate and sodium but in insufficient amounts to qualify as calcic or natric horizons.

The dominant soil-forming processes in Entisols are weak forms of humification, calcificiation, silicification, and gleization. These are discussed in Chap. 14.

10.3 Use and Management

The more abundant Orthents are used primarily for open range, but some extensive soil series (Kyler, Bluewing, Wardenot, and Katelana) are used for livestock grazing. A few extensive Psamments are used for livestock grazing, including the Stumble and Kawich soil series, but most are

Fig. 10.3 The Bricone soil series, a Lithic Cryorthents, is formed from residuum and colluvium derived from limestone and dolomite. The knife is resting on fractured limestone bedrock. *Source* NRCS

Table 10.1 Analytical properties of representative Entisols found in Nevada

Horizon	Depth (cm)	Clay (%)	Silt (%)	Sand (%)	SOC (%)	CEC7 (cmolc/kg)	Base sat. (%)	pH H2O	CaCO3 (%)	CaSO4 (%)	EC (dS/m)	Ex. Na (%)	SAR	Tot. salts (%)	Al + 1/2Fe (%)	1.5 mPa H2O/ clay
Cagwin; Dystric Xeropsamments; Douglas, NV; pedon no. 00P0306																
A1	3–5	1.4	10.0	88.6	2	7.8	47	5.4				3			0.28	3.21
A2	5–10	1.6	10.9	87.5	1	7.5	51	5.8				3			0.30	3.06
A3	10–18	0.9	11.3	87.8	2	6.4	39	5.8				3			0.25	
AC	18–43	0.9	12.3	86.8	tr	3.6	42	6.0				6			0.17	
C	43–64	0.3	10.3	89.4	tr	3.4	44	6.1				6			0.26	
Cr	64–145	0.0	5.4	94.6		1.4	50	6.0				14			0.07	
Cirac; Typic Torrifluvents; Nye, NV; pedon no. 92P0571																
A	0–8	4.6	13.5	81.9	0.33	14.5	100	8.4	3		0.37	8	4	tr	0.10	1.74
C1	8–21	2.2	11.2	86.6	0.11	11.6	100	8.6	2		0.22	9			0.08	2.95
C2	21–52	7.0	29.2	63.8	0.28	17.1	100	8.7	4		0.34	11	6	tr	0.11	1.30
C3	52–94	3.0	16.8	80.2	0.11	13.6	100	8.7	3		0.52	22	12	tr	0.07	2.20
Sonoma; Aeric Fluvaquents; Pershing, NY; pedon no. 40A3220								sat								
Ap	3–20	31.2	54.2	14.6	1.76		100	7.9	8		1.7		6	0.1		0.57
C1	20–32	30.8	55.3	13.9	1.00		100	8.0	8		1.5		7	tr		0.71
C2	32–53	25.8	66.9	7.3	0.55		100	8.0	7		1.5		11	tr		0.63
C3	53–84	27.4	56.2	16.4	0.50		100	7.8	7		2.1		14	0.1		0.60
2Ab	84–117	47.2	49.3	3.5	0.61		100	7.7	3		2.2		15	0.1		0.53
Stumble; Typic Torripsamments; Nye, NV; pedon no. 87P0272																
A	0–10	5.1	10.6	84.3	0.31	8.3	100	8.9			0.14	2				0.75
C1	10–30	6.4	11.1	82.5	0.28	9.5	100	8.9			0.21	2				0.69
C2	30–43	6.3	12.6	81.1	0.13	9.5	100	8.9	tr		0.2	3				0.70
C3	43–107	5.8	11.1	83.1	0.17	8.9	100	8.8	1		0.21	4				0.76
C4	107–145	6.1	11.1	82.8	0.18	10.4	100	8.9	8		0.23	14				1.16
Trocken; Typic Torriorthents; Churchill, NY; pedon no. 72C0013																
A	0–7	4.3	22.4	73.3	0.25	9.1	100	8.0	1		0.7	11	3	tr		0.84
Bw	7–16	7.4	21.4	71.2	0.11	11.4	100	8.7	1		0.64	11	5	tr		0.61
Bk1	16–27	10.7	17.5	71.8	0.08	13.0	100	8.6	2		0.83	15	8	tr		0.50

(continued)

Table 10.1 (continued)

Horizon	Depth (cm)	Clay (%)	Silt (%)	Sand (%)	SOC (%)	CEC7 (cmolc/kg)	Base sat. (%)	pH H2O	CaCO3 (%)	CaSO4 (%)	EC (dS/m)	Ex. Na (%)	SAR	Tot. salts (%)	Al + 1/2Fe (%)	1.5 mPa H2O/ clay
Bk2	27–39	10.0	14.6	75.4	0.11	12.0	100	9.1	1		1.22	24	11	tr		0.63
Bk3	39–58	7.1	14.6	78.3	0.07	10.8	100	9.2	1		0.91	32	11	tr		0.49
Bk4	58–73	5.8	11.4	82.8	0.21	9.7	100	9.3	1		2.3	37	24	tr		0.79
C1	73–87	5.9	14.4	79.7	0.51	10.7	100	9.0	1		5.9	43	35	0.1		0.80
Temo; Typic Cryopsamments; Washoe, NV; pedon no. 79P0472																
A1	0–2	2.5	4.7	92.8	2.09	6.5	18	4.8				2				1.28
A2	2–9	2.1	4.2	93.7	1.58	6.0	17	4.4				2				1.14
A3	9–21	2.4	5.4	92.2	0.57	4.4	2	5.0								0.88
C	21–39	3.7	12.6	83.7	0.60	5.8	2	5.0								0.70
Cr	39–50															.
Graylock; Typic Cryorthents; Douglas, NV; pedon no. 00P0194																
A	3–8	2.6	8.5	88.9	4	14.5	80	6.2				1			0.32	2.35
Bw1	8–20	2.3	9.4	88.3	1	6.0	50	6.0				5			0.36	1.26
Bw2	20–46	2.5	10.0	87.5	1	5.4	44	5.8				4			0.25	1.20
C1	46–69	3.2	9.7	87.1	tr	4.6	43	5.6				4			0.22	0.84

used for wildlife habitat. Many of the Fluvents, including the Cirac, Slaw, Stargo, Alhambra soil series) and the Sonoma soil series (Aquents) are used for livestock grazing. The Sonoma and Alhambra soil series are irrigated for cropland.

10.4 Conclusions

Entisols are the third most common soil order in Nevada, accounting for 17% of the soil series and 22% of the land area. Entisols occur in all 10 Major Land Resource Areas in Nevada, but are most common in MLRA 30 (33% of all soil series), MLRA 27 (32%), and MLRA 29 (25%). On an area basis, Entisols are ranked as follows: Orthents (84%), Psamments (10%), Fluvents (4.4%), and Aquents (0.1%). More than three-quarters (84%) of the Entisol soil series are Torriorthents. Orthents occur in mountain ranges and basins, Fluvents along the rivers, Psamments in the former pluvial Lake Lahontan basin, and Aquents along the Humboldt River and its tributaries. Entisols are the least developed soils in Nevada usually containing only an ochric epipedon. Most of the Entisols are used for range and wildlife habitat, but a few of the soil series are irrigated for cropland.

11 Inceptisols

Abstract

Inceptisols occur in all 10 of the Major Land Resource Areas of Nevada. This chapter discusses the distribution, properties and processes, and use and management of Inceptisols in Nevada.

11.1 Distribution

Inceptisols comprise only 3.6% of the soil series and 1.9% of the land area of Nevada (Table 7.2). Inceptisols occur in all 10 of the MLRAs in Nevada but are most pronounced in the Nevada portion of MLRA 22A, the Sierra Nevada Mountains, where they comprise 20% of the soil series (Table 4.1). More than three-quarters (79%) of the Inceptisols are Aquepts, followed by Cryepts (17%), Xerepts (2.3%), and Ustepts (1.0%) (Table 7.2). Halaquepts comprise 78% of the Inceptisol great groups in Nevada. The Wendane and Ocala soil series are Halaquepts and occupy the greatest areas of Inceptisols in Nevada at 794 and 547 km^2, respectively.

In Nevada, Inceptisols are commonly separated into high-elevation (cryic and frigid soil-temperature regimes) and low-elevation (mesic and thermic soil-temperature regimes) types. High-elevation types include the Osditch and Meiss soil series. The Osditch soil series, a Lamellic Haplocryepts, is derived from in colluvium derived from quartzite and argillite in the mountains of Nevada (Fig. 11.1). The soil has an ochric epipedon (A horizon) to 8 cm, an albic (E) horizon to 46 cm, and a cambic horizon with lamellae to 152 cm and has abundant coarse fragments. The Meiss soil series, a Lithic Humicryepts, is a shallow, somewhat excessively drained soil formed in material from weathered andesitic bedrock in the Tahoe area (Fig. 11.2). The soil contains a brown umbric epipedon that overlies bedrock at 33 cm.

Low-elevation Inceptisols include the Gefo and Tahoe soil series. The Gefo soil series, a Humic Dystroxerepts, is formed on glacial outwash and alluvium in the Tahoe Basin (Fig. 11.3). This soil has an umbric epipedon (A horizon) to 38 cm that is underlain by C horizons. The Tahoe soil series, a Cumulic Humaquepts, only occurs around the rim of Lake Tahoe (Fig. 11.4). This soil is very poorly drained, has formed in alluvium derived from mixed material, and is in the floodplain. The soil contains an umbric epipedon (A horizon) from 8 to 76 cm that is underlain by gleyed C horizons.

Inceptisols in Nevada form in a comparatively cool, moist climate, with an average annual temperature averaging 5°C and mean annual precipitation averaging 620 mm. The vegetation commonly is greasewood, rabbitbrush, saltgrass, and alkali sacaton on Halaquepts at the lower elevations and coniferous forests in Cryepts at the higher elevations. Halaquepts are derived from alluvium and lacustrine sediments and Cryepts originate from colluvium and residuum.

Inceptisols in Nevada occur in a variety of particle-size classes, tend to have a mixed mineralogy (65% of soil series) and a superactive CEC activity class (49%), are most abundant in cryic-frigid (57%) and mesic (40%) soil-temperature classes, and aquic (51%) and xeric (45%) soil-moisture classes.

P. W. Blackburn et al., *The Soils of Nevada*, World Soils Book Series,
https://doi.org/10.1007/978-3-030-53157-7_11

Fig. 11.1 The Osditch soil series, a Lamellic Haplocryepts, is derived from in colluvium derived from quartzite and argillite in the mountains of Nevada. The soil has an ochric epipedon (A horizon) to 8 cm, an albic (E) horizon to 46 cm, and a cambic horizon with lamellae to 152 cm. The soil contains abundant coarse fragments. The scale is in decimeters. Source: NRCS

Fig. 11.2 The Meiss soil series, a Lithic Humicryepts, is a shallow, somewhat excessively drained soil formed in material from weathered andesitic bedrock in the Tahoe area. The soil contains a brown umbric epipedon that overlies bedrock at 33 cm. The scale is in decimeters. *Source* NRCS

11.2 Properties and Processes

The definition of Inceptisols is perhaps unnecessarily complex. Inceptisols must contain a cambic, calcic, petrocalcic, gypsic, petrogypsic, or a duripan within 100 cm of the mineral soil surface, or a folistic, histic, mollic, plaggen, or umbric epipedon, or a salic horizon. Soils with an exchangeable Na percentage of 15 or more and groundwater within 100 cm of the mineral soil surface also qualify as Inceptisols. For this reason, Halaquepts in Nevada qualify as Inceptisols even though they contain only an ochric epipedon. About half (56%) Nevada's Inceptisols have an ochric epipedon and 40% have an umbric epipedon, with no diagnostic subsurface horizon. Cambic horizons are very uncommon.

The representative Inceptisols shown in Table 11.1 are more strongly developed than Entisols (see the previous chapter), but do not meet the requirements for other soil orders. The Bakerpeak and Robbersfire soil series have calcic horizons; the Strawbcrek, Umpa, and Robbersfire soil series have a cambic horizon; and the Tahoe and Meiss soil series have an umbric epipedon. Two of the three Cryepts (Strawbcrek and Meiss) are acidic and have lower base saturations than the other soil series. The Settlement soil series (Halaquepts) contains abundant sodium salts.

The dominant soil-forming processes in Nevada Inceptisols include gleization, melanization, and salinization. Information about soil-forming processes is given in Chap. 14.

11.3 Use and Management

Several of the more extensive Inceptisols are used for livestock grazing, including the Ocala, Nuyobe, and Dianev soil series. The Argenta soil series and a portion of the Ocala are irrigated for cropland.

Fig. 11.3 The Gefo soil series, a Humic Dystroxerepts, is derived from glacial outwash and alluvium in the Tahoe Basin. The soil contains an umbric epipedon (A horizon) to 38 cm and is underlain by C horizons. The scale is in decimeters. *Source* NRCS

Fig. 11.4 The Tahoe soil series, a Cumulic Humaquepts, only occurs around the rim of Lake Tahoe. The soil is very poorly drained, has formed in alluvium derived from mixed material, and is in the floodplain. The soil contains an umbric epipedon (A horizon) from 8 to 76 cm that is underlain by gleyed C horizons. *Source* NRCS

Table 11.1 Analytical properties of representative Inceptisols found in Nevada

Horizon1	Depth	Clay	Silt	Sand	SOC	CEC7	Base sat.	pH	CaCO3	CaSO4	EC	Ex. Na		Tot. salts	Al + 1/2Fe	1.5 mPa H2O/
	(cm)	(%)	(%)	(%)	(%)	(cmolc/kg)	(%)	H2O	(%)	(%)	(dS/m)	(%)	SAR	(%)	(%)	clay
Settlement; Aeric Halaquepts; Churchill, NV; pedon no. 72C0015																
A	0–3	42.6	44.4	13.0	0.27	42.3	100	8.5			9.5	47	44	0.4		0.46
Bk1	3–15	60.7	32.6	6.7	0.24	51.1	100	8.5			10.3	60	62	0.6		0.48
Bk2	15–31	64.4	31.2	4.4	0.19	52.6	100	8.2			15.6	58	59	1.0		0.49
Bk3	31–60	62.9	34.1	3.0	0.12	49.6	100	8.2			8.0	36	36	0.5		0.49
Strawbcrek; Lamellic Haplocryepts; White Pine, NY; pedon no. 92P0147																
A	0–3															
E1	3–23	5.1	25.2	69.7	0.79	6.0	73	5.7								0.69
E2	23–61	4.8	22.4	72.8	0.30	4.3	79	5.9								0.77
E & Bt	61–152	4.1	19.2	76.7	0.27	3.3	76	5.6				3				0.61
Bakerpeak; Xeric Calcicryepts; White Pine, NV; pedon no. 13N0353																
A1	0–5	22.6	51.1	26.3	6.8	27.2	100	7.4	27		1.20			0.1		0.58
A2	5–19	22.0	46.9	31.1	2.4	21.1	100	7.7	21		0.49			tr		0.45
Bk1	19–39	19.5	41.3	39.2	1.6	16.8	100	7.9	37		0.37			tr		0.48
Bk2	39–65	8.1	28.4	63.5	0.2	4.9	100	8.2	59		0.13					0.43
Bk3	65–123	7.7	20.8	71.5	0.1	3.8	100	8.3	64		0.12					0.39
Umpa; Andic Dystroxerepts; Washoe, NV; pedon no. 00P0308																
A	3–5	8.5	27.0	64.5	4.74	31.1	67	6.6				1			1.17	1.94
AB	5–18	9.4	26.6	64.0	2.78	24.4	59	6.5				1			1.17	2.01
BA1	18–38	9.9	26.4	63.7	2.24	22.6	54	6.4				1			1.06	1.33
BA2	38–58	11.4	27.1	61.5	1.40	21.2	54	6.2				1			0.84	1.48
BA3	58–71	11.9	28.6	59.5	1.33	20.0	65	6.1				2			0.52	1.14
BA4	71–81	14.3	28.4	57.3	0.89	18.4	66	6.1				2			0.39	1.05
Bw	81–109	10.5	27.2	62.3	0.60	18.6	60	6.4				3			0.41	1.55
Crt	109–142	16.7	28.8	54.5	0.66	21.0	62	6.4				3			0.38	1.36
Overton; Aeric Endoaquepts; Clark, NV; pedon no. 40A3221								sat								
Ap	0–18	44.5	42.8	12.7	1.47		100	7.8	15		1.3		3	tr		0.46
A	18–41	45.2	44.3	10.5	1.18		100	7.9	14		1.1		4	tr		0.48
Cg1	41–48	45.8	42.0	12.2	0.63		100	7.7	16		1.2		4	tr		0.48
Cg2	48–63	22.7	41.6	35.7	0.21		100	8.0	13		1.3		4	tr		0.51
Robbersfire; Calcic Haplustepts; Clark, NV; pedon no. 10N0514																

(continued)

Table 11.1 (continued)

Horizon1	Depth	Clay	Silt	Sand	SOC	CEC7	Base sat.	pH	CaCO3	CaSO4	EC	Ex. Na		Tot. salts	Al + 1/2Fe	1.5 mPa H2O/
	(cm)	(%)	(%)	(%)	(%)	(cmolc/kg)	(%)	H2O	(%)	(%)	(dS/m)	(%)	SAR	(%)	(%)	clay
A	0–16	21.4	44.8	33.8	4.80	24.5	100	7.0	3			tr			0.25	0.54
Bw	16–30	26.1	44.7	29.2	2.40	25.1	100	7.7	11			tr			0.20	0.43
Bk1	30–48	22.3	40.4	37.3	1.40	18.1	100	7.7	20			tr			0.20	0.39
Bk2	48–100	20.1	43.6	36.3	1.70	17.1	100	7.9	24			tr			0.16	0.46
Tahoe; Cumulic Humaquepts; Douglas, NV; pedon no. 00P0305																
A1	0–4	10.9	15.5	73.6	16.00	35.3	78	5.5				2			0.54	4.84
A2	4–20	6.9	8.6	84.5	2.55	14.4	65	5.9				3			0.32	1.88
A3	20–38	11.3	10.7	78.0	1.62	14.4	74	6.0				2			0.49	0.91
A4	38–58	9.8	14.2	76.0	0.46	11.0	69	6.0				3			1.12	1.06
A5	58–79	21.6	19.3	59.1	1.92	22.8	56	5.2				2			0.68	0.92
Cg1	79–91	23.8	13.2	63.0	1.06	14.4	37	4.7				2			1.32	0.87
Meiss; Lithic Humicryepts; Washoe, NV; pedon no. 00P0309																
A1	0–10	4.1	25.0	70.9	2.23	11.7	21	5.5				2			1.15	2.68
A2	10–20	4.5	26.2	69.3	1.62	12.8	20	5.6				2			1.35	3.09
AC	20–50	4.0	26.9	69.1	1.97	13.0	18	5.6							1.46	3.93
R	50															

[a]Boldface indicates the presence of an albic horizon (E), cambic (E and Bt; Bw), calcic (Bk), and umbric (A)

11.4 Conclusions

Inceptisols are more strongly developed than Entisols (see the previous chapter), but they do not have properties that meet the requirements for other soil orders. Inceptisols comprise only 3.6% of the soil series and 1.8% of the land area of Nevada. Inceptisols occur in all 10 of the MLRAs in Nevada but are most pronounced in the Nevada portion of MLRA 22A, the Sierra Nevada Mountains, where they comprise 20% of the soil series. More than three-quarters (78%) of the Inceptisols are Aquepts, followed by Cryepts (17%), Xerepts (2.3%), and Ustepts (1.0%). In Nevada, Inceptisols are commonly separated into high-elevation (cryic and frigid soil-temperature regimes) and low-elevation (mesic and thermic soil-temperature regimes) types. Most of the Inceptisols in Nevada contain either an umbric epipedon or an ochric epipedon overlying materials with an exchangeable Na percentage of 15 or more and groundwater within 100 cm of the mineral soil surface.

12 Alfisols, Vertisols, and Andisols

Abstract

Collectively, Alfisols, Vertisols, and Andisols comprise a small proportion of the soil series and the land area in Nevada. This chapter discusses the distribution, properties and processes, and use and management of soils in these three orders in Nevada.

12.1 Distribution

Collectively, Alfisols, Vertisols, and Andisols comprise less than 3% of the soil series (Table 6.1) and 1.3% of the land area of Nevada. Alfisols occur primarily in MLRA 22A, the Sierra Nevada Mountains, but exist to a limited extent in MLRAs 25, 26, 28A, 28B, 29, and 30. Vertisols are most abundant in MLRA 23, the Malheur High Plateau, but are found to a limited extent in MLRAs 25, 26, 27, and 28B. Andisols occur only in MLRA 22A, the Sierra Nevada Mountains, but compose only 5.7% of the soil area in the Nevada portion of the Sierra.

12.1.1 Alfisols

There are only three Alfisols that compose more than 100 km^2 in Nevada, including the Bulake (321 km^2), Hardzem (245 km^2), and Borealis (152 km^2) soil series. Alfisols in Nevada develop in a cool, dry climate. The mean annual precipitation ranges between 330 and 760 mm, and the mean annual temperature from 5 to 7°C. Alfisols in Nevada usually support forest vegetation, including Jeffrey pine, red fir, white fir, and lodgepole pine, but they also occur at the upper elevation limits of sagebrush and pinyon pine. Common parent materials include colluvium over residuum, commonly with a volcanic ash component; and typical landforms are mountains and plateaus.

12.1.2 Vertisols

Nearly three-quarters (73%) of the Vertisols in Nevada are Haploxererts. Nevada Vertisols have mainly a fine or very-fine particle-size class and have a smectitic mineralogy class; nearly all have a mesic or frigid soil-temperature regime; and 73% have a cracking pattern that is associated with adjacent soils with a xeric soil-moisture regime. Only two Vertisols are moderately extensive in Nevada, which are the Carson (308 km^2) soil series in MLRA 27 and the Tunnison (~ 200 km^2) soil series in MLRA 23. Most Vertisols in Nevada receive very low amounts of precipitation (127 to 280 mm/yr). The mean annual air temperature ranges between 8 and 12°C. Vertisols in Nevada support plant communities of big sagebrush, rabbitbrush, black greasewood, inland saltgrass, horsebrush, spiny hopsage, and related species. Common parent materials are alluvium and colluvium on flood plains and plateaus, respectively.

12.1.3 Andisols

Four Andisol soil series compose less than 18 km^2 in Nevada. These soils occur between 2,000 and 3,000 m in the easternmost portion of the Sierra Nevada Mountains. This area receives 1,200 mm/yr of precipitation and has a mean annual temperature of 5°C. Andisols in Nevada occur under white fir, red fir, and mountain hemlock and are formed in

P. W. Blackburn et al., *The Soils of Nevada*, World Soils Book Series,
https://doi.org/10.1007/978-3-030-53157-7_12

weathered parent materials derived from volcanic rocks such as the andesitic tuff and lahar mudflows of the late Tertiary-age Mehrton Formation.

Nevada Andisols have the ashy-skeletal or medial-skeletal substitutes for particle-size class and amorphic or mixed mineralogy classes. The soil temperature regimes are frigid or cryic and the soil moisture regime is xeric. Andisols require the presence of a diagnostic soil characteristic known as andic soil properties. The partial weathering of tephra or other parent materials containing a significant content of volcanic glass is one genetic pathway for the formation of the short-range-order minerals, which are necessary for the presence of andic soil properties. Although Andisols comprise only a small portion of Nevada, there are more than 100 additional soil series that are derived from volcaniclastic materials or have been affected by volcanic ash deposition. These soils classify in Andic and related intergrade subgroups (i.e., Aquandic, Vitrandic, Vitritorrandic, Vitrixerandic) and often have the glassy mineralogy class and the ashy or ashy-skeletal substitutes for particle-size class. Volcanic parent materials, such as eolian volcanic ash deposits, are abundant in Nevada, but the State's arid and semiarid climates preclude the formation of short-range-order minerals. As such, no Andisols with an aridic soil moisture regime (i.e., Torrands) have been recognized in Nevada.

12.2 Properties and Processes

12.2.1 Alfisols

Alfisols in Nevada tend to be in loamy and skeletal particle-size classes and have an isotic (36%) or smectitic (32%) mineralogy class. All of the Alfisols have a frigid or cryic soil temperature regime, and all have a xeric soil moisture regime. Alfisols in Nevada have an ochric epipedon averaging 28 cm in thickness over an argillic horizon averaging 53 cm in thickness. Half of the Alfisols in Nevada are classified as Haploxeralfs and contain primarily an argillic horizon. However, the Borealis soil series also has a duripan and is classified as a Durixeralf. Alfisols must also have a high base saturation (ca. 50% or more). The presence of an argillic horizon and a high base saturation is evident in analytical data for the Borealis and Bulake soil series (Table 12.1). Additionally, the Borealis soil series has a duripan below 58 cm; and, although it is not designated on the OSD, the Carioca soil series may have an umbric epipedon epipedon based on the thickness ranges of the A horizons and the range in characteristics for the series.

Three examples of Alfisols follow. The Ceebee soil series (Fig. 12.1) is a Lamellic Haplocryalfs that occurs in Great Basin National Park. This pedon has an ochric epipedon to 42 cm underlain by poorly visible lamellae (alternating E and Bt horizons). The profile contains abundant stones in the light gray E horizon. The Deerhill soil series (Fig. 12.2), an Ultic Palexeralfs, occurs in the Tahoe Basin Area of California and Nevada. The soil contains brown A1, A2, Bw1, and Bw2 horizons to 90 cm, which are underlain by yellowish brown 2Bt horizons. The Jorge soil series, an Andic Haploxeralfs, occurs in the Tahoe Basin Area of California and Nevada. This soil has a monochromatic brown color (7.5YR 5/4) to a depth of 126 cm but contains an argillic horizon from 85 to 126 cm. (Figure 12.3)

12.2.2 Vertisols

Nearly three–quarters (73%) of the Vertisols in Nevada have an ochric epipedon averaging 15 cm in thickness over a cambic horizon averaging 57 cm. By definition, Vertisols must contain a layer 25 cm or more in thickness, within 100 cm of the mineral soil surface, that has either slickensides or wedge–shaped peds with their long axes tilted 10 to 60 degrees from the horizontal, and a clay content of 30% or more within the upper 50 cm. These conditions are evident in analytical data for the Carson and Tunnison soil series (Table 12.1). The Carson soil series has a mollic epipedon.

12.2.3 Andisols

Andisols in Nevada either have an umbric epipedon with no diagnostic subsurface horizon or an ochric epipedon over a cambic horizon. Andisols must have andic properties within the upper 60 cm; andic soil properties commonly form during weathering of tephra or other parent materials with a significant content of volcanic glass. The Melody and Waca soil series occur to a very limited extent in Nevada but reflect these properties. Both soils have a wide 1.5 mPa H_2O/clay ratio; they also have high melanic indices (33–44%) and contain from 27 to 36% glass shards in the fine–sand (0.25–0.10 mm) fraction (data not shown). The Wardcreek series (Fig. 12.4), a Xeric Vitricryands, occurs in the Tahoe Basin Area of California and Nevada. This soil has an ochric epipedon to 30 cm that is underlain by a slightly clay-enriched cambic horizon (Bt horizon). Because the Bt horizon does not meet the required increase in clay content, it does qualify as an argillic horizon. The stones are from weathered volcanic bedrock.

The dominant soil-forming processes in Nevada Alfisols are base accumulation and argilluviation; Vertisols feature vertisolization; and Andisols reflect andosolization. Chapter 14 discusses these processes in detail.

Table 12.1 Analytical properties of representative Alfisols, Vertisols, and Andisols found in Nevada

	Depth	Clay	Silt	Sand	SOC	CEC7	Base sat.	pH	$CaCO_3$	$CaSO_4$	EC	Ex. Na		Volcanic	Al+1/2Fe	1.5 mPa H_2O/
Horizon[1]	(cm)	(%)	(%)	(%)	(%)	($cmol_c$/kg)	(%)	H_2O	(%)	(%)	(dS/m)	(%)	SAR	glass (%)	(%)	clay
Borealis; Abruptic Durixeralfs; Mineral, NV; pedon no. 82P0627																
A1	0–5	4.6	17.3	78.1	0.77	6.6	100	6.6			0.14	2		71	0.11	1.13
A2	5–14	10.0	21.8	68.2	0.68	9.9	100	7.0			0.12	1		65	0.1	0.68
Bt1	14–22	26.2	24.3	49.5	0.64	20.0	100	7.2			0.11	1		41	0.14	0.46
Bt2	22–42	32.6	23.9	43.5	0.49	23.9	100	7.3			0.11	1		38	0.17	0.46
Bt3	42–51	34.9	26.3	38.8	0.48	28.7	100	7.5			0.12	5		38	0.18	0.50
Bt4	51–58	28.0	31.1	40.9	0.69	33.3	100	7.7	tr	tr	0.18	5		36	0.20	0.73
Bqkm	58–79	2.0	12.9	85.1	0.14	7.1	100	8.6	3		0.23	7		16		
Bulake; Lithic Mollic Haploxeralfs; Elko, NV; pedon no. 81P0244																
A1	0–5	14.4	50.8	34.8	2.13	16.3	96	7.0	tr			1		17		0.65
A2	5–15	12.5	56.1	31.4	1.08	13.5	96	7.1	tr			1		20		0.62
Bt1	15–23	26.7	46.7	26.6	0.60	18.3	93	7.1	tr			3		11		0.42
Bt2	23–41	66.6	16.0	17.4	0.63	54.8	96	7.2	tr			2		11		0.48
Bt3	41–48	47.9	10.0	42.1	0.54	56.7	100	7.6	tr			3		16		0.62
R	48															
Carioca; Andic Haplocryalfs; Washoe, NV; pedon no. 40A3415																
O	0–3															
A	3–21	10.5	34.1	55.4	4.25	24.9	14	5.1								1.13
E	21–49	7.6	34.4	58.0	1.84	17.2	14	5.6								1.45
A2	49–79	7.0	37.1	55.9	0.60	12.8	27	5.8								1.24
Bt1	79–115	15.3	44.3	40.4	0.15	14.8	66	5.6				1				0.73
Bt2	115–145	19.0	37.6	43.4	0.12	14.3	73	5.9				1				0.59
C	145–168	20.8	39.7	39.5	0.18	16.3	66	5.8				1				0.69
Cr	168–193	5.3	12.4	82.3	0.04											
Carson; Halic Haploxererts; Churchill, NV; pedon no. 40A3236								sat								
Ap	0–18	64.7	26.4	8.9	0.93		100	7.5			4.40		11			0.42
Ayz	18–43	64.5	26.4	9.1	0.87		100	7.4			6.00		10			0.43
Bssz1	43–69	65.3	25.8	8.9	0.56		100	7.1			8.60		12			0.42
Bssz2	69–86	67.2	23.5	9.3	0.47		100	6.9			11.10		12			0.42
Bkssz	86–127	70.7	20.8	8.5	0.48		100	7.2			11.50		12			0.42
Bkz	127–163	78.2	16.0	5.8	0.42		100	7.3			9.70		18			0.39

(continued)

Table 12.1 (continued)

	Depth	Clay	Silt	Sand	SOC	CEC7	Base sat.	pH	$CaCO_3$	$CaSO_4$	EC	Ex. Na		Volcanic	Al+1/2Fe	1.5 mPa H_2O/
Tunnison; Aridic Haploxererts; Washoe, NV; pedon no. 69C0094																
A1	0–3	56.2	36.0	7.8	1.13	52.0	99	6.8				2				0.44
A2	3–14	57.5	33.9	8.6	0.59	53.7	97	6.7				2				0.48
Bw1	14–26	58.0	33.9	8.1	0.55	53.3	100	7.0				2				0.47
Bss1	26–54	57.7	34.0	8.3	0.60	50.6	100	7.0				4				0.54
Bss2	54–76	57.3	35.8	6.9	0.44	54.0	100	7.9	1		0.41	7				0.48
Bk	76–107	48.3	42.8	8.9	0.26	54.9	100	8.3	2		0.56	10				0.61
Bkq	107–130	12.3	27.2	60.5		63.7	100	8.3	9		1.01	13				2.21
Melody; Lithic Vitricryands; Placer, CA; 00P0302																
A	3–15	3.3	34.8	61.9	5.75	21.7	59	6.0				1			0.96	3.58
AC	15–33	2.8	30.6	66.6	3.34	9.9	49	6.0				2		1	0.99	3.46
C	33–43	2.7	31.2	66.1	2.45	9.0	41	5.9							1.15	3.15
R	43															
Waca; Humic Vitrixerands; Placer, CA; pedon no. 00P0358																
A1	3–10	4.3	17.5	78.2	9.96	32.6	52	5.5				1			0.59	3.93
A2	10–41	4.7	20.1	75.2	0.88	8.4	46	5.9				5		tr	0.52	1.26
A3	41–56	4.4	24.1	71.5	1.22	10.0	38	5.5							0.47	1.34
Cr	56															

[1]Boldface indicates diagnostic horizon: mollic or umbric (A horizon); argillic (Bt)

Fig. 12.1 The Ceebee soil series is a Lamellic Haplocryalfs that occurs in the Great Basin National Park. This pedon has an ochric epipedon to 42 cm underlain by poorly visible lamellae (alternating E and Bt horizons). The profile contains abundant stones in the light gray E horizon. The scale is in decimeters. Photo by USDA, NRCS (2009)

Fig. 12.2 The Deerhill soil series, an Ultic Palexeralfs, occurs in the Tahoe Basin Area of California and Nevada. The soil contains brown A1, A2, Bw1, and Bw2 horizons to 90 cm, which are underlain by yellowish brown 2Bt horizons. The scale is in decimeters. Photo by USDA, NRCS (2007)

12.3 Use and Management

The Alfisols and Andisols occur in mountains in the Lake Tahoe area and, where managed, are used for woodlands, recreation, and urban development. Vertisols in floodplains are used for irrigated cropland and livestock grazing.

12.4 Conclusions

Alfisols are base-enriched and have an argillic horizon; Vertisols are enriched in smectitic clays with high shrink-swell potential; and Andisols have abundant volcanic

Fig. 12.3 The Jorge soil series, an Andic Haploxeralfs, occurs in the Tahoe Basin Area of California and Nevada. This soil has a brown color (7.5YR 5/4) to a depth of 126 cm but contains an argillic horizon from 85 to 126 cm. The scale is in decimeters. Photo by USDA, NRCS (2007)

Fig. 12.4 The Wardcreek series, a Xeric Vitricryands, occurs in the Tahoe Basin Area of California and Nevada. This soil has an ochric epipedon to 30 cm that is underlain by a slightly clay-enriched cambic horizon. Because the Bt horizon does not meet the required increase in clay content, it does qualify as an argillic horizon. The stones are from weathered volcanic bedrock. The scale is in decimeters. Photo by USDA, NRCS (2007)

glass and soil OC and retain phosphates due to the presence of short-range-order minerals. Collectively, Alfisols, Vertisols, and Andisols comprise less than 3% of the soil and 1.3% of the land area of Nevada. Alfisols occur primarily in MLRA 22A, the Sierra Nevada Mountains; Vertisols are most abundant in MLRA 23, the Malheur High Plateau and Andisols occur only in MLRA 22A, the Sierra Nevada Mountains. Alfisols in Nevada develop in a cool, dry climate and usually support forest vegetation. Vertisols. Most Vertisols in Nevada receive very low amounts of precipitation (127 to 280 mm/yr) and are formed in alluvium on flood plains and colluvium on plateaus. Andisols in Nevada occur under coniferous vegetation and are formed in parent materials derived from weathered volcanic rocks such as andesitic tuff and lahar mudflows.

13 Soil-Forming Processes in Nevada

Abstract

This chapter discusses the role of these processes in the development of soils in Nevada. The 17 generalized soil-forming processes occurring in soils of Nevada include argilluviation, melanization, silicification, calcification, gleization, cambisolization, vertization, solonization, gypsification, andisolization, and paludization.

13.1 Introduction

Specific soil processes are determined by the soil-forming factors and are expressed in diagnostic horizons, properties, and materials, which are then used to classify soils: soil-forming factors → soil-forming processes → diagnostic horizons, properties, materials → soil taxonomic system (Bockheim and Gennadiyev 2000). They investigators identified 17 generalized soil-forming processes. One additional process is added here—cambisolization—which will be defined forthwith. The dominant soil-forming processes in Nevada are argilluviation, melanization, silicification, calcification, gleization, cambisolization, vertization, solonization, and salinization (Fig. 13.1). Gypsification, andisolization, and paludization occur to a very limited extent.

13.2 Argilluviation

Argilluviation (lessivage) refers to the movement and accumulation of clays in the solum. Argilluviation is a dominant process in many Aridisols, including the Argids suborder and the Argidurids and Natridurids great groups; many Mollisols, including the Argixerolls, Argicryolls, and Argiustolls; and in the Alfisols, the Xeralfs. The evidence for argilluviation in Nevada soils is the presence of argillans, i.e., clay skins and bridges and abrupt increases in the clay content from the eluvial horizons (A or E) to the illuvial horizons (Bt or Btn). This process is favored by strong desiccation of subsurface horizons followed by wetting during cool-season rains, melting snows, and occasional intense summer thunderstorms; parent materials enriched in exchangeable sodium and carbonate-free clays (Alexander and Nettleton 1977); high rock fragment content on soil surfaces; stable landforms (e.g., fan remnants) and hillslope positions (e.g., foot slopes and back slopes, rather than eroding shoulders); and a time interval of >2,000 years (Bockheim and Hartemink 2013).

13.3 Melanization

Melanization (humification) refers to the accumulation of well-humified organic compounds in the upper mineral soil. Soils reflecting melanization include the Mollisols and Humi-great groups in Inceptisols. Common native grassland species that enhance melanization include Nevada bluegrass (*Poa nevadensis*), Idaho fescue (*Festuca idahoensis*), bluejoint grass (*Calamagrostis canadensis*), desert needlegrass (*Stipa speciosa*), Indian ricegrass (*Oryzopsis hymenoides*), saltgrass (*Distichlis spicata*), wheatgrass (*Agropyron cristatums*), and Great Basin wild rye (*Leymus cinereus*). These species commonly occur under a canopy of big sagebrush (*Artemisia tridentata*). This process is favored in Nevada by grassland vegetation, base-rich parent materials (in the case of Mollisols), and a time interval of >500 years.

13.4 Silicification

Silicification refers to the secondary accumulation of silica in the form of durinodes or a duripan (Chadwick et al. 1987). This process is common in Durids, Durixerolls, Durixeralfs, Duric subgroups of Halaquepts, Haplocalcids, Palexerolls, Petroargids, Torrifluvents, and Torriorthents, Duridic

P. W. Blackburn et al., *The Soils of Nevada*, World Soils Book Series,
https://doi.org/10.1007/978-3-030-53157-7_13

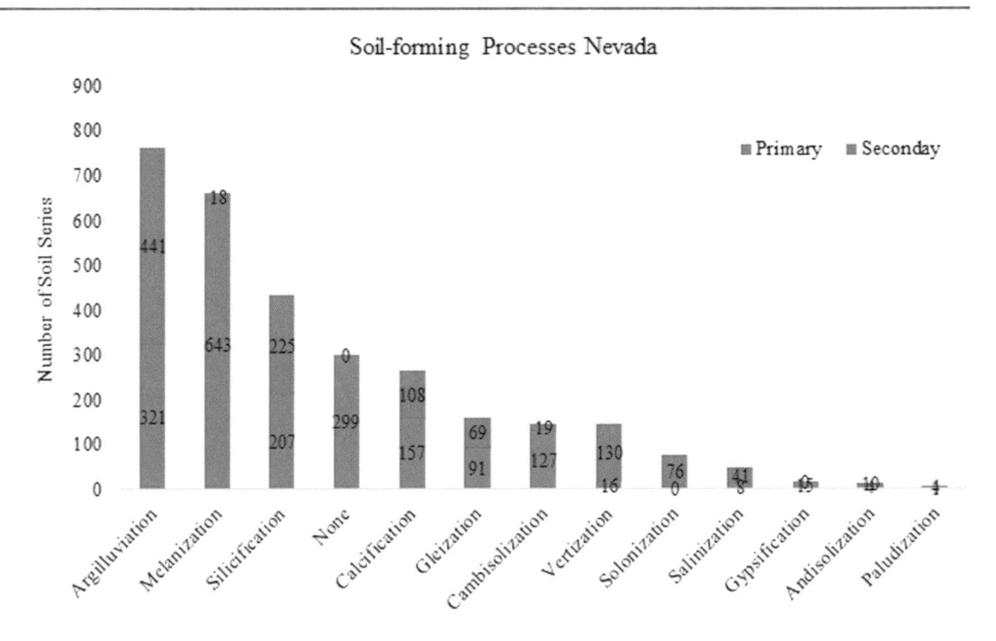

Fig. 13.1 Soil-forming processes in soils of Nevada. Primary processes are identified at the order and suborder taxonomic levels; and secondary processes are identified at the great-group and subgroup levels

subgroups of Haploxerolls, and Durinodic and Durinodic Xeric subgroups of Calciargids, Haplargids, Haplocalcids, Haplocambids, and Natrargids. Silicification is favored by an aridic-torric soil-moisture regime, parent materials enriched in opaline Si, eolian additions of loess containing volcanic ash, and possibly secondary $CaCO_3$ (Chadwick et al. 1987, 1989).

13.5 Calcification

Calcification refers to the accumulation of secondary carbonates in semiarid and arid soils (Harper 1957). The $CaCO_3$ initially fills micropores but over millennia may result in a strongly cemented to indurated petrocalcic horizon (Harper 1957; Gile et al. 1965, 1966, Gardner 1972; Brock and Buck 2009). In Nevada, calcification occurs in the suborder of Calcids, Calcic great groups of Aridisols (e.g., Calciargids) and Mollisols (e.g., Calcixerolls), and in many taxa of Entisols (e.g., Torriorthents) which lack diagnostic calcic or petrocalcic horizons. Calcification is related to mean annual precipitation, the presence of calcareous parent materials, dust inputs, and, in the case of petrocalcic horizons, long periods of pedogenesis (Amundson et al. 1989a, b; Marion 1989; Reheis et al. 1992; Robins et al. 2012).

13.6 Gleization

Gleization (hydromorphism) refers to the presence of aquic conditions often evidenced by reductimorphic or redoximorphic features such as mottles (i.e., redox concentrations and redox depletions) and gleying. In Nevada, gleization occurs in Aquic ("aqu") suborders of Mollisols, Inceptisols, and Vertisols. Gleization in Nevada is favored by bolsons, depressions in the landscape, and proximity to lakes and by parent materials that restrict drainage by virtue of texture, a layer that restricts moisture movement, or the presence of bedrock.

13.7 Cambisolization

Cambisolization refers to a collection of weak soil-forming processes that leads to the formation of a cambic horizon, which is present largely in Cambids but also in some Haplodurids, Vitricryands, and a few Haploxerolls. Cambic horizons in Nevada are favored by processes not leading to the accumulation of clay or salts; a cambic horizon may form in 1–2 ky on gravelly parent materials (Gile 1975).

13.8 Vertization

Vertization represents a collection of subprocesses occurring in soils with >60% smectitic clay, which enables soils to undergo shrinking and swelling that leads to reversible cracking of surface and subsurface horizons, tilted, wedge-shaped peds, and slickensides on faces of peds. In Nevada, vertization is limited to 15 soil series in the Vertisols order and 28 soil series in mainly Vertic ("ert") intergrade subgroups of Mollisols, Aridisols, and Entisols that classify in fine or very-fine particle-size classes and have the smectitic mineralogy class.

13.9 Solonization

Also referred to as alkalization, this process occurs when soils subject to salinization are naturally leached by soil water or purposely drained by humans. The excess soluble salts are leached out, the colloids under the influence of exchangeable Na become dispersed, and a strongly alkaline reaction develops. Solonization is dominant Natrargids, Natridurids, and Halaquepts. Solonization in Nevada is favored by the presence of basin fill in bolsons and semi-bolsons and may develop in less than 6.6 ky (Alexander and Nettleton 1977).

13.10 Salinization

Nowadays, salinization often is used to describe human-caused increases in soluble salts in soils and surface waters as a result of "desertification". From a soil genesis standpoint, salinization refers to the collection of subprocesses that enable the accumulation of soluble salts of Na, Ca, Mg, and K as chlorides, sulfates, carbonates, and bicarbonates. In general, these salts are more soluble than gypsum in cold water, and may become concentrated in a salic horizon. Salinization is a dominant process in Salids, Halaquepts, and Sodic subgroups of Calcids, Cambids, Gypsids, and Torrerts. Salinization in Nevada is favored by depressions in the landscape, seasonally high water tables, and proximity to the edge of playas.

13.11 Gypsification

Gypsification refers to the secondary accumulation of gypsum. The $CaSO_4$ initially fills micropores but over millennia may result in a strongly cemented petrogypsic horizon (references). In Nevada, gypsification occurs mainly in Gypsids. These soils occur almost exclusively in MLRA 30, the Mojave Desert, and comprise an area of less than 300 km^2. The soils occur on dissected rock pediments, fan remnants, alluvial flats, and hills formed in parent materials derived from gypsiferous sedimentary rocks.

13.12 Andisolization

Andisolization results in soils whose fine-earth fraction is dominated by amorphous compounds. Andisols must have andic properties, which include high amounts of acid-oxalate-extractable Al and Fe, a low bulk density, a high phosphate retention, and in vitric (allophanic) soils an abundance of volcanic glass. This process occurs in Andisols, of which there are four soil series in Nevada, and in Andic, Aquandic, Vitrandic, and related subgroups. Andisolization is especially important in MLRAs 22A, 23, 25, 26, and 28A. In addition to Andisols, Argixerolls, Haploxerolls, and Argicryolls often feature andic properties in Nevada.

13.13 Paludization

This term pertains primarily to the deep (>40 cm) accumulation of organic matter (histic materials) on the landscape usually in marshy areas. Most soils featuring paludization are in the Histosol order. There are no Histosols in Nevada; however, there are three soil series in Histic subgroups that cover only 7 km^2.

13.14 Conclusions

The dominant soil-forming processes in Nevada are argilluviation, the transfer of clay into the subsoil, melanization, the accumulation of well-humified materials in the topsoil, silicification, the plugging of soil pores by secondary Si in the form of opaline materials, and calcification, the plugging of pores by secondary carbonates. Other key processes include gleization, reducing conditions from restricted drainage, cambisolization, the development of B horizons with weak color and structure, vertization, the development of cracking and slickensides in parent materials enriched in smectitic clays, solonization, the accumulation of Na salts that lead to the formation of natric horizons, salinization, the accumulation of soluble salts in depressions of the landscape, and gypsification, the accumulation of gypsum in soils in depressions that are derived from gypsiferous rocks.

References

Alexander EB, Nettleton WD (1977) Post-Mazama Natrargids in Dixie Valley. Nevada. Soil Sci. Soc. Am. J. 41:1210–1212

Amundson RG, Chadwick OA, Sowers JM, Doner HE (1989a) Soil evolution along an altitudinal transect in the eastern Mojave Desert of Nevada, U.S.A. Geoderma 43:349–371

Amundson RG, Chadwick OA, Sowers JM, Doner HE (1989b) The stable isotope chemistry of pedogenic carbonates at Kyle Canyon. Nevada. Soil Sci. Soc. Am. J. 53:201–210

Bockheim JG, Gennadiyev AN (2000) The role of soil-forming processes in the definition of taxa in Soil Taxonomy and the World Soil Reference Base. Geoderma 95:53–72

Bockheim JG, Hartemink AE (2013) Distribution and classification of soils with clay-enriched horizons in the USA. Geoderma 209–210:153–160

Brock AL, Buck BJ (2009) Polygenetic development of the Mormon Mesa, NV petrocalcic horizons: geomorphic and paleoenvironmental interpretations. CATENA 77:65–75

Chadwick OA, Hendricks DM, Nettleton WD (1987) Silica in duric soils: I. A depositional model. Soil Sci Soc Am J 51:975–982

Chadwick OA, Hendricks DM, Nettleton WD (1989) Silicification of Holocene soils, in northern Monitor Valley. Nevada. Soil Sci. Soc. Am. J. 53:158–164

Gardner R (1972) Origin of the Mormon Mesa caliche, Clark County. Nevada. GSA Bull. 83:143–156

Gile LH (1975) Holocene soils and soil-geomorphic relations in an arid region of southern New Mexico. Quat. Res. 5:321–360

Gile LH, Peterson FF, Grossman RB (1965) The K horizon: a master soil horizon of carbonate accumulation. Soil Sci 99:74–82

Gile LH, Peterson FF, Grossman RB (1966) Morphological and genetic sequences of carbonate accumulation in desert soils. Soil Sci 101:347–360

Harper WG (1957) Morphology and genesis of Calcisols. Soil Sci. Soc. Am. 21:420–424

Marion GM (1989) Correlation between long-term pedogenic $CaCO_3$ formation rate and modern precipitation in deserts of the American Southwest. Quat. Res. 32:291–295

Reheis MC, Sowers JM, Taylor EM, McFadden LD, Harden JW (1992) Morphology and genesis of carbonate soils on the Kyle Canyon fan, Nevada, U.S.A. Geoderma 52:303–342

Robins CR, Brock-Hon AL, Buck BJ (2012) Conceptual mineral genesis models for calcic pendants and petrocalcic horizons. Nevada. Soil Sci. Soc. Am. J. 76:1887–1903

14 Benchmark, Endemic, Rare, and Endangered Soils in Nevada

Abstract

This chapter discusses four broad kinds of soils in Nevada in terms of their need for protection. Benchmark soils are those that (i) have a large extent within one or more Major Land Resource Areas, (ii) hold a key position in the Soil Taxonomy, (iii) have a large amount of data, (iv) have special importance to one or more significant land uses, (v) or are of significant ecological importance. An endemic soil is defined as the only soil in a family. Rare soils are those with an area less than 10,000 ha. Endangered soils are those that are endemic and rare. In Nevada, 6.4% of the soils are designated as benchmark soils, 30% are endemic, 62% are rare, and 21% are endangered.

14.1 Introduction

Benchmark soils are those that (i) have a large extent within one or more MLRAs, (ii) hold a key position in the *Soil Taxonomy*, (iii) have a large amount of data, (iv) have special importance to one or more significant land uses, (v) or are of significant ecological importance. About 6.4% of the soil series in the US have been designated as benchmark soils (Table 14.1).

An endemic soil is defined as the only soil in a family (Bockheim 2005). About 30% of the soil series identified in the USA are endemic (Table 14.1). Rare soils are those with an area less than 10,000 ha (Ditzler 2003). About 62% of the soil series in the USA occupy less than 10,000 ha (100 km^2). Endangered soils are those that are endemic and rare. About 21% of the soil series in the USA are endangered. However, there are certainly endemic soils with an area exceeding 10,000 ha that, depending on land use, have become endangered.

14.2 Benchmark Soils

About 7.8% of the soils of Nevada have been identified as benchmark soils (Table 14.1). This is slightly larger than the proportion of soils recognized as benchmark soils in other states. This is probably due to the large size and the special importance of many of the soils in Nevada for significant land uses and the ecological significance of many Nevada soils. Benchmark soils are listed in Appendix D.

14.3 Endemic Soils

About 18% of the soil series recognized in Nevada are the only soil in the family and, therefore, may be considered endemic (Table 14.1). This is comparable to the 19% of endemic soils reported for Wisconsin (Bockheim and Hartemink 2017) but is less than the 30% value for the USA.

Nearly three-quarters (71%) of the soil series in Nevada occur only within the confines of the state. Nevada ranks third in the USA in the number of soil series occurring only in the state and sixth in the proportion (percentage) of the total soil series identified only in the state (Table 7.3).

14.4 Rare Soils

About 62% of the soil series recognized in Nevada occupy less than 100 km^2 (10,000 ha) each and, therefore, may be considered rare (Fig. 7.5; Table 14.1). This is the same as the 62% value for the nation as a whole.

P. W. Blackburn et al., *The Soils of Nevada*, World Soils Book Series,
https://doi.org/10.1007/978-3-030-53157-7_14

Table 14.1 Proportion of benchmark, endemic, rare, endangered, shallow, and lithic soils in Nevada and the USA

Soil Class	NV (%)	USA (%)
Benchmark	7.8	6.3
Endemic	18	30
Rare	62	62
Endangered	8.5	21
Shallow	19	5.6
Lithic	13	5.7

14.5 Endangered Soils

About 8.5% of the soil series in Nevada are rare and endemic to the state, i.e., are considered "endangered". This is less than the 21% value for all soil series in the USA, primarily because of the lower proportion of endemic soils in Nevada (Table 14.1). Endemic, rare, and endangered soils are listed in Appendix E.

14.6 Shallow Soils

Nineteen percent of Nevada's soils classify in the shallow family class, due mainly to the presence of a duripan or a petrocalcic horizon occurring within 50 cm of the mineral soil surface. Another 13% classify in lithic subgroups, i.e., a lithic contact of hard bedrock occurs within 50 cm of the mineral soil surface).

14.7 Conclusions

Nevada has a comparably high pedodiversity compared to other states in the USA. This is reflected by the large number of soil series (1,800; exceeded only by CA), the high proportion (71%) of soil series that only occur in Nevada, the large number of endemic soils (317; exceeded only by CA, TX, WA, and OR), and the abundance of soils that occupy less than 100 km^2 in the state.

References

Bockheim JG (2005) Soil endemism and its relation to soil formation theory. Geoderma 129:109–124

Bockheim JG, Hartemink AE (2017) The Soils of Wisconsin. Springer, NY, p 393

Ditzler C (2003) Endangered soils. National Coop. Soil Surv. Newsletter No. 25, Nov., 2003, pp 1–2

Land Use in Nevada

15

Abstract

About 85% of the land in Nevada is federally managed. This chapter discusses the use of soils for rangeland, pasture, agricultural crops, forest products, wildlife habitat, and urban/suburban development.

15.1 Introduction

About 85% of the land in Nevada is federally managed (see Fig. 1.2). Some of the federally managed land is open to public use; other lands have restrictions; and some lands, notably that of the U.S. Department of Defense, are highly restricted (Table 15.1).

For the land area in Nevada as a whole, about 85% is rangeland suited for livestock grazing, recreation, and wildlife habitat, 8.5% is suitable for irrigated cropland or pasture, 0.5% is forested, and 6% is urbanized (Fig. 15.1). For private land, 83% is range, 5.6% is cropland, 5.2% is developed, 3.0% is forest, and 2.2% is in pasture (Fig. 15.2).

The State Land Use Planning Agency (SLUPA) provides technical planning assistance to local governments and other agencies in Nevada and represents the state on a variety of federal land management issues.

15.2 Rangeland

In soil survey reports for Nevada, the term "rangeland" refers to a kind of land rather than land use. Rangelands provides many important resource values, including watershed protection, habitat for wildlife, livestock forage, and opportunities for recreation. About 10% of the rangeland in Nevada is privately owned. Public rangeland use is allowed via permits on most of the federal land in the state. Over 43 million acres of grazing land is managed by the Bureau of Land Management in Nevada.

Rangeland has degraded over the years in many areas of the state due to wildfires and erosion. Management plans evaluating rangeland quality have only partially included soils. Historically, rangelands throughout Nevada experience heavy grazing pressures. In the last half of the twentieth century, the sagebrush ecosystem has experienced shortened disturbance return intervals due primarily to the introduction and spread of a non-native cool-season annual grass, cheatgrass (*Bromus tectorum*). Researchers and land management agencies are working on novel ideas to deal with soil erosion, wildfires, and the spread of undesirable species across the state. Soil survey interpretations provide an important tool for managing rangelands. Concepts of resistance and resilience based on soil properties described by the soil survey help managers control soil erosion, manage plant productivity, and implement restoration. Soil survey area reports prepared by the NRCS often provide tables evaluating suitability for rangeland seeding and/or rangeland production and species composition.

15.3 Pasture

Only 2.2% of the total nonfederal (i.e., private) land area in Nevada is irrigated pasture used for livestock production, including cattle, calves, dairy cows, sheep, lambs, and hogs (Fig. 15.3). However, this land accounts for nearly $300 million in revenue for the state, which constitutes 54% of all agricultural commodities produced in Nevada (Fig. 15.4).

P. W. Blackburn et al., *The Soils of Nevada*, World Soils Book Series,
https://doi.org/10.1007/978-3-030-53157-7_15

Table 15.1 Federal management of land in Nevada

Agency	Area (km^2)	%
Bureau of Indian Affairs	5441	1.9
Bureau of Land Management	195017	68.1
Bureau of Reclamation	2577	0.9
Department of Defense	1145	0.4
Department of Energy	3436	1.2
Fish and Wildlife Service	9450	3.3
Forest Service	23482	8.2
National Park Service	3150	1.1
Private, state, other	42669	14.9
Total	286368	100

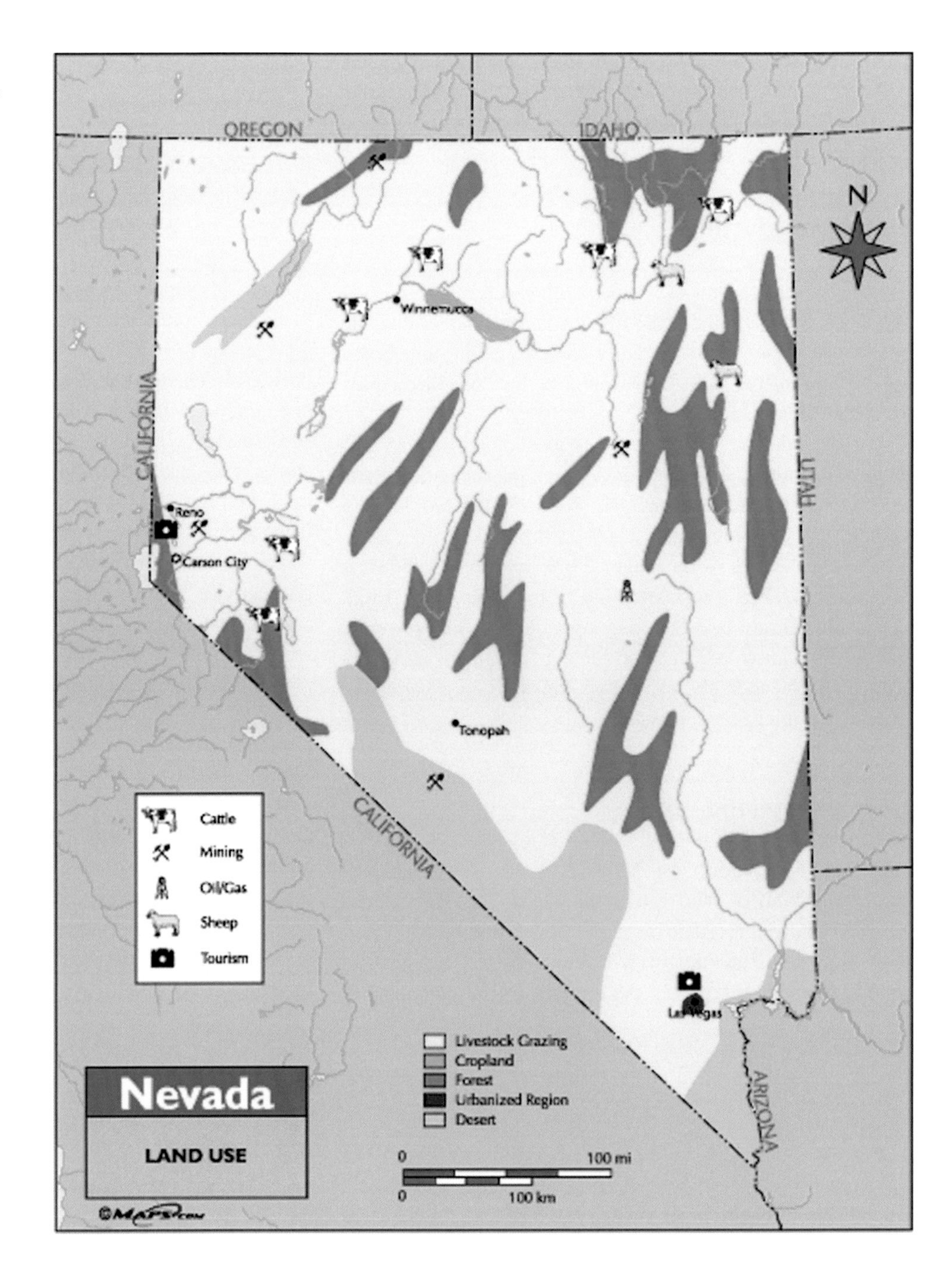

Fig. 15.1 Nevada land use includes open rangeland, forest, desert, cropland, and urbanized land. (*Source* Maps.com)

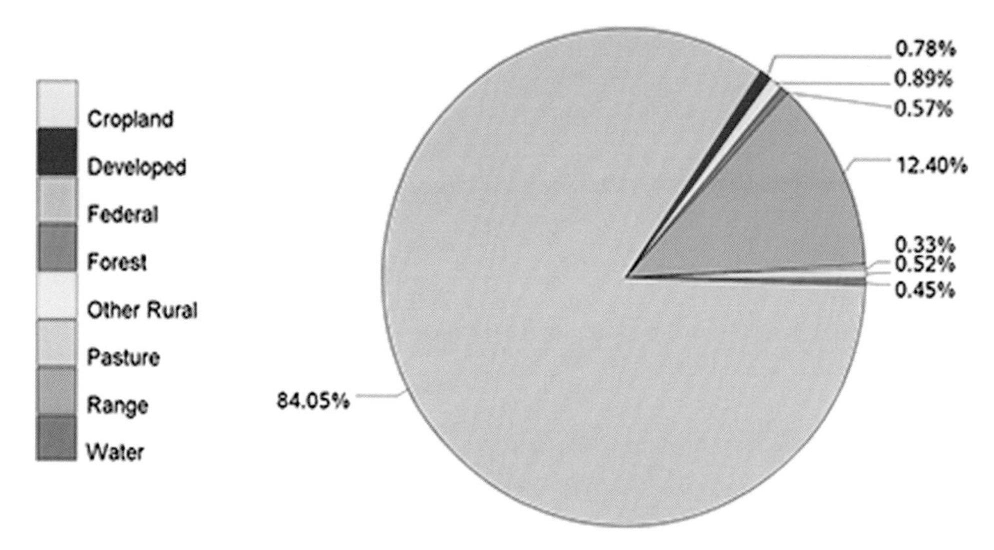

Fig. 15.2 2012 Natural Resources Inventory of nonfederal land in Nevada (*Source* USDA Natural Resources Conservation Service)

Fig. 15.3 Livestock grazing at the University of Nevada-Reno's Gund Ranch near Austin, Nevada. (*Source* Todd Klassy; https://www.toddklassy.com)

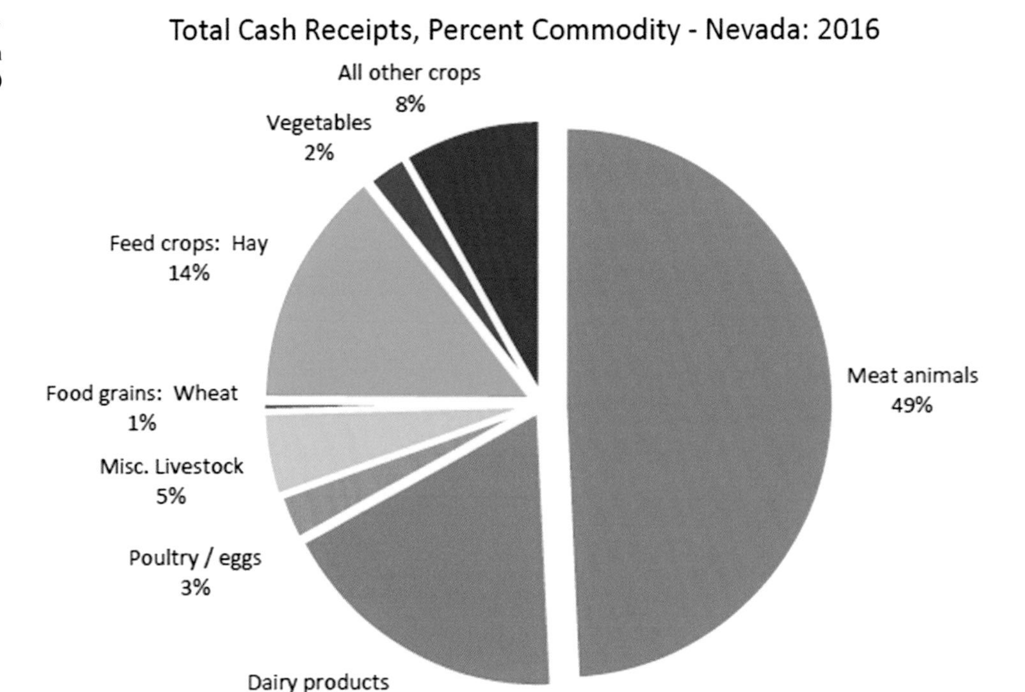

Fig. 15.4 Total cash receipts by agricultural commodity in 2016 in Nevada. (*Source* Rumberg, 2016)

15.4 Agricultural Crops

Nevada has 25,495 km^2 (6.3 million acres) of farmland. Because of the aridity of the state, most of the agriculture conducted in Nevada requires irrigation (Fig. 15.5). About 70% of the state's total water supply originates from the Colorado River, particularly Lake Mead, and other surface water sources (U.S EPA 2016; https://www.epa.gov/sites/proudction/files/2017-02/documents/ws-ourwater-nevada-state-fact-sheet.pdf). Groundwater supplies the remaining 30% of Nevada's water supply.

There are approximately 4,100 farms in Nevada, with an average size of 1,400 ha (3,500 acres) (Rumberg 2016). Farms in Elko County account for 36% of the total in Nevada. Other key agricultural counties include Churchill, Washoe, Humboldt, Pershing, and Lyon. Much of this land is in the Humboldt River, Carson River, and Truckee River basins.

The most important crops in terms of revenue generated are alfalfa hay, alfalfa seed, potatoes, barley, winter and spring wheat, rye, oats, vegetables, garlic and onions, and some fruits (Fig. 15.6).

Agriculture in Nevada is regulated by the Nevada Department of Agriculture, the U.S. Department of Agriculture, and other agencies.

Soil surveys play an important role in Nevada's agriculture by providing land capability designations, identifying prime farmland, and showing crop-yield estimates.

15.5 Forest Products

Forests (also referred to in soil survey area reports as "woodlands") account for 42,896 km^2 (10.6 million acres) of the land area in Nevada (Fig. 15.7). Unique among the western states, Nevada has some 300 forested mountain "islands" separated by wide non-forested basins. More than half (63%) of Nevada's forests are managed by the Bureau of Land Management; national forests compose 30%, and 3.7% are privately owned.

The Nevada Division of Forestry, the Bureau of Land Management, the U.S. Forest Service, and other agencies manage forests in Nevada. The Humboldt-Toiyabe National Forest is the largest national forest in the "Lower 48". From a state-wide perspective, 88% of Nevada's forests are single-leaf pinyon (*Pinus monophylla*) or Utah juniper (*Juniperus osteosperma*), or a combination of these two key species (Fig. 15.8). However, most mountain ranges include high elevation enclaves of other conifers and hardwoods that augment the diversity of vegetation types in Nevada.

Fig. 15.5 Irrigation of pasture land in Spring Valley, Nevada. (*Source* Las Vegas Review-Journal)

Fig. 15.6 Hay harvesting in Nevada. (*Source* Nevada Farm Bureau Federation)

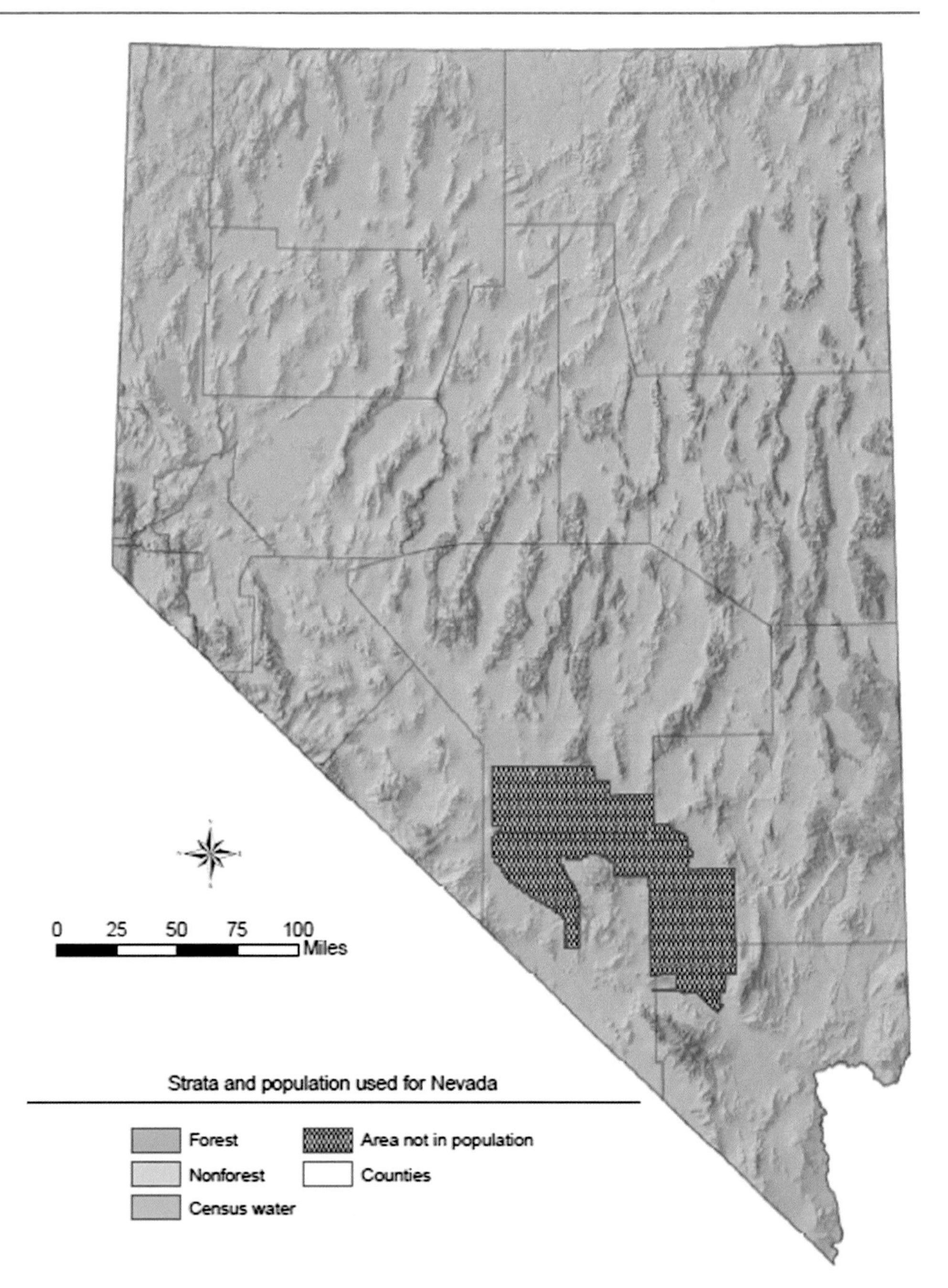

Fig. 15.7 Forest land in Nevada. (*Source* Menlove et al. 2016)

Though the area of forest land is relatively small, the value of this resource is immeasurable in terms of commodities, recreational uses, and aesthetic properties. Key forest products in Nevada include Christmas trees, firewood, pinyon nuts (Fig. 15.9), fence posts, and wildings for transplant. Preferred species for firewood include Utah juniper, single-leaf pinyon pine, quaking aspen (*Populus tremuloides*), and curl-leaf mountain mahogany (*Cercocarpus ledifolius*). Juniper is also used for fence posts. County soil surveys report prospects for woodland management and products where it is appropriate. One of the key concerns with forest management in Nevada is wildfire (Fig. 15.10).

15.6 Wildlife Habitat

A key use of Nevada's soils is preserving wildlife habitat. The Nevada Department of Wildlife is the state agency and the Fish and Wildlife Service is the federal agency responsible for the restoration and management of wildlife

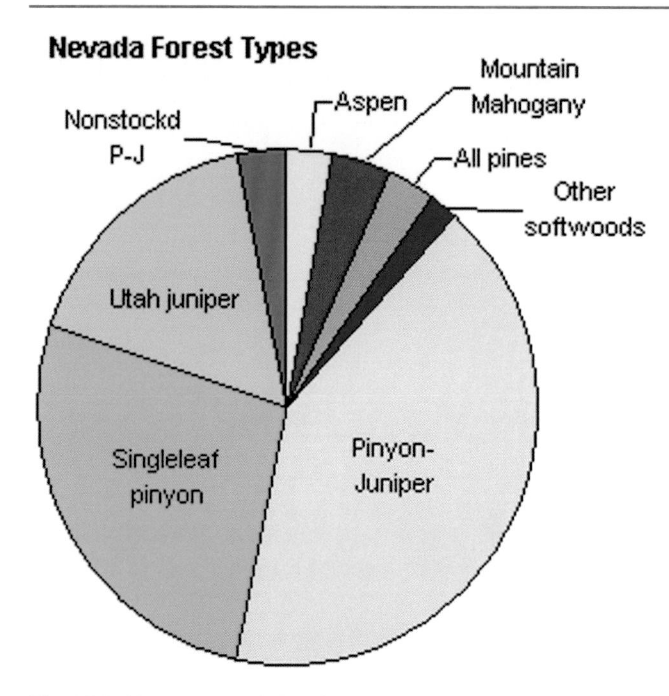

Fig. 15.8 Forest types of Nevada. (*Source* Nevada Forest Overview, U.S. Forest Service)

resources in Nevada. The Desert NWR (6,540 km^2; 1.615 million acres) is the largest national wildlife refuge in the "Lower 48". The fauna of Nevada contains mostly species adapted to high desert. Animals in Nevada include scorpions, mountain lions, snakes, lizards, spiders, wolves, coyotes, foxes, ground squirrels, rabbits, falcons, ravens, desert tortoises, hawks, eagles, bobcats, sheep, deer, pronghorns, geckos, owls, bats, horned toads, and more. The desert bighorn sheep *(Ovis canadensis nelsoni)* is the official state animal and is found in most of Nevada's mountainous desert. County soil surveys often identify wildlife suitability groups.

15.7 Development

Due to its lower population (under 3 million), only a small part of Nevada's land is urbanized. Urbanization is mainly in the Carson City area (410 km^2; 157 mi^2), the Reno-Sparks area (367 km^2; 142 mi^2), the greater Las Vegas area (352 km^2; 136 mi^2), and the Henderson area (279 km^2; 108 mi^2). In addition to urban development, land in Nevada is used for mining and

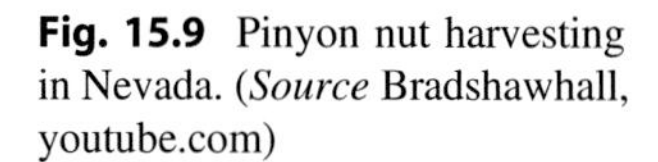

Fig. 15.9 Pinyon nut harvesting in Nevada. (*Source* Bradshawhall, youtube.com)

Fig. 15.10 Prescribed burn in the Lake Tahoe Basin. (*Source* Alan H. Taylor)

energy production (Fig. 15.11). Nevada's early history as a US territory and later a State was strongly shaped by the mining industry and extraction of minerals remains an important industry.

Lands in Nevada which are managed by the U.S. Department of Defense (DOD) are used for military training, munitions storage and disposal, and weapons development and testing. The largest area in the state, which lacks soil survey information, is in southern Nevada in parts of Clark, Lincoln, and Nye Counties and constitutes 4.1 million acres (16853 km^2; 6507 mi^2) of federal land controlled by the DOD. These lands are highly restricted and not open to public visitation or use.

15.8 Conclusions

About 85% of the land in Nevada is federally managed. Current land use includes livestock grazing on rangeland (85%), irrigated cropland or pasture (8.5%), urban (6%), and forest (0.5%). Pastures that are often irrigated account for 54% of all agricultural commodities produced in Nevada. Degradation of rangeland with the introduction of exotic species, wildfires, and accelerated soil erosion is a key issue in Nevada. Around 8.5% of the land in Nevada is cropped, primarily in river basins. The most important crops in terms of revenue generated are alfalfa hay, alfalfa seed, potatoes,

Fig. 15.11 Oil and gas lease on BLM land in Nevada. (*Source* Bureau of Land Management)

barley, winter and spring wheat, rye, oats, vegetables, garlic and onions, and some fruits. Despite their small area, forests are important from a recreational and aesthetic standpoint in that they occur primarily as "sky islands" on north-south trending mountain ranges surrounded by fault-block valleys. Key forest products in Nevada include Christmas trees, firewood, pinyon pine nuts, fence posts, and wildings for transplant. Nevada's high desert environment provides a unique wildlife habitat that includes mountain frogs, spinedace and bull trout fish, sage grouse, Sierra Nevada bighorn sheep, feral horses and donkeys, other fauna.

References

Menlove J, Shaw JD, Witt C et al (2016) Nevada's forest resources, 2004–2013. U.S. For. Serv., Resour. Bull. RMRS-RB-22. 167 pp

Rumberg S (2016) Nevada agricultural statistics annual bulletin 2016 crop year. www.nass.usda.gov/nv/

U.S. Environmental Protection Agency (2016) Saving water in Nevada. EPA-832-F-16–001, May 2016

Conclusions

16

- The soil cover in Nevada is considered here as "a natural body comprised of solids (minerals and organic matter), liquid, and gases that occurs on the land surface, occupies space, and is characterized by one or both of the following: horizons, or layers, that are distinguishable from the initial material as a result of additions, losses, transfers, and transformations of energy and matter *or* the ability to support rooted plants in a natural environment" (Soil Survey Staff 2014).
- Nevada is divided into 10 Major Land Resource Areas that reflect differences in physiography, geology, climate, water, soils, biological resources, and land use.
- In that the soils of the Fallon Area were mapped in 1909, Nevada was one of the earliest states to have a soil survey.
- The soils of Nevada are classified using *Soil Taxonomy* (Soil Survey Staff, 1999) and the *Keys to Soil Taxonomy* (Soil Survey Staff, 2014). Soil taxonomy is a hierarchical system based mainly on natural soil properties defined operationally in diagnostic surface and subsurface horizons and diagnostic soil characteristics. This taxonomic system divides the soils of the world into orders, suborders, great groups, subgroups, families, and soil series. Phases of soil series and miscellaneous areas are the fundamental map unit components used at the typical detailed map scale of 1:24,000 in the National Cooperative Soil Survey (NCSS) of the USA and its territories.
- Nevada has benefited from considerable soil research by university and NRCS investigators over the past 60 years.
- The Nevada state legislature has recognized by importance of soils in the state by approving the Orovada soil series as the state soil.
- The cumulative number of soil series and soil surveys has increased exponentially in Nevada, since the early 1970s. To date, 89% of Nevada's land area in its 16 countics has been mapped according to NCSS standards and made available for public use.
- The soils of Nevada result from five soil-forming factors operating collectively: climate, organisms, relief, parent material, and time. Most of Nevada is included in the Great Basin Section of the Basin and Range Province of the western USA. The presence of the Sierra Nevada Mountain Range and more than 100 north-south trending mountain ranges separated by deep fault-block valleys (basins) enable Nevada to experience a diversity of climates. The soils of the state are classified into four soil moisture regimes and five soil temperature regimes that have contributed to a high pedodiversity in the state. Although the state has not experienced continental glaciation, large pluvial lakes formed during alpine glaciations have impacted soil formation during the Quaternary Period. The vegetation of Nevada is divided into six elevation zones: salt desert, sagebrush-grassland, pinyon-juniper woodland, mountain shrubs, subalpine forest, and alpine tundra. Nevada is composed of a variety of lithologic units, ranging in age from Early Proterozoic (2,500 million years ago) to Holocene (last 10,000 years) and in composition that includes igneous, metamorphic, and sedimentary rocks. The Basin and Range Province is due to widespread volcanism, crustal extension, and block faulting since the early Miocene Epoch (beginning about 17 million years ago). The soil parent materials of Nevada are dominated by colluvium and residuum in the mountain ranges, alluvium and lacustrine deposits in the basins, and subsequent deposition of loess and/or volcanic ash. Soils in Nevada range from $\sim$ 5 million years to less than 1,000 years in age. Humans have impacted the soils of Nevada by mining, weapons testing, deforestation, wildfire, irrigation, and urban development.
- An analysis of the soils of Nevada by Major Land Resources Regions suggests that Torriorthents, Haplargids, Argixerolls, Haplocalcids, Haplocambids, Haplodurids, and Argidurids are the most conspicuous great groups in Nevada. Argixerolls, some Haplargids, and Torriorthents are dominant in mountain ranges; Haplocambids, Natrargids, Torriorthents, Haplocalcids, Haplodurids, Argidurids, and some Haplargids are most prevalent in the basins; Halaquepts and Aquisalids

P. W. Blackburn et al., *The Soils of Nevada*, World Soils Book Series,
https://doi.org/10.1007/978-3-030-53157-7_16

occupy basin floors or lake plains; and Torripsamments exist on dunes and sand sheets.

- Prof. F. F. Peterson of the University of Nevada-Reno defined the landforms of the Basin and Range Province for soil survey. Subsequent block diagrams of soil associations contained in soil survey reports illustrate soil-landform relationships in the state.
- Six of the eight epipedons (diagnostic surface horizons) recognized in *Soil Taxonomy* (Soil Survey Staff, 1999) and ten of the 20 diagnostic subsurface horizons occur in soils of Nevada. The state contains soils representative of seven of the 12 orders, 29 of the 68 suborders, 69 of the 337 great groups, 300 subgroups, over 1,000 families, and over 1,800 soil series.
- Soils of Nevada contain primarily ochric and mollic epipedons (99% of soil series) overlying argillic, calcic, cambic, natric, or petrocalcic horizons and duripans (89% of soil series). Albic, gypsic, petrogypsic, and salic horizons occur only minimally in the soils of Nevada, but are locally important. On an area basis, the dominant orders represented in Nevada are Aridisols, Mollisols, and Entisols (97%). The dominant suborders are Xerolls, Argids, Orthents, Durids, and Calcids (84% of land area); the dominant great groups are Torriorthents, Haplargids, Argixerolls, Haplocalcids, Argidurids, Haplocambids, and Haplodurids (77% of land area). Of the more than 1,800 soil series in Nevada, 72% occur only in the state.
- Nevada is divided here into 17 soil great group associations. The dominant great groups are Torriorthents, Haplargids, Argixerolls, Haplocalcids, Argidurids, Haplocambids, Haplodurids, Haploxerolls, Petrocalcids, Halaquepts, Argicryolls, Haplocryolls, and Argiustolls. These soil map units are displayed in a wall-size, color map at a scale of 1:1.2 million.
- The dominant soil-forming processes in Nevada are argilluviation, the transfer of clay into the subsoil; melanization, the accumulation of well-humified organic matter in the topsoil; silicification, the plugging of soil pores by secondary silica (SiO_2) in the form of opaline materials; and calcification, the plugging of pores by secondary calcium carbonate. Other key processes include gleization, reducing conditions from restricted drainage; cambisolization, the weak development of B horizons with color and structure; vertization, the development of reversible cracking, slickensides, and wedge-shaped peds in parent materials enriched in smectitic clays; solonization, the accumulation of Na salts that may lead to the formation of natric horizons; salinization, the accumulation of soluble salts of Na, Ca, Mg, and K as chlorides, sulfates, carbonates, and bicarbonates in depressions in the landscape, landforms with seasonally high water tables, and proximity to the edge of playas; and gypsification, the accumulation of gypsum in soils that formed in parent materials derived from gypsiferous rocks on dissected rock pediments, fan remnants, alluvial flats, and hills.
- Nevada has a comparably high pedodiversity compared to other states in the USA. This is reflected by the large number of soil series (1,800; exceeded only by CA), the high proportion (71%) of soil series that only occur in Nevada, the large number of endemic soils (317; exceeded only by CA, TX, WA, and OR), and the abundance of soils that occupy less than 100 km^2 in the state.
- Nearly 85% of the land in Nevada is managed by the federal government. The dominant type of land in Nevada is rangeland typically used for livestock grazing, recreation, and wildlife habitat. Only 5.6% of the privately owned land in Nevada is irrigated cropland which is used primarily for livestock production and contributes $750 million to the state's economy.

References

Soil Survey Staff (1999) Soil Taxonomy: a Basic System of Soil Classification for Making and Interpreting Soil Surveys. 2nd ed. Agric. Handbook, vol 436. U.S. Govt. Print. Office, Washington, DC. 869 pp

Soil Survey Staff (2014) Keys to Soil Taxonomy. 12th ed. U.S. Department of Agricultural Natural Resource Conservation Service, Linoln, NE

Appendix A
Taxonomy of Nevada Soils

P. W. Blackburn et al., *The Soils of Nevada*, World Soils Book Series,
https://doi.org/10.1007/978-3-030-53157-7

Soil series	Area (km²)	Bench-mark	Year Est-ablished	TL State	Other States Using	MLRAs Using	Order	Suborder	Great Group	Subgroup	Particle-size Class	Mineralogy Class	CEC Activity Class	Reaction Class	Soil temp. Class	SMR	Other Family
ABALAN	2.4	NO	2004	NV		28A, 28B	Aridisols	Argids	Haplargids	Xeric Haplargids	loamy-skeletal	Mixed	superactive		mesic	aridic	shallow
ABGESE	75.4	NO	1974	NV	ID	11, 28B	Aridisols	Argids	Haplargids	Xeric Haplargids	fine-loamy	Mixed	superactive		mesic	aridic	
ABOTEN	149.7	NO	1988	NV		27	Aridisols	Durids	Natridurids	Natrargidic Natridurids	loamy	Mixed	superactive		mesic	aridic	shallow
ACANA	100.3	NO	1971	NV	UT	28A, 28B, 29	Aridisols	Durids	Argidurids	Haploxeralfi c Argidurids	loamy	Mixed	superactive		mesic	aridic	shallow
ACKETT	92.6	NO	1986	ID	NV	25	Aridisols	Durids	Argidurids	Xeric Argidurids	clayey-skeletal	Smectitic			mesic	aridic	shallow
ACKLEY	17.0	NO	1981	NV	ID	11, 26	Aridisols	Argids	Haplargids	Xeric Haplargids	fine-loamy	mixed	superactive		mesic	aridic	
ACOMA	103.2	NO	1971	NV		28B, 29	Aridisols	Argids	Paleargids	Calcic Paleargids	fine	smectitic			mesic	aridic	
ACRELANE	598.5	NO	1980	NV	ID	23, 25, 26, 27	Mollisols	Xerolls	Argixerolls	Aridic Argixerolls	loamy-skeletal	mixed	superactive		mesic	xeric	shallow
ACTI	122.2	NO	1992	NV		29	Mollisols	Ustolls	Argiustolls	Aridic Lithic Argiustolls	clayey-skeletal	smectitic			mesic	ustic	
ADAMATT	2.3	NO	2010	CA	NV	26	Mollisols	Xerolls	Argixerolls	Vitrandic Argixerolls	ashy-skeletal	glassy			frigid	xeric	shallow
ADAVEN	1.5	NO	1965	NV		29	Aridisols	Calcids	Petrocalcids	Aquic Petrocalcids	coarse-loamy	mixed	superactive		mesic	aridic	
ADELAIDE	153.9	NO	1974	NV		24	Aridisols	Durids	Haplodurids	Cambidic Haplodurids	loamy	mixed	superactive		mesic	aridic	shallow
ADOBE	189.3	NO	1990	NV		28A, 28B	Mollisols	Cryolls	C alcicryolls	Lithic Calcicryolls	loamy-skeletal	carbonatic			cryic	xeric-aridic	
ADOS	40.4	NO	1983	NV		28B	Aridisols	Calcids	Petrocalcids	Xeric Petrocalcids	loamy-skeletal	carbonatic			frigid	aridic	
ADVOKAY	96.3	NO	1984	NV		29	Aridisols	Argids	Haplargids	Typic Haplargids	loamy	mixed	superactive		mesic	aridic	shallow
AFFEY	7.8	NO	1986	NV		25	Mollisols	Xerolls	Argixerolls	Argiduridic Argixerolls	clayey-skeletal	smectitic			frigid	xeric	
AGASSIZ	10.1	NO	1969	UT	ID, NV	25, 28A, 47	Mollisols	Xerolls	Haploxerolls	Lithic Haploxerolls	loamy-skeletal	mixed	superactive		frigid	xeric	
AGON	8.9	NO	1996	NV		30	Aridisols	Durids	Haplodurids	Typic Haplodurids	sandy	mixed			thermic	aridic	
AGORT	35.3	NO	1986	NV		25	Mollisols	Xerolls	Haploxerolls	Entic Haploxerolls	loamy	mixed	superactive		frigid	xeric	shallow
AGUACHIQUITA	16.8	NO	2006	NV		30	Aridisols	Durids	Haplodurids	Cambidic Haplodurids	loamy-skeletal	mixed	superactive		thermic	aridic	
AHCHEW	21.8	NO	1991	NV		26	Aridisols	Argids	Haplargids	Vitrixerandic Haplargids	loamy-skeletal	mixed	superactive		mesic	aridic	shallow
AKELA	106.7		1970	NM	AZ	40, 41, 42	Entisols	Orthents	Torriorthents	Lithic Torriorthents	loamy-skeletal	mixed	superactive	calcareous	thermic	torric	
AKERCAN	13.8	NO	1983	NV		28B	Aridisols	Cambids	Haplocambids	Xeric Haplocambids	fine-loamy	mixed	superactive		frigid	aridic	
AKERUE	64.2	NO	1983	NV		24, 28B	Aridisols	Durids	Argidurids	Xeric Argidurids	clayey-skeletal	smectitic			frigid	aridic	shallow
AKLER	138.7	NO	1986	NV		25	Aridisols	Argids	Haplargids	Xeric Haplargids	clayey	smectitic			frigid	aridic	shallow
ALADSHI	27.1	NO	1980	NV		26	Aridisols	Argids	Haplargids	Durinodic Xeric Haplargids	fine-loamy	mixed	superactive		mesic	aridic	
ALAMOROAD	40.3	NO	2011	NV		30	Aridisols	Calcids	Haplocalcids	Petronodic Haplocalcids	loamy-skeletal	carbonatic			thermic	aridic	
ALBURZ	37.1	NO	1986	NV		25	Mollisols	Aquolls	Endoaquolls	Fluvaquentic Endoaquolls	sandy-skeletal	mixed			frigid	aquic	
ALCAN	37.8	NO	1984	NV		29	Aridisols	Argids	Haplargids	Xeric Haplargids	loamy-skeletal	mixed	superactive		mesic	aridic	shallow
ALDAX	6.7	NO	1973	NV	CA	21, 26	Mollisols	Xerolls	Haploxerolls	Aridic Lithic Haploxerolls	loamy-skeletal	mixed	superactive		mesic	xeric	
ALHAMBRA	142.0	NO	1971	NV		28B	Entisols	Fluvents	Torrifluvents	Duric Xeric Torrifluvents	coarse-loamy	mixed	superactive	calcareous	frigid	torric	
ALKO	173.3	NO	1976	NV	AZ, CA	30	Aridisols	Durids	Haplodurids	Typic Haplodurids	loamy	mixed	superactive		thermic	aridic	shallow
ALLEY	467.2	YES	1979	NV		24, 25, 28B	Aridisols	Argids	Haplargids	Durinodic Xeric Haplargids	fine-loamy	mixed	superactive		mesic	aridic	
ALLKER	201.2	NO	1983	NV		24, 25, 29	Aridisols	Argids	Haplargids	Durinodic Xeric Haplargids	fine-loamy over sandy or sandy-skeletal	mixed	superactive		mesic	aridic	

(continued)

(continued)

Soil series	Area (km^2)	Bench-mark	Year Est-ablished	TL State	Other States Using	MLRAs Using	Order	Suborder	Great Group	Subgroup	Particle-size Class	Mineralogy Class	CEC Activity Class	Reaction Class	Soil temp. Class	SMR	Other Family
ALLOR	177.2	NO	1985	NV		24, 28B	Aridisols	Argids	Haplargids	Durinodic Xeric Haplargids	fine-loamy	mixed	superactive		mesic	aridic	
ALPHA	11.2	NO	1971	NV		28B	Mollisols	Xerolls	Haploxerolls	Duridic Haploxerolls	fine-loamy	mixed	superactive		frigid	xeric	
ALTA	40.0	NO	1994	NV		23	Mollisols	Cryolls	Haplocryolls	Pachic Haplocryolls	sandy-skeletal	mixed			cryic	xeric-aridic	
ALVODEST	6.4	NO	1991	OR	NV	24	Aridisols	Cambids	Aquicambids	Sodic Aquicambids	fine	smectitic			mesic	aridic	
ALYAN	228.7	NO	1986	NV	OR	23, 24, 25	Mollisols	Xerolls	Argixerolls	Aridic Argixerolls	fine	smectitic			frigid	xeric	
AMBUSH	34.7	NO	2006	NV		29	Aridisols	Calcids	Haplocalcids	Petronodic Haplocalcids	coarse-loamy	mixed	superactive		mesic	aridic	
AMELAR	121.4	NO	1990	NV		28B	Mollisols	Xerolls	Argixerolls	Calcic Argixerolls	loamy-skeletal	mixed	superactive		frigid	xeric	
AMENE	44.5	NO	1986	NV		25, 28B	Mollisols	Xerolls	Calcixerolls	Aridic Lithic Calcixerolls	loamy-skeletal	carbonatic			frigid	xeric	
AMTOFT	354.8	NO	1971	UT	NV	25, 28A, 28B, 47	Aridisols	Calcids	Haplocalcids	Lithic Xeric Haplocalcids	loamy-skeletal	carbonatic			mesic	aridic	
ANAUD	60.5	NO	2006	NV		28A	Mollisols	Xerolls	Argixerolls	Aridic Lithic Argixerolls	loamy-skeletal	mixed	superactive		frigid	xeric	
ANAWALT	265.6	NO	1991	OR	CA, NV	10, 23, 25	Aridisols	Argids	Haplargids	Lithic Xeric Haplargids	clayey	smectitic			frigid	aridic	
ANED	25.1	NO	1971	NV	UT	28A	Mollisols	Xerolls	Durixerolls	Argidic Durixerolls	loamy	mixed	superactive		mesic	xeric	shallow
ANNAW	331.8	NO	1984	NV		27, 29	Aridisols	Cambids	Haplocambids	Typic Haplocambids	sandy-skeletal	mixed			mesic	aridic	
ANOWELL	31.4	NO	1986	NV		25	Aridisols	Argids	Haplargids	Xeric Haplargids	loamy	mixed	superactive		mesic	aridic	shallow
ANSPING	108.2	NO	1974	NV		28B	Mollisols	Xerolls	Calcixerolls	Aridic Calcixerolls	loamy-skeletal	carbonatic			frigid	xeric	
ANTEL	26.1	NO	1974	NV		24	Aridisols	Cambids	Haplocambids	Durinodic Haplocambids	fine-silty	mixed	superactive		mesic	aridic	
ANTENNAPEAK	28.3	NO	2010	NV		28A	Mollisols	Cryolls	Haplocryolls	Vitrandic Haplocryolls	loamy-skeletal	mixed	superactive		cryic	typic-xeric	
ANTHOLOP	15.8	NO	1985	NV		26	Aridisols	Durids	Argidurids	Abruptic Xeric Argidurids	clayey	smectitic			mesic	aridic	shallow
ANTHONY	3.7		1912	AN	NM, TX	40, 41, 42	Entisols	Fluvents	Torrifluvents	Typic Torrifluvents	coarse-loamy	mixed	superactive	calcareous	thermic	torric	
APMAT	11.3	NO	1980	NV		26	Mollisols	Xerolls	Argixerolls	Alfic Argixerolls	loamy-skeletal	mixed	active		frigid	xeric	
APPIAN	422.6	YES	1971	NV		24, 26, 27, 28A, 28B	Aridisols	Argids	Natrargids	Typic Natrargids	fine-loamy over sandy or sandy-skeletal	mixed	superactive		mesic	aridic	
AQUINAS	31.4	NO	1980	NV		26	Aridisols	Durids	Argidurids	Haploxeralfic Argidurids	fine-loamy	mixed	superactive		mesic	aridic	
ARADA	180.0	NO	1970	NV	AZ	30	Aridisols	Calcids	Haplocalcids	Typic Haplocalcids	sandy	mixed			thermic	aridic	
ARCIA	123.5	NO	1986	NV	OR	10, 25	Mollisols	Xerolls	Argixerolls	Vitrandic Argixerolls	fine	smectitic			frigid	xeric	
ARCLAY	193.7	NO	1988	NV		23, 27	Mollisols	Xerolls	Argixerolls	Aridic Argixerolls	loamy	mixed	superactive		mesic	xeric	shallow
ARDIVEY	475.6	NO	1972	NV		27, 29	Aridisols	Argids	Haplargids	Durinodic Haplargids	loamy-skeletal	mixed	superactive		mesic	aridic	
ARGALT	96.3	NO	1985	NV		27, 29	Aridisols	Durids	Argidurids	Xeric Argidurids	loamy	mixed	superactive		mesic	aridic	shallow
ARGENTA	163.1	NO	1974	NV		24	Inceptisols	Aquepts	Halaquepts	Duric Halaquepts	coarse-loamy	mixed	superactive	calcareous	mesic	aquic	
ARIZO	1204.8	YES	1971	NV	AZ, CA, NM	29, 30, 40, 41, 42	Entisols	Orthents	Torriorthents	Typic Torriorthents	sandy-skeletal	mixed			thermic	torric	
ARKSON	2.7	NO	1975	NV		22A	Mollisols	Cryolls	Haplocryolls	Xeric Haplocryolls	coarse-loamy	mixed	superactive		cryic		

(continued)

(continued)

Soil series	Area (km^2)	Bench-mark	Year Est-ablished	TL State	Other States Using	MLRAs Using	Order	Suborder	Great Group	Subgroup	Particle-size Class	Mineralogy Class	CEC Activity Class	Reaction Class	Soil temp. Class	SMR	Other Family
ARMESPAN	1248.1	YES	1984	NV		27, 28A, 28B, 29	Aridisols	Calcids	Haplocalcids	Durinodic Xeric Haplocalcids	loamy-skeletal	mixed	superactive		mesic	aridic	
ARMOINE	113.0	NO	1984	NV		29	Aridisols	Argids	Haplargids	Xeric Haplargids	loamy-skeletal	mixed	superactive		mesic	aridic	shallow
ARMPUP	19.2	NO	1996	NV		30	Aridisols	Argids	Natrargids	Typic Natrargids	fine	smectitic			thermic	aridic	
ARMYDRAIN	6.6	NO	1985	NV		27	Mollisols	Aquolls	Endoaquolls	Fluvaquentic Endoaquolls	clayey over loamy	smectitic OVER mixed	superactive	calcareous	mesic	aquic	
ARROLIME	19.9	NO	1970	NV		30	Aridisols	Gypsids	Calcigypsids	Typic Calcigypsids	loamy-skeletal	mixed	active		thermic	aridic	
ARVA	28.8	NO	1986	NV		25	Mollisols	Xerolls	Argixerolls	Pachic Argixerolls	fine	smectitic			frigid	xeric	
ARZO	41.3	NO	1980	NV		26, 27	Mollisols	Xerolls	Argixerolls	Calciargidic Argixerolls	fine	smectitic			mesic	xeric	
ASH SPRINGS	7.1	NO	1940	NV		29, 30	Aridisols	Calcids	Haplocalcids	Aquic Haplocalcids	coarse-loamy	mixed	superactive		mesic	aridic	
ASHART	45.3	NO	1986	NV		25	Aridisols	Argids	Haplargids	Vitrixerandic Haplargids	ashy	glassy			mesic	aridic	shallow
ASHCAMP	26.4	NO	1995	NV		23	Mollisols	Xerolls	Argixerolls	Vitritorrandic Argixerolls	ashy	glassy			mesic	xeric	shallow
ASHDOS	24.4	NO	1995	NV	CA	23	Mollisols	Xerolls	Argixerolls	Vitritorrandic Argixerolls	ashy	glassy			frigid	xeric	
ASHFLAT	2.1	NO	2005	CA	NV	26	Mollisols	Cryolls	Argicryolls	Vitrandic Argicryolls	ashy-skeletal	glassy			cryic	xeric	
ASHMED	53.4	NO	1994	NV		30	Aridisols	Argids	Haplargids	Typic Haplargids	loamy-skeletal	mixed	superactive		thermic	aridic	
ASHONE	15.5	YES	1995	NV		23	Mollisols	Xerolls	Argixerolls	Vitritorrandic Argixerolls	ashy	glassy			mesic	xeric	
ASHTRE	27.6	NO	1995	NV	CA	23	Mollisols	Xerolls	Argixerolls	Vitritorrandic Argixerolls	ashy	glassy			frigid	xeric	
ATLANTA	36.3	NO	2004	NV		28A	Aridisols	Calcids	Haplocalcids	Xeric Haplocalcids	coarse-loamy	mixed	superactive		mesic	aridic	
ATLOW	479.6	NO	1985	NV	OR	10, 24, 27, 28B	Aridisols	Argids	Haplargids	Lithic Xeric Haplargids	loamy-skeletal	mixed	superactive		mesic	aridic	
ATRYPA	121.6	NO	1971	NV		25, 28B	Mollisols	Xerolls	Haploxerolls	Calcidic Haploxerolls	loamy	mixed	superactive		frigid	xeric	shallow
ATTELLA	33.4	NO	1985	NV		24, 28B	Entisols	Orthents	Torriorthents	Lithic Xeric Torriorthents	loamy-skeletal	mixed	superactive	calcareous	frigid	torric	
AURUM	3.6	NO	2013	NV		28B	Mollisols	Xerolls	Durixerolls	Haplic Durixerolls	fine-loamy	mixed	superactive		frigid	xeric	
AUTOMAL	402.8	NO	1990	NV		28B	Aridisols	Calcids	Haplocalcids	Durinodic Xeric Haplocalcids	loamy-skeletal	mixed	superactive		mesic	aridic	
AYCAB	64.9	NO	1993	NV		23, 25	Mollisols	Cryolls	Haplocryolls	Pachic Haplocryolls	coarse-loamy	mixed	superactive		cryic	xeric	
AYMATE	64.7	NO	1992	NV		30	Aridisols	Calcids	Petrocalcids	Ustalfic Petrocalcids	fine-loamy	mixed	superactive		thermic	aridic	
AYSEES	9.0	NO	1974	NV	ID	11, 28A, 28B, 29	Aridisols	Calcids	Haplocalcids	Typic Haplocalcids	sandy-skeletal	mixed			mesic	aridic	
AZSAND	11.2	NO	2006	NV		30	Aridisols	Calcids	Haplocalcids	Typic Haplocalcids	sandy-skeletal	mixed			hyperthermic	aridic	
AZTEC	30.0	NO	1964	NV	NM	30, 42	Aridisols	Gypsids	Calcigypsids	Typic Calcigypsids	loamy-skeletal	mixed	active		thermic	aridic	
AZURERIDGE	33.4	NO	2006	NV		30	Aridisols	Durids	Haplodurids	Cambidic Haplodurids	loamy-skeletal	mixed	superactive		thermic	aridic	shallow
BABERWIT	15.9	NO	2004	NV		28A	Aridisols	Argids	Natrargids	Typic Natrargids	clayey-skeletal	carbonatic			mesic	aridic	
BACHO	40.0	NO	1996	NV		30	Aridisols	Durids	Argidurids	Typic Argidurids	clayey-skeletal	smectitic			thermic	aridic	shallow
BADENA	72.5	NO	2004	NV		28A, 28B	Mollisols	Xerolls	Argixerolls	Aridic Argixerolls	loamy-skeletal	mixed	superactive		mesic	xeric	
BADGERCAMP	40.7	NO	1994	NV		23	Mollisols	Cryolls	Argicryolls	Xeric Argicryolls	loamy-skeletal	mixed	superactive		cryic	xeric	shallow

(continued)

(continued)

Soil series	Area (km^2)	Bench-mark	Year Est-ablished	TL State	Other States Using	MLRAs Using	Order	Suborder	Great Group	Subgroup	Particle-size Class	Mineralogy Class	CEC Activity Class	Reaction Class	Soil temp. Class	SMR	Other Family
BADHAP	64.2	NO	2004	NV		28A	Mollisols	Cryolls	Haplocryolls	Pachic Haplocryolls	loamy-skeletal	mixed	superactive		cryic		
BAGARD	0.1	NO	1971	UT	NV	28A, 47	Mollisols	Xerolls	Argixerolls	Aridic Argixerolls	clayey-skeletal	smectitic			frigid	xeric	
BAKERPEAK	52.5	NO	2009	NV		28A	Inceptisols	Cryepts	Calcicryepts	Xeric Calcicryepts	loamy-skeletal	carbonatic			cryic	xeric	
BAKSCRATCH	3.8	NO	2006	NV	CA	22A, 26	Mollisols	Cryolls	Argicryolls	Xeric Argicryolls	loamy-skeletal	mixed	superactive		cryic	xeric	shallow
BAMOS	0.3	NO	1997	UT	NV	28A	Mollisols	Xerolls	Argixerolls	Calciargidic Argixerolls	fine-loamy	mixed	superactive		mesic	xeric	
BANCY	23.3	NO	1986	ID	NV	25	Mollisols	Xerolls	Durixerolls	Typic Durixerolls	clayey	smectitic			frigid	xeric	shallow
BANGO	205.7	NO	1971	NV		26, 27, 28B	Aridisols	Argids	Natrargids	Haplic Natrargids	fine-loamy	mixed	superactive		mesic	aridic	
BARD	555.5	NO	1970	NV	AZ	30	Aridisols	Calcids	Petrocalcids	Calcic Petrocalcids	loamy	carbonatic			thermic	aridic	shallow
BARERANCH	1.5	NO	2006	CA	NV	23	Mollisols	Xerolls	Argixerolls	Vitritorrandic Argixerolls	ashy-skeletal	glassy			mesic	xeric	
BARFAN	34.8	NO	1990	NV		28B	Entisols	Orthents	Torriorthents	Lithic Xeric Torriorthents	ashy	glassy		calcareous	mesic	torric	
BARNARD	28.1	NO	1941	OR	CA, ID, NV	10, 21, 24, 25, 26, 27	Mollisols	Xerolls	Durixerolls	Argiduridic Durixerolls	fine	smectitic			mesic	xeric	
BARNMOT	47.0	NO	1985	NV		27, 29	Entisols	Orthents	Torriorthents	Typic Torriorthents	fine	smectitic		calcareous	mesic	torric	
BARRIER	66.2	NO	1983	NV		28B	Aridisols	Durids	Haplodurids	Xereptic Haplodurids	loamy	mixed	superactive		frigid	aridic	shallow
BARSHAAD	26.2	NO	1980	NV	CA	22A, 26	Mollisols	Xerolls	Palexerolls	Aridic Palexerolls	fine	smectitic			mesic	xeric	
BARTINE	74.4	NO	1971	NV		28B	Mollisols	Xerolls	Calcixerolls	Aridic Calcixerolls	loamy-skeletal	carbonatic			frigid	xeric	
BARTOME	265.3	NO	1986	NV	OR	25	Aridisols	Durids	Argidurids	Vitrixerandic Argidurids	loamy	mixed	superactive		mesic	aridic	shallow
BASELINE	76.8	NO	2006	NV		30	Aridisols	Calcids	Haplocalcids	Typic Haplocalcids	loamy-skeletal	carbonatic			hyperthermic	aridic	
BASINPEAK	3.3	NO	2004	NV		28A	Mollisols	Cryolls	Haplocryolls	Xeric Haplocryolls	loamy-skeletal	mixed	superactive		cryic	xeric	
BASKET	168.6	NO	1971	NV		28A, 29	Aridisols	Argids	Haplargids	Xeric Haplargids	loamy-skeletal	mixed	superactive		mesic	aridic	
BASTIAN	1.9	NO	1996	NV		29	Entisols	Orthents	Torriorthents	Oxyaquic Torriorthents	fine-loamy	mixed	superactive	calcareous	mesic	torric	
BATAN	427.2	NO	1974	NV		24, 25, 28B	Entisols	Orthents	Torriorthents	Duric Torriorthents	fine-silty	mixed	superactive	calcareous	mesic	torric	
BEANFLAT	73.6	NO	1983	NV		28B	Inceptiso ls	Aquepts	Halaquepts	Aeric Halaquepts	coarse-loamy	mixed	superactive	calcareous	frigid	aquic	
BEANO	18.7	NO	1984	NV		27, 29	Aridisols	Durids	Argidurids	Argidic Argidurids	loamy-skeletal	mixed	superactive		mesic	aridic	shallow
BEARBUTTE	41.8	NO	1993	NV		23	Mollisols	Xerolls	Argixerolls	Pachic Argixerolls	coarse-loamy	mixed	superactive		frigid	xeric	
BEARSKIN	0.1	NO	1971	UT	NV	25, 28A, 47	Mollisols	Xerolls	Argixerolls	Lithic Argixerolls	loamy	mixed	superactive		frigid	xeric	
BEDELL	22.3	NO	1980	NV		25, 26	Mollisols	Xerolls	Argixerolls	Aridic Argixerolls	coarse-loamy	mixed	superactive		mesic	xeric	
BEDWYR	34.1	NO	1986	NV		27	Aridisols	Argids	Natrargids	Typic Natrargids	clayey	smectitic			mesic	aridic	shallow
BEDZEE	11.7	NO	1987	NV		27	Aridisols	Argids	Haplargids	Xeric Haplargids	clayey	smectitic			mesic	aridic	shallow
BEELEM	359.7	NO	1984	NV		27, 29	Entisols	Orthents	Torriorthents	Lithic Xeric Torriorthents	loamy	mixed	superactive	calcareous	mesic	torric	
BEEOX	30.1	NO	1992	NV		24	Aridisols	Argids	Natrargids	Durinodic Natrargids	fine	smectitic			mesic	aridic	
BEERBO	12.8	NO	2006	NV		30	Mollisols	Ustolls	Argiustolls	Aridic Argiustolls	loamy-skeletal	mixed	superactive		mesic	ustic	shallow
BELATE	83.8	NO	1985	NV		24, 27	Mollisols	Xerolls	Argixerolls	Aridic Argixerolls	loamy-skeletal	mixed	superactive		frigid	xeric	
BELCHER	52.2	NO	1972	NV		29	Aridisols	Durids	Haplodurids	Cambidic Haplodurids	loamy	mixed	superactive		mesic	aridic	shallow
BELLEHELEN	877.3	YES	1984	NV		28A, 29	Mollisols	Xerolls	Argixerolls	Aridic Lithic Argixerolls	loamy-skeletal	mixed	superactive		mesic	xeric	
BELLENMINE	29.7	NO	2004	NV		28A, 28B, 29	Mollisols	Xerolls	Argixerolls	Lithic Argixerolls	loamy-skeletal	mixed	superactive		frigid	xeric	

(continued)

(continued)

Soil series	Area (km^2)	Bench-mark	Year Est-ablished	TL State	Other States Using	MLRAs Using	Order	Suborder	Great Group	Subgroup	Particle-size Class	Mineralogy Class	CEC Activity Class	Reaction Class	Soil temp. Class	SMR	Other Family
BELMILL	67.0	NO	1974	NV		28B	Mollisols	Xerolls	Argixerolls	Calciargidic Argixerolls	loamy-skeletal	mixed	superactive		mesic	xeric	
BELSAC	49.2	NO	1986	NV		25, 28B	Mollisols	Cryolls	Haplocryolls	Pachic Haplocryolls	loamy-skeletal	mixed	superactive		cryic		
BELTED	328.2	NO	1984	NV		27, 28B, 29	Aridisols	Durids	Argidurids	Argidic Argidurids	loamy	mixed	superactive		mesic	aridic	shallow
BENDASTIK	10.1	NO	2016	NV		25	Mollisols	Cryolls	Argicryolls	Aquic Argicryolls	fine-loamy	mixed	superactive		cryic		
BENIN	321.5	NO	1974	NV		24, 27, 28B	Entisols	Orthents	Torriorthents	Vertic Torriorthents	fine	smectitic		calcareous	mesic	torric	
BEOSKA	614.9	YES	1974	NV		24	Aridisols	Argids	Natrargids	Durinodic Natrargids	fine-loamy	mixed	superactive		mesic	aridic	
BEOWAWE	23.5	NO	1974	NV		24	Aridisols	Argids	Natrargids	Durinodic Natrargids	fine-loamy	mixed	superactive		mesic	aridic	
BERIT	67.7	NO	1981	NV		23, 26	Aridisols	Argids	Haplargids	Xeric Haplargids	loamy-skeletal	mixed	superactive		mesic	aridic	shallow
BERNING	38.0	NO	1974	NV		24, 25	Aridisols	Argids	Haplargids	Xeric Haplargids	clayey-skeletal	smectitic			mesic	aridic	
BERRYCREEK	39.1	NO	2009	NV		28B	Mollisols	Cryolls	Haplocryolls	Xeric Haplocryolls	loamy-skeletal	mixed	superactive		cryic	xeric	
BERZATIC	79.1	NO	1996	NV		29	Entisols	Orthents	Torriorthents	Lithic Torriorthents	loamy-skeletal	mixed	superactive	calcareous	mesic	torric	
BESHERM	94.5	NO	1985	NV		30	Aridisols	Calcids	Haplocalcids	Typic Haplocalcids	fine	carbonatic			thermic	aridic	
BETRA	50.2	NO	1986	NV		25, 28B	Mollisols	Xerolls	Durixerolls	Paleargidic Durixerolls	clayey-skeletal	smectitic			frigid	xeric	
BEZO	46.4	NO	1986	NV		27	Inceptisols	Aquepts	Halaquepts	Aeric Halaquepts	fine-silty	mixed	superactive	calcareous	mesic	aquic	
BICONDOA	23.2	NO	1971	CA	NV, OR	21, 23, 25	Mollisols	Aquolls	Endoaquolls	Fluvaquentic Vertic Endoaquolls	fine	smectitic		calcareous	frigid	aquic	
BIDART	0.8	NO	2007	CA	NV	22A	Inceptisols	Aquepts	Cryaquepts	Fluvaquentic Cryaquepts	coarse-loamy	mixed	superactive	acid	cryic	aquic	
BIDDLEMAN	291.7	NO	1971	NV		27	Aridisols	Argids	Natrargids	Typic Natrargids	fine-loamy over sandy or sandy-skeletal	mixed	superactive		mesic	aridic	
BIDRIM	22.4	NO	1995	NV	CA	23	Mollisols	Xerolls	Argixerolls	Aridic Lithic Argixerolls	clayey	smectitic			mesic	xeric	
BIEBER	45.9	NO	1920	CA	ID, NV	21, 23, 25, 26	Mollisols	Xerolls	Durixerolls	Argiduridic Durixerolls	clayey	smectitic			mesic	xeric	shallow
BIENFAIT	59.6	NO	2004	NV		28A	Aridisols	Cambids	Haplocambids	Sodic Haplocambids	sandy	mixed			mesic	aridic	
BIGA	255.3	NO	1988	NV		27	Aridisols	Argids	Natrargids	Durinodic Natrargids	clayey over loamy	smectitic OVER mixed	superactive		mesic	aridic	
BIGHAT	5.3	NO	1995	NV	CA	23	Aridisols	Argids	Natrargids	Typic Natrargids	fine-loamy over sandy or sandy-skeletal	mixed	superactive		mesic	aridic	
BIGMEADOW	10.9	NO	1985	NV		27	Mollisols	Aquolls	Endoaquolls	Fluvaquentic Vertic Endoaquolls	fine	smectitic		calcareous	mesic	aquic	
BIGSPRING	12.9	NO	2004	NV		28A	Mollisols	Xerolls	Calcixerolls	Aridic Calcixerolls	fine-loamy	mixed	superactive		mesic	xeric	
BIGWASH	38.5	NO	2009	NV		28A	Mollisols	Xerolls	Haploxerolls	Cumulic Haploxerolls	coarse-loamy	mixed	superactive		frigid	xeric	
BIJI	61.5	NO	2004	NV		28A	Aridisols	Calcids	Haplocalcids	Vertic Haplocalcids	fine	carbonatic			mesic	aridic	
BIJORJA	9.7	NO	1985	NV		27, 28B, 29	Aridisols	Cambids	Haplocambids	Xeric Haplocambids	coarse-loamy	mixed	superactive		mesic	aridic	
BIKEN	143.0	NO	1990	NV		28B	Aridisols	Calcids	Haplocalcids	Xeric Haplocalcids	loamy-skeletal	mixed	superactive		mesic	aridic	shallow
BILBO	256.3	NO	1986	NV		25	Aridisols	Argids	Haplargids	Xeric Haplargids	clayey-skeletal	smectitic			mesic	aridic	

(continued)

(continued)

Soil series	Area (km^2)	Bench-mark	Year Est-ablished	TL State	Other States Using	MLRAs Using	Order	Suborder	Great Group	Subgroup	Particle-size Class	Mineralogy Class	CEC Activity Class	Reaction Class	Soil temp. Class	SMR	Other Family
BIMMER	12.5	NO	1995	NV		27	Entisols	Orthents	Torriorthents	Typic Torriorthents	loamy	mixed	superactive	nonacid	mesic	torric	shallow
BIOYA	471.0	NO	1986	NV		24, 25	Aridisols	Durids	Haplodurids	Xeric Haplodurids	fine-loamy	mixed	superactive		mesic	aridic	
BIRCHCREEK	109.8	NO	1987	ID	NV	25, 28B	Mollisols	Xerolls	Argixerolls	Typic Argixerolls	clayey-skeletal	smectitic			frigid	xeric	
BIRDSPRING	218.9	NO	2006	NV		30	Entisols	Orthents	Torriorthents	Lithic Torriorthents	loamy-skeletal	carbonatic			thermic	torric	
BISHOP	4.3	NO	1924	CA	NV	20, 26, 29	Mollisols	Aquolls	Endoaquolls	Cumulic Endoaquolls	fine-loamy	mixed	superactive	calcareous	mesic	aquic	
BITNER	11.8	NO	1995	NV	CA	23	Mollisols	Xerolls	Haploxerolls	Vitritorrandic Haploxerolls	ashy	glassy			mesic	xeric	
BITTER SPRING	118.6	NO	1970	NV		30, 41	Aridisols	Argids	Calciargids	Typic Calciargids	sandy-skeletal	mixed			thermic	aridic	
BITTERRIDGE	8.7	NO	2006	NV		30	Aridisols	Calcids	Haplocalcids	Typic Haplocalcids	loamy-skeletal	carbonatic			thermic	aridic	shallow
BLACK BUTTE	6.5	NO	1970	NV		30	Aridisols	Salids	Haplosalids	Typic Haplosalids	fine-silty over sandy or sandy-skeletal	mixed	active		thermic	aridic	
BLACKA	16.6	NO	1974	NV		24	Aridisols	Durids	Haplodurids	Cambidic Haplodurids	coarse-loamy	mixed	superactive		mesic	aridic	
BLACKCAN	58.0	NO	2006	NV		29	Aridisols	Durids	Haplodurids	Xeric Haplodurids	loamy-skeletal	mixed	superactive		mesic	aridic	shallow
BLACKHAWK	245.6	NO	1972	NV		24, 27	Aridisols	Durids	Haplodurids	Cambidic Haplodurids	loamy	mixed	superactive		mesic	aridic	shallow
BLACKMESA	5.2	NO	2006	NV		30	Aridisols	Durids	Haplodurids	Typic Haplodurids	loamy	mixed	active		hyperthermic	aridic	shallow
BLACKNAT	1.0	NO		NV		30	Aridisols	Calcids	Petrocalcids	Typic Petrocalcids	sandy	mixed			thermic	aridic	shallow
BLACKTOP	865.3	NO	1980	CA	NV	29	Entisols	Orthents	Torriorthents	Lithic Torriorthents	loamy-skeletal	mixed	superactive	calcareous	mesic	torric	
BLACKWELL	3.9	NO	1973	ID	CO, MT, NV, WY	25, 26, 43B, 48A, 48B	Mollisols	Aquolls	Cryaquolls	Typic Cryaquolls	fine-loamy	mixed	superactive		cryic	aquic	
BLAPPERT	84.8	NO	1984	NV		29	Aridisols	Argids	Haplargids	Typic Haplargids	loamy-skeletal	mixed	superactive		mesic	aridic	shallow
BLIMO	321.9	YES	1990	NV		28B	Entisols	Orthents	Torriorthents	Duric Torriorthents	coarse-loamy	mixed	superactive	calcareous	mesic	torric	
BLISS	245.0	NO	1974	NV		24	Aridisols	Durids	Haplodurids	Xereptic Haplodurids	coarse-loamy	mixed	superactive		mesic	aridic	
BLITZEN	16.2	NO	1986	NV		25	Mollisols	Xerolls	Argixerolls	Aridic Argixerolls	clayey-skeletal	smectitic			frigid	xeric	
BLIZZARD	0.3	NO	1991	OR	NV	23	Mollisols	Cryolls	Argicryolls	Lithic Argicryolls	clayey	smectitic			cryic		
BLOOR	53.9	NO	1986	NV	CA, ID	24, 25, 28A	Aridisols	Argids	Natrargids	Durinodic Xeric Natrargids	fine-silty	mixed	superactive		mesic	aridic	
BLUDIAMOND	19.3	NO	2006	NV		30	Aridisols	Calcids	Petrocalcids	Argic Petrocalcids	loamy-skeletal	mixed	superactive		thermic	aridic	
BLUEAGLE	31.4	NO	1993	NV		29	Entisols	Orthents	Torriorthents	Typic Torriorthents	fine-silty	mixed	superactive	calcareous	mesic	torric	
BLUEGYP	3.3	NO	2006	NV		30	Aridisols	Gypsids	Haplogypsids	Leptic Haplogypsids	coarse-loamy	gypsic			hyperthermic	aridic	
BLUEHILL	56.5	NO	1985	ID	NV, UT	25	Inceptisols	Xerepts	Haploxerepts	Vitrandic Haploxerepts	ashy	glassy			mesic	xeric	
BLUEMASS	13.3	NO	2004	NV		28A	Mollisols	Xerolls	Durixerolls	Argidic Durixerolls	loamy	mixed	superactive		mesic	xeric	shallow
BLUEPOINT	198.4	NO	1964	NV	AZ, CA, NM, TX	30, 41, 42	Entisols	Psamments	Torripsamments	Typic Torripsamments		mixed			thermic	torric	
BLUEWING	953.8	YES	1963	NV	UT	27, 28A, 28B, 29	Entisols	Orthents	Torriorthents	Typic Torriorthents	sandy-skeletal	mixed			mesic	torric	
BOBNBOB	35.5	NO	1996	NV		30	Entisols	Fluvents	Torrifluvents	Aquic Torrifluvents	fine-silty	mixed	superactive	calcareous	thermic	torric	
BOBS	301.0	NO	1971	NV		25, 28B	Mollisols	Xerolls	Palexerolls	Petrocalcidic Palexerolls	loamy	carbonatic			frigid	xeric	shallow
BOBZBULZ	3.2	NO	2001	AZ	NV	30	Aridisols	Cambids	Haplocambids	Typic Haplocambids	loamy-skeletal	mixed	superactive		thermic	aridic	
BODIEHILL	2.7	NO	2010	CA	NV	26	Mollisols	Xerolls	Argixerolls	Vitrandic Argixerolls	ashy-skeletal over loamy-skeletal	glassy OVER mixed	superactive		frigid	xeric	
BOGER	61.8	NO	1992	NV	OR	23, 24	Aridisols	Durids	Haplodurids	Vitrixerandic Haplodurids	loamy-skeletal	mixed	superactive		mesic	aridic	shallow

(continued)

(continued)

Soil series	Area (km^2)	Bench-mark	Year Est-ablished	TL State	Other States Using	MLRAs Using	Order	Suborder	Great Group	Subgroup	Particle-size Class	Mineralogy Class	CEC Activity Class	Reaction Class	Soil temp. Class	SMR	Other Family
BOJO	155.8	NO	1986	NV		24	Aridisols	Argids	Haplargids	Lithic Haplargids	loamy	mixed	superactive		mesic	aridic	
BOLTZ	7.0	NO	1995	NV		23	Mollisols	Xerolls	Haploxerolls	Vitritorrandic Haploxerolls	ashy	glassy			frigid	xeric	
BOMBADIL	165.7	NO	1980	NV	CA	23, 26, 27	Aridisols	Argids	Haplargids	Lithic Xeric Haplargids	loamy	mixed	superactive		mesic	aridic	
BOOFORD	21.4	NO	1980	NV	ID	22A, 25	Mollisols	Xerolls	Argixerolls	Typic Argixerolls	fine	smectitic			frigid	xeric	
BOOFUSS	96.1	NO	1990	NV		28B	Inceptisols	Aquepts	Halaquepts	Typic Halaquepts	clayey over loamy	smectitic OVER mixed	superactive	calcareous	mesic	aquic	
BOOMSTICK	123.1	NO	1988	NV		27	Aridisols	Argids	Haplargids	Lithic Xeric Haplargids	loamy-skeletal	mixed	superactive		mesic	aridic	
BOOMTOWN	44.7	NO	1980	NV		22A	Alfisols	Xeralfs	Haploxeralfs	Ultic Haploxeralfs	fine	kaolinitic			frigid	xeric	
BOONDOCK	54.3	NO	2007	NV		26	Entisols	Orthents	Torriorthents	Xeric Torriorthents	loamy	mixed	superactive	nonacid	mesic	torric	shallow
BORDA	4.6	NO	1981	NV	CA	26	Aridisols	Argids	Paleargids	Vertic Paleargids	fine	smectitic			mesic	aridic	
BOREALIS	152.2	YES	1985	NV		26	Alfisols	Xeralfs	Durixeralfs	Abruptic Durixeralfs	ashy	glassy			frigid	xeric	
BORVANT	11.2	YES	1971	UT	NV	28A, 28B, 47	Mollisols	Xerolls	Palexerolls	Petrocalcic Palexerolls	loamy-skeletal	carbonatic			mesic	xeric	shallow
BOSCO	8.8	NO	1974	NV		25, 27	Mollisols	Xerolls	Haploxerolls	Torriorthentic Haploxerolls	loamy-skeletal	mixed	superactive		frigid	xeric	
BOSO	5.9	NO	1986	NV		25	Mollisols	Xerolls	Palexerolls	Petrocalcidic Palexerolls	loamy-skeletal	mixed	superactive		frigid	xeric	shallow
BOTLEG	4.2	NO	2006	NV		30	Aridisols	Argids	Haplargids	Typic Haplargids	loamy-skeletal	mixed	superactive		thermic	aridic	shallow
BOTON	1105.2	NO	1986	NV		24, 27	Entisols	Orthents	Torriorthents	Duric Torriorthents	fine-silty	mixed	superactive	calcareous	mesic	torric	
BOULDER LAKE	13.6	NO	1974	NV	CA, ID, OR	23, 25, 28B	Vertisols	Aquerts	Epiaquerts	Xeric Epiaquerts	fine	smectitic			frigid	aquic	
BOULFLAT	55.1	NO	1979	NV		24, 25	Aridisols	Durids	Argidurids	Haploxeralfic Argidurids	fine-loamy	mixed	superactive		mesic	aridic	
BOUNCER	8.0	NO	1985	NV		26	Aridisols	Argids	Haplargids	Xeric Haplargids	loamy-skeletal	mixed	superactive		mesic	aridic	shallow
BOXSPRING	817.4	NO	1992	NV		29, 30	Entisols	Orthents	Torriorthents	Lithic Ustic Torriorthents	loamy-skeletal	carbonatic			mesic	torric	
BRACKEN	36.2	NO	1923	NV		30	Aridisols	Gypsids	Haplogypsids	Leptic Haplogypsids	coarse-loamy	gypsic			thermic	aridic	
BRADSHAW	10.3	NO	1971	UT	ID, NV	26, 47	Mollisols	Xerolls	Haploxerolls	Typic Haploxerolls	loamy-skeletal	mixed	superactive		frigid	xeric	
BRAMWELL	0.1	NO	1949	ID	NV, UT	11, 28A, 28B	Aridisols	Calcids	Haplocalcids	Aquic Haplocalcids	fine-silty	mixed	superactive		mesic	aridic	
BRAWLEY	42.6	NO	1985	NV		26	Alfisols	Xeralfs	Palexeralfs	Vitrandic Palexeralfs	clayey-skeletal	smectitic			frigid	xeric	
BREGAR	287.6	NO	1974	NV	CA, ID	23, 24, 25	Aridisols	Argids	Haplargids	Lithic Xeric Haplargids	loamy-skeletal	mixed	superactive		frigid	aridic	
BREKO	148.3	NO	1984	NV		29, 30	Aridisols	Argids	Haplargids	Xeric Haplargids	loamy-skeletal	mixed	superactive		mesic	aridic	
BRICONE	90.1	NO	2009	NV		28A	Entisols	Orthents	Cryorthents	Lithic Cryorthents	loamy-skeletal	carbonatic			cryic		
BRIER	287.0	NO	1984	NV		29, 30	Mollisols	Xerolls	Argixerolls	Aridic Lithic Argixerolls	loamy-skeletal	mixed	superactive		mesic	xeric	
BRINKER	10.1	NO	1985	NV		27	Mollisols	Aquolls	Endoaquolls	Fluvaquentic Endoaquolls	fine	mixed	superactive	calcareous	mesic	aquic	
BRINNUM	20.2	NO	1983	NV		28B	Inceptisols	Aquepts	Halaquepts	Typic Halaquepts	fine-silty	mixed	superactive	calcareous	mesic	aquic	
BROCK	45.5	NO	1968	NV	OR	24	Aridisols	Durids	Argidurids	Xeric Argidurids	loamy-skeletal	mixed	superactive		mesic	aridic	shallow
BROCKLISS	4.2	NO	1975	NV	CA	26	Mollisols	Xerolls	Haploxerolls	Torriorthentic Haploxerolls	sandy-skeletal	mixed			mesic	xeric	
BROE	21.8	NO	1972	NV		29	Aridisols	Cambids	Haplocambids	Durinodic Haplocambids	sandy	mixed			mesic	aridic	

(continued)

(continued)

Soil series	Area (km^2)	Bench-mark	Year Est-ablished	TL State	Other States Using	MLRAs Using	Order	Suborder	Great Group	Subgroup	Particle-size Class	Mineralogy Class	CEC Activity Class	Reaction Class	Soil temp. Class	SMR	Other Family
BROKIT	15.2	NO	2009	NV		28A	Mollisols	Cryolls	Haplocryolls	Aquic Cumulic Haplocryolls	loamy-skeletal over sandy or sandy-skeletal	mixed	superactive		cryic	aquic	
BROLAND	205.5	NO	1990	NV		28B	Aridisols	Durids	Argidurids	Haploxeralfic Argidurids	loamy-skeletal	mixed	superactive		mesic	aridic	shallow
BROWNSBOWL	1.9	NO	2006	NV	CA	23	Mollisols	Cryolls	Haplocryolls	Vitrandic Haplocryolls	ashy	glassy			cryic		
BROYLES	742.7	YES	1972	NV		24, 28B	Aridisols	Cambids	Haplocambids	Durinodic Haplocambids	ashy over loamy	glassy OVER mixed	superactive		mesic	aridic	
BRUBECK	22.9	NO	1990	CA	NV	23	Vertisols	Xererts	Haploxererts	Aridic Haploxererts	fine	smectitic			mesic	xeric	
BRUFFY	13.2	NO	1971	NV		25, 28B	Entisols	Fluvents	Torrifluvents	Xeric Torrifluvents	fine-loamy	mixed	superactive	calcareous	frigid	torric	
BUBUS	517.0	YES	1974	NV		24, 28B	Entisols	Orthents	Torriorthents	Duric Torriorthents	coarse-loamy	mixed	superactive	calcareous	mesic	torric	
BUCAN	277.4	NO	1979	NV		24, 25, 28B	Aridisols	Argids	Haplargids	Xeric Haplargids	fine	smectitic			frigid	aridic	
BUCKAROO	180.3	NO	1985	NV		27	Aridisols	Argids	Natrargids	Typic Natrargids	fine	smectitic			mesic	aridic	
BUCKLAKE	354.5	NO	1986	CA	NV, OR	10, 21, 23	Mollisols	Xerolls	Argixerolls	Aridic Argixerolls	fine	smectitic			mesic	xeric	
BUCKSPRING	130.5	NO	2006	NV		30	Mollisols	Ustolls	Argiustolls	Aridic Lithic Argiustolls	loamy-skeletal	mixed	superactive		mesic	ustic	
BUDIHOL	99.3	NO	1985	NV		27, 29	Entisols	Orthents	Torriorthents	Xeric Torriorthents	loamy	mixed	superactive	nonacid	mesic	torric	shallow
BUFFARAN	388.0	NO	1985	NV	CA, OR	23, 24, 25, 27, 28B	Aridisols	Durids	Argidurids	Xeric Argidurids	clayey	smectitic			mesic	aridic	shallow
BULAKE	335.0	NO	1984	NV	ID	25	Alfisols	Xeralfs	Haploxeralfs	Lithic Mollic Haploxeralfs	clayey	smectitic			frigid	xeric	
BULLFOR	23.7	NO	1996	NV		30	Aridisols	Durids	Haplodurids	Typic Haplodurids	sandy	mixed			thermic	aridic	
BULLUMP	275.6	NO	1986	NV	CA, OR, UT	21, 23, 25, 28A	Mollisols	Xerolls	Argixerolls	Pachic Argixerolls	loamy-skeletal	mixed	superactive		frigid	xeric	
BULLVARO	16.1	NO	1986	NV		25	Mollisols	Xerolls	Argixerolls	Pachic Argixerolls	loamy-skeletal	mixed	superactive		frigid	xeric	
BULLVILLE	10.6	YES	2006	NV	CA	22A, 26	Mollisols	Xerolls	Argixerolls	Ultic Argixerolls	loamy-skeletal	mixed	superactive		frigid	xeric	
BUNDORF	50.8	NO	1978	NV		27	Aridisols	Durids	Argidurids	Typic Argidurids	clayey	smectitic			mesic	aridic	shallow
BUNEJUG	33.6	NO	1971	NV		24, 27	Mollisols	Aquolls	Endoaquolls	Fluvaquentic Endoaquolls	coarse-loamy	mixed	superactive		mesic	aquic	
BUNKY	30.8	NO	1975	NV		24, 25	Aridisols	Durids	Haplodurids	Xereptic Haplodurids	fine-loamy	mixed	superactive		mesic	aridic	
BURNBOROUGH	74.8	NO	1980	NV		23, 24, 26, 27	Mollisols	Xerolls	Argixerolls	Aridic Argixerolls	loamy-skeletal	mixed	superactive		frigid	xeric	
BURRITA	540.5	NO	1985	NV		23, 24, 25	Aridisols	Argids	Haplargids	Lithic Xeric Haplargids	clayey-skeletal	smectitic			mesic	aridic	
BUZZTAIL	37.3	NO	2004	NV		28A	Mollisols	Xerolls	Haploxerolls	Aridic Lithic Haploxerolls	loamy-skeletal	carbonatic			frigid	xeric	
BYLO	76.0	NO	1974	NV		24, 27, 28B	Aridisols	Cambids	Haplocambids	Typic Haplocambids	fine-silty	mixed	superactive		mesic	aridic	
CABINPINE	6.2	NO	2014	NV		29	Mollisols	Ustolls	Haplustolls	Fluventic Haplustolls	sandy-skeletal	mixed			mesic	ustic	
CAFETAL	23.4	NO	2006	NV		30	Aridisols	Argids	Calciargids	Durinodic Calciargids	loamy-skeletal	mixed	superactive		thermic	aridic	
CAGAS	17.7	NO	2006	NV		28A	Mollisols	Xerolls	Argixerolls	Vitritorrandic Argixerolls	ashy-skeletal	glassy			frigid	xeric	
C AGLE	45.9	NO	1975	NV	CA	10, 23, 26, 27	Mollisols	Xerolls	Argixerolls	Aridic Argixerolls	fine	smectitic			mesic	xeric	
CAGWIN	4.1	YES	1970	CA	NV	22A	Entisols	Psamments	Xeropsamments	Dystric Xeropsamments		mixed			frigid	xeric	

(continued)

(continued)

Soil series	Area (km²)	Bench-mark	Year Est-ablished	TL State	Other States Using	MLRAs Using	Order	Suborder	Great Group	Subgroup	Particle-size Class	Mineralogy Class	CEC Activity Class	Reaction Class	Soil temp. Class	SMR	Other Family
CALICO	13.7	NO	1970	NV		30	Entisols	Fluvents	Xerofluvents	Aquic Xerofluvents	coarse-loamy over clayey	mixed OVER smectitic	superactive	calcareous	thermic	xeric	
CALIZA	19.5		1970	NM	NV	42	Aridisols	Calcids	Haplocalcids	Typic Haplocalcids	sandy-skeletal	mixed			thermic	aridic	
CALLAT	0.1		xx	CA		22A	Inceptisols	Cryepts	Humicryepts	Xeric Humicryepts	loamy-skeletal	isotic			cryic	xeric	
CALLVILLE	50.6	NO	2006	NV		30	Aridisols	Gypsids	Haplogypsids	Leptic Haplogypsids	coarse-loamy	mixed	active		hyperthermic	aridic	
CALPEAK	36.6	NO	1985	NV		27	Entisols	Orthents	Torriorthents	Xeric Torriorthents	loamy-skeletal	mixed	superactive	calcareous	mesic	torric	shallow
CALPINE	8.6	YES	1973	CA	NV	19, 20, 21, 22A, 26	Mollisols	Xerolls	Haploxerolls	Aridic Haploxerolls	coarse-loamy	mixed	superactive		mesic	xeric	
CALWASH	7.4	NO	2006	NV		30	Entisols	Orthents	Torriorthents	Typic Torriorthents	loamy	mixed	superactive	calcareous	thermic	torric	shallow
CAMEEK	55.3	NO	1986	NV		25	Mollisols	Xerolls	Durixerolls	Argiduridic Durixerolls	clayey	smectitic			frigid	xeric	shallow
CANDELARIA	582.3	NO	1984	NV		27, 28B, 29	Aridisols	Calcids	Haplocalcids	Durinodic Haplocalcids	sandy-skeletal	mixed			mesic	aridic	
CANFIRE	0.0	NO	2006	CA	NV	22A, 26	Mollisols	Xerolls	Argixerolls	Lithic Argixerolls	loamy-skeletal	mixed	superactive		mesic	xeric	
CANIWE	7.8	NO	1985	NV		28B	Mollisols	Xerolls	Haploxerolls	Duridic Haploxerolls	fine-silty	mixed	superactive		mesic	xeric	
CANOTO	177.3	NO	2000	NV		30	Entisols	Orthents	Torriorthents	Typic Torriorthents	loamy-skeletal	mixed	superactive	calcareous	thermic	torric	
CANUTIO	28.3	NO	1970	TX	NM, NV	30, 42	Entisols	Orthents	Torriorthents	Typic Torriorthents	loamy-skeletal	mixed	superactive	calcareous	thermic	torric	
CANYONFORK	13.3	NO	2009	NV		28A	Mollisols	Xerolls	Calcixerolls	Typic Calcixerolls	loamy-skeletal	carbonatic			frigid	xeric	
CANYOUNG	43.3	NO	2009	NV		28A	Mollisols	Cryolls	Calcicryolls	Xeric Calcicryolls	loamy-skeletal	carbonatic			cryic	xeric	
CAPHOR	42.7	NO	1985	NV		28B	Aridisols	Calcids	Haplocalcids	Durinodic Haplocalcids	coarse-loamy	mixed	superactive		mesic	aridic	
CAPSUS	62.6	NO	1992	NV		29	Mollisols	Ustolls	Argiustolls	Aridic Lithic Argiustolls	clayey	smectitic			mesic	ustic	
CARCITY	9.6	NO	1971	NV		27	Mollisols	Aquolls	Endoaquolls	Cumulic Vertic Endoaquolls	clayey over sandy or sandy-skeletal	smectitic OVER mixed			mesic	aquic	
CARIOCA	12.9	NO	1980	NV		22A	Alfisols	Cryalfs	Haplocryalfs	Andic Haplocryalfs	loamy-skeletal	isotic			cryic	udic	
CARRIZO	174.8	NO	1918	CA	AZ, NV	30, 31, 40	Entisols	Orthents	Torriorthents	Typic Torriorthents	sandy-skeletal	mixed			hyperthermic	torric	
CARRWASH	36.2	NO	2005	NV	AZ	30	Entisols	Orthents	Torriorthents	Typic Torriorthents	sandy-skeletal	mixed			hyperthermic	torric	
CARSON	308.0	YES	1909	NV		27	Vertisols	Xererts	Haploxererts	Halic Haploxererts	very-fine	smectitic			mesic	xeric	
CARSTUMP	163.3	NO	1968	NV		25, 47	Mollisols	Xerolls	Argixerolls	Calciargidic Argixerolls	clayey-skeletal	smectitic			frigid	xeric	
CARWALKER	15.7	NO	2005	NV		26, 27	Entisols	Fluvents	Torrifluvents	Oxyaquic Torrifluvents	sandy	mixed			mesic	torric	
CASAGA	79.6	NO	1982	NV		30	Aridisols	Argids	Natrargids	Typic Natrargids	fine-loamy	mixed	superactive		thermic	aridic	
CASLO	2.9	NO	1996	NV		30	Entisols	Aquents	Fluvaquents	Typic Fluvaquents	fine-loamy	carbonatic			thermic	aquic	
CASSENAI	32.4	NO	2007	NV	CA	22A	Entisols	Psamments	Xeropsamments	Dystric Xeropsamments		mixed			frigid	xeric	
CASSIRO	67.8	NO	1980	NV	CA	26, 28B	Mollisols	Xerolls	Argixerolls	Aridic Argixerolls	clayey-skeletal	smectitic			mesic	xeric	
CATH	386.5	NO	1971	NV		28B, 29	Aridisols	Argids	Calciargids	Durinodic Xeric Calciargids	fine-loamy	mixed	superactive		mesic	aridic	
CAUDLE	13.4	NO	1972	NV	CA	23, 26, 28B	Aridisols	Argids	Haplargids	Durinodic Haplargids	fine-loamy	mixed	superactive		mesic	aridic	
CAVANAUGH	16.7	NO	1984	NV	ID	25	Mollisols	Xerolls	Argixerolls	Ultic Argixerolls	clayey-skeletal	smectitic			frigid	xeric	
CAVE	341.3		1936	AZ	NM		Aridisols	Calcids	Petrocalcids	Typic Petrocalcids	loamy	mixed	superactive		thermic	aridic	
CAVEHILL	1065.9	YES	1983	NV		25, 28B	Mollisols	Xerolls	Calcixerolls	Typic Calcixerolls	loamy-skeletal	carbonatic			frigid	xeric	

(continued)

(continued)

Soil series	Area (km^2)	Bench-mark	Year Est-ablished	TL State	Other States Using	MLRAs Using	Order	Suborder	Great Group	Subgroup	Particle-size Class	Mineralogy Class	CEC Activity Class	Reaction Class	Soil temp. Class	SMR	Other Family
CAVEMOUNTAIN	9.8	NO	2013	NV		28B	Mollisols	Cryolls	Argicryolls	Calcic Argicryolls	loamy-skeletal	mixed	superactive		cryic	xeric	
CAVEROCK	1.5	NO	2007	NV		22A	Inceptisols	Xerepts	Dystroxerepts	Humic Dystroxerepts	coarse-loamy	isotic			frigid	xeric	
CAVIN	6.5	NO	2006	NV	CA	23	Mollisols	Xerolls	Haploxerolls	Vitritorrandic Haploxerolls	ashy-skeletal	glassy			frigid	xeric	
CEDARAN	126.6	NO	1971	NV		28B, 29	Mollisols	Ustolls	Argiustolls	Aridic Lithic Argiustolls	clayey-skeletal	smectitic			mesic	ustic	
CEDARCABIN	21.2	NO	2009	NV		28A	Mollisols	Xerolls	Calcixerolls	Typic Calcixerolls	loamy-skeletal	carbonatic			frigid	xeric	
CEEBEE	5.4	NO	2009	NV		28A	Alfisols	Cryalfs	Haplocryalfs	Lamellic Haplocryalfs	sandy-skeletal	mixed			cryic		
CEEJAY	315.5	NO	1990	NV	CA	23, 27	Aridisols	Argids	Haplargids	Lithic Xeric Haplargids	clayey	smectitic			mesic	aridic	
CELETON	96.1	NO	1971	NV		27	Entisols	Orthents	Torriorthents	Typic Torriorthents	loamy	mixed	superactive	calcareous	mesic	torric	shallow
CETREPAS	23.5	NO	2006	NV		29	Aridisols	Argids	Haplargids	Ustic Haplargids	loamy-skeletal	mixed	superactive		mesic	aridic	shallow
CEWAT	11.3	NO	1990	NV	CA	23, 27	Aridisols	Cambids	Haplocambids	Xeric Haplocambids	loamy-skeletal	mixed	superactive		mesic	aridic	
CHAD	79.0	NO	1983	NV		28B	Mollisols	Xerolls	Argixerolls	Aridic Argixerolls	fine	mixed	superactive		frigid	xeric	
CHAINLINK	83.9	NO	2004	NV		28A	Mollisols	Xerolls	Durixerolls	Cambidic Durixerolls	loamy	mixed	superactive		mesic	xeric	shallow
CHALCO	115.4	NO	1980	NV	CA	23, 26	Aridisols	Argids	Haplargids	Xeric Haplargids	clayey	smectitic			mesic	aridic	shallow
CHANYBUCK	7.5	NO	1992	NV		29	Mollisols	Xerolls	Haploxerolls	Lithic Haploxerolls	ashy-skeletal	glassy			frigid	xeric	
CHAPPUIS	7.7	NO	1990	CA	NV	23	Aridisols	Argids	Natrargids	Xeric Natrargids	fine	smectitic			mesic	aridic	
CHARKILN	5.9	NO	2006	NV		30	Mollisols	Ustolls	Argiustolls	Aridic Argiustolls	fine-loamy	mixed	superactive		mesic	ustic	
CHARLEBOIS	14.4	NO	1981	NV		26	Mollisols	Xerolls	Argixerolls	Argiduridic Argixerolls	fine-loamy	mixed	superactive		mesic	xeric	
CHARNOCK	17.8	NO	1972	NV		28B, 29	Inceptisols	Aquepts	Halaquepts	Duric Halaquepts	fine-loamy	mixed	superactive	calcareous	mesic	aquic	
CHARPEAK	202.9	NO	2006	NV		30	Inceptisols	Cryepts	Calcicryepts	Typic Calcicryepts	loamy-skeletal	mixed	superactive		cryic		
CHARWELL	8.0	NO	1993	NV		25	Mollisols	Xerolls	Durixerolls	Abruptic Argiduridic Durixerolls	very-fine	smectitic			frigid	xeric	
CHAYSON	49.5	NO	1986	ID	NV	25	Mollisols	Xerolls	Durixerolls	Typic Durixerolls	fine-loamy	mixed	superactive		frigid	xeric	
CHECKETT	130.3	NO	1970	UT	NV	28A	Aridisols	Argids	Haplargids	Lithic Xeric Haplargids	loamy-skeletal	mixed	superactive		mesic	aridic	
CHEDEHAP	17.0	NO	1973	ID	NV, WA	11, 24, 28B, 7	Aridisols	Cambids	Haplocambids	Xeric Haplocambids	coarse-loamy	mixed	superactive		mesic	aridic	
CHEME	95.2	NO	2006	NV		30	Aridisols	Durids	Haplodurids	Typic Haplodurids	loamy-skeletal	mixed	superactive		hyperthermic	aridic	shallow
CHEN	884.3	YES	1979	NV	ID, OR	23, 24, 25, 26, 28B	Mollisols	Xerolls	Argixerolls	Aridic Lithic Argixerolls	clayey-skeletal	smectitic			frigid	xeric	
CHERRY SPRING	310.5	NO	1977	NV		24, 25	Aridisols	Durids	Argidurids	Haploxeralfic Argidurids	fine-loamy	mixed	superactive		mesic	aridic	
CHIARA	2197.0	YES	1974	NV	ID	11, 24, 25, 28B	Aridisols	Durids	Haplodurids	Vitrixerandic Haplodurids	loamy	mixed	superactive		mesic	aridic	shallow
CHIEFPAN	14.6	NO	2010	NV		29	Aridisols	Durids	Argidurids	Xeric Argidurids	clayey-skeletal	smectitic			mesic	aridic	
CHIEFRANGE	12.4	NO	2010	NV		29	Aridisols	Argids	Haplargids	Xeric Haplargids	loamy-skeletal	mixed	superactive		mesic	aridic	
CHILL	98.9	NO	1981	NV		23, 26, 27	Aridisols	Argids	Haplargids	Xeric Haplargids	loamy	mixed	superactive		mesic	aridic	shallow

(continued)

(continued)

Soil series	Area (km^2)	Bench-mark	Year Est-ablished	TL State	Other States Using	MLRAs Using	Order	Suborder	Great Group	Subgroup	Particle-size Class	Mineralogy Class	CEC Activity Class	Reaction Class	Soil temp. Class	SMR	Other Family
CHILPER	143.6	NO	1986	NV		27	Aridisols	Argids	Natrargids	Durinodic Natrargids	clayey over loamy-skeletal	smectitic OVER mixed	superactive		mesic	aridic	
CHIME	26.1	NO	1986	NV	CA	23, 25	Aridisols	Argids	Haplargids	Durinodic Xeric Haplargids	fine-loamy	mixed	superactive		mesic	aridic	
CHINKLE	31.9	NO	1992	NV		30	Entisols	Orthents	Torriorthents	Typic Torriorthents	loamy	mixed	superactive	calcareous	thermic	torric	shallow
CHRISTOPHER	0.9	NO	2007	CA	NV	22A	Entisols	Psamments	Xeropsamments	Dystric Xeropsamments		mixed			frigid	xeric	
CHUBARD	609.3	NO	2006	NV		29	Aridisols	Argids	Haplargids	Lithic Xeric Haplargids	loamy-skeletal	mixed	superactive		mesic	aridic	
CHUCKLES	54.8	NO	1990	NV	CA	23, 27	Aridisols	Cambids	Haplocambids	Sodic Haplocambids	fine-silty	mixed	superactive		mesic	aridic	
CHUCKMILL	197.3	NO	2006	NV		28A	Aridisols	Durids	Argidurids	Vitrixerandic Argidurids	loamy	mixed	superactive		mesic	aridic	shallow
CHUCKRIDGE	205.7	NO	1985	NV		27, 28A, 28B, 29	Aridisols	Durids	Argidurids	Xeric Argidurids	loamy	mixed	superactive		mesic	aridic	shallow
CHUFFA	193.2	NO	2004	NV		28A	Aridisols	Cambids	Haplocambids	Xeric Haplocambids	fine-silty	mixed	superactive		mesic	aridic	
CHUG	101.4	NO	2015	NV		25	Mollisols	Xerolls	Haploxerolls	Vitrandic Haploxerolls	fine-loamy	mixed	superactive		frigid	xeric	
CHURCHILL	25.4	NO	1909	NV		27	Aridisols	Argids	Natrargids	Vertic Natrargids	fine	smectitic			mesic	aridic	
CHUSKA	85.0	NO	1986	ID	NV, UT	11, 25, 28A	Aridisols	Durids	Argidurids	Xeric Argidurids	loamy	mixed	superactive		mesic	aridic	shallow
CIRAC	512.9	YES	1984	NV		27, 29	Entisols	Fluvents	Torrifluvents	Typic Torrifluvents	coarse-loamy	mixed	superactive	calcareous	mesic	torric	
CLANALPINE	243.0	NO	1985	NV		27, 28B	Mollisols	Xerolls	Argixerolls	Typic Argixerolls	loamy-skeletal	mixed	superactive		frigid	xeric	
CLEAVAGE	1679.0	YES	1983	NV	ID, OR	23, 24, 25, 27, 28B	Mollisols	Xerolls	Argixerolls	Aridic Lithic Argixerolls	loamy-skeletal	mixed	superactive		frigid	xeric	
CLEAVER	416.6	YES	1941	NV		26, 27	Aridisols	Durids	Argidurids	Typic Argidurids	loamy	mixed	superactive		mesic	aridic	shallow
CLEAVMOR	22.9	NO	1986	NV		25	Mollisols	Xerolls	Argixerolls	Aridic Lithic Argixerolls	loamy-skeletal	mixed	superactive		frigid	xeric	
CLEMENTINE	149.2	NO	1986	NV		23, 24, 25	Mollisols	Aquolls	Endoaquolls	Cumulic Endoaquolls	fine-silty	mixed	superactive		mesic	aquic	
CLIFFDOWN	177.0	NO	1971	NV	UT	28A, 28B, 29	Entisols	Orthents	Torriorthents	Typic Torriorthents	loamy-skeletal	mixed	superactive	calcareous	mesic	torric	
CLIMINE	7.0	NO	1992	NV		24	Mollisols	Xerolls	Haploxerolls	Pachic Haploxerolls	loamy-skeletal	mixed	superactive		frigid	xeric	
CLOSKEY	8.8	NO	2004	NV		28A	Mollisols	Xerolls	Argixerolls	Aridic Argixerolls	loamy-skeletal	mixed	superactive		frigid	xeric	
CLOWFIN	88.2	NO	1983	NV		28B, 29	Entisols	Orthents	Torriorthents	Typic Torriorthents	loamy-skeletal	mixed	superactive	calcareous	mesic	torric	
CLURDE	123.0	NO	1979	NV	OR	23, 24, 25	Aridisols	Cambids	Haplocambids	Durinodic Xeric Haplocambids	fine-loamy	mixed	superactive		mesic	aridic	
CLURO	39.3	NO	1979	NV		24	Aridisols	Cambids	Haplocambids	Durinodic Xeric Haplocambids	fine-loamy	mixed	superactive		mesic	aridic	
COBATUS	20.2	NO	1996	NV		30	Inceptisols	Aquepts	Halaquepts	Aeric Halaquepts	fine-loamy	mixed	superactive	calcareous	thermic	aquic	
COBBLYWHEEL	17.8	NO	2009	NV		28A, 28B	Mollisols	Cryolls	Haplocryolls	Xeric Haplocryolls	loamy-skeletal	mixed	active		cryic	xeric	
COBRE	145.3	YES	1986	NV		25, 28B	Aridisols	Cambids	Haplocambids	Durinodic Xeric Haplocambids	ashy	glassy			mesic	aridic	
COFF	11.0	NO	1968	NV		24	Aridisols	Calcids	Petrocalcids	Calcic Petrocalcids	loamy-skeletal	carbonatic			frigid	aridic	
COFFEPOT	19.3	NO		NV		25	Mollisols	Cryolls	Haplocryolls	Pachic Haplocryolls	coarse-loamy	mixed	superactive		cryic		
COILS	96.3	NO	1983	NV		24, 28B	Aridisols	Durids	Argidurids	Haploxeralfic Argidurids	fine	smectitic			frigid	aridic	

(continued)

(continued)

Soil series	Area (km^2)	Bench-mark	Year Est-ablished	TL State	Other States Using	MLRAs Using	Order	Suborder	Great Group	Subgroup	Particle-size Class	Mineralogy Class	CEC Activity Class	Reaction Class	Soil temp. Class	SMR	Other Family
COIT	8.1	NO	1980	NV		24	Mollisols	Aquolls	Endoaquolls	Cumulic Endoaquolls	fine-loamy	mixed	superactive		mesic	aquic	
COLADO	1.7	NO	2009	NV		27	Entisols	Orthents	Torriorthents	Oxyaquic Torriorthents	sandy over clayey	mixed OVER smectitic		calcareous	mesic	torric	
COLBAR	265.5	NO	1985	NV		24, 27, 28B	Aridisols	Argids	Haplargids	Xeric Haplargids	fine-loamy	mixed	superactive		mesic	aridic	
COLDENT	49.9	NO	1988	NV		27	Entisols	Orthents	Torriorthents	Duric Torriorthents	coarse-loamy	mixed	superactive	calcareous	mesic	torric	
COLOLAG	6.2	NO	2006	NV		30	Aridisols	Argids	Calciargids	Typic Calciargids	loamy-skeletal	mixed	superactive		hyperthermic	aridic	
COLOROCK	66.6	NO	1970	NV		30	Aridisols	Calcids	Petrocalcids	Argic Petrocalcids	loamy-skeletal	mixed	superactive		thermic	aridic	shallow
COLTROOP	66.0	NO	1986	NV		25	Aridisols	Durids	Haplodurids	Xeric Haplodurids	loamy	mixed	superactive		mesic	aridic	shallow
COLVAL	114.8	NO	2006	NV		29	Aridisols	Argids	Calciargids	Durinodic Calciargids	fine-silty	mixed	superactive		mesic	aridic	
COMMSKI	390.9	NO	1996	NV		30	Aridisols	Calcids	Haplocalcids	Typic Haplocalcids	loamy-skeletal	carbonatic			thermic	aridic	
CONNEL	243.4	NO	1986	NV		24, 25	Aridisols	Cambids	Haplocambids	Durinodic Xeric Haplocambids	coarse-loamy over sandy or sandy-skeletal	mixed	superactive		mesic	aridic	
CONTACT	7.8	NO	1986	NV		25	Mollisols	Xerolls	Haploxerolls	Torripsammentic Haploxerolls		mixed			frigid	xeric	
COOPERWASH	1.1	NO	2013	NV		28B	Mollisols	Cryolls	Argicryolls	Vertic Argicryolls	fine	smectitic			cryic		
COPPEREID	112.4	NO	1990	NV		27	Entisols	Orthents	Torriorthents	Xeric Torriorthents	loamy	mixed	superactive	calcareous	mesic	torric	shallow
COPPERSMITH	1.7	NO	2006	CA	NV	23	Mollisols	Xerolls	Argixerolls	Vitritorrandic Argixerolls	ashy	glassy			mesic	xeric	
CORBETT	18.4	NO	1975	NV	CA	22A, 5	Entisols	Psamments	Xeropsamments	Typic Xeropsamments		mixed			frigid	xeric	
CORBILT	116.1	NO	1996	NV		30	Aridisols	Calcids	Haplocalcids	Duric Haplocalcids	coarse-loamy	mixed	superactive		thermic	aridic	
CORMOL	20.2	NO	2006	CA	NV	23	Mollisols	Xerolls	Argixerolls	Vitritorrandic Argixerolls	ashy	glassy			mesic	xeric	shallow
CORNCREEK	27.2	NO	2006	NV		30	Aridisols	Calcids	Haplocalcids	Sodic Haplocalcids	loamy-skeletal	carbonatic			thermic	aridic	
CORNFLAT	3.4	NO	2011	NV		30	Aridisols	Argids	Calciargids	Lithic Calciargids	loamy	carbonatic			thermic	aridic	
CORRAL	40.9	NO	1990	CA	NV, OR	21, 23, 26, 27	Aridisols	Argids	Haplargids	Xeric Haplargids	loamy	mixed	superactive		mesic	aridic	shallow
CORTEZ	108.0	NO	1979	NV		24, 25	Aridisols	Durids	Natridurids	Vitrixerandic Natridurids	fine	smectitic			mesic	aridic	
COSER	130.0	NO	1986	NV	ID	25	Mollisols	Xerolls	Palexerolls	Typic Palexerolls	fine	smectitic			frigid	xeric	
COTANT	370.4	NO	1986	NV	OR	10, 23, 24, 25, 28B	Mollisols	Xerolls	Argixerolls	Aridic Argixerolls	clayey	smectitic			frigid	xeric	shallow
COUCH	40.5	NO	1971	CA	NV	23	Aridisols	Argids	Natrargids	Vitrixerandic Natrargids	fine	smectitic			mesic	aridic	
COUTIS	11.6		1976	WY	CO		Mollisols	Cryolls	Haplocryolls	Pachic Haplocryolls	coarse-loamy	mixed	superactive		cryic		
COWBELL	0.2	NO	2006	CA	NV	23	Mollisols	Cryolls	Argicryolls	Vitrandic Argicryolls	ashy-skeletal	glassy			cryic	xeric	
COWGIL	73.8	NO	1986	NV	ID	11, 25, 28B	Aridisols	Argids	Haplargids	Xeric Haplargids	loamy-skeletal	mixed	superactive		mesic	aridic	
COZTUR	27.8	NO	1985	NV	OR	23, 24, 25	Aridisols	Argids	Haplargids	Lithic Xeric Haplargids	loamy	mixed	superactive		frigid	aridic	
CRADLEBAUGH	7.0	NO	1940	NV		26	Mollisols	Aquolls	Endoaquolls	Duric Endoaquolls	fine-loamy	mixed	superactive	calcareous	mesic	aquic	
CREDO	35.6	NO	1971	NV		25, 26, 28B	Aridisols	Argids	Haplargids	Xeric Haplargids	fine-loamy	mixed	superactive		frigid	aridic	
CREEMON	290.9	NO	1974	NV		24, 25	Aridisols	Cambids	Haplocambids	Durinodic Haplocambids	coarse-silty	mixed	superactive		mesic	aridic	
CREN	47.7	NO	1974	NV		24, 28B	Entisols	Orthents	Torriorthents	Duric Torriorthents	coarse-silty	mixed	superactive	calcareous	mesic	torric	

(continued)

(continued)

Soil series	Area (km^2)	Bench-mark	Year Est-ablished	TL State	Other States Using	MLRAs Using	Order	Suborder	Great Group	Subgroup	Particle-size Class	Mineralogy Class	CEC Activity Class	Reaction Class	Soil temp. Class	SMR	Other Family
CRESAL	67.2	NO	1988	NV		24, 27	Entisols	Orthents	Torriorthents	Duric Torriorthents	coarse-silty	mixed	superactive	calcareous	mesic	torric	
CRESTLINE	59.1	NO	1942	UT	NV	28A, 47	Aridisols	Calcids	Haplocalcids	Xeric Haplocalcids	coarse-loamy	mixed	superactive		mesic	aridic	
CRETHERS	4.4	NO	2013	NV		28B	Mollisols	Cryolls	Argicryolls	Vitrandic Argicryolls	clayey-skeletal	smectitic			cryic		
CREVA	42.4	NO	1968	NV		25	Aridisols	Argids	Haplargids	Lithic Ruptic-Entic Haplargids	clayey-skeletal	smectitic			frigid	aridic	
CRISPY	8.9	NO	2006	CA	NV	22A, 26	Mollisols	Xerolls	Argixerolls	Typic Argixerolls	loamy-skeletal	mixed	superactive		frigid	xeric	shallow
CROCAN	25.1	NO	1995	NV	CA	23	Mollisols	Xerolls	Argixerolls	Aridic Lithic Argixerolls	clayey	smectitic			frigid	xeric	
CROESUS	41.5	NO	1971	NV		23, 25, 28B	Mollisols	Cryolls	Haplocryolls	Pachic Haplocryolls	loamy-skeletal	mixed	superactive		cryic		
CROOKED CREEK	312.2	NO	1971	UT	CO, ID, NV	12, 25, 28A, 47, 48A	Mollisols	Aquolls	Endoaquolls	Cumulic Endoaquolls	fine	smectitic			frigid	aquic	
CROPPER	458.5	YES	1990	NV		28B, 29	Mollisols	Xerolls	Argixerolls	Aridic Lithic Argixerolls	loamy-skeletal	mixed	superactive		frigid	xeric	
CROSGRAIN	176.4	NO	2000	NV	CA	30	Aridisols	Durids	Haplodurids	Typic Haplodurids	loamy-skeletal	mixed	superactive		thermic	aridic	shallow
CRUNKER	43.4	NO	1985	NV		26, 27	Entisols	Orthents	Torriorthents	Duric Torriorthents	sandy-skeletal	mixed			mesic	torric	
CRUNKVAR	4.6	NO	1985	NV		26, 28A, 29	Entisols	Orthents	Torriorthents	Xeric Torriorthents	sandy-skeletal	mixed			mesic	torric	
CRUTCHER	4.0	NO	1974	NV	CA	23	Aridisols	Cambids	Haplocambids	Sodic Xeric Haplocambids	ashy	glassy			mesic	aridic	
CRUZSPRING	46.3	NO	2001	NV		30	Aridisols	Argids	Haplargids	Typic Haplargids	loamy-skeletal	mixed	superactive		mesic	aridic	shallow
CRYSTAL SPRINGS	45.4	NO	1970	NV		29, 30	Aridisols	Calcids	Petrocalcids	Typic Petrocalcids	loamy	carbonatic			mesic	aridic	shallow
CUCAMUNGO	69.8	NO	1984	NV		28B, 29	Mollisols	Xerolls	Argixerolls	Typic Argixerolls	loamy-skeletal	mixed	superactive		frigid	xeric	shallow
DAB	13.5	NO	2006	NV	CA	22A, 26	Mollisols	Cryolls	Argicryolls	Pachic Argicryolls	loamy-skeletal	mixed	superactive		cryic		
DACKER	508.8	YES	1986	NV		24, 25	Aridisols	Durids	Argidurids	Xeric Argidurids	fine-loamy	mixed	superactive		mesic	aridic	
DAGGET	12.0	NO	2007	NV	CA	22A	Entisols	Orthents	Cryorthents	Typic Cryorthents	sandy-skeletal	mixed			cryic		
DAICK	34.8	NO	1986	NV		24, 27	Entisols	Orthents	Torriorthents	Typic Torriorthents	clayey	smectitic		calcareous	mesic	torric	shallow
DAKENT	16.8	NO	1985	NV		27	Aridisols	Calcids	Haplocalcids	Durinodic Xeric Haplocalcids	loamy-skeletal	mixed	superactive		mesic	aridic	
DALIAN	11.4		1972	NM		42	Entisols	Orthents	Torriorthents	Typic Torriorthents	loamy-skeletal	carbonatic			thermic	torric	
DALZELL	16.5	NO	1940	NV		26, 27	Aridisols	Durids	Natridurids	Natrargidic Natridurids	fine-loamy	mixed	superactive		mesic	aridic	
DANGBERG	12.8	NO	1975	NV		26	Aridisols	Durids	Natridurids	Aquic Natrargidic Natridurids	fine	smectitic			mesic	aridic	
DAPHSUE	51.1	NO	2016	NV		25	Mollisols	Xerolls	Haploxerolls	Typic Haploxerolls	loamy-skeletal	mixed	superactive		frigid	xeric	
DAVEY	305.1	NO	1974	NV	CA, ID, OR	11, 23, 24, 26, 27	Aridisols	Cambids	Haplocambids	Xeric Haplocambids	sandy	mixed			mesic	aridic	
DEADYON	153.7	NO	1988	NV		27	Aridisols	Argids	Haplargids	Xeric Haplargids	coarse-loamy	mixed	superactive		mesic	aridic	
DEANRAN	10.1	NO	1990	NV		23	Mollisols	Xerolls	Argixerolls	Aridic Argixerolls	loamy-skeletal	mixed	superactive		frigid	xeric	shallow
DEARBUSH	39.0	NO	2016	NV		25	Mollisols	Xerolls	Haploxerolls	Pachic Haploxerolls	loamy-skeletal	mixed	superactive		frigid	xeric	
DECAN	222.7	YES	1971	NV		29	Mollisols	Xerolls	Durixerolls	Argiduridic Durixerolls	fine	smectitic			mesic	xeric	
DECATHON	37.5	NO	1971	NV		28A	Aridisols	Durids	Argidurids	Haploxeralfic Argidurids	clayey	smectitic			mesic	aridic	shallow
DECRAM	42.4	NO	1983	NV		28B	Mollisols	Cryolls	Haplocryolls	Xeric Haplocryolls	loamy-skeletal	mixed	superactive		cryic	xeric	
DEDAS	39.4	NO	1996	NV		30	Aridisols	Durids	Argidurids	Typic Argidurids	loamy-skeletal	mixed	superactive		thermic	aridic	shallow

(continued)

(continued)

Soil series	Area (km^2)	Bench-mark	Year Est-ablished	TL State	Other States Using	MLRAs Using	Order	Suborder	Great Group	Subgroup	Particle-size Class	Mineralogy Class	CEC Activity Class	Reaction Class	Soil temp. Class	SMR	Other Family
DEDMOUNT	30.6	NO	1985	NV		24, 27	Entisols	Orthents	Torriorthents	Vertic Torriorthents	fine	smectitic		calcareous	mesic	torric	
DEEFAN	25.9	NO	1985	NV		27	Aridisols	Durids	Argidurids	Abruptic Argidurids	clayey	smectitic			mesic	aridic	shallow
DEEPEEK	44.5	NO	1986	NV		25	Aridisols	Durids	Argidurids	Xeric Argidurids	loamy-skeletal	mixed	superactive		mesic	aridic	shallow
DEERHILL	3.0	NO	2007	NV		22A	Alfisols	Xeralfs	Palexeralfs	Ultic Palexeralfs	fine-loamy	isotic			frigid	xeric	
DEERLODGE	4.2	NO	1971	UT	NV	28A	Aridisols	Durids	Argidurids	Xeric Argidurids	fine-loamy	mixed	superactive		mesic	aridic	
DEFLER	91.2	NO	1985	NV		28B	Entisols	Orthents	Torriorthents	Typic Torriorthents	loamy-skeletal	mixed	superactive	calcareous	mesic	torric	
DEKOOM	12.0	NO	1986	NV		24	Mollisols	Rendolls	Cryrendolls	Typic Cryrendolls	loamy-skeletal	carbonatic			cryic		
DELACIT	4.7	NO	1993	NV		29	Aridisols	Argids	Natrargids	Typic Natrargids	loamy-skeletal	mixed	superactive		mesic	aridic	shallow
DELAMAR	333.0	NO	1992	NV		29	Aridisols	Durids	Argidurids	Typic Argidurids	fine-loamy	mixed	superactive		mesic	aridic	
DELEPLAIN	3.8	NO	1986	NV		25	Inceptisols	Aquepts	Endoaquepts	Aquandic Endoaquepts	ashy over sandy or sandy-skeletal	glassy OVER mixed		calcareous	mesic	aquic	
DELHEW	14.9	YES	2006	NV	CA	22A, 26	Mollisols	Cryolls	Argicryolls	Pachic Argicryolls	loamy-skeletal	mixed	superactive		cryic		
DELMO	12.5	NO	2012	NV		26	Mollisols	Xerolls	Palexerolls	Aridic Palexerolls	clayey-skeletal	smectitic			frigid	xeric	
DELP	29.0	NO	1981	NV		26, 27	Aridisols	Argids	Haplargids	Typic Haplargids	coarse-loamy	mixed	superactive		mesic	aridic	
DELVADA	85.8	NO	1993	NV		24	Mollisols	Aquolls	Endoaquolls	Cumulic Vertic Endoaquolls	fine	smectitic		calcareous	mesic	aquic	
DEMILL	2.4	NO	1991	NV		29	Entisols	Orthents	Torriorthents	Duric Torriorthents	sandy	mixed			mesic	torric	
DENAY	38.5	NO	1980	NV		24, 25	Mollisols	Xerolls	Calcixerolls	Aridic Calcixerolls	loamy-skeletal	mixed	superactive		frigid	xeric	
DENIHLER	55.2	NO	2016	NV		25	Mollisols	Xerolls	Haploxerolls	Typic Haploxerolls	loamy-skeletal	mixed	superactive		frigid	xeric	
DENIO	30.2	NO	1994	NV		23	Entisols	Orthents	Torriorthents	Xeric Torriorthents	sandy-skeletal	mixed			mesic	torric	
DENMARK	0.1	YES	1947	UT	NV	28A, 29, 47	Aridisols	Calcids	Petrocalcids	Calcic Petrocalcids	loamy	carbonatic			mesic	aridic	shallow
DENPARK	92.3	NO	2010	NV		28A	Mollisols	Cryolls	Argicryolls	Vitrandic Argicryolls	ashy	glassy			cryic		
DEPPY	114.9	NO	1991	OR	NV	24	Aridisols	Durids	Argidurids	Argidic Argidurids	loamy	mixed	superactive		mesic	aridic	shallow
DESATOYA	98.3	NO	1985	NV		24, 27	Aridisols	Argids	Haplargids	Durinodic Xeric Haplargids	clayey over loamy-skeletal	smectitic OVER mixed	superactive		mesic	aridic	
DESEED	13.4	NO	1986	NV	OR	23, 25	Aridisols	Argids	Haplargids	Xeric Haplargids	fine	smectitic			frigid	aridic	
DESTAZO	27.7	NO	1977	CA	NV	30	Aridisols	Calcids	Haplocalcids	Petronodic Haplocalcids	loamy-skeletal	carbonatic			thermic	aridic	
DEUNAH	43.2	NO	1984	ID	NV	25	Alfisols	Xeralfs	Durixeralfs	Abruptic Durixeralfs	very-fine	smectitic			frigid	xeric	
DEVADA	1311.0	NO	1981	NV	CA, OR	21, 23, 24, 25, 26	Mollisols	Xerolls	Argixerolls	Aridic Lithic Argixerolls	clayey	smectitic			mesic	xeric	
DEVEN	8.4	NO	1974	CA	ID, NV	21, 26	Mollisols	Xerolls	Argixerolls	Lithic Argixerolls	clayey	smectitic			mesic	xeric	
DEVILDOG	139.5	NO	2006	NV		29	Aridisols	Cambids	Haplocambids	Vitrixerandic Haplocambids	loamy-skeletal	mixed	superactive		mesic	aridic	
DEVILS	18.6	NO	1981	NV		26	Mollisols	Xerolls	Argixerolls	Aridic Argixerolls	loamy-skeletal	mixed	superactive		frigid	xeric	
DEVILSGAIT	249.0	YES	1986	NV		25, 28B	Mollisols	Aquolls	Endoaquolls	Cumulic Endoaquolls	fine-silty	mixed	superactive	calcareous	mesic	aquic	
DEVILSTHUMB	5.3	NO	2006	NV		30	Inceptisols	Ustepts	Calciustepts	Aridic Calciustepts	loamy-skeletal	mixed	superactive		frigid	ustic	
DEVOY	24.3	NO	1971	NV	OR	23, 28B	Mollisols	Cryolls	Argicryolls	Xeric Argicryolls	clayey-skeletal	smectitic			cryic	xeric	
DEWAR	1059.8	NO	1985	NV		24, 25, 28B	Aridisols	Durids	Argidurids	Xeric Argidurids	loamy	mixed	superactive		mesic	aridic	shallow
DEWRUST	27.7	NO	1992	NV		29	Aridisols	Durids	Argidurids	Xeric Argidurids	fine	smectitic			mesic	aridic	

(continued)

(continued)

Soil series	Area (km^2)	Bench-mark	Year Est-ablished	TL State	Other States Using	MLRAs Using	Order	Suborder	Great Group	Subgroup	Particle-size Class	Mineralogy Class	CEC Activity Class	Reaction Class	Soil temp. Class	SMR	Other Family
DIA	114.8	YES	1971	NV		27	Mollisols	Xerolls	Haploxerolls	Oxyaquic Haploxerolls	fine-loamy over sandy or sandy-skeletal	mixed	superactive		mesic	xeric	
DIAMONDHIL	5.5	NO	2006	NV		29	Aridisols	Durids	Argidurids	Ustic Argidurids	loamy-skeletal	mixed	superactive		mesic	aridic	
DIANEV	182.1	NO	1971	NV		25, 28B	Inceptisols	Aquepts	Halaquepts	Aeric Halaquepts	fine-silty	mixed	superactive	calcareous	frigid	aquic	
DIAZ	0.2	NO	1991	C A	NV, OR	23	Aridisols	Argids	Haplargids	Xeric Haplargids	fine	smectitic			mesic	aridic	
DITHOD	157.0	NO	1971	NV		27	Mollisols	Xerolls	Haploxerolls	Oxyaquic Haploxerolls	coarse-loamy	mixed	superactive		mesic	xeric	
DOBEL	116.3	NO	1972	NV		29	Aridisols	Durids	Argidurids	Argidic Argidurids	loamy	mixed	superactive		mesic	aridic	shallow
DOESPRING	21.4	NO	2006	NV		30	Mollisols	Ustolls	Calciustolls	Petrocalcic Calciustolls	loamy-skeletal	carbonatic			mesic	ustic	shallow
DOMEHILL	14.5	NO	2005	CA	NV	26	Mollisols	Xerolls	Argixerolls	Aridic Lithic Argixerolls	ashy-skeletal	glassy			frigid	xeric	
DOMEZ	11.5	NO	1972	NV		29	Entisols	Orthents	Torriorthents	Duric Torriorthents	fine-loamy	mixed	superactive	calcareous	mesic	torric	
DONNA	503.6	YES	1975	NV		25	Mollisols	Xerolls	Durixerolls	Abruptic Argiduridic Durixerolls	very-fine	smectitic			frigid	xeric	
DOOH	12.3	NO	2012	NV		26	Mollisols	Xerolls	Argixerolls	Pachic Argixerolls	loamy-skeletal	mixed	superactive		frigid	xeric	
DOORKISS	167.7	NO	2009	NV		27	Aridisols	Argids	Haplargids	Lithic Xeric Haplargids	loamy-skeletal	mixed	superactive		mesic	aridic	
DOOWAK	8.9	NO	1985	NV		24	Entisols	Orthents	Torriorthents	Xeric Torriorthents	sandy-skeletal	mixed			mesic	torric	
DORPER	520.4	NO	1988	NV		27	Aridisols	Argids	Natrargids	Durinodic Natrargids	fine	smectitic			mesic	aridic	
DOSIE	125.3	NO	1990	NV	CA	23	Mollisols	Xerolls	Argixerolls	Pachic Argixerolls	clayey-skeletal	smectitic			mesic	xeric	
DOTEN	60.0	NO	1980	NV		26, 28B	Vertisols	Xererts	Haploxererts	Aridic Haploxererts	fine	smectitic			mesic	xeric	
DOTSOLOT	20.7	NO	2012	NV		26	Mollisols	Xerolls	Argixerolls	Aridic Argixerolls	loamy-skeletal	mixed	superactive		frigid	xeric	shallow
DOUHIDE	280.9	NO	1995	NV		28B	Mollisols	Xerolls	Argixerolls	Aridic Lithic Argixerolls	clayey-skeletal	smectitic			frigid	xeric	
DOWNEYVILLE	1253.9	YES	1984	NV		27, 29	Aridisols	Argids	Haplargids	Lithic Haplargids	loamy-skeletal	mixed	superactive		mesic	aridic	
DRESSLER	9.0	NO	1974	NV		26	Mollisols	Xerolls	Haploxerolls	Torrifluventic Haploxerolls	coarse-loamy	mixed	superactive		mesic	xeric	
DRESSLEWET	6.8	NO	2010	NV		26	Mollisols	Xerolls	Haploxerolls	Oxyaquic Haploxerolls	coarse-loamy	mixed	superactive		mesic	xeric	
DREWING	30.3	NO	1974	NV		28B	Mollisols	Xerolls	Durixerolls	Argiduridic Durixerolls	loamy-skeletal	mixed	superactive		mesic	xeric	shallow
DRIT	17.1	NO	1981	NV		26	Mollisols	Xerolls	Haploxerolls	Pachic Haploxerolls	loamy-skeletal	mixed	superactive		mesic	xeric	
DRYGYP	24.6	YES	2006	NV		30	Aridisols	Gypsids	Petrogypsids	Typic Petrogypsids	loamy	gypsic			hyperthermic	aridic	shallow
DUCKHILL	18.6	NO	1980	NV		22A, 26	Alfisols	Xeralfs	Haploxeralfs	Ultic Haploxeralfs	loamy-skeletal	mixed	superactive		frigid	xeric	shallow
DUCO	307.0	YES	1980	NV	CA, ID	10, 26, 27, 28B	Mollisols	Xerolls	Argixerolls	Aridic Lithic Argixerolls	loamy-skeletal	mixed	superactive		mesic	xeric	
DUFF	44.6	NO	1983	NV	OR	23, 25	Mollisols	Cryolls	Haplocryolls	Pachic Haplocryolls	fine-loamy	mixed	superactive		cryic		
DUFFER	421.4	NO	1974	NV		24, 28B	Aridisols	Calcids	Haplocalcids	Aquic Haplocalcids	fine-silty	carbonatic			mesic	aridic	
DUGCHIP	86.2	NO	1992	NV		24	Aridisols	Durids	Natridurids	Xeric Natridurids	fine-loamy	mixed	superactive		mesic	aridic	
DUGWAY	60.0	NO	1995	NV		23	Aridisols	Durids	Natridurids	Natrixeralfic Natridurids	fine	smectitic			mesic	aridic	

(continued)

(continued)

Soil series	Area (km²)	Bench-mark	Year Est-ablished	TL State	Other States Using	MLRAs Using	Order	Suborder	Great Group	Subgroup	Particle-size Class	Mineralogy Class	CEC Activity Class	Reaction Class	Soil temp. Class	SMR	Other Family
DUN GLEN	287.4	YES	1974	NV		24, 26, 27, 28B	Aridisols	Cambids	Haplocambids	Sodic Haplocambids	coarse-loamy	mixed	superactive		mesic	aridic	
DUNPHY	88.6	NO	1979	NV		24, 28B	Inceptisols	Aquepts	Halaquepts	Duric Halaquepts	coarse-loamy	mixed	superactive	calcareous	mesic	aquic	
DUTCHJOHN	13.1	NO	1992	NV	OR	25	Mollisols	Xerolls	Argixerolls	Vitritorrandic Argixerolls	loamy-skeletal	mixed	superactive		frigid	xeric	
EAGLEPASS	477.1	NO	1984	NV		28A, 29	Entisols	Orthents	Torriorthents	Lithic Xeric Torriorthents	loamy-skeletal	carbonatic			mesic	torric	
EAGLEROCK	112.5	NO	1988	NV		23, 26, 27	Mollisols	Xerolls	Argixerolls	Aridic Argixerolls	loamy-skeletal	mixed	superactive		mesic	xeric	
EARCREE	8.1	NO	1977	ID	MT, NV	10, 23, 25, 48A	Mollisols	Cryolls	Haplocryolls	Pachic Haplocryolls	coarse-loamy	mixed	superactive		cryic		
EAST FORK	94.4	YES	1940	NV		26, 27	Mollisols	Xerolls	Haploxerolls	Oxyaquic Haploxerolls	fine-loamy	mixed	superactive		mesic	xeric	
EASTE	2.7	NO	1995	CA	NV	21, 22B	Inceptisols	Xerepts	Haploxerepts	Vitrandic Haploxerepts	loamy-skeletal	isotic			frigid	xeric	
EASTGATE	197.9	NO	1985	NV		27, 29	Aridisols	Cambids	Haplocambids	Typic Haplocambids	sandy	mixed			mesic	aridic	
EASTLAND	0.1	NO	1964	NV	AZ	30	Aridisols	Calcids	Haplocalcids	Typic Haplocalcids	sandy-skeletal	mixed			thermic	aridic	
EASTMORE	166.7	NO	2004	NV		28A	Aridisols	Durids	Haplodurids	Xereptic Haplodurids	loamy-skeletal	mixed	superactive		mesic	aridic	shallow
EASTVAL	23.0	NO	1999	NV		26	Aridisols	Durids	Argidurids	Haploxeralfic Argidurids	fine-loamy	mixed	superactive		mesic	aridic	
EASTWELL	65.2	NO	1985	NV	OR	24, 28A, 28B	Aridisols	Durids	Haplodurids	Xereptic Haplodurids	loamy-skeletal	mixed	superactive		mesic	aridic	shallow
EASYCHAIR	42.4	NO	1993	NV		29	Entisols	Orthents	Torriorthents	Typic Torriorthents	fine-silty	mixed	superactive	calcareous	mesic	torric	
EBIC	44.8	NO	1983	NV		25	Mollisols	Xerolls	Palexerolls	Typic Palexerolls	clayey-skeletal	smectitic			frigid	xeric	
EBODA	104.5	NO	1986	NV		25	Mollisols	Xerolls	Argixerolls	Aridic Argixerolls	fine-loamy	mixed	superactive		frigid	xeric	
EDNAGREY	10.0	NO	2006	NV		30	Entisols	Orthents	Torriorthents	Lithic Ustic Torriorthents	loamy-skeletal	mixed	superactive	calcareous	mesic	torric	
EENREED	22.1	NO	2004	NV		28A	Mollisols	Xerolls	Calcixerolls	Aridic Calcixerolls	loamy-skeletal	mixed	superactive		frigid	xeric	
EGANROC	96.9	NO	1990	NV		28B	Mollisols	Cryolls	Haplocryolls	Calcic Pachic Haplocryolls	loamy-skeletal	mixed	superactive		cryic		
EIGHTMILE	30.5	NO	1983	NV		27, 28B	Entisols	Orthents	Torriorthents	Xeric Torriorthents	loamy-skeletal	mixed	superactive	nonacid	frigid	torric	shallow
EKIM	34.1	NO	1986	NV		25	Mollisols	Xerolls	Calcixerolls	Aridic Calcixerolls	loamy-skeletal	carbonatic			frigid	xeric	
ELAERO	0.1	NO	2006	CA	NV	26	Mollisols	Xerolls	Argixerolls	Typic Argixerolls	loamy-skeletal	mixed	superactive		frigid	xeric	
ELBOWCANYON	52.4	NO	2011	NV		30	Entisols	Orthents	Torriorthents	Typic Torriorthents	loamy-skeletal	carbonatic			thermic	torric	
ELHINA	7.6	NO	1986	ID	NV	25	Aridisols	Durids	Argidurids	Abruptic Xeric Argidurids	fine	smectitic			frigid	aridic	
ELOCIN	13.6	NO	1986	NV		25	Mollisols	Xerolls	Palexerolls	Duric Palexerolls	clayey-skeletal	smectitic			frigid	xeric	
EMAGERT	5.0	NO	1995	NV	CA	23	Mollisols	Xerolls	Haploxerolls	Vitritorrandic Haploxerolls	ashy	glassy			mesic	xeric	
EMAMOUNT	1.1	NO	2006	NV	CA	23	Mollisols	Xerolls	Haploxerolls	Vitritorrandic Haploxerolls	ashy	glassy			frigid	xeric	
ENKO	778.4	YES	1983	NV	ID, OR	24, 25, 28B	Aridisols	Cambids	Haplocambids	Durinodic Xeric Haplocambids	coarse-loamy	mixed	superactive		mesic	aridic	
ENTERO	73.6	NO	1984	NV		29	Aridisols	Argids	Haplargids	Xeric Haplargids	loamy-skeletal	mixed	superactive		mesic	aridic	shallow
ENVOL	69.7	NO	1988	NV		27	Aridisols	Argids	Haplargids	Lithic Haplargids	loamy	mixed	superactive		mesic	aridic	
EOJ	34.5	NO	1990	NV		28B	Mollisols	Xerolls	Palexerolls	Typic Palexerolls	fine	smectitic			frigid	xeric	

(continued)

(continued)

Soil series	Area (km^2)	Bench-mark	Year Est-ablished	TL State	Other States Using	MLRAs Using	Order	Suborder	Great Group	Subgroup	Particle-size Class	Mineralogy Class	CEC Activity Class	Reaction Class	Soil temp. Class	SMR	Other Family
EPVIP	11.0	NO	1985	NV	CA	26	Mollisols	Xerolls	Argixerolls	Vitrandic Argixerolls	ashy-skeletal	glassy			frigid	xeric	shallow
EQUIS	124.4	NO	1990	NV		28B	Inceptisols	Aquepts	Halaquepts	Typic Halaquepts	fine	carbonatic			mesic	aquic	
ERAKATAK	21.4	NO	1986	OR	NV	10, 23, 25, 43C	Mollisols	Xerolls	Argixerolls	Vitrandic Argixerolls	clayey-skeletal	smectitic			frigid	xeric	
ERASTRA	73.2	NO	1999	NV		26	Mollisols	Xerolls	Argixerolls	Aridic Argixerolls	loamy-skeletal	mixed	superactive		mesic	xeric	shallow
ERBER	31.5	NO	1971	NV		24, 27	Mollisols	Aquolls	Endoaquolls	Fluvaquentic Endoaquolls	sandy	mixed			mesic	aquic	
ESCALANTE	253.6	NO	1942	UT	ID, NV	11, 25, 28A	Aridisols	Calcids	Haplocalcids	Xeric Haplocalcids	coarse-loamy	mixed	superactive		mesic	aridic	
ESMOD	73.0	NO	1995	NV		23	Aridisols	Durids	Argidurids	Abruptic Xeric Argidurids	clayey	smectitic			mesic	aridic	shallow
ESPINT	40.2	NO	1984	NV		29	Aridisols	Argids	Haplargids	Xeric Haplargids	clayey	smectitic			mesic	aridic	shallow
ESSAL	62.2	NO	1992	NV	UT	24, 27	Entisols	Orthents	Torriorthents	Typic Torriorthents	coarse-loamy over sandy or sandy-skeletal	mixed	superactive	calcareous	mesic	torric	
EWELAC	89.5	NO	2004	NV		28A	Aridisols	Cambids	Haplocambids	Vertic Haplocambids	fine	smectitic			mesic	aridic	
FADOLL	45.7	NO	1985	NV		26	Entisols	Orthents	Torriorthents	Vitrandic Torriorthents	ashy	glassy		nonacid	mesic	torric	
FAIRYDELL	35.3	NO	1971	NV		28B	Mollisols	Cryolls	Haplocryolls	Xeric Haplocryolls	loamy-skeletal	mixed	superactive		cryic	xeric	
FALERIA	32.9	NO	1992	NV		29	Alfisols	Ustalfs	Haplustalfs	Vitrandic Haplustalfs	ashy-skeletal	glassy			frigid	ustic	
FALLON	74.3	NO	1960	NV		27	Entisols	Fluvents	Torrifluvents	Oxyaquic Torrifluvents	coarse-loamy	mixed	superactive	nonacid	mesic	torric	
FANG	125.8	NO	1965	NV		28B, 29	Entisols	Orthents	Torriorthents	Typic Torriorthents	coarse-loamy	mixed	superactive	calcareous	mesic	torric	
FANU	66.0	NO	1971	NV		28B, 29	Mollisols	Xerolls	Haploxerolls	Cumulic Haploxerolls	fine-loamy	mixed	superactive		mesic	xeric	
FAREPEAK	45.9	NO	2006	NV		28A	Mollisols	Xerolls	Argixerolls	Aridic Lithic Argixerolls	ashy-skeletal	glassy			frigid	xeric	
FAWIN	28.0	NO	1985	NV		27	Aridisols	Cambids	Haplocambids	Typic Haplocambids	sandy	mixed			mesic	aridic	
FAX	76.9	NO	1990	NV		28B	Mollisols	Xerolls	Durixerolls	Argidic Durixerolls	loamy-skeletal	mixed	superactive		mesic	xeric	
FENELON	24.4	NO	1986	NV		25	Mollisols	Xerolls	Argixerolls	Calciargidic Argixerolls	fine-loamy	mixed	superactive		frigid	xeric	
FENSTER	42.2	NO	1983	NV	CO	28B, 34A	Entisols	Orthents	Torriorthents	Typic Torriorthents	fine-silty	mixed	superactive	calcareous	frigid	torric	
FERA	58.9	NO	1971	NV		25, 28B, 47	Mollisols	Xerolls	Argixerolls	Aridic Argixerolls	clayey-skeletal	smectitic			frigid	xeric	
FERDELFORD	29.1	NO	1980	NV		24, 25	Aridisols	Cambids	Haplocambids	Xeric Haplocambids	fine-loamy	mixed	superactive		mesic	aridic	
FERNLEY	37.5	NO	1971	NV		27	Entisols	Psamments	Torripsamments	Oxyaquic Torripsamments		mixed			mesic	torric	
FERNPOINT	26.0	NO	1995	NV		23	Mollisols	Xerolls	Argixerolls	Aridic Argixerolls	fine-loamy	mixed	superactive		mesic	xeric	
FERROGOLD	130.8	NO	2001	NV		30	Aridisols	Calcids	Petrocalcids	Calcic Petrocalcids	loamy-skeletal	carbonatic			thermic	aridic	shallow
FERTALINE	20.7	NO	1983	NV	CA, OR	23, 28B	Aridisols	Durids	Argidurids	Abruptic Xeric Argidurids	fine	smectitic			frigid	aridic	
FERVER	40.8	NO	1995	NV	CA	23	Aridisols	Durids	Argidurids	Vertic Argidurids	very-fine	smectitic			mesic	aridic	
FETTIC	8.7	NO	1975	NV		26	Mollisols	Xerolls	Natrixerolls	Aridic Natrixerolls	fine-silty	mixed	superactive		mesic	xeric	
FEZ	39.6	NO	1986	NV		25	Mollisols	Xerolls	Haploxerolls	Vitrandic Haploxerolls	ashy	glassy			frigid	xeric	
FIDDLER	6.7	NO	1986	CA	NV	21, 23	Mollisols	Xerolls	Argixerolls	Typic Argixerolls	clayey-skeletal	smectitic			mesic	xeric	

(continued)

(continued)

Soil series	Area (km^2)	Bench-mark	Year Est-ablished	TL State	Other States Using	MLRAs Using	Order	Suborder	Great Group	Subgroup	Particle-size Class	Mineralogy Class	CEC Activity Class	Reaction Class	Soil temp. Class	SMR	Other Family
FIFTEENMILE	58.6	NO	2010	NV		28A	Aridisols	Cambids	Haplocambids	Xeric Haplocambids	fine-silty	mixed	superactive		mesic	aridic	
FILAREE	48.5	NO	2005	AZ	NV	30	Aridisols	Cambids	Haplocambids	Typic Haplocambids	coarse-loamy	mixed	superactive		thermic	aridic	
FILIRAN	33.6	NO	1985	NV		24, 28B	Aridisols	Durids	Natridurids	Natrixeralfic Natridurids	fine	smectitic			mesic	aridic	
FINDOUT	59.7	NO	1986	NV		27	Aridisols	Calcids	Haplocalcids	Lithic Haplocalcids	loamy-skeletal	carbonatic			mesic	aridic	
FIREBALL	56.0	NO	1980	NV		27	Aridisols	Argids	Haplargids	Typic Haplargids	loamy-skeletal	mixed	superactive		mesic	aridic	
FIVEMILE	18.1	NO	1969	WY	NM, NV	28B, 29, 32, 36	Entisols	Fluvents	Torrifluvents	Typic Torrifluvents	fine-silty	mixed	superactive	calcareous	mesic	torric	
FLATNOSEWASH	23.5	NO	2010	NV		28A, 29	Entisols	Fluvents	Torrifluvents	Xeric Torrifluvents	fine-loamy	mixed	superactive	calcareous	mesic	torric	
FLATTOP	86.9	NO	1970	NV		30	Aridisols	Argids	Natrargids	Typic Natrargids	loamy-skeletal	mixed	superactive		thermic	aridic	
FLEISCHMANN	10.7	NO	1980	NV		26	Mollisols	Xerolls	Durixerolls	Argidic Durixerolls	fine	smectitic			mesic	xeric	
FLETCHERPEAK	80.6	NO	2006	NV		30	Mollisols	Ustolls	Argiustolls	Aridic Lithic Argiustolls	loamy-skeletal	mixed	superactive		frigid	ustic	
FLEX	39.4	NO	1980	NV	CA	26, 27	Aridisols	Argids	Haplargids	Xeric Haplargids	loamy-skeletal	mixed	superactive		mesic	aridic	shallow
FLOER	5.1	NO	1985	NV		23, 24	Mollisols	Xerolls	Palexerolls	Aridic Palexerolls	clayey-skeletal	smectitic			frigid	xeric	
FLUE	139.7	NO	1993	NV		24	Aridisols	Durids	Natridurids	Vitrixerandic Natridurids	fine	smectitic			mesic	aridic	
FORTANK	79.3	NO	1983	NV		25, 28B	Aridisols	Argids	Haplargids	Xeric Haplargids	fine	smectitic			frigid	aridic	
FORVIC	14.6	NO	1986	ID	NV	25	Mollisols	Xerolls	Durixerolls	Typic Durixerolls	fine	smectitic			frigid	xeric	
FOUR STAR	11.3	NO	1971	CA	NV	21, 23, 25	Mollisols	Aquolls	Endoaquolls	Aquandic Endoaquolls	ashy	glassy			frigid	aquic	
FOXCAN	28.7	NO	1990	NV		27	Entisols	Orthents	Torriorthents	Xeric Torriorthents	loamy	mixed	superactive	calcareous	mesic	torric	shallow
FOXMOUNT	20.8	NO	1983	NV	CA, ID	23, 25, 28B	Mollisols	Cryolls	Haplocryolls	Xeric Haplocryolls	loamy-skeletal	mixed	superactive		cryic	xeric	
FOXVIRE	12.6	NO	1991	NV		29	Mollisols	Cryolls	Haplocryolls	Pachic Haplocryolls	coarse-loamy	mixed	superactive		cryic		
FRANKTOWN	7.0	NO	1974	NV	CA	22A	Mollisols	Xerolls	Haploxerolls	Lithic Ultic Haploxerolls	loamy-skeletal	mixed	superactive		frigid	xeric	
FRAVAL	31.9	NO	1980	NV	CA	22A, 26	Mollisols	Xerolls	Argixerolls	Ultic Argixerolls	loamy-skeletal	mixed	superactive		frigid	xeric	
FRENTERA	46.5	NO	1994	NV		23	Mollisols	Xerolls	Haploxerolls	Vitritorrandic Haploxerolls	ashy	glassy			frigid	xeric	
FREWA	11.2	NO	1986	NV		24	Entisols	Psamments	Torripsamments	Haploduridic Torripsamments		mixed			mesic	torric	
FREZNIK	4.0	NO	1983	NV	CA, OR	23, 25	Aridisols	Argids	Paleargids	Xeric Paleargids	fine	smectitic			frigid	aridic	
FRINES	8.3	NO	1988	NV		27	Aridisols	Argids	Haplargids	Typic Haplargids	fine	smectitic			mesic	aridic	
FRODO	38.0	NO	1980	NV		26	Mollisols	Xerolls	Durixerolls	Paleargidic Durixerolls	clayey	smectitic			mesic	xeric	shallow
FUBBLE	26.1	NO	1995	NV		27	Aridisols	Argids	Haplargids	Lithic Xeric Haplargids	loamy	mixed	superactive		mesic	aridic	
FUEGOSTA	26.2	NO	1984	NV		29	Aridisols	Durids	Argidurids	Abruptic Argidurids	clayey	smectitic			mesic	aridic	shallow
FUGAWEE	14.9	NO	1970	NV	CA	22A	Alfisols	Xeralfs	Haploxeralfs	Andic Haploxeralfs	fine-loamy	isotic			frigid	xeric	
FULSTONE	440.0	NO	1981	NV	CA	23, 25, 26	Aridisols	Durids	Argidurids	Abruptic Xeric Argidurids	clayey	smectitic			mesic	aridic	shallow
FUSULINA	32.9	NO	1971	NV		28B	Mollisols	Cryolls	Haplocryolls	Xeric Haplocryolls	loamy	mixed	superactive		cryic	xeric	shallow
FUSUVAR	3.4	NO	1985	NV		26	Mollisols	Cryolls	Haplocryolls	Xeric Haplocryolls	loamy	mixed	superactive		cryic	xeric	shallow

(continued)

(continued)

Soil series	Area (km^2)	Bench-mark	Year Est-ablished	TL State	Other States Using	MLRAs Using	Order	Suborder	Great Group	Subgroup	Particle-size Class	Mineralogy Class	CEC Activity Class	Reaction Class	Soil temp. Class	SMR	Other Family
GABBS	11.0	NO	1972	NV	UT	29	Aridisols	Durids	Haplodurids	Typic Haplodurids	loamy-skeletal	mixed	superactive		mesic	aridic	
GABBVALLY	1233.3	NO	1984	NV		26, 27, 29	Aridisols	Argids	Haplargids	Lithic Xeric Haplargids	loamy-skeletal	mixed	superactive		mesic	aridic	
GABEL	3.7	NO	1971	NV		25, 28B	Aridisols	Argids	Haplargids	Xeric Haplargids	loamy-skeletal	mixed	superactive		frigid	aridic	
GABICA	32.1	YES	1975	NV		26	Mollisols	Xerolls	Argixerolls	Lithic Argixerolls	loamy-skeletal	mixed	superactive		frigid	xeric	
GAIA	2.5	NO	2009	NV		28A	Mollisols	Cryolls	Haplocryolls	Xeric Haplocryolls	loamy-skeletal	mixed	superactive		cryic	xeric	
GALEHILLS	40.1	NO	2006	NV		30	Entisols	Orthents	Torriorthents	Lithic Torriorthents	loamy-skeletal	mixed	superactive	calcareous	thermic	torric	
GALEPPI	36.3	NO	1973	CA	NV	23, 26	Mollisols	Xerolls	Argixerolls	Argiduridic Argixerolls	fine-loamy	mixed	superactive		mesic	xeric	
GAMGEE	26.2	NO	1981	NV		27	Aridisols	Argids	Natrargids	Haplic Natrargids	fine-loamy	mixed	superactive		mesic	aridic	
GANAFLAN	54.3	NO	1990	NV		27, 29	Entisols	Orthents	Torriorthents	Typic Torriorthents	coarse-loamy	mixed	superactive	calcareous	mesic	torric	
GANCE	264.7	NO	1986	NV		24, 25, 28B	Aridisols	Argids	Haplargids	Durinodic Xeric Haplargids	clayey-skeletal	smectitic			mesic	aridic	
GANDO	59.3	NO	1983	NV		24, 25, 28B	Mollisols	Xerolls	Haploxerolls	Aridic Lithic Haploxerolls	loamy-skeletal	mixed	superactive		frigid	xeric	
GARDELLA	9.3	NO	1971	NV		27	Aridisols	Durids	Haplodurids	Cambidic Haplodurids	sandy	mixed			mesic	aridic	shallow
GARDENVALLEY	25.9	NO	2006	NV		29	Aridisols	Cambids	Haplocambids	Durinodic Haplocambids	coarse-loamy	mixed	superactive		mesic	aridic	
GARDNERVILLE	5.7	NO	1941	NV		26	Aridisols	Argids	Natrargids	Durinodic Xeric Natrargids	fine	smectitic			mesic	aridic	
GARFAN	31.1	NO	1990	NV		28B	Aridisols	Argids	Paleargids	Xeric Paleargids	clayey-skeletal	smectitic			frigid	aridic	
GARHILL	221.8	NO	1984	NV		29	Aridisols	Durids	Haplodurids	Typic Haplodurids	loamy	mixed	superactive		mesic	aridic	shallow
GARNEL	22.1	NO	2004	NV		28A, 28B	Mollisols	Xerolls	Argixerolls	Aridic Argixerolls	loamy-skeletal	mixed	superactive		frigid	xeric	shallow
GEER	461.6	YES	1940	NV		26, 28A, 28B, 29	Entisols	Orthents	Torriorthents	Typic Torriorthents	coarse-loamy	mixed	superactive	calcareous	mesic	torric	
GEFO	1.5	NO	1970	CA	NV	22A	Inceptisols	Xerepts	Dystroxerepts	Humic Dystroxerepts	sandy	mixed			frigid	xeric	
GENAW	130.6	NO	1985	NV		23, 24, 28B	Aridisols	Argids	Haplargids	Xeric Haplargids	loamy	mixed	superactive		mesic	aridic	shallow
GENEGRAF	399.6	NO	1986	NV		27	Aridisols	Argids	Natrargids	Durinodic Natrargids	fine-loamy	mixed	superactive		mesic	aridic	
GENOA	5.9	NO	1940	NV		26	Mollisols	Xerolls	Argixerolls	Lithic Argixerolls	loamy-skeletal	mixed	superactive		frigid	xeric	
GENOAPEAK	0.6	NO	2007	NV		22A	Entisols	Orthents	Xerorthents	Dystric Xerorthents	fragmental	mixed			frigid	xeric	
GETA	92.5	NO	1992	NV		30	Aridisols	Calcids	Haplocalcids	Ustic Haplocalcids	coarse-loamy	mixed	superactive		thermic	aridic	
GEYSEN	31.7	NO	1980	NV		24, 25	Aridisols	Argids	Natrargids	Durinodic Xeric Natrargids	fine-loamy	mixed	superactive		mesic	aridic	
GILA	0.1	YES	1900	AZ	NM, NV, TX	40, 41, 42	Entisols	Fluvents	Torrifluvents	Typic Torrifluvents	coarse-loamy	mixed	superactive	calcareous	thermic	torric	
GINEX	7.9	NO	1985	NV		24	Aridisols	Argids	Haplargids	Xeric Haplargids	loamy-skeletal	mixed	superactive		mesic	aridic	shallow
GITAKUP	28.5	NO	1990	NV		27	Aridisols	Cambids	Haplocambids	Sodic Haplocambids	fine-silty	mixed	superactive		mesic	aridic	
GLASSHAWK	7.9	NO	2006	NV		23	Aridisols	Durids	Haplodurids	Cambidic Haplodurids	ashy	glassy			mesic	aridic	shallow
GLEAN	141.9	NO	1973	CA	NV	21, 22A, 23, 24, 26, 28B	Mollisols	Xerolls	Haploxerolls	Pachic Haploxerolls	loamy-skeletal	mixed	superactive		frigid	xeric	
GLENBROOK	72.6	NO	1973	NV	CA	26, 27	Entisols	Psamments	Torripsamments	Xeric Torripsamments		mixed			mesic	torric	shallow

(continued)

(continued)

Soil series	Area (km^2)	Bench-mark	Year Est-ablished	TL State	Other States Using	MLRAs Using	Order	Suborder	Great Group	Subgroup	Particle-size Class	Mineralogy Class	CEC Activity Class	Reaction Class	Soil temp. Class	SMR	Other Family
GLENCARB	96.2	NO	1982	NV		30	Entisols	Fluvents	Torrifluvents	Typic Torrifluvents	fine-silty	carbonatic			thermic	torric	
GLENDALE	21.6		1946	AZ	NV	40, 41, 42	Entisols	Fluvents	Torrifluvents	Typic Torrifluvents	fine-silty	mixed	superactive	calcareous	thermic	torric	
GLIDESKI	5.5	NO	2009	NV		28A	Mollisols	Xerolls	Argixerolls	Typic Argixerolls	loamy-skeletal	mixed	superactive		frigid	xeric	
GLOTRAIN	82.5	NO	2006	NV		29	Aridisols	Argids	Haplargids	Typic Haplargids	coarse-loamy	mixed	superactive		mesic	aridic	
GLYPHS	74.2	NO	1985	NV		28B	Aridisols	Argids	Haplargids	Durinodic Xeric Haplargids	fine-loamy	mixed	superactive		mesic	aridic	
GOCHEA	190.8	NO	1986	NV	OR	23, 25, 29	Mollisols	Xerolls	Argixerolls	Argiduridic Argixerolls	fine-loamy	mixed	superactive		frigid	xeric	
GODECKE	12.2	NO	1975	NV		26	Aridisols	Argids	Natrargids	Durinodic Xeric Natrargids	fine-loamy	mixed	superactive		mesic	aridic	
GOL	58.2	NO	1986	NV		24, 27	Aridisols	Argids	Haplargids	Xeric Haplargids	loamy-skeletal	mixed	superactive		frigid	aridic	shallow
GOLCONDA	195.4	NO	1974	NV		24, 25	Aridisols	Durids	Natridurids	Natrargidic Natridurids	fine-loamy	mixed	superactive		mesic	aridic	
GOLDBUTTE	74.7	NO	2006	NV		30	Entisols	Orthents	Torriorthents	Typic Torriorthents	loamy-skeletal	mixed	superactive	nonacid	mesic	torric	shallow
GOLDROAD	324.0	NO	2000	AZ	CA, NV	30	Entisols	Orthents	Torriorthents	Lithic Torriorthents	loamy-skeletal	mixed	superactive	calcareous	hyperthermic	torric	
GOLDRUN	340.0	YES	1974	NV	OR, UT	24, 27, 28A	Entisols	Psamments	Torripsamments	Xeric Torripsamments		mixed			mesic	torric	
GOLDYKE	113.5	NO	1985	NV		27, 29	Entisols	Orthents	Torriorthents	Typic Torriorthents	loamy	mixed	superactive	calcareous	mesic	torric	shallow
GOLLAHER	325.7	NO	1986	NV		25	Entisols	Orthents	Xerorthents	Lithic Xerorthents	loamy-skeletal	carbonatic			frigid	xeric	
GOLSUM	56.0	NO	1974	NV		24, 25, 28A, 47	Mollisols	Xerolls	Argixerolls	Calciargidic Argixerolls	clayey-skeletal	smectitic			frigid	xeric	
GOMINE	0.1	NO	1997	UT	NV	47	Mollisols	Xerolls	Haploxerolls	Lithic Haploxerolls	loamy-skeletal	mixed	superactive		frigid	xeric	
GOODSKI	23.3	NO	2009	NV		28A	Mollisols	Cryolls	Haplocryolls	Pachic Haplocryolls	loamy-skeletal	mixed	superactive		cryic		
GOODSPRINGS	72.0	NO	1964	NV	AZ	30	Aridisols	Calcids	Petrocalcids	Typic Petrocalcids	loamy	mixed	superactive		thermic	aridic	shallow
GOODWATER	68.5	NO	2006	NV		30	Aridisols	Calcids	Petrocalcids	Calcic Petrocalcids	loamy-skeletal	carbonatic			mesic	aridic	shallow
GOOSEL	120.2	NO	1992	NV		24, 25	Aridisols	Durids	Argidurids	Xeric Argidurids	fine	smectitic			mesic	aridic	
GORZELL	0.5	NO	1971	CA	NV	23	Aridisols	Argids	Haplargids	Durinodic Xeric Haplargids	fine-loamy over sandy or sandy-skeletal	mixed	superactive		mesic	aridic	
GOSUMI	46.6	NO	1974	NV		23, 24	Mollisols	Xerolls	Argixerolls	Aridic Argixerolls	clayey-skeletal	smectitic			frigid	xeric	
GOVWASH	3.6	NO	2006	NV		30	Aridisols	Gypsids	Haplogypsids	Leptic Haplogypsids	coarse-loamy	gypsic			hyperthermic	aridic	
GOWJAI	41.3	NO	1992	NV		24, 25	Mollisols	Xerolls	Argixerolls	Aridic Argixerolls	loamy-skeletal	mixed	superactive		frigid	xeric	
GRALEY	332.7	NO	1974	NV		24, 25, 28B	Mollisols	Xerolls	Argixerolls	Aridic Lithic Argixerolls	clayey-skeletal	smectitic			frigid	xeric	
GRANDEPOSIT	52.4	NO	2013	NV		28B	Mollisols	Xerolls	Argixerolls	Lithic Argixerolls	loamy-skeletal	mixed	superactive		frigid	xeric	
GRANDRIDGE	3.9	NO	2006	CA	NV	22A, 26	Mollisols	Xerolls	Argixerolls	Typic Argixerolls	loamy-skeletal	mixed	superactive		frigid	xeric	shallow
GRANIPEAK	8.9	NO	1990	NV		23	Mollisols	Xerolls	Argixerolls	Aridic Argixerolls	loamy-skeletal	mixed	superactive		frigid	xeric	
GRANMOUNT	23.1	NO	1985	NV		26	Mollisols	Cryolls	Argicryolls	Xeric Argicryolls	clayey-skeletal	mixed	superactive		cryic	xeric	
GRANQUIN	16.1	NO	2006	NV		29	Mollisols	Xerolls	Argixerolls	Aridic Lithic Argixerolls	loamy-skeletal	mixed	superactive		frigid	xeric	
GRANSHAW	362.0	NO	1988	NV		24, 27	Aridisols	Argids	Haplargids	Typic Haplargids	coarse-loamy	mixed	superactive		mesic	aridic	
GRANZAN	12.2	NO	1983	NV		25, 27, 28B	Mollisols	Xerolls	Calcixerolls	Typic Calcixerolls	loamy-skeletal	carbonatic			frigid	xeric	
GRAPEVINE	51.0	NO	1964	NV	AZ, CA	30	Aridisols	Calcids	Haplocalcids	Typic Haplocalcids	coarse-loamy	mixed	superactive		thermic	aridic	

(continued)

(continued)

Soil series	Area (km^2)	Bench-mark	Year Est-ablished	TL State	Other States Using	MLRAs Using	Order	Suborder	Great Group	Subgroup	Particle-size Class	Mineralogy Class	CEC Activity Class	Reaction Class	Soil temp. Class	SMR	Other Family
GRASSVAL	350.0	YES	1985	NV	OR	24, 28B	Aridisols	Durids	Argidurids	Xeric Argidurids	loamy	mixed	superactive		mesic	aridic	shallow
GRASSYCAN	69.1	NO	1995	NV		23	Aridisols	Durids	Argidurids	Abruptic Xeric Argidurids	clayey	smectitic			mesic	aridic	shallow
GRAUFELS	98.0	NO	1980	NV	CA	26	Mollisols	Xerolls	Haploxerolls	Torripsammentic Haploxerolls		mixed			mesic	xeric	
GRAVIER	336.6	NO	1986	NV		28A, 28B	Aridisols	Calcids	Haplocalcids	Sodic Haplocalcids	loamy-skeletal	mixed	superactive		mesic	aridic	
GRAYLOCK	3.1	NO	1969	ID	CA, NV	22A, 25, 43B	Entisols	Orthents	Cryorthents	Typic Cryorthents	sandy-skeletal	mixed			cryic		
GREATDAY	2.1	NO	2004	NV		28A	Aridisols	Calcids	Haplocalcids	Petronodic Xeric Haplocalcids	fine-loamy	mixed	superactive		mesic	aridic	
GREENBRAE	69.7	NO	1975	NV		23, 26	Aridisols	Argids	Haplargids	Xeric Haplargids	fine-loamy	mixed	superactive		mesic	aridic	
GREENGROVE	18.8	NO	2010	NV		28A	Mollisols	Cryolls	Haplocryolls	Vitrandic Haplocryolls	ashy-skeletal	glassy			cryic		
GREMMERS	31.9	NO	2004	NV		28A	Aridisols	Durids	Haplodurids	Xereptic Haplodurids	loamy	mixed	active		mesic	aridic	shallow
GREYEAGLE	196.3	NO	1980	CA	AZ, NV	29, 30, 40	Aridisols	Durids	Haplodurids	Typic Haplodurids	loamy-skeletal	mixed	superactive		thermic	aridic	shallow
GRIFFY	8.9	NO	1969	WY	NV	27, 29, 32, 34A, 43B	Aridisols	Argids	Haplargids	Typic Haplargids	fine-loamy	mixed	superactive		mesic	aridic	
GRIFLEYS	29.6	NO	2004	NV		28A	Aridisols	Argids	Calciargids	Xeric Calciargids	loamy-skeletal	mixed	superactive		mesic	aridic	
GRIMLAKE	3.8	NO	2006	NV	CA	23	Vertisols	Xererts	Haploxererts	Aquic Haploxererts	fine	smectitic			frigid	xeric	
GRINA	155.7	NO	1985	NV		25, 28B	Entisols	Orthents	Torriorthents	Xeric Torriorthents	loamy	mixed	superactive	calcareous	mesic	torric	shallow
GRINK	67.7	NO	1990	NV		28B	Mollisols	Xerolls	Haploxerolls	Lithic Haploxerolls	loamy-skeletal	mixed	superactive		frigid	xeric	
GRIVER	20.0	NO	1979	NV	UT	25, 28A	Entisols	Fluvents	Torrifluvents	Aquic Torrifluvents	coarse-loamy	mixed	superactive	calcareous	mesic	torric	
GROSSCHAT	8.2	NO	2004	NV		28A	Mollisols	Xerolls	Argixerolls	Aridic Lithic Argixerolls	loamy-skeletal	mixed	superactive		frigid	xeric	
GROWSET	8.9	NO	2004	NV		28A	Mollisols	Xerolls	Haploxerolls	Typic Haploxerolls	loamy	mixed	superactive		frigid	xeric	shallow
GRUBE	49.0	NO	2004	NV		28A, 28B	Mollisols	Xerolls	Calcixerolls	Aridic Calcixerolls	loamy-skeletal	mixed	superactive		frigid	xeric	
GRUMBLEN	340.3	YES	1988	NV		27	Aridisols	Argids	Haplargids	Lithic Xeric Haplargids	clayey-skeletal	smectitic			mesic	aridic	
GUARDIAN	49.6	NO	2006	NV		30	Aridisols	Gypsids	Haplogypsids	Leptic Haplogypsids	loamy	gypsic			hyperthermic	aridic	shallow
GUISER	61.9	NO	1990	NV		28B	Mollisols	Cryolls	Argicryolls	Xeric Argicryolls	loamy-skeletal	mixed	superactive		cryic	xeric	
GUMBLE	321.0	NO	1986	NV	OR	10, 25	Aridisols	Argids	Haplargids	Xeric Haplargids	clayey	smectitic			mesic	aridic	shallow
GUND	63.1	NO	1985	NV		28B	Entisols	Orthents	Torriorthents	Xerertic Torriorthents	fine-silty over clayey	mixed OVER smectitic	superactive	nonacid	mesic	torric	
GURDUGEE	9.4	NO	2010	NV	CA	26	Aridisols	Argids	Natrargids	Aquic Natrargids	fine-loamy	mixed	superactive		mesic	aridic	
GWENA	55.4	NO	1986	NV		24	Aridisols	Durids	Natridurids	Xeric Natridurids	loamy	mixed	superactive		mesic	aridic	shallow
GYNELLE	697.7	NO	1984	NV		27, 29	Entisols	Orthents	Torriorthents	Typic Torriorthents	sandy-skeletal	mixed			mesic	torric	
GYPWASH	26.9	NO	2006	NV		30	Aridisols	Gypsids	Calcigypsids	Typic Calcigypsids	loamy-skeletal	carbonatic			hyperthermic	aridic	
HAAR	44.6	NO	1982	NV		26, 29	Entisols	Orthents	Torriorthents	Xeric Torriorthents	loamy	mixed	superactive	nonacid	mesic	torric	shallow
HAARVAR	20.3	NO	1985	NV		29	Entisols	Orthents	Torriorthents	Xeric Torriorthents	clayey	smectitic		calcareous	mesic	torric	shallow
HACKWOOD	345.5	NO	1983	NV	OR	23, 25, 28B	Mollisols	Cryolls	Haplocryolls	Pachic Haplocryolls	fine-loamy	mixed	superactive		cryic		

(continued)

(continued)

Soil series	Area (km^2)	Bench-mark	Year Est-ablished	TL State	Other States Using	MLRAs Using	Order	Suborder	Great Group	Subgroup	Particle-size Class	Mineralogy Class	CEC Activity Class	Reaction Class	Soil temp. Class	SMR	Other Family
HALACAN	149.1	NO	1985	NV		28B	Mollisols	Rendolls	Cryrendolls	Lithic Cryrendolls	loamy-skeletal	carbonatic			cryic		
HALEBURU	447.0	NO	2000	CA	NV	30	Entisols	Orthents	Torriorthents	Lithic Torriorthents	loamy-skeletal	mixed	superactive	calcareous	thermic	torric	
HALFASH	7.1	NO	2005	CA	NV	26	Mollisols	Xerolls	Argixerolls	Vitritorrandic Argixerolls	ashy-skeletal	glassy			frigid	xeric	shallow
HALLECK	49.2	NO	1986	NV		25	Mollisols	Aquolls	Endoaquolls	Cumulic Endoaquolls	fine-silty	mixed	superactive	calcareous	frigid	aquic	
HALVERT	7.7	NO	1995	NV	CA	23	Mollisols	Xerolls	Durixerolls	Vertic Durixerolls	very-fine	smectitic			mesic	xeric	
HAMACER	22.0	NO	1971	NV		28B	Entisols	Psamments	Torripsamments	Haploduridic Torripsamments		mixed			frigid	torric	
HAMTAH	144.0	NO	1971	NV	UT	28A, 47	Mollisols	Xerolls	Argixerolls	Vitrandic Argixerolls	clayey-skeletal	smectitic			frigid	xeric	
HANDPAH	657.4	YES	1984	NV		29	Aridisols	Durids	Argidurids	Xeric Argidurids	loamy	mixed	superactive		mesic	aridic	shallow
HANDY	121.5	NO	1971	NV		25, 28B	Aridisols	Argids	Haplargids	Xeric Haplargids	clayey over loamy	smectitic OVER mixed	superactive		frigid	aridic	
HANGROCK	118.5	NO	1995	NV		23	Aridisols	Durids	Argidurids	Haploxeralfic Argidurids	ashy	glassy			mesic	aridic	shallow
HAPGOOD	432.0	YES	1971	NV	CA, CO, ID, MT, OR	21, 23, 24, 25, 27, 28A, 28B	Mollisols	Cryolls	Haplocryolls	Pachic Haplocryolls	loamy-skeletal	mixed	superactive		cryic		
HARCANY	64.6	NO	1974	NV	OR	23, 24	Mollisols	Cryolls	Haplocryolls	Pachic Haplocryolls	loamy-skeletal	mixed	superactive		cryic		
HARDBASIN	5.2	NO	2006	NV		30	Aridisols	Gypsids	Petrogypsids	Typic Petrogypsids	loamy	mixed	superactive		thermic	aridic	shallow
HARDHAT	273.3	NO	1986	NV		28A, 28B, 29	Aridisols	Calcids	Haplocalcids	Durinodic Haplocalcids	coarse-loamy	mixed	superactive		mesic	aridic	
HARDNUT	16.6	NO	2005	CA	NV	26	Mollisols	Xerolls	Argixerolls	Aridic Lithic Argixerolls	ashy-skeletal	glassy			frigid	xeric	
HARDOL	194.4	NO	1990	NV		28B	Mollisols	Cryolls	Calcicryolls	Pachic Calcicryolls	loamy-skeletal	carbonatic			cryic		
HARDZEM	245.4	NO	1990	NV		28B	Alfisols	Cryalfs	Haplocryalfs	Xeric Haplocryalfs	loamy-skeletal	mixed	superactive		cryic	xeric	
HARSKEL	2.0	NO	2006	CA	NV	23	Mollisols	Xerolls	Argixerolls	Vitritorrandic Argixerolls	ashy-skeletal	glassy			frigid	xeric	shallow
HART CAMP	5.1	NO	1974	NV	CA, OR	10, 23, 25, 27	Mollisols	Xerolls	Argixerolls	Aridic Argixerolls	loamy	mixed	superactive		frigid	xeric	shallow
HARTIG	48.8	NO	1981	NV	CA	23, 26	Mollisols	Xerolls	Haploxerolls	Aridic Haploxerolls	loamy-skeletal	mixed	superactive		frigid	xeric	
HARVAN	8.0	NO	2010	NV		29	Aridisols	Argids	Haplargids	Xeric Haplargids	loamy-skeletal	mixed	superactive		mesic	aridic	shallow
HASHWOODS	5.5	NO	2006	NV	CA	23	Mollisols	Cryolls	Argicryolls	Vitrandic Argicryolls	ashy	glassy			cryic		
HASTEE	13.3	NO	1990	NV		23	Mollisols	Cryolls	Haplocryolls	Pachic Haplocryolls	sandy-skeletal	mixed			cryic		
HATPEAK	85.7	NO	1984	NV	ID	25	Mollisols	Xerolls	Durixerolls	Typic Durixerolls	fine	smectitic			frigid	xeric	
HATUR	8.9	NO	1983	NV		28B	Mollisols	Rendolls	Cryrendolls	Typic Cryrendolls	loamy-skeletal	carbonatic			cryic		
HAUNCHEE	623.6	YES	1983	NV		28A, 28B	Mollisols	Rendolls	Cryrendolls	Lithic Cryrendolls	loamy-skeletal	carbonatic			cryic		
HAVINGDON	97.1	NO	1974	NV		24	Aridisols	Argids	Haplargids	Xeric Haplargids	clayey-skeletal	smectitic			mesic	aridic	
HAWSLEY	1533.0	YES	1980	NV	ID	11, 24, 26, 27	Entisols	Psamments	Torripsamments	Typic Torripsamments		mixed			mesic	torric	
HAYBOURNE	192.2	NO	1974	NV	CA, UT	25, 26, 27, 28A	Aridisols	Cambids	Haplocambids	Xeric Haplocambids	coarse-loamy	mixed	superactive		mesic	aridic	

(continued)

(continued)

Soil series	Area (km^2)	Bench-mark	Year Est-ablished	TL State	Other States Using	MLRAs Using	Order	Suborder	Great Group	Subgroup	Particle-size Class	Mineralogy Class	CEC Activity Class	Reaction Class	Soil temp. Class	SMR	Other Family
HAYESTON	181.8	NO	1971	NV		11, 25, 28B	Entisols	Orthents	Torriorthents	Xeric Torriorthents	coarse-loamy	mixed	superactive	calcareous	frigid	torric	
HAYMONT	128.9	NO	1996	NV		30	Entisols	Orthents	Torriorthents	Typic Torriorthents	coarse-silty	mixed	superactive	calcareous	thermic	torric	
HAYPEAK	16.2	NO	2011	NV		30	Alfisols	Ustalfs	Haplustalfs	Oxyaquic Haplustalfs	loamy-skeletal	mixed	superactive		frigid	ustic	
HAYPRESS	49.7	NO	1973	CA	NV	22A, 26	Mollisols	Xerolls	Haploxerolls	Psammentic Haploxerolls		mixed			frigid	xeric	
HAYSPUR	15.6	NO	1984	NV	ID	25	Mollisols	Aquolls	Endoaquolls	Fluvaquentic Endoaquolls	fine-loamy	mixed	superactive		frigid	aquic	
HECKISON	14.0	NO	1986	ID	NV	25	Mollisols	Xerolls	Durixerolls	Argiduridic Durixerolls	fine-loamy	mixed	superactive		frigid	xeric	
HEECHEE	122.1	NO	1986	NV		25	Mollisols	Xerolls	Argixerolls	Typic Argixerolls	loamy-skeletal	mixed	superactive		frigid	xeric	
HEFED	24.0	NO	1980	NV		26	Aridisols	Argids	Haplargids	Xeric Haplargids	loamy-skeletal	mixed	superactive		mesic	aridic	
HEIDTMAN	6.3	NO	1975	NV		26	Mollisols	Xerolls	Haploxerolls	Cumulic Haploxerolls	fine-loamy	mixed	superactive		mesic	xeric	
HEIST	690.0	NO	1942	UT	NV	28A, 28B, 29	Aridisols	Cambids	Haplocambids	Xeric Haplocambids	coarse-loamy	mixed	superactive		mesic	aridic	
HELEWEISER	69.0	NO	2006	NV		30	Aridisols	Calcids	Haplocalcids	Typic Haplocalcids	loamy-skeletal	mixed	superactive		hyperthermic	aridic	
HELKITCHEN	16.2	NO	2001	NV	AZ	30	Aridisols	Calcids	Haplocalcids	Lithic Haplocalcids	loamy-skeletal	carbonatic			thermic	aridic	
HENDAP	5.7	NO	1994	NV		28B	Mollisols	Xerolls	Haploxerolls	Aridic Lithic Haploxerolls	loamy-skeletal	mixed	superactive		mesic	xeric	
HENNINGSEN	10.9	NO	1970	NV		26	Mollisols	Xerolls	Haploxerolls	Oxyaquic Haploxerolls	coarse-loamy over sandy or sandy-skeletal	mixed	superactive		mesic	xeric	
HESSING	153.1	NO	1985	NV		24, 27, 28B	Aridisols	Cambids	Haplocambids	Typic Haplocambids	coarse-loamy	mixed	superactive		mesic	aridic	
HEUSSER	16.0	NO	2004	NV		28A	Mollisols	Xerolls	Palexerolls	Aridic Palexerolls	clayey-skeletal	smectitic			frigid	xeric	
HIDDENFOREST	4.2	NO	2011	NV		30	Mollisols	Ustolls	Argiustolls	Aridic Lithic Argiustolls	loamy-skeletal	mixed	superactive		mesic	ustic	
HIDDENSUN	40.0	NO	2006	NV		30	Aridisols	Calcids	Haplocalcids	Lithic Haplocalcids	loamy-skeletal	mixed	superactive		thermic	aridic	
HIGHAMS	17.4	NO	1973	ID	NV, UT, WY	12, 13, 25, 34A, 47	Entisols	Orthents	Torriorthents	Lithic Xeric Torriorthents	loamy-skeletal	carbonatic			frigid	torric	
HIGHLAND	49.1	NO	2006	NV		30	Aridisols	Argids	Haplargids	Typic Haplargids	loamy-skeletal	mixed	superactive		thermic	aridic	
HIGHUP	36.4	NO	2004	NV		28A	Mollisols	Xerolls	Calcixerolls	Typic Calcixerolls	loamy-skeletal	carbonatic			frigid	xeric	
HIKO PEAK	2.7	YES	1977	UT	NV	28A	Aridisols	Calcids	Haplocalcids	Xeric Haplocalcids	loamy-skeletal	mixed	active		mesic	aridic	
HIKO SPRINGS	6.4	NO	1940	UT	NV	28A, 28B, 34B	Aridisols	Calcids	Haplocalcids	Typic Haplocalcids	coarse-loamy	mixed	superactive		mesic	aridic	
HILLER	66.6	NO	2006	NV		30	Aridisols	Calcids	Haplocalcids	Durinodic Haplocalcids	loamy-skeletal	mixed	superactive		thermic	aridic	
HIRIDGE	57.0	NO	1984	NV		26, 27, 29	Mollisols	Cryolls	Argicryolls	Xeric Argicryolls	loamy-skeletal	mixed	superactive		cryic	xeric	shallow
HIRSCHDALE	26.2	NO	1980	NV		22A	Alfisols	Xeralfs	Haploxeralfs	Mollic Haploxeralfs	fine	smectitic			frigid	xeric	
HOBOG	5.2	NO	1972	UT	NV	30	Aridisols	Calcids	Haplocalcids	Lithic Haplocalcids	loamy-skeletal	mixed	superactive		thermic	aridic	
HOCAR	7.9	NO	1975	NV		26	Mollisols	Xerolls	Haploxerolls	Calcidic Haploxerolls	loamy-skeletal	mixed	superactive		mesic	xeric	shallow

(continued)

(continued)

Soil series	Area (km^2)	Bench-mark	Year Est-ablished	TL State	Other States Using	MLRAs Using	Order	Suborder	Great Group	Subgroup	Particle-size Class	Mineralogy Class	CEC Activity Class	Reaction Class	Soil temp. Class	SMR	Other Family
HODEDO	77.5	NO	1983	NV		28B	Mollisols	Xerolls	Durixerolls	Argiduridic Durixerolls	fine	smectitic			frigid	xeric	
HOGMALAT	10.1	NO	1986	ID	NV	25	Mollisols	Cryolls	Argicryolls	Lithic Argicryolls	loamy-skeletal	mixed	superactive		cryic		
HOGUM	4.3	NO	2004	NV		28A	Inceptisols	Aquepts	Endoaquepts	Typic Endoaquepts	fine-loamy	mixed	superactive	calcareous	mesic	aquic	
HOLBORN	43.6	NO	1986	NV		25, 28A	Entisols	Orthents	Torriorthents	Xeric Torriorthents	loamy	mixed	superactive	calcareous	mesic	torric	shallow
HOLBROOK	82.0	NO	1940	NV	CA	23, 26	Mollisols	Xerolls	Haploxerolls	Torriorthentic Haploxerolls	loamy-skeletal	mixed	superactive		mesic	xeric	
HOLLACE	56.1	NO	1992	NV		29	Aridisols	Calcids	Petrocalcids	Ustalfic Petrocalcids	loamy-skeletal	mixed	superactive		mesic	aridic	shallow
HOLLYWELL	219.0	NO	1993	NV		27, 29	Aridisols	Cambids	Haplocambids	Typic Haplocambids	loamy-skeletal	mixed	superactive		mesic	aridic	
HOLSINE	23.3	NO	1971	NV		28A	Aridisols	Calcids	Haplocalcids	Xeric Haplocalcids	sandy-skeletal over loamy	mixed	superactive		mesic	aridic	
HOLTLE	22.9	NO	1971	NV		28B	Mollisols	Xerolls	Haploxerolls	Duridic Haploxerolls	coarse-loamy	mixed	superactive		mesic	xeric	
HOLTVILLE	5.8	YES	1918	CA	AZ, NV	30, 31, 40	Entisols	Fluvents	Torrifluvents	Typic Torrifluvents	clayey over loamy	smectitic OVER mixed	superactive	calcareous	hyperthermic	torric	
HOME CAMP	5.1	YES	1971	NV	CA	21, 23	Mollisols	Xerolls	Argixerolls	Vitrandic Argixerolls	clayey-skeletal	smectitic			frigid	xeric	
HOMESTAKE	103.9	NO	1971	NV		28A, 28B, 29	Mollisols	Xerolls	Argixerolls	Argiduridic Argixerolls	clayey-skeletal	smectitic			frigid	xeric	
HOOPLITE	205.0	NO	1985	NV		27, 28B	Aridisols	Argids	Haplargids	Lithic Xeric Haplargids	loamy-skeletal	mixed	superactive		mesic	aridic	
HOOT	348.7	NO	1975	NV	WY	24, 32	Aridisols	Argids	Haplargids	Lithic Haplargids	loamy-skeletal	mixed	superactive		mesic	aridic	
HOOTEN	18.0	NO	1971	NV		27	Aridisols	Durids	Haplodurids	Cambidic Haplodurids	loamy-skeletal	mixed	superactive		mesic	aridic	shallow
HOPEKA	413.6	NO	1971	NV		25, 27, 28A, 28B	Entisols	Orthents	Torriorthents	Lithic Xeric Torriorthents	loamy-skeletal	carbonatic			frigid	torric	
HOPPSWELL	93.6	NO	2006	NV		30	Aridisols	Argids	Haplargids	Ustic Haplargids	loamy-skeletal	mixed	superactive		thermic	aridic	
HORSECAMP	5.1	NO	1990	CA	NV	23	Vertisols	Xererts	Haploxererts	Aridic Haploxererts	fine	smectitic			mesic	xeric	
HORSETRACK	18.3	NO	2013	NV		28B	Mollisols	Xerolls	Argixerolls	Aridic Lithic Argixerolls	loamy-skeletal	mixed	superactive		frigid	xeric	
HOTSPRINGS	21.9	NO	1981	NV		26	Entisols	Psamments	Torripsamments	Xeric Torripsamments		mixed			mesic	torric	
HOUGH	116.0	NO	1981	NV		27	Aridisols	Argids	Haplargids	Typic Haplargids	fine-loamy over sandy or sandy-skeletal	mixed	superactive		mesic	aridic	
HOURLAND	11.1	NO	2016	NV		25	Mollisols	Xerolls	Haploxerolls	Pachic Haploxerolls	sandy-skeletal	mixed			frigid	xeric	
HUEVI	415.0	NO	1994	NV	AZ	30	Aridisols	Calcids	Haplocalcids	Durinodic Haplocalcids	loamy-skeletal	mixed	superactive		hyperthermic	aridic	
HUILEPASS	90.7	NO	2004	NV		28A	Aridisols	Argids	Haplargids	Xeric Haplargids	loamy-skeletal	mixed	superactive		mesic	aridic	
HULDERMAN	6.4	NO	1994	NV		28B	Mollisols	Aquolls	Endoaquolls	Duric Endoaquolls	fine-loamy over sandy or sandy-skeletal	mixed	superactive		mesic	aquic	
HUMBOLDT	288.0	YES	1957	NV	CA	23, 24, 25, 27, 28B	Mollisols	Aquolls	Endoaquolls	Fluvaquentic Vertic Endoaquolls	fine	smectitic		calcareous	mesic	aquic	
HUMDUN	77.0	NO	1979	NV	ID	24, 25	Aridisols	Cambids	Haplocambids	Durinodic Xeric Haplocambids	coarse-loamy	mixed	superactive		frigid	aridic	

(continued)

(continued)

Soil series	Area (km²)	Bench-mark	Year Est-ablished	TL State	Other States Using	MLRAs Using	Order	Suborder	Great Group	Subgroup	Particle-size Class	Mineralogy Class	CEC Activity Class	Reaction Class	Soil temp. Class	SMR	Other Family
HUNDRAW	335.0	NO	1986	NV	UT	25, 28A, 28B	Entisols	Orthents	Torriorthents	Xeric Torriorthents	loamy	mixed	superactive	calcareous	mesic	torric	shallow
HUNEWILL	115.7	NO	1981	NV		25, 26	Aridisols	Argids	Haplargids	Xeric Haplargids	loamy-skeletal	mixed	superactive		mesic	aridic	
HUNNTON	945.0	YES	1986	NV	CA, ID	23, 24, 25, 28B	Aridisols	Durids	Argidurids	Xeric Argidurids	fine	smectitic			mesic	aridic	
HUSSA	112.0	NO	1971	CA	NV	21, 23, 25	Mollisols	Aquolls	Endoaquolls	Aquandic Endoaquolls	ashy	glassy		calcareous	frigid	aquic	
HUSSELL	19.8	NO	1986	NV		25	Aridisols	Argids	Haplargids	Durinodic Xeric Haplargids	coarse-loamy	mixed	superactive		mesic	aridic	
HUSSMAN	7.2	NO	1975	NV		26	Mollisols	Xerolls	Haploxerolls	Torrertic Haploxerolls	fine	smectitic			mesic	xeric	
HUTCHLEY	148.0	NO	1987	ID	CA, NV, OR	10, 12, 13, 23, 25, 27, 28B	Mollisols	Xerolls	Argixerolls	Lithic Argixerolls	loamy-skeletal	mixed	superactive		frigid	xeric	
HUXLEY	12.2	NO	1971	NV		27	Aridisols	Argids	Natrargids	Typic Natrargids	clayey-skeletal over sandy or sandy-skeletal	smectitic OVER mixed			mesic	aridic	
HYLOC	111.2	NO	1981	NV		26	Mollisols	Xerolls	Argixerolls	Aridic Argixerolls	clayey	smectitic			mesic	xeric	shallow
HYMAS	138.8	NO	1972	ID	NV	10, 12, 13, 25, 28A, 28B, 47	Mollisols	Xerolls	Haploxerolls	Lithic Haploxerolls	loamy-skeletal	carbonatic			frigid	xeric	
HYPOINT	49.2	NO	2000	NV	CA	30	Entisols	Orthents	Torriorthents	Typic Torriorthents	sandy	mixed			thermic	torric	
HYZEN	474.8	YES	1990	NV		28A, 28B, 29	Mollisols	Xerolls	Haploxerolls	Aridic Lithic Haploxerolls	loamy-skeletal	carbonatic			frigid	xeric	
ICEBERG	36.2	NO	2001	NV	AZ	30	Aridisols	Calcids	Haplocalcids	Lithic Haplocalcids	loamy-skeletal	carbonatic			hyperthermic	aridic	
ICHBOD	8.2	NO	1986	NV		10, 25	Mollisols	Xerolls	Argixerolls	Aridic Argixerolls	clayey	smectitic			frigid	xeric	shallow
IDLEWILD	9.2	NO	1980	NV		26	Mollisols	Xerolls	Argixerolls	Aquic Argixerolls	fine	smectitic			mesic	xeric	
IDWAY	109.0	NO	1994	NV		28B	Aridisols	Calcids	Haplocalcids	Durinodic Xeric Haplocalcids	coarse-loamy over sandy or sandy-skeletal	mixed	superactive		mesic	aridic	
IFTEEN	12.6	NO	2006	NV		30	Aridisols	Calcids	Haplocalcids	Durinodic Haplocalcids	coarse-loamy	carbonatic			thermic	aridic	
IGDELL	144.7	NO	1986	NV		25	Mollisols	Xerolls	Durixerolls	Abruptic Argiduridic Durixerolls	fine	smectitic			frigid	xeric	
ILTON	2.7	NO	1974	NV		28B	Aridisols	Calcids	Haplocalcids	Durinodic Xeric Haplocalcids	coarse-loamy	mixed	superactive		mesic	aridic	
INCY	54.7	NO	1975	NV	CA	21, 23, 26	Entisols	Psamments	Torripsamments	Xeric Torripsamments		mixed			mesic	torric	
INDIAN CREEK	164.0	YES	1974	NV	CA	23, 26	Aridisols	Durids	Argidurids	Xeric Argidurids	clayey	smectitic			mesic	aridic	shallow
INDIANO	112.0	NO	1975	NV	CA	21, 26, 29	Mollisols	Xerolls	Argixerolls	Aridic Argixerolls	fine-loamy	mixed	superactive		mesic	xeric	
INDICOVE	52.5	NO	2010	NV		29	Aridisols	Durids	Argidurids	Xeric Argidurids	loamy-skeletal	mixed	superactive		mesic	aridic	shallow
INMO	116.7	NO	1985	NV		27, 29	Entisols	Orthents	Torriorthents	Typic Torriorthents	sandy-skeletal	mixed			mesic	torric	
INPENDENCE	19.7	NO	1986	NV		25	Inceptisols	Xerepts	Humixerepts	Pachic Humixerepts	loamy-skeletal	mixed	superactive		frigid	xeric	
INVILLE	2.4	NO	1970	NV	CA	22A	Alfisols	Xeralfs	Haploxeralfs	Ultic Haploxeralfs	loamy-skeletal	isotic			frigid	xeric	
IRETEBA	15.0	NO	1964	NV	AZ	30	Entisols	Fluvents	Torrifluvents	Typic Torrifluvents	coarse-loamy	mixed	superactive	calcareous	thermic	torric	
IRON BLOSSOM	22.4	NO	1968	NV		24	Entisols	Fluvents	Torrifluvents	Duric Torrifluvents	fine-loamy	mixed	superactive	calcareous	mesic	torric	
IRONGOLD	844.0	NO	2006	NV		30	Aridisols	Calcids	Petrocalcids	Typic Petrocalcids	loamy	mixed	superactive		thermic	aridic	shallow

(continued)

(continued)

Soil series	Area (km²)	Bench-mark	Year Est-ablished	TL State	Other States Using	MLRAs Using	Order	Suborder	Great Group	Subgroup	Particle-size Class	Mineralogy Class	CEC Activity Class	Reaction Class	Soil temp. Class	SMR	Other Family
ISOLDE	1267.0	NO	1980	NV	CA	23, 24, 27	Entisols	Psamments	Torripsamments	Typic Torripsamments		mixed			mesic	torric	
ISTER	218.4	NO	1981	NV		26	Mollisols	Xerolls	Argixerolls	Aridic Argixerolls	loamy-skeletal	mixed	superactive		mesic	xeric	
ITCA	383.6	YES	1971	NV	ID, OR	11, 21, 24, 25, 27, 28B, 29	Mollisols	Xerolls	Argixerolls	Lithic Argixerolls	clayey-skeletal	smectitic			frigid	xeric	
ITME	185.0	NO	1984	NV		29	Entisols	Orthents	Torriorthents	Typic Torriorthents	sandy-skeletal	mixed			mesic	torric	
IVER	81.1	NO	1974	NV		24	Mollisols	Xerolls	Haploxerolls	Pachic Haploxerolls	coarse-loamy	mixed	superactive		frigid	xeric	
IXIAN	28.9	NO	1986	NV		28A	Entisols	Orthents	Torriorthents	Aquic Torriorthents	fine-silty	mixed	superactive	calcareous	mesic	torric	
IZAMATCH	329.0	NO	1994	NV	UT	28A, 28B	Entisols	Orthents	Torriorthents	Typic Torriorthents	sandy-skeletal	mixed			mesic	torric	
IZAR	370.1	NO	1986	NV		25, 28A, 28B	Entisols	Orthents	Torriorthents	Lithic Xeric Torriorthents	loamy-skeletal	mixed	superactive	calcareous	mesic	torric	
IZO	838.0	NO	1984	NV		27, 28A, 29	Entisols	Orthents	Torriorthents	Typic Torriorthents	sandy-skeletal	mixed			mesic	torric	
IZOD	124.7	NO	1985	NV		24, 25, 28B	Entisols	Orthents	Torriorthents	Lithic Xeric Torriorthents	loamy-skeletal	carbonatic			mesic	torric	
JABU	0.6	NO	1970	CA	NV	22A	Alfisols	Xeralfs	Haploxeralfs	Ultic Haploxeralfs	coarse-loamy	mixed	superactive		frigid	xeric	
JACARANDA	2.0	NO	2009	NV		27	Entisols	Orthents	Torriorthents	Lithic Torriorthents	sandy-skeletal	mixed			mesic	torric	
JACK CREEK	26.5	NO	1968	NV	UT	25	Mollisols	Xerolls	Haploxerolls	Torriorthentic Haploxerolls	sandy-skeletal	mixed			frigid	xeric	
JACKMORE	11.9	NO	2016	NV		25	Entisols	Orthents	Cryorthents	Typic Cryorthents	loamy-skeletal	mixed	superactive	nonacid	cryic		
JACKPOT	36.5	NO	1986	NV		25, 28B	Aridisols	Cambids	Haplocambids	Vitrixerandic Haplocambids	ashy	glassy			mesic	aridic	shallow
JACKROCK	36.2	NO	2013	NV		28B	Mollisols	Xerolls	Argixerolls	Lithic Argixerolls	clayey-skeletal	smectitic			frigid	xeric	
JACRATZ	30.5	NO	1995	NV		27	Entisols	Orthents	Torriorthents	Xeric Torriorthents	loamy	mixed	superactive	calcareous	mesic	torric	shallow
JAMCANVAR	2.2	NO	2012	NV		26	Mollisols	Xerolls	Haploxerolls	Aquic Haploxerolls	fine-loamy	mixed	superactive		mesic	xeric	
JAMES CANYON	45.0	NO	1969	NV	CA, UT	21, 26, 28A, 28B	Mollisols	Aquolls	Endoaquolls	Cumulic Endoaquolls	fine-loamy	mixed	superactive		mesic	aquic	
JAMESCANNY	3.3	NO	2010	NV	CA	26	Mollisols	Xerolls	Haploxerolls	Pachic Haploxerolls	fine-loamy	mixed	superactive		mesic	xeric	
JARAB	141.1	NO	1971	NV		29	Mollisols	Xerolls	Durixerolls	Haploduridic Durixerolls	loamy-skeletal	mixed	superactive		mesic	xeric	shallow
JARBOE	18.7	NO	1974	NV		29	Aridisols	Calcids	Haplocalcids	Petronodic Haplocalcids	fine-loamy	carbonatic			mesic	aridic	
JAYBEE	315.0	NO	1990	NV	CA	23	Aridisols	Argids	Haplargids	Lithic Xeric Haplargids	loamy	mixed	superactive		mesic	aridic	
JEAN	85.5	NO	1964	NV		30	Entisols	Orthents	Torriorthents	Typic Torriorthents	sandy-skeletal	mixed			thermic	torric	
JENNESS	61.3		1944	ID			Entisols	Orthents	Torriorthents	Xeric Torriorthents	coarse-loamy	mixed	superactive		mesic	torric	
JENOR	21.9	NO	1985	NV	ID	24, 25	Aridisols	Durids	Haplodurids	Typic Haplodurids	coarse-loamy	mixed	superactive		mesic	aridic	
JERICHO	559.4	NO	1981	UT	NV	25, 28A, 28B	Aridisols	Durids	Haplodurids	Xeric Haplodurids	loamy-skeletal	mixed	superactive		mesic	aridic	shallow
JERVAL	487.4	YES	1986	NV		24, 27	Aridisols	Argids	Natrargids	Durinodic Natrargids	fine-loamy	mixed	superactive		mesic	aridic	

(continued)

(continued)

Soil series	Area (km^2)	Bench-mark	Year Est-ablished	TL State	Other States Using	MLRAs Using	Order	Suborder	Great Group	Subgroup	Particle-size Class	Mineralogy Class	CEC Activity Class	Reaction Class	Soil temp. Class	SMR	Other Family
JESAYNO	38.0	NO	2006	NV	CA	23	Aridisols	Cambids	Haplocambids	Vitrixerandic Haplocambids	ashy	glassy			mesic	aridic	
JESSE CAMP	94.0	NO	1971	NV	CA, OR	23, 28B	Aridisols	Cambids	Haplocambids	Xeric Haplocambids	fine-silty	mixed	superactive		frigid	aridic	
JETCOP	64.6	NO	1985	NV		29	Aridisols	Durids	Argidurids	Abruptic Xeric Argidurids	clayey	mixed	superactive		mesic	aridic	shallow
JETMINE	8.9	NO	2006	NV		30	Aridisols	Durids	Haplodurids	Cambidic Haplodurids	loamy	mixed	superactive		thermic	aridic	shallow
JEVETS	56.2	NO	1996	NV		29	Aridisols	Durids	Haplodurids	Typic Haplodurids	sandy	mixed			mesic	aridic	
JIVAS	50.4	NO	1983	NV		25	Mollisols	Xerolls	Argixerolls	Aridic Argixerolls	loamy-skeletal	mixed	superactive		frigid	xeric	
JOB	3.4	NO	1981	NV	UT	26, 28A	Entisols	Fluvents	Torrifluvents	Oxyaquic Torrifluvents	fine-loamy	mixed	superactive	calcareous	mesic	torric	
JOBPEAK	96.3	NO	1986	NV		27	Entisols	Orthents	Torriorthents	Lithic Xeric Torriorthents	loamy-skeletal	mixed	superactive	nonacid	mesic	torric	
JOBSIS	1.2	NO	2006	CA	NV	22A	Entisols	Orthents	Cryorthents	Typic Cryorthents	sandy-skeletal	mixed			cryic	xeric	shallow
JOEMAY	118.2	NO	2011	NV		30	Aridisols	Calcids	Haplocalcids	Sodic Haplocalcids	loamy-skeletal	carbonatic			thermic	aridic	
JOLAN	27.9	NO	1972	NV		28B, 29	Aridisols	Durids	Haplodurids	Typic Haplodurids	coarse-loamy	mixed	superactive		mesic	aridic	
JONLAKE	30.2	NO	2004	NV		28A	Mollisols	Cryolls	Haplocryolls	Lithic Haplocryolls	loamy-skeletal	mixed	superactive		cryic		
JONNIC	34.8	NO	1994	NV		30	Aridisols	Durids	Argidurids	Xeric Argidurids	clayey-skeletal	smectitic			thermic	aridic	
JORGE	15.7	NO	1970	CA	NV	22A	Alfisols	Xeralfs	Haploxeralfs	Andic Haploxeralfs	loamy-skeletal	isotic			frigid	xeric	
JOTAVA	23.0	NO	1993	NV		29	Aridisols	Salids	Aquisalids	Calcic Aquisalids	fine-loamy	mixed	active		mesic	aridic	
JOWEC	19.4	NO	1980	NV		26	Aridisols	Argids	Paleargids	Xeric Paleargids	fine	smectitic			mesic	aridic	
JUBILEE	30.9	NO	1975	NV		26	Mollisols	Aquolls	Endoaquolls	Typic Endoaquolls	coarse-loamy	mixed	superactive		mesic	aquic	
JUMBLE	10.2	NO	2009	NV		28A	Inceptisols	Cryepts	Dystrocryepts	Lamellic Dystrocryepts	loamy-skeletal	mixed	superactive		cryic		
JUMBO	17.0	NO	1980	NV	CA	26	Mollisols	Xerolls	Argixerolls	Pachic Ultic Argixerolls	loamy-skeletal	mixed	superactive		frigid	xeric	
JUMBOPEAK	19.4	NO	2006	NV		30	Mollisols	Ustolls	Argiustolls	Aridic Argiustolls	loamy-skeletal	mixed	superactive		mesic	ustic	
JUNG	326.4	NO	1985	NV		24, 27, 28B	Aridisols	Argids	Haplargids	Lithic Xeric Haplargids	clayey-skeletal	smectitic			mesic	aridic	
JUNGO	41.0	NO	1988	NV		27, 28A	Aridisols	Argids	Haplargids	Xeric Haplargids	loamy-skeletal	mixed	superactive		mesic	aridic	
JURADO	17.9	NO	2004	NV		28A	Aridisols	Argids	Haplargids	Typic Haplargids	loamy-skeletal	mixed	superactive		mesic	aridic	
JUVA	94.1	NO	1971	NV		27, 28A	Entisols	Fluvents	Torrifluvents	Typic Torrifluvents	coarse-loamy	mixed	superactive	calcareous	mesic	torric	
KAFFUR	14.4	NO	1990	NV		23	Entisols	Orthents	Torriorthents	Xeric Torriorthents	loamy-skeletal	mixed	superactive	calcareous	mesic	torric	shallow
KANACKEY	78.4	NO	1992	NV		30	Aridisols	Argids	Haplargids	Lithic Haplargids	clayey-skeletal	smectitic			thermic	aridic	
KANESPRINGS	127.3	NO	1992	NV		30	Aridisols	Durids	Argidurids	Typic Argidurids	loamy	mixed	superactive		thermic	aridic	shallow
KARLO	21.2	NO	1971	NV	CA	23	Vertisols	Xererts	Haploxererts	Leptic Haploxererts	very-fine	smectitic			frigid	xeric	
KARPP	49.6	NO	1986	NV		25	Aridisols	Durids	Haplodurids	Xeric Haplodurids	loamy-skeletal	mixed	superactive		mesic	aridic	shallow
KASPAL	77.0	NO	1992	NV		30	Aridisols	Argids	Petroargids	Typic Petroargids	fine	smectitic			thermic	aridic	
KATELANA	949.4	YES	1990	NV		28B	Entisols	Orthents	Torriorthents	Typic Torriorthents	fine-silty	carbonatic			mesic	torric	
KAWICH	312.0	NO	1984	NV	OR	24, 28A, 28B, 29	Entisols	Psamments	Torripsamments	Typic Torripsamments		mixed			mesic	torric	

(continued)

(continued)

Soil series	Area (km^2)	Bench-mark	Year Est-ablished	TL State	Other States Using	MLRAs Using	Order	Suborder	Great Group	Subgroup	Particle-size Class	Mineralogy Class	CEC Activity Class	Reaction Class	Soil temp. Class	SMR	Other Family
KAYO	53.4	NO	1980	NV		26	Aridisols	Argids	Haplargids	Xeric Haplargids	loamy-skeletal	mixed	superactive		mesic	aridic	
KAZUL	3.3	NO	2012	NV		26	Mollisols	Xerolls	Haploxerolls	Torrifluventic Haploxerolls	fine-loamy	mixed	superactive		mesic	xeric	
KEEFA	463.5	YES	1984	NV		28A, 29	Aridisols	Cambids	Haplocambids	Durinodic Haplocambids	coarse-loamy	mixed	superactive		mesic	aridic	
KELK	618.9	NO	1985	NV		24, 25, 28B	Aridisols	Cambids	Haplocambids	Durinodic Xeric Haplocambids	fine-silty	mixed	superactive		mesic	aridic	
KEMAN	79.0	NO	1986	ID	NV	25	Mollisols	Cryolls	Argicryolls	Pachic Argicryolls	loamy-skeletal	mixed	superactive		cryic		
KEYOLE	63.8	NO	2009	NV		28A	Inceptisols	Cryepts	Haplocryepts	Xeric Haplocryepts	loamy-skeletal	mixed	active		cryic	xeric	
KIDWELL	98.7	NO	2006	NV		30	Aridisols	Argids	Calciargids	Typic Calciargids	fine-loamy	mixed	superactive		thermic	aridic	
KIMMERLING	22.1	YES	1975	NV		26	Mollisols	Aquolls	Endoaquolls	Cumulic Endoaquolls	fine-loamy	mixed	superactive		mesic	aquic	
KINGINGHAM	77.0	NO	1985	NV		24	Aridisols	Durids	Natridurids	Typic Natridurids	fine	smectitic			mesic	aridic	
KINGSBEACH	0.6	NO	2007	CA	NV	22A	Alfisols	Xeralfs	Palexeralfs	Ultic Palexeralfs	fine-loamy	isotic			frigid	xeric	
KINGSRIVER	21.9	NO	1992	NV		24, 25	Mollisols	Aquolls	Endoaquolls	Cumulic Endoaquolls	coarse-loamy	mixed	superactive		mesic	aquic	
KIOTE	20.0	NO	1984	NV	CA	26	Mollisols	Cryolls	Argicryolls	Vitrandic Argicryolls	loamy-skeletal	mixed	superactive		cryic		
KIOUS	28.6	NO	2004	NV		28A	Mollisols	Cryolls	Haplocryolls	Pachic Haplocryolls	loamy-skeletal	mixed	superactive		cryic		shallow
KITGRAM	20.4	NO	2006	NV		30	Mollisols	Ustolls	Calciustolls	Pachic Calciustolls	loamy-skeletal	carbonatic			frigid	ustic	
KLECK	37.1	NO	1992	NV		24	Aridisols	Cambids	Haplocambids	Xeric Haplocambids	loamy	mixed	superactive		mesic	aridic	shallow
KLECKNER	159.2	NO	1986	NV		25	Mollisols	Xerolls	Argixerolls	Aridic Argixerolls	clayey-skeletal	smectitic			frigid	xeric	
KLEINBUSH	24.3	NO	1980	NV		27	Aridisols	Argids	Natrargids	Typic Natrargids	fine	smectitic			mesic	aridic	
KNIESLEY	3.3	NO	1985	NV		27	Entisols	Fluvents	Xerofluvents	Oxyaquic Xerofluvents	fine-silty over clayey	mixed OVER smectitic	superactive	calcareous	mesic	xeric	
KNOB HILL	40.0	NO	1964	NV	CA	30	Aridisols	Calcids	Haplocalcids	Typic Haplocalcids	sandy	mixed			thermic	aridic	
KNOSS	23.2	NO	1986	NV		24, 27	Aridisols	Durids	Natridurids	Typic Natridurids	clayey	smectitic			mesic	aridic	shallow
KNOTT	55.0	NO	1986	NV		23, 24	Aridisols	Durids	Natridurids	Typic Natridurids	clayey	smectitic			mesic	aridic	shallow
KOBEH	218.0	NO	1971	NV		25, 28B	Aridisols	Cambids	Haplocambids	Durinodic Xeric Haplocambids	loamy-skeletal	mixed	superactive		frigid	aridic	
KODAK	1.8	NO	1963	NV		27	Mollisols	Xerolls	Haploxerolls	Torrifluventic Haploxerolls	coarse-loamy	mixed	superactive		mesic	xeric	
KODRA	43.8	NO	1983	NV		24, 25	Aridisols	Durids	Haplodurids	Xereptic Haplodurids	coarse-loamy	mixed	superactive		mesic	aridic	
KOLCHECK	0.9	NO	2013	NV		28B	Mollisols	Cryolls	Argicryolls	Oxyaquic Argicryolls	fine	smectitic			cryic		
KOLDA	177.9	NO	1990	NV		27, 28B	Mollisols	Aquolls	Endoaquolls	Typic Endoaquolls	fine	smectitic		calcareous	mesic	aquic	
KOONTZ	88.7	NO	1975	NV	CA	21, 26	Mollisols	Xerolls	Argixerolls	Aridic Argixerolls	loamy-skeletal	mixed	superactive		mesic	xeric	shallow
KORTTY	14.6	NO	1986	NV		25	Aridisols	Argids	Petroargids	Duric Petroargids	fine-loamy	mixed	superactive		mesic	aridic	
KOYEN	1215.2	NO	1972	NV		28A, 28B, 29	Aridisols	Cambids	Haplocambids	Typic Haplocambids	coarse-loamy	mixed	superactive		mesic	aridic	

(continued)

(continued)

Soil series	Area (km^2)	Bench-mark	Year Est-ablished	TL State	Other States Using	MLRAs Using	Order	Suborder	Great Group	Subgroup	Particle-size Class	Mineralogy Class	CEC Activity Class	Reaction Class	Soil temp. Class	SMR	Other Family
KOYNIK	21.8	NO	1985	NV		24, 28B	Entisols	Orthents	Torriorthents	Lithic Torriorthents	loamy-skeletal	carbonatic			mesic	torric	
KRAM	274.7	NO	1981	NV		24, 25, 26, 27, 28B	Entisols	Orthents	Torriorthents	Lithic Xeric Torriorthents	loamy-skeletal	carbonatic			mesic	torric	
KRENKA	9.7	NO	1994	NV		25	Mollisols	Xerolls	Argixerolls	Pachic Argixerolls	loamy-skeletal	mixed	superactive		frigid	xeric	
KREZA	9.7	NO	1990	NV		27	Aridisols	Argids	Haplargids	Lithic Xeric Haplargids	loamy	mixed	superactive		mesic	aridic	
KUMIVA	111.4	NO	1988	NV		27	Entisols	Orthents	Torriorthents	Typic Torriorthents	coarse-loamy	mixed	superactive	calcareous	mesic	torric	
KUNZLER	982.0	YES	1985	UT	NV	28A, 28B, 29	Aridisols	Calcids	Haplocalcids	Durinodic Xeric Haplocalcids	coarse-loamy	mixed	superactive		mesic	aridic	
KURSTAN	127.8	NO	1992	NV		30	Aridisols	Calcids	Haplocalcids	Durinodic Haplocalcids	coarse-loamy	mixed	superactive		thermic	aridic	
KYLECANYON	7.9	NO	2006	NV		30	Mollisols	Ustolls	Calciustolls	Petrocalcic Calciustolls	loamy-skeletal	carbonatic			mesic	ustic	
KYLER	1341.2	YES	1971	NV		28A, 28B, 29	Entisols	Orthents	Torriorthents	Lithic Xeric Torriorthents	loamy-skeletal	carbonatic			mesic	torric	
KZIN	114.2	NO	1986	NV		25, 28A, 28B	Entisols	Orthents	Torriorthents	Xeric Torriorthents	loamy-skeletal	mixed	superactive	calcareous	mesic	torric	shallow
LABKEY	203.8	NO	1988	NV		27	Aridisols	Cambids	Haplocambids	Typic Haplocambids	sandy-skeletal	mixed			mesic	aridic	
LABOU	14.3	NO	1970	NV		27	Aridisols	Argids	Natrargids	Lithic Natrargids	loamy-skeletal	mixed	superactive		mesic	aridic	
LABSHAFT	105.5	NO	1971	NV		25, 28B	Mollisols	Cryolls	Haplocryolls	Lithic Haplocryolls	loamy-skeletal	mixed	superactive		cryic		
LADYOFSNOW	14.5	NO	2006	NV		30	Inceptisols	Cryepts	Calcicryepts	Oxyaquic Calcicryepts	loamy-skeletal	carbonatic			cryic	ustic	
LAHONTAN	126.9	NO	1909	NV	ID, UT	27, 28A, 29	Entisols	Orthents	Torriorthents	Vertic Torriorthents	fine	smectitic		calcareous	mesic	torric	
LAKASH	21.5	NO	1991	NV		29	Entisols	Orthents	Torriorthents	Vitrandic Torriorthents	ashy	glassy		calcareous	mesic	torric	
LAMADRE	7.6	NO	2006	NV		30	Mollisols	Ustolls	Haplustolls	Torriorthentic Haplustolls	loamy-skeletal	mixed	superactive		frigid	ustic	
LAND	39.5	YES	1923	NV		30	Aridisols	Salids	Aquisalids	Typic Aquisalids	fine-silty	mixed	superactive		thermic	aridic	
LANDCO	29.4	NO	1974	NV		24	Entisols	Orthents	Torriorthents	Typic Torriorthents	coarse-silty over clayey	mixed OVER smectitic	superactive	calcareous	mesic	torric	
LANDERMEYER	13.8	NO		NV		25, 28B	Mollisols	Xerolls	Haploxerolls	Vitritorrandic Haploxerolls	fine-loamy	mixed	superactive		mesic	xeric	
LANFAIR	53.1	NO	2006	NV		30	Aridisols	Cambids	Haplocambids	Ustic Haplocambids	sandy-skeletal	mixed			thermic	aridic	
LANGSTON	88.0	NO	1985	NV	CA	23, 27	Aridisols	Argids	Haplargids	Xeric Haplargids	fine-loamy over sandy or sandy-skeletal	mixed	superactive		mesic	aridic	
LANIP	247.2	NO	2006	NV		30	Aridisols	Argids	Calciargids	Typic Calciargids	fine-loamy	mixed	superactive		thermic	aridic	
LAPED	86.2	NO	1985	NV	ID	23, 24, 25	Aridisols	Durids	Argidurids	Typic Argidurids	loamy	mixed	superactive		mesic	aridic	shallow
LAPON	138.2	NO	1940	NV		24, 26, 27	Aridisols	Durids	Argidurids	Xeric Argidurids	loamy-skeletal	mixed	superactive		mesic	aridic	shallow
LAROSS	37.5	NO	1992	NV		29	Mollisols	Ustolls	Haplustolls	Vitrandic Haplustolls	ashy-skeletal	glassy			mesic	ustic	
LARYAN	4.1	NO	2016	NV		25	Inceptisols	Aquepts	Cryaquepts	Histic Cryaquepts	clayey	kaolinitic		nonacid	cryic	aquic	
LAS VEGAS	135.5	NO	1923	NV		30	Aridisols	Calcids	Petrocalcids	Typic Petrocalcids	loamy	carbonatic			thermic	aridic	shallow

(continued)

(continued)

Soil series	Area (km^2)	Bench-mark	Year Est-ablished	TL State	Other States Using	MLRAs Using	Order	Suborder	Great Group	Subgroup	Particle-size Class	Mineralogy Class	CEC Activity Class	Reaction Class	Soil temp. Class	SMR	Other Family
LASTCHANCE	136.4	NO	2001	NV		30	Aridisols	Calcids	Petrocalcids	Calcic Petrocalcids	loamy-skeletal	carbonatic			thermic	aridic	
LASTONE	45.2	NO	2006	NV		30	Aridisols	Argids	Haplargids	Ustic Haplargids	loamy-skeletal	mixed	superactive		mesic	aridic	shallow
LASTSUMMER	2.4	NO	2010	CA	NV	26	Mollisols	Cryolls	Argicryolls	Vitrandic Argicryolls	ashy-skeletal over loamy-skeletal	glassy OVER mixed	superactive		cryic	xeric	
LATHROP	187.7	NO	1972	NV		29	Aridisols	Argids	Haplargids	Durinodic Haplargids	fine-loamy over sandy or sandy-skeletal	mixed	superactive		mesic	aridic	
LAXAL	163.7	NO	1972	NV		28B, 29	Aridisols	Calcids	Haplocalcids	Durinodic Haplocalcids	loamy-skeletal	mixed	superactive		mesic	aridic	
LAYVIEW	171.6	NO	1985	NV		23, 24, 27, 28B	Mollisols	Cryolls	Argicryolls	Lithic Argicryolls	loamy-skeletal	mixed	superactive		cryic		
LAZAN	182.4	NO	1984	NV		26, 29	Entisols	Orthents	Torriorthents	Xeric Torriorthents	sandy-skeletal	mixed			mesic	torric	shallow
LEALANDIC	37.4	NO	1996	NV		30	Aridisols	Durids	Argidurids	Typic Argidurids	clayey-skeletal	smectitic			thermic	aridic	
LEECANYON	17.0	NO	2006	NV		30	Mollisols	Ustolls	Calciustolls	Petrocalcic Calciustolls	loamy-skeletal	carbonatic			mesic	ustic	shallow
LEEVAN	46.0	NO	1986	NV	OR	23, 25	Mollisols	Xerolls	Argixerolls	Typic Argixerolls	clayey-skeletal	smectitic			frigid	xeric	
LEHMANDOW	5.3	NO	2009	NV		28A	Mollisols	Aquolls	Endoaquolls	Typic Endoaquolls	loamy-skeletal	mixed	superactive		frigid	aquic	
LEMCAVE	6.1	NO	2009	NV		28A	Inceptisols	Cryepts	Haplocryepts	Oxyaquic Haplocryepts	sandy-skeletal	mixed			cryic		
LEMM	32.1	NO	1980	NV		26	Mollisols	Xerolls	Argixerolls	Aridic Argixerolls	loamy-skeletal	mixed	superactive		mesic	xeric	
LEO	467.0	NO	1984	NV	UT	28A, 29	Entisols	Orthents	Torriorthents	Typic Torriorthents	sandy-skeletal	mixed			mesic	torric	
LERROW	175.0	NO	1986	NV	OR	25	Mollisols	Xerolls	Argixerolls	Vitritorrandic Argixerolls	fine	smectitic			frigid	xeric	
LEVIATHAN	64.4	NO	1980	NV	CA	23, 26	Mollisols	Xerolls	Argixerolls	Aridic Argixerolls	loamy-skeletal	mixed	superactive		mesic	xeric	
LEWDLAC	17.2	NO	1996	NV		30	Aridisols	Durids	Haplodurids	Cambidic Haplodurids	loamy	mixed	superactive		thermic	aridic	shallow
LIDAN	16.8	NO	1996	NV		29	Aridisols	Durids	Argidurids	Abruptic Argidurids	clayey-skeletal	smectitic			mesic	aridic	
LIEN	243.7	NO	1971	NV		28A, 28B, 29	Aridisols	Durids	Haplodurids	Xeric Haplodurids	loamy-skeletal	mixed	superactive		mesic	aridic	shallow
LIMEWASH	6.3	NO	2006	NV		30	Aridisols	Gypsids	Haplogypsids	Leptic Haplogypsids	loamy	mixed	active		thermic	aridic	shallow
LINCO	134.8	NO	1971	NV		28B, 29	Aridisols	Cambids	Haplocambids	Durinodic Xeric Haplocambids	coarse-loamy	mixed	superactive		mesic	aridic	
LINHART	29.1	NO	1980	NV		26	Mollisols	Xerolls	Haploxerolls	Torriorthentic Haploxerolls	sandy-skeletal	mixed			mesic	xeric	
LINKUP	298.3	NO	1986	NV		25	Aridisols	Argids	Haplargids	Lithic Xeric Haplargids	clayey	smectitic			frigid	aridic	
LINOYER	700.0	YES	1971	UT	NV	28A, 28B, 47	Entisols	Orthents	Torriorthents	Xeric Torriorthents	coarse-silty	mixed	superactive	calcareous	mesic	torric	
LINPEAK	14.7	NO	2009	NV		28A	Inceptisols	Cryepts	Calcicryepts	Xeric Calcicryepts	loamy-skeletal	carbonatic			cryic	xeric	
LINROSE	83.4	NO	1986	NV		24	Mollisols	Xerolls	Haploxerolls	Calcidic Haploxerolls	loamy-skeletal	mixed	superactive		frigid	xeric	
LITTLEAILIE	257.8	NO	2006	NV		29	Aridisols	Durids	Argidurids	Xeric Argidurids	loamy-skeletal	mixed	superactive		mesic	aridic	shallow
LITTLESPRING	9.1	NO	2004	NV		28A	Aridisols	Calcids	Haplocalcids	Typic Haplocalcids	fine-loamy over sandy or sandy-skeletal	mixed	superactive		mesic	aridic	
LOCANE	125.0	NO	1971	NV	OR	23, 24, 25, 27, 28B	Aridisols	Argids	Haplargids	Lithic Xeric Haplargids	clayey-skeletal	smectitic			frigid	aridic	
LODAR	13.2	NO	1971	UT	NV	28A, 28B, 47	Mollisols	Xerolls	Calcixerolls	Lithic Calcixerolls	loamy-skeletal	carbonatic			mesic	xeric	
LOFFTUS	0.1	NO	1991	OR	NV	23	Aridisols	Durids	Haplodurids	Aquicambidic Haplodurids	ashy	glassy			mesic	aridic	
LOGAN	49.3	YES	1913	UT	ID, NV	13, 28A, 28B	Mollisols	Aquolls	Calciaquolls	Typic Calciaquolls	fine-silty	mixed	superactive		mesic	aquic	

(continued)

(continued)

Soil series	Area (km^2)	Bench-mark	Year Est-ablished	TL State	Other States Using	MLRAs Using	Order	Suborder	Great Group	Subgroup	Particle-size Class	Mineralogy Class	CEC Activity Class	Reaction Class	Soil temp. Class	SMR	Other Family
LOGRING	406.5	NO	1984	NV		28A, 29	Entisols	Orthents	Torriorthents	Lithic Xeric Torriorthents	loamy-skeletal	carbonatic			mesic	torric	
LOJET	375.2	NO	2006	NV		29	Aridisols	Durids	Argidurids	Xeric Argidurids	fine-loamy	mixed	superactive		mesic	aridic	
LOMOINE	159.5	NO	1984	NV		27, 28B, 29	Entisols	Orthents	Torriorthents	Lithic Xeric Torriorthents	loamy-skeletal	mixed	superactive	calcareous	mesic	torric	
LONCAN	224.6	NO	1983	NV		23, 24, 25, 28B	Mollisols	Xerolls	Haploxerolls	Aridic Haploxerolls	loamy-skeletal	mixed	superactive		frigid	xeric	
LONE	22.9	NO	1971	NV		25, 28B	Aridisols	Durids	Haplodurids	Xeric Haplodurids	loamy-skeletal	mixed	superactive		frigid	aridic	
LONGCREEK	72.1	NO	1992	CA	ID, NV, OR	10, 21, 23, 25, 26	Mollisols	Xerolls	Argixerolls	Aridic Lithic Argixerolls	clayey-skeletal	smectitic			mesic	xeric	
LONGDAY	5.3	NO	2006	CA	NV	22A, 26	Mollisols	Xerolls	Argixerolls	Typic Argixerolls	loamy-skeletal	mixed	superactive		frigid	xeric	
LONGDIS	124.6	NO	1995	NV		23, 24	Aridisols	Argids	Natrargids	Xeric Natrargids	fine	smectitic			mesic	aridic	
LONGJIM	146.2	NO	1992	NV		30	Aridisols	Durids	Haplodurids	Typic Haplodurids	loamy-skeletal	mixed	superactive		thermic	aridic	shallow
LOOMER	190.6	NO	1981	NV		26, 27, 28B	Mollisols	Xerolls	Argixerolls	Aridic Lithic Argixerolls	clayey-skeletal	smectitic			mesic	xeric	
LOOMIS	88.0	NO	1986	NV	CA, ID	23, 25	Aridisols	Argids	Haplargids	Lithic Xeric Haplargids	clayey-skeletal	smectitic			mesic	aridic	
LOPWASH	30.9	NO	1983	NV		28B	Aridisols	Cambids	Haplocambids	Typic Haplocambids	loamy-skeletal	mixed	superactive		frigid	aridic	
LORAY	183.0	NO	1986	NV	ID	11, 28A, 28B	Aridisols	Calcids	Haplocalcids	Typic Haplocalcids	sandy-skeletal	mixed			mesic	aridic	
LOSTLEADER	3.0	NO	2011	NV		30	Aridisols	Argids	Haplargids	Ustic Haplargids	loamy-skeletal	mixed	superactive		mesic	aridic	
LOUDERBACK	65.6	NO	1984	NV		27, 29	Entisols	Orthents	Torriorthents	Oxyaquic Torriorthents	sandy	mixed			mesic	torric	
LOVAMP	28.6	NO	2017	NV		25	Mollisols	Cryolls	Haplocryolls	Xeric Haplocryolls	loamy-skeletal	mixed	superactive		cryic	xeric	
LOVEBOLDT	0.8	NO	2009	NV		27	Entisols	Fluvents	Torrifluvents	Typic Torrifluvents	sandy	mixed			mesic	torric	
LOVELOCK	79.9	NO	1963	NV		27	Mollisols	Aquolls	Endoaquolls	Fluvaquentic Endoaquolls	fine	mixed	superactive	calcareous	mesic	aquic	
LOWEMAR	55.7	NO	2016	NV		25	Inceptisols	Cryepts	Humicryepts	Xeric Humicryepts	sandy-skeletal	mixed			cryic	xeric	
LOX	10.8	NO	1981	NV		27	Aridisols	Argids	Natrargids	Typic Natrargids	loamy-skeletal	mixed	superactive		mesic	aridic	
LUAP	41.2	NO	1986	NV		28A	Aridisols	Calcids	Petrocalcids	Typic Petrocalcids	loamy-skeletal	mixed	superactive		mesic	aridic	
LUCKYSTRIKE	7.5	NO	2006	NV		30	Mollisols	Ustolls	Argiustolls	Calcidic Argiustolls	loamy-skeletal	mixed	superactive		mesic	ustic	
LUNDER	62.0	NO	1981	NV	CA	23, 25, 26	Mollisols	Xerolls	Durixerolls	Abruptic Argiduridic Durixerolls	clayey	smectitic			mesic	xeric	shallow
LUNING	315.5	NO	1984	NV		27, 28B, 29	Entisols	Orthents	Torriorthents	Typic Torriorthents	sandy	mixed			mesic	torric	
LUPPINO	32.5	NO	1980	NV		26	Mollisols	Xerolls	Argixerolls	Aridic Argixerolls	loamy	mixed	superactive		mesic	xeric	shallow
LUSET	91.1	NO	2013	NV		28B	Mollisols	Cryolls	Argicryolls	Vitrandic Argicryolls	loamy-skeletal	mixed	superactive		cryic		
LYDA	276.8	YES	1972	NV		29	Aridisols	Durids	Argidurids	Typic Argidurids	loamy-skeletal	mixed	superactive		mesic	aridic	shallow
LYKAL	5.3	NO	1994	NV		28B	Aridisols	Cambids	Aquicambids	Fluventic Aquicambids	coarse-silty	carbonatic			mesic	aridic	
LYNNBOW	21.1	NO	1986	NV		25, 28B	Mollisols	Xerolls	Palexerolls	Typic Palexerolls	fine	smectitic			frigid	xeric	

(continued)

(continued)

Soil series	Area (km^2)	Bench-mark	Year Est-ablished	TL State	Other States Using	MLRAs Using	Order	Suborder	Great Group	Subgroup	Particle-size Class	Mineralogy Class	CEC Activity Class	Reaction Class	Soil temp. Class	SMR	Other Family
LYRA	8.5	NO	1986	NV		25	Mollisols	Xerolls	Argixerolls	Aridic Argixerolls	loamy-skeletal	mixed	superactive		frigid	xeric	shallow
LYX	220.3	NO	1993	NV		29	Aridisols	Cambids	Haplocambids	Typic Haplocambids	sandy-skeletal	mixed			mesic	aridic	
MACAREENO	13.6	NO	1980	NV		22A	Mollisols	Cryolls	Argicryolls	Oxyaquic Argicryolls	fine-loamy	isotic			cryic		
MACKERLAKE	44.7	NO	2009	NV		27	Mollisols	Aquolls	Endoaquolls	Vertic Endoaquolls	fine	smectitic			mesic	aquic	
MACKEY	0.5	NO	1976	ID	NV	10, 25	Aridisols	Calcids	Haplocalcids	Xeric Haplocalcids	loamy-skeletal	mixed	superactive		mesic	aridic	
MACKRANCH	4.5	NO	2012	NV		26	Mollisols	Xerolls	Haploxerolls	Aquic Haploxerolls	fine-loamy over sandy or sandy-skeletal	mixed	superactive		mesic	xeric	
MACKSCANYON	58.7	NO	2006	NV		30	Mollisols	Ustolls	Calciustolls	Aridic Calciustolls	loamy-skeletal	carbonatic			mesic	ustic	
MACNOT	114.5	NO	2006	NV	CA	23	Aridisols	Calcids	Haplocalcids	Vitrixerandic Haplocalcids	ashy-skeletal	glassy			mesic	aridic	
MACYFLET	23.1	NO	1991	NV	OR	23	Aridisols	Argids	Paleargids	Vertic Paleargids	very-fine	smectitic			frigid	aridic	
MADELINE	63.6	NO	1971	NV	CA, OR	10, 21, 23, 24, 25, 27	Mollisols	Xerolls	Argixerolls	Aridic Lithic Argixerolls	clayey	smectitic			frigid	xeric	
MADERBAK	37.7	NO	1990	NV		28B	Aridisols	Argids	Haplargids	Xeric Haplargids	clayey-skeletal	smectitic			mesic	aridic	
MAGGIE	20.6	NO	1972	NV		29	Aridisols	Durids	Argidurids	Typic Argidurids	loamy-skeletal	mixed	superactive		mesic	aridic	shallow
MAGHILLS	11.4	NO	1983	NV		28B	Entisols	Orthents	Torriorthents	Typic Torriorthents	loamy-skeletal	carbonatic			frigid	torric	
MAHALA	35.5	NO	1986	NV	CA	23, 25	Aridisols	Argids	Paleargids	Vertic Paleargids	fine	smectitic			mesic	aridic	
MAHOGEE	4.2	NO	2017	NV		25	Mollisols	Cryolls	Argicryolls	Lithic Argicryolls	loamy	mixed	superactive		cryic		
MAJORSPLACE	63.4	NO	2009	NV		28A	Mollisols	Xerolls	Argixerolls	Lithic Argixerolls	loamy-skeletal	mixed	superactive		frigid	xeric	
MAJUBA	86.7	NO	1988	NV		27	Mollisols	Xerolls	Argixerolls	Calciargidic Argixerolls	loamy-skeletal	mixed	superactive		frigid	xeric	
MALMESA	41.7	NO	1984	NV		29	Aridisols	Durids	Argidurids	Xeric Argidurids	loamy-skeletal	mixed	superactive		mesic	aridic	shallow
MALPAIS	213.4	NO	1972	NV		24, 25, 26, 27, 29	Aridisols	Cambids	Haplocambids	Typic Haplocambids	loamy-skeletal	mixed	superactive		mesic	aridic	
MANARD	10.0	NO	1977	ID	NV	10, 25	Mollisols	Xerolls	Durixerolls	Typic Durixerolls	fine	smectitic			frigid	xeric	
MANOGUE	81.6	NO	1980	NV		23, 26, 27	Vertisols	Xererts	Haploxererts	Aridic Haploxererts	fine	smectitic			mesic	xeric	
MAREPAS	10.7	NO	2006	NV		23	Mollisols	Xerolls	Argixerolls	Aridic Lithic Argixerolls	ashy-skeletal	glassy			mesic	xeric	
MARLA	3.4	NO	1980	NV	CA	22A	Inceptisols	Xerepts	Dystroxerepts	Aquic Dystroxerepts	sandy	mixed			frigid	xeric	
MARMOTHILL	0.2	NO	2010	CA	NV	26	Mollisols	Xerolls	Palexerolls	Vertic Palexerolls	ashy-skeletal over clayey	glassy OVER smectitic			frigid	xeric	
MARYJANE	81.1	NO	2006	NV		30	Mollisols	Ustolls	Calciustolls	Pachic Calciustolls	loamy-skeletal	carbonatic			frigid	ustic	
MASCAMP	12.0	NO	1971	NV	CA, OR	10, 23, 25	Mollisols	Xerolls	Argixerolls	Aridic Lithic Argixerolls	loamy-skeletal	mixed	superactive		frigid	xeric	
MASTLY	13.6	NO	1990	NV		27	Entisols	Fluvents	Torrifluvents	Duric Xeric Torrifluvents	coarse-loamy	mixed	superactive	calcareous	mesic	torric	
MATTIER	130.7	NO	2013	NV		28B	Mollisols	Xerolls	Argixerolls	Lithic Argixerolls	loamy-skeletal	mixed	superactive		frigid	xeric	
MAU	86.0	NO	1971	NV		25, 28A, 28B	Aridisols	Argids	Haplargids	Durinodic Xeric Haplargids	clayey-skeletal	smectitic			frigid	aridic	
MAYGAL	3.8	NO	2012	NV		26	Mollisols	Cryolls	Argicryolls	Lithic Argicryolls	loamy-skeletal	mixed	superactive		cryic		
MAYNARD LAKE	8.6	NO	1940	NV		30	Entisols	Psamments	Torripsamments	Typic Torripsamments		mixed			thermic	torric	

(continued)

(continued)

Soil series	Area (km^2)	Bench-mark	Year Est-ablished	TL State	Other States Using	MLRAs Using	Order	Suborder	Great Group	Subgroup	Particle-size Class	Mineralogy Class	CEC Activity Class	Reaction Class	Soil temp. Class	SMR	Other Family
MAZUMA	1601.0	YES	1963	NV	CA, ID	11, 23, 24, 27, 28B, 29	Entisols	Orthents	Torriorthents	Typic Torriorthents	coarse-loamy	mixed	superactive	calcareous	mesic	torric	
MCCARRAN	298.2	YES	1964	NV	TX	30, 42	Aridisols	Gypsids	Haplogypsids	Typic Haplogypsids	coarse-loamy	mixed	superactive		thermic	aridic	
MCCLANAHAN	30.7	NO	2006	NV		29	Aridisols	Argids	Haplargids	Ustic Haplargids	loamy-skeletal	mixed	superactive		mesic	aridic	shallow
MCCLEARY	17.2	NO	1986	NV		25	Inceptisols	Aquepts	Epiaquepts	Vertic Epiaquepts	fine	smectitic		nonacid	mesic	aquic	
MCCONNEL	277.0	YES	1974	NV	CA, OR	23, 24, 25, 27, 28B	Aridisols	Cambids	Haplocambids	Xeric Haplocambids	sandy-skeletal	mixed			mesic	aridic	
MCCULLOUGH	18.0	NO	1964	NV	NM	30, 42	Entisols	Orthents	Torriorthents	Typic Torriorthents	coarse-loamy	mixed	superactive	calcareous	thermic	torric	
MCCUTCHEN	7.6	NO	1965	NV		29	Aridisols	Calcids	Haplocalcids	Typic Haplocalcids	coarse-loamy	mixed	superactive		mesic	aridic	
MCIVEY	714.8	YES	1986	NV		25, 28B	Mollisols	Xerolls	Argixerolls	Typic Argixerolls	clayey-skeletal	smectitic			frigid	xeric	
MCTOM	0.3	NO	2006	CA	NV	22A	Inceptisols	Cryepts	Humicryepts	Xeric Humicryepts	sandy-skeletal	mixed			cryic	xeric	
MCVEGAS	34.9	NO	1985	NV		24, 28B	Aridisols	Durids	Natridurids	Natrargidic Natridurids	clayey-skeletal	smectitic			mesic	aridic	shallow
MCWATT	36.0	NO	1995	NV	CA	23	Aridisols	Cambids	Haplocambids	Xeric Haplocambids	sandy-skeletal	mixed			mesic	aridic	
MEADVIEW	8.9	NO	2001	AZ	NV	30	Aridisols	Calcids	Haplocalcids	Durinodic Haplocalcids	sandy-skeletal	mixed			thermic	aridic	
MEDBURN	104.7	YES	1981	UT	NV	28A, 28B	Entisols	Orthents	Torriorthents	Xeric Torriorthents	coarse-loamy	mixed	superactive	calcareous	mesic	torric	
MEDLAVAL	4.4	NO	2004	NV		28A	Mollisols	Xerolls	Calcixerolls	Vertic Calcixerolls	fine	smectitic			mesic	xeric	
MEDVED	3.8	NO	2006	NV		23	Entisols	Orthents	Torriorthents	Lithic Xeric Torriorthents	loamy-skeletal	mixed	superactive	nonacid	mesic	torric	
MEEKS	35.0	NO	1970	CA	NV	22A	Inceptisols	Xerepts	Dystroxerepts	Humic Dystroxerepts	sandy-skeletal	mixed			frigid	xeric	
MEISS	10.0	NO	1970	CA	NV	22A	Inceptisols	Cryepts	Humicryepts	Lithic Humicryepts	loamy	isotic			cryic	?	
MELLOR	13.1	NO	1947	UT	ID, NV	11, 26, 28A, 47	Aridisols	Argids	Natrargids	Xeric Natrargids	fine-silty	mixed	superactive		mesic	aridic	
MELODY	6.7	NO	2007	CA	NV	22A	Andisols	Cryands	Vitricryands	Lithic Vitricryands	ashy-skeletal	mixed			cryic	xeric	
MENBO	46.5	NO	1992	OR	CA, NV	10, 23	Mollisols	Xerolls	Argixerolls	Vitrandic Argixerolls	clayey-skeletal	smectitic			frigid	xeric	
MESABASE	22.1	NO	2006	NV		30	Aridisols	Calcids	Haplocalcids	Typic Haplocalcids	sandy-skeletal	mixed			hyperthermic	aridic	
MEZZER	142.8	NO	2006	NV		29	Aridisols	Cambids	Haplocambids	Xeric Haplocambids	loamy-skeletal	carbonatic			mesic	aridic	
MICKEY	63.6	NO	1985	NV		27	Aridisols	Durids	Argidurids	Haploxeralfic Argidurids	loamy	mixed	superactive		mesic	aridic	shallow
MIDAS	49.2	NO	1968	NV		24	Aridisols	Cambids	Haplocambids	Durinodic Haplocambids	loamy-skeletal	mixed	superactive		mesic	aridic	
MIDRAW	162.0	NO	1985	NV	ID	23, 24, 25	Aridisols	Durids	Argidurids	Xeric Argidurids	clayey	smectitic			mesic	aridic	shallow
MIGERN	15.9	NO	1996	NV		30	Aridisols	Argids	Haplargids	Durinodic Haplargids	fine-loamy over sandy or sandy-skeletal	mixed	superactive		thermic	aridic	
MIJAY	12.4	NO	1993	NV		29	Aridisols	Durids	Argidurids	Haploxeralfic Argidurids	loamy-skeletal	mixed	superactive		mesic	aridic	
MIJOYSEE	157.1	NO	2013	NV		28B	Entisols	Orthents	Torriorthents	Lithic Xeric Torriorthents	loamy-skeletal	carbonatic			frigid	torric	

(continued)

(continued)

Soil series	Area (km^2)	Bench-mark	Year Est-ablished	TL State	Other States Using	MLRAs Using	Order	Suborder	Great Group	Subgroup	Particle-size Class	Mineralogy Class	CEC Activity Class	Reaction Class	Soil temp. Class	SMR	Other Family
MILKIWAY	11.3	NO	2012	NV		26	Aridisols	Argids	Haplargids	Xeric Haplargids	fine	smectitic			mesic	aridic	
MILLAN	4.4	NO	2004	NV		28A	Mollisols	Xerolls	Argixerolls	Aridic Argixerolls	loamy-skeletal	mixed	superactive		frigid	xeric	
MILLERLUX	10.2	NO	1986	NV		23, 24, 26, 27	Aridisols	Argids	Haplargids	Lithic Xeric Haplargids	clayey	smectitic			frigid	aridic	
MIMENTOR	20.7	NO	2010	NV	CA	26	Aridisols	Argids	Haplargids	Durinodic Xeric Haplargids	fine-loamy	mixed	superactive		mesic	aridic	
MINA	5.1	NO	1972	NV		29	Aridisols	Cambids	Haplocambids	Durinodic Xeric Haplocambids	loamy-skeletal	mixed	superactive		mesic	aridic	
MINAT	32.6	NO	1985	NV		24	Aridisols	Cambids	Haplocambids	Xeric Haplocambids	loamy-skeletal	mixed	superactive		mesic	aridic	
MINDLEBAUGH	38.9	NO	2010	NV		26	Mollisols	Xerolls	Haploxerolls	Fluvaquentic Haploxerolls	fine-loamy	mixed	superactive		mesic	xeric	
MINNEHA	43.7	NO	1981	NV		26, 27	Mollisols	Xerolls	Haploxerolls	Aridic Haploxerolls	loamy-skeletal	mixed	superactive		mesic	xeric	shallow
MINNYE	36.4	NO	1996	NV		27, 29	Aridisols	Argids	Haplargids	Typic Haplargids	loamy-skeletal	mixed	superactive		mesic	aridic	
MINU	187.5	NO	1971	NV		28A, 28B, 29	Aridisols	Durids	Argidurids	Xeric Argidurids	loamy	mixed	superactive		mesic	aridic	shallow
MIRKWOOD	130.6	NO	1981	NV		27	Aridisols	Argids	Haplargids	Lithic Haplargids	loamy-skeletal	mixed	superactive		mesic	aridic	
MISAD	161.8	NO	1985	NV		24	Entisols	Orthents	Torriorthents	Duric Torriorthents	loamy-skeletal	mixed	superactive	calcareous	mesic	torric	
MIZEL	60.3	NO	1980	NV		26, 27	Entisols	Orthents	Torriorthents	Lithic Xeric Torriorthents	loamy-skeletal	mixed	superactive	nonacid	mesic	torric	
MIZPAH	10.0	NO	1994	NV		28B	Aridisols	Argids	Paleargids	Calcic Paleargids	fine	smectitic			mesic	aridic	
MOAPA	52.7	NO	1970	NV		30	Entisols	Psamments	Torripsamments	Typic Torripsamments		mixed			thermic	torric	
MOBL	34.5	NO	1996	NV		30	Aridisols	Argids	Natrargids	Typic Natrargids	coarse-loamy	mixed	superactive		thermic	aridic	
MODEM	6.8	NO	2006	NV		28A	Mollisols	Xerolls	Durixerolls	Vitritorrandic Durixerolls	ashy-skeletal	glassy			mesic	xeric	shallow
MOENTRIA	76.4	NO	2006	NV		29	Entisols	Orthents	Torriorthents	Typic Torriorthents	loamy-skeletal	mixed	superactive	calcareous	mesic	torric	shallow
MOHOCKEN	22.3	NO	1984	NV		29	Mollisols	Xerolls	Palexerolls	Typic Palexerolls	fine	smectitic			frigid	xeric	
MOLION	98.2	NO	1983	NV		28B, 29	Aridisols	Durids	Haplodurids	Xereptic Haplodurids	loamy-skeletal	mixed	superactive		mesic	aridic	shallow
MONARCH	94.8	NO	2004	NV		28A, 29	Mollisols	Xerolls	Calcixerolls	Lithic Calcixerolls	loamy-skeletal	carbonatic			frigid	xeric	
MONGER	16.4	NO	2007	NV		30	Aridisols	Calcids	Haplocalcids	Petronodic Haplocalcids	coarse-loamy	mixed	superactive		thermic	aridic	
MONTE CRISTO	46.7	NO	1972	NV		28B, 29	Aridisols	Durids	Natridurids	Natrargidic Natridurids	loamy	mixed	superactive		mesic	aridic	shallow
MOPANA	19.0	NO	1985	NV	CA	26	Mollisols	Xerolls	Durixerolls	Vitritorrandic Durixerolls	clayey	smectitic			frigid	xeric	shallow
MORANCH	27.7	NO	1986	NV		25	Entisols	Orthents	Torriorthents	Duric Torriorthents	coarse-silty	mixed	superactive	calcareous	mesic	torric	
MORBENCH	18.6	NO	1993	NV		29	Aridisols	Calcids	Haplocalcids	Durinodic Xeric Haplocalcids	loamy-skeletal	mixed	superactive		mesic	aridic	
MORMON MESA	604.0	YES	1939	NV	AZ	30	Aridisols	Calcids	Petrocalcids	Calcic Petrocalcids	loamy	carbonatic			thermic	aridic	shallow
MORMONWELL	40.1	NO	2011	NV		30	Mollisols	Ustolls	Haplustolls	Aridic Lithic Haplustolls	loamy-skeletal	mixed	superactive		mesic	ustic	
MORMOUNT	144.8	NO	1992	NV		30	Aridisols	Calcids	Petrocalcids	Argic Petrocalcids	loamy	mixed	superactive		thermic	aridic	shallow
MORWEN	2.4	NO	2012	NV		26	Mollisols	Xerolls	Haploxerolls	Oxyaquic Haploxerolls	coarse-loamy over sandy or sandy-skeletal	mixed	superactive		mesic	xeric	
MOSIDA	64.0	NO	1958	UT	NV	28A, 28B	Entisols	Fluvents	Torrifluvents	Xeric Torrifluvents	coarse-loamy	mixed	superactive	calcareous	mesic	torric	
MOSQUET	13.8	NO	1974	NV	CA	23, 25, 26, 28B	Mollisols	Cryolls	Argicryolls	Lithic Argicryolls	clayey	smectitic			cryic		

(continued)

(continued)

Soil series	Area (km^2)	Bench-mark	Year Est-ablished	TL State	Other States Using	MLRAs Using	Order	Suborder	Great Group	Subgroup	Particle-size Class	Mineralogy Class	CEC Activity Class	Reaction Class	Soil temp. Class	SMR	Other Family
MOTOQUA	120.6	NO	1972	UT	NM, NV, WY	29, 35, 36, 67	Mollisols	Ustolls	Argiustolls	Aridic Lithic Argiustolls	loamy-skeletal	mixed	superactive		mesic	ustic	
MOTTSKEL	7.7	NO	2006	NV	CA	26	Mollisols	Xerolls	Haploxerolls	Torriorthentic Haploxerolls	sandy-skeletal	mixed			mesic	xeric	
MOTTSVILLE	78.0	YES	1940	NV	CA	26	Mollisols	Xerolls	Haploxerolls	Torripsammentic Haploxerolls		mixed			mesic	xeric	
MOUNTMCULL	109.0	NO	2006	NV		30	Entisols	Orthents	Torriorthents	Lithic Ustic Torriorthents	loamy-skeletal	mixed	superactive	nonacid	mesic	torric	
MOUNTMUMMY	43.4	NO	2006	NV		30	Mollisols	Ustolls	Calciustolls	Pachic Calciustolls	loamy-skeletal	carbonatic			frigid	ustic	
MOUNTROSE	10.5	NO	2007	NV	CA	22A	Andisols	Cryands	Vitricryands	Xeric Vitricryands	medial-skeletal	amorphic			cryic	xeric	
MUIRAL	114.2	NO	1990	NV		28B	Inceptisols	Cryepts	Haplocryepts	Calcic Haplocryepts	loamy-skeletal	mixed	superactive		cryic		
MULHOP	24.7	NO	1986	NV		24	Aridisols	Calcids	Haplocalcids	Lithic Xeric Haplocalcids	loamy-skeletal	mixed	superactive		frigid	aridic	
MUNI	107.3	NO	1985	NV		28B	Aridisols	Durids	Argidurids	Haploxeralfic Argidurids	loamy	mixed	superactive		mesic	aridic	shallow
MYSOL	49.3	NO	1994	NV		28B	Aridisols	Calcids	Haplocalcids	Durinodic Haplocalcids	fine-loamy over sandy or sandy-skeletal	mixed	superactive		mesic	aridic	
NADRA	100.7	NO	1983	NV		28B	Aridisols	Durids	Haplodurids	Cambidic Haplodurids	loamy	mixed	superactive		frigid	aridic	shallow
NALL	14.8	NO	1981	NV		26	Mollisols	Xerolls	Haploxerolls	Aridic Haploxerolls	loamy	mixed	superactive		mesic	xeric	shallow
NAYE	49.7	NO	1992	NV		30	Aridisols	Calcids	Petrocalcids	Typic Petrocalcids	loamy-skeletal	carbonatic			thermic	aridic	
NAYFAN	27.1	NO	1995	NV		27	Mollisols	Xerolls	Haploxerolls	Calcidic Haploxerolls	fine-loamy	mixed	superactive		frigid	xeric	
NAYPED	47.1	NO	1971	NV		28B	Aridisols	Cambids	Haplocambids	Durinodic Xeric Haplocambids	fine-loamy	mixed	superactive		frigid	aridic	
NEEDAHOE	0.4	NO	2010	CA	NV	26	Mollisols	Xerolls	Durixerolls	Vitrandic Durixerolls	ashy over clayey	glassy OVER smectitic			frigid	xeric	
NEEDLE PEAK	151.7	NO	1974	NV	UT	24, 28A, 28B	Entisols	Orthents	Torriorthents	Oxyaquic Torriorthents	fine-silty	mixed	superactive	calcareous	mesic	torric	
NELLSPRING	15.4	NO	1995	NV		23	Aridisols	Durids	Argidurids	Vertic Argidurids	fine	smectitic			mesic	aridic	
NEMICO	82.6	NO	1981	NV		27	Aridisols	Durids	Natridurids	Typic Natridurids	clayey	smectitic			mesic	aridic	shallow
NETTI	102.4	NO	2017	NV		25	Mollisols	Xerolls	Argixerolls	Typic Argixerolls	clayey-skeletal	smectitic			frigid	xeric	
NEVADANILE	4.1	NO	1985	NV		27	Entisols	Fluvents	Torrifluvents	Oxyaquic Torrifluvents	fine-loamy over sandy or sandy-skeletal	mixed	superactive	calcareous	mesic	torric	
NEVADASH	56.8	NO	2006	NV	CA	23	Aridisols	Argids	Haplargids	Durinodic Xeric Haplargids	ashy	glassy			mesic	aridic	
NEVADOR	130.3	NO	1981	NV	OR	23, 24, 25	Aridisols	Argids	Haplargids	Durinodic Xeric Haplargids	fine-loamy	mixed	superactive		mesic	aridic	
NEVKA	14.0	NO	1971	NV	UT	25, 28B, 34A	Aridisols	Calcids	Haplocalcids	Aquic Haplocalcids	fine-loamy	mixed	superactive		frigid	aridic	
NEVOYER	15.4	NO	1971	NV		29	Aridisols	Durids	Haplodurids	Xeric Haplodurids	loamy	mixed	superactive		mesic	aridic	shallow
NEVTAH	15.8	NO	1971	NV		28A	Mollisols	Xerolls	Haploxerolls	Vitrandic Haploxerolls	loamy-skeletal	mixed	superactive		frigid	xeric	
NEVU	87.5	NO	1971	NV		29	Mollisols	Xerolls	Durixerolls	Vitritorrandic Durixerolls	ashy	glassy			mesic	xeric	

(continued)

(continued)

Soil series	Area (km^2)	Bench-mark	Year Est-ablished	TL State	Other States Using	MLRAs Using	Order	Suborder	Great Group	Subgroup	Particle-size Class	Mineralogy Class	CEC Activity Class	Reaction Class	Soil temp. Class	SMR	Other Family
NEWERA	141.6	NO	2000	NV	CA	30	Aridisols	Argids	Haplargids	Lithic Haplargids	loamy-skeletal	mixed	superactive		thermic	aridic	
NEWLANDS	69.0	NO	1971	CA	NV, OR	21, 23, 28B	Mollisols	Cryolls	Argicryolls	Xeric Argicryolls	fine-loamy	mixed	superactive		cryic	xeric	
NEWPASS	76.6	NO	1985	NV		24, 28B	Aridisols	Durids	Natridurids	Natrixeralfic Natridurids	fine	smectitic			mesic	aridic	
NEWVIL	105.3	NO	2006	NV		28A	Mollisols	Xerolls	Durixerolls	Argiduridic Durixerolls	loamy	mixed	superactive		mesic	xeric	shallow
NIAVI	40.8	NO	2001	NV		30	Aridisols	Calcids	Haplocalcids	Typic Haplocalcids	sandy-skeletal	mixed			thermic	aridic	
NICANOR	13.3	NO	1993	NV		27	Entisols	Orthents	Torriorthents	Xeric Torriorthents	loamy	mixed	superactive	nonacid	mesic	torric	shallow
NICKEL	71.0	NO	1939	NV	AZ, NM, TX	30, 40, 41, 42	Aridisols	Calcids	Haplocalcids	Typic Haplocalcids	loamy-skeletal	mixed	superactive		thermic	aridic	
NILESVAL	5.2	NO	2011	NV		30	Aridisols	Cambids	Haplocambids	Ustic Haplocambids	coarse-loamy	carbonatic			mesic	aridic	
NINCH	2.3	NO	1974	NV		24	Entisols	Fluvents	Torrifluvents	Duric Xeric Torrifluvents	sandy	mixed			mesic	torric	
NINEMILE	1534.0	YES	1985	NV	CA, OR	23, 25, 27, 28B	Mollisols	Xerolls	Argixerolls	Aridic Lithic Argixerolls	clayey	smectitic			frigid	xeric	
NIPPENO	132.1	NO	2006	NV		30	Aridisols	Argids	Haplargids	Lithic Ustic Haplargids	loamy-skeletal over fragmental	mixed	superactive		mesic	aridic	
NIPTON	228.8	NO	1994	NV		30	Entisols	Orthents	Torriorthents	Lithic Torriorthents	loamy-skeletal	mixed	superactive	nonacid	thermic	torric	
NIRAC	63.0	NO	1986	NV		25	Mollisols	Xerolls	Calcixerolls	Aridic C alcixerolls	loamy-skeletal	mixed	superactive		frigid	xeric	
NIRE	38.4	NO	1985	NV		26	Mollisols	Cryolls	Argicryolls	Vitrandic Argicryolls	loamy-skeletal over clayey	isotic OVER smectitic			cryic		
NITPAC	53.2	NO	1995	NV	CA	23	Mollisols	Xerolls	Durixerolls	Vertic Durixerolls	fine	smectitic			mesic	xeric	
NOBUCK	9.6	NO	1985	NV		28B	Aridisols	Argids	Haplargids	Xeric Haplargids	loamy-skeletal	mixed	superactive		frigid	aridic	
NODUR	11.6	NO	1988	NV		27	Aridisols	Argids	Natrargids	Durinodic Xeric Natrargids	fine	smectitic			mesic	aridic	
NOFET	5.9	NO	2012	NV		26	Aridisols	Argids	Natrargids	Xeric Natrargids	fine-silty	mixed	superactive		mesic	aridic	
NOLENA	175.1	NO	2006	NV		30	Entisols	Orthents	Torriorthents	Typic Torriorthents	loamy-skeletal	mixed	superactive	nonacid	thermic	torric	shallow
NOMARA	17.1	NO	1974	NV		24	Mollisols	Xerolls	Argixerolls	Calcic Pachic Argixerolls	loamy-skeletal	mixed	superactive		frigid	xeric	
NOMAZU	8.5	NO	2006	NV	CA	23	Aridisols	Calcids	Haplocalcids	Durinodic Haplocalcids	ashy	glassy			mesic	aridic	
NONAMEWASH	9.6	NO	2006	NV		30	Entisols	Fluvents	Torrifluvents	Typic Torrifluvents	sandy	mixed			hyperthermic	torric	
NOPAH	69.1	NO	1996	NV		30	Entisols	Orthents	Torriorthents	Typic Torriorthents	fine-silty	carbonatic			thermic	torric	
NOPEG	13.2	NO	2006	NV	CA	23	Aridisols	Cambids	Haplocambids	Sodic Haplocambids	ashy	glassy			mesic	aridic	shallow
NORFORK	40.0	NO	1985	NV		24, 25	Aridisols	Durids	Argidurids	Xeric Argidurids	clayey	smectitic			mesic	aridic	shallow
NORTHMORE	15.8	NO	1980	NV		26	Mollisols	Xerolls	Argixerolls	Aridic Argixerolls	fine	smectitic			mesic	xeric	
NOSAVVY	13.3	NO	2006	NV		23	Aridisols	Argids	Haplargids	Vitrixerandic Haplargids	ashy	glassy			mesic	aridic	
NOSKI	38.8	NO	2009	NV		28A	Mollisols	Xerolls	Calcixerolls	Lithic Calcixerolls	loamy-skeletal	carbonatic			frigid	xeric	

(continued)

(continued)

Soil series	Area (km^2)	Bench-mark	Year Est-ablished	TL State	Other States Using	MLRAs Using	Order	Suborder	Great Group	Subgroup	Particle-size Class	Mineralogy Class	CEC Activity Class	Reaction Class	Soil temp. Class	SMR	Other Family
NOSLO	7.7	NO	1990	NV		23	Mollisols	Xerolls	Argixerolls	Aridic Argixerolls	fine-loamy	mixed	superactive		mesic	xeric	
NOSRAC	69.0	NO	1975	NV	CA	26	Mollisols	Xerolls	Argixerolls	Aridic Argixerolls	loamy-skeletal	mixed	superactive		mesic	xeric	
NOTELLUMCREEK	49.1	NO	2013	NV		28B	Mollisols	Xerolls	Argixerolls	Lithic Argixerolls	ashy-skeletal	glassy			frigid	xeric	
NOTUS	6.0	NO	1962	ID	NV, OR	10, 11, 25, 26	Entisols	Fluvents	Xerofluvents	Aquic Xerofluvents	sandy-skeletal	mixed			mesic	xeric	
NOVACAN	10.9	NO	1985	NV		28B	Aridisols	Durids	Argidurids	Abruptic Xeric Argidurids	fine	smectitic			mesic	aridic	
NOWOY	74.1	NO	1996	NV		30	Aridisols	Calcids	Haplocalcids	Typic Haplocalcids	fine-loamy	carbonatic			thermic	aridic	
NOYSON	75.4	NO	1972	NV		29	Aridisols	Durids	Haplodurids	Cambidic Haplodurids	coarse-loamy	mixed	superactive		mesic	aridic	
NUAHS	19.1	NO	1985	NV		27	Aridisols	Calcids	Haplocalcids	Typic Haplocalcids	coarse-loamy	mixed	superactive		mesic	aridic	
NUC	52.2	NO	1974	NV		28B	Aridisols	Calcids	Haplocalcids	Durinodic Xeric Haplocalcids	loamy-skeletal	carbonatic			mesic	aridic	
NUHELEN	71.6	NO	2006	NV		28A	Mollisols	Xerolls	Argixerolls	Aridic Lithic Argixerolls	loamy-skeletal	mixed	superactive		mesic	xeric	
NUMANA	0.0	NO	2009	NV	UT	27, 28A	Entisols	Fluvents	Torrifluvents	Oxyaquic Torrifluvents	sandy-skeletal	mixed			mesic	torric	
NUPART	121.1	NO	1985	NV		26	Mollisols	Xerolls	Haploxerolls	Torriorthentic Haploxerolls	sandy-skeletal	mixed			frigid	xeric	shallow
NUPPER	13.1	NO	2006	NV		29, 30	Entisols	Orthents	Torriorthents	Lithic Ustic Torriorthents	loamy-skeletal	mixed	superactive	nonacid	mesic	torric	
NUTVAL	2.2	NO	2012	NV		26	Mollisols	Xerolls	Argixerolls	Pachic Argixerolls	loamy-skeletal	mixed	superactive		frigid	xeric	
NUTZAN	2.1	NO	1995	NV	CA	23	Mollisols	Xerolls	Haploxerolls	Vitritorrandic Haploxerolls	ashy-skeletal	glassy			frigid	xeric	
NUYOBE	226.3	NO	1984	NV		27, 29	Inceptisols	Aquepts	Halaquepts	Aeric Halaquepts	fine-silty	mixed	superactive	calcareous	mesic	aquic	
NYAK	30.1	NO	1990	NV		28B	Entisols	Orthents	Torriorthents	Duric Torriorthents	coarse-loamy	mixed	superactive	calcareous	mesic	torric	
NYALA	17.9	NO	1990	NV		28B, 29	Aridisols	Argids	Haplargids	Durinodic Haplargids	fine-loamy	mixed	superactive		mesic	aridic	
NYSERVA	111.4	NO	1972	NV	CA	29	Aridisols	Argids	Natrargids	Durinodic Natrargids	fine-loamy	mixed	superactive		mesic	aridic	
OBANION	15.5	NO	1981	NV		26, 27	Inceptisols	Aquepts	Halaquepts	Aeric Halaquepts	fine-loamy	mixed	superactive	nonacid	mesic	aquic	
OCALA	547.0	NO	1963	NV		24, 25, 28B	Inceptisols	Aquepts	Halaquepts	Duric Halaquepts	fine-silty	mixed	superactive	calcareous	mesic	aquic	
OCASHE	11.4	NO	2005	CA	NV	26	Mollisols	Xerolls	Argixerolls	Aridic Lithic Argixerolls	ashy-skeletal	glassy			mesic	xeric	
OCUD	9.7	NO	2012	NV		26	Mollisols	Xerolls	Argixerolls	Aridic Lithic Argixerolls	loamy-skeletal	mixed	superactive		mesic	xeric	
OEST	45.0	NO	1980	NV	CA	26	Mollisols	Xerolls	Argixerolls	Aridic Argixerolls	loamy-skeletal	mixed	superactive		mesic	xeric	
OKAN	268.9	NO	1994	NV		28B	Entisols	Orthents	Torriorthents	Duric Torriorthents	coarse-loamy	mixed	superactive	calcareous	mesic	torric	
OKAYVIEW	42.9	NO	2006	NV		28A	Mollisols	Xerolls	Argixerolls	Vitritorrandic Argixerolls	ashy	glassy			mesic	xeric	shallow
OLA	103.5	YES	1962	ID	NV	10, 11, 23, 25	Mollisols	Xerolls	Haploxerolls	Pachic Haploxerolls	coarse-loamy	mixed	superactive		frigid	xeric	
OLAC	148.5	NO	1981	NV	CA, OR	24, 25, 26, 27	Aridisols	Argids	Haplargids	Lithic Xeric Haplargids	loamy-skeletal	mixed	superactive		mesic	aridic	
OLD CAMP	1448.0	YES	1972	NV	CA, OR	23, 24, 25, 26, 27, 29	Aridisols	Argids	Haplargids	Lithic Xeric Haplargids	loamy-skeletal	mixed	superactive		mesic	aridic	

(continued)

(continued)

Soil series	Area (km^2)	Bench-mark	Year Est-ablished	TL State	Other States Using	MLRAs Using	Order	Suborder	Great Group	Subgroup	Particle-size Class	Mineralogy Class	CEC Activity Class	Reaction Class	Soil temp. Class	SMR	Other Family
OLDSPAN	107.7	NO	2006	NV		30	Aridisols	Calcids	Haplocalcids	Sodic Haplocalcids	loamy-skeletal	carbonatic			thermic	aridic	
OLEMAN	94.9	NO	1992	NV		29	Aridisols	Durids	Argidurids	Ustic Argidurids	loamy-skeletal	mixed	superactive		mesic	aridic	shallow
ONEIDAS	1.2	NO	2007	CA	NV	22A	Alfisols	Xeralfs	Haploxeralfs	Fragiaquic Haploxeralfs	coarse-loamy	mixed	active		frigid	xeric	
ONKEYO	178.7	NO	1986	NV		25, 28B	Mollisols	Xerolls	Calcixerolls	Lithic Calcixerolls	loamy-skeletal	mixed	active		frigid	xeric	
OPHIR	17.7	NO	1980	NV		26	Mollisols	Xerolls	Haploxerolls	Aquic Haploxerolls	sandy	mixed			mesic	xeric	
OPPIO	118.8	NO	1975	NV		26	Aridisols	Argids	Haplargids	Xeric Haplargids	fine	smectitic			mesic	aridic	
ORENEVA	0.2	NO	1991	OR	NV	23	Aridisols	Cambids	Haplocambids	Xeric Haplocambids	loamy-skeletal	mixed	superactive		frigid	aridic	
ORICTO	408.6	NO	1984	NV		27, 29	Aridisols	Argids	Calciargids	Typic Calciargids	sandy-skeletal	mixed			mesic	aridic	
ORIZABA	97.3	NO	1972	NV		26, 27, 28B	Inceptisols	Aquepts	Halaquepts	Aeric Halaquepts	fine-loamy	mixed	superactive	calcareous	mesic	aquic	
ORMSBY	6.8	NO	1973	NV	CA	21, 26	Entisols	Psamments	Torripsamments	Haploduridic Torripsamments		mixed			mesic	torric	
OROVADA	983.0	YES	1972	NV	ID, OR	24, 25, 28B	Aridisols	Cambids	Haplocambids	Durinodic Xeric Haplocambids	coarse-loamy	mixed	superactive		mesic	aridic	
ORPHANT	43.9	NO	1972	NV		29	Aridisols	Durids	Argidurids	Argidic Argidurids	loamy	mixed	superactive		mesic	aridic	shallow
ORR	78.5	NO	1980	NV	CA	23, 26, 28B	Mollisols	Xerolls	Argixerolls	Aridic Argixerolls	fine-loamy	mixed	superactive		mesic	xeric	
ORRUBO	0.3	NO	2005	AZ	NV	30	Aridisols	Calcids	Petrocalcids	Calcic Petrocalcids	loamy-skeletal	carbonatic			thermic	aridic	shallow
ORUPA	44.5	NO	1990	NV		28B	Entisols	Orthents	Torriorthents	Xeric Torriorthents	fine	smectitic		calcareous	mesic	torric	
ORWASH	121.0	NO	1984	NV	AZ	30	Entisols	Orthents	Torriorthents	Typic Torriorthents	sandy	mixed			thermic	torric	
OSDITCH	23.5	NO	2004	NV		28A	Inceptisols	Cryepts	Haplocryepts	Lamellic Haplocryepts	loamy-skeletal	mixed	superactive		cryic		
OSOBB	175.9	NO	1971	NV		27, 29	Aridisols	Durids	Haplodurids	Typic Haplodurids	loamy-skeletal	mixed	superactive		mesic	aridic	shallow
OSOLL	29.7	NO	1985	NV	OR	23, 24	Aridisols	Durids	Haplodurids	Typic Haplodurids	loamy-skeletal	mixed	superactive		mesic	aridic	shallow
OTOMO	13.9	NO	1981	NV		27	Aridisols	Durids	Haplodurids	Typic Haplodurids	loamy-skeletal	mixed	superactive		mesic	aridic	shallow
OUPICO	82.4	NO	1986	NV		25, 28B	Aridisols	Durids	Haplodurids	Xeric Haplodurids	coarse-loamy	mixed	superactive		mesic	aridic	
OUTERKIRK	39.0	NO	1993	OR	NV	24	Aridisols	Calcids	Haplocalcids	Durinodic Haplocalcids	coarse-loamy	mixed	superactive		mesic	aridic	
OVERLAND	21.2	NO	1971	NV	UT	28A, 28B	Aridisols	Calcids	Haplocalcids	Xeric Haplocalcids	loamy-skeletal	carbonatic			frigid	aridic	
OVERTON	10.8	NO	1970	NV		30	Inceptisols	Aquepts	Endoaquepts	Aeric Endoaquepts	fine	smectitic		calcareous	thermic	aquic	
OXCOREL	819.5	YES	1985	NV		24, 28B	Aridisols	Argids	Natrargids	Durinodic Natrargids	fine	smectitic			mesic	aridic	
OXVALLEY	26.9	NO	2010	NV		28A	Mollisols	Cryolls	Argicryolls	Pachic Argicryolls	clayey-skeletal	smectitic			cryic		
PACKER	173.9	YES	1979	NV		24, 25, 27, 28B, 29	Mollisols	Cryolls	Argicryolls	Xeric Argicryolls	loamy-skeletal	mixed	superactive		cryic	xeric	
PAGECREEK	27.6	NO	2010	NV		28A	Mollisols	Xerolls	Argixerolls	Vitrandic Argixerolls	fine-loamy	mixed	superactive		frigid	xeric	
PAHRANAGAT	28.0	NO	1940	NV	UT	28A, 28B, 29, 30	Mollisols	Aquolls	Endoaquolls	Fluvaquentic Endoaquolls	fine-silty	mixed	superactive	calcareous	mesic	aquic	
PAHRANGE	30.1	NO	1980	NV		26	Mollisols	Xerolls	Argixerolls	Aridic Argixerolls	fine-loamy	mixed	superactive		mesic	xeric	
PAHROC	164.3	NO	1940	NV		29	Aridisols	Durids	Haplodurids	Typic Haplodurids	loamy-skeletal	mixed	superactive		mesic	aridic	shallow
PAHRUMP	55.2	NO	1996	NV		30	Aridisols	Calcids	Haplocalcids	Petronodic Haplocalcids	loamy-skeletal	carbonatic			thermic	aridic	
PALINOR	2508.2	YES	1990	NV		28B	Aridisols	Durids	Haplodurids	Xeric Haplodurids	loamy-skeletal	carbonatic			mesic	aridic	shallow
PAMISON	39.5	NO	1986	NV		25	Mollisols	Xerolls	Calcixerolls	Aridic Calcixerolls	loamy-skeletal	mixed	superactive		frigid	xeric	

(continued)

(continued)

Soil series	Area (km^2)	Bench-mark	Year Est-ablished	TL State	Other States Using	MLRAs Using	Order	Suborder	Great Group	Subgroup	Particle-size Class	Mineralogy Class	CEC Activity Class	Reaction Class	Soil temp. Class	SMR	Other Family
PAMSDEL	20.5	NO	1971	NV		28A	Mollisols	Xerolls	Durixerolls	Haploduridic Durixerolls	loamy-skeletal	carbonatic			mesic	xeric	shallow
PANACKER	16.7	NO	2006	NV		29	Aridisols	Calcids	Haplocalcids	Sodic Haplocalcids	fine-loamy	mixed	semiactive		mesic	aridic	
PANLEE	217.0	NO	1992	NV		24	Aridisols	Cambids	Petrocambids	Xeric Petrocambids	loamy-skeletal	mixed	superactive		mesic	aridic	
PANOR	10.7	NO	1996	NV		30	Entisols	Orthents	Torriorthents	Typic Torriorthents	fine-loamy	mixed	superactive	calcareous	thermic	torric	
PAPOOSE	130.8	NO	1974	NV	UT	27, 29	Aridisols	Argids	Haplargids	Typic Haplargids	fine-loamy	mixed	superactive		mesic	aridic	
PARADISE	5.3	NO	1964	NV		30	Mollisols	Aquolls	C alciaquolls	Typic Calciaquolls	coarse-loamy				thermic	aquic	
PARANAT	193.6	NO	1983	NV		24, 25, 28B	Mollisols	Aquolls	Endoaquolls	Fluvaquentic Endoaquolls	fine-silty	mixed	superactive	calcareous	mesic	aquic	
PARISA	346.3	NO	1990	NV		28B	Aridisols	Durids	Haplodurids	Xeric Haplodurids	loamy-skeletal	carbonatic			mesic	aridic	
PARRAN	383.1	YES	1971	NV		27, 28B	Aridisols	Salids	Aquisalids	Typic Aquisalids	fine	smectitic			mesic	aridic	
PATNA	247.8	YES	1971	NV		27	Aridisols	Argids	Haplargids	Typic Haplargids	sandy	mixed			mesic	aridic	
PATTANI	13.2	NO	1968	NV		25	Vertisols	Xererts	Haploxererts	Leptic Haploxererts	fine	smectitic			frigid	xeric	
PATTER	34.2	NO	1971	NV		28A, 28B, 29	Aridisols	Cambids	Haplocambids	Durinodic Xeric Haplocambids	coarse-loamy	mixed	superactive		mesic	aridic	
PAYPOINT	83.6	YES	1995	NV		23	Aridisols	Argids	Haplargids	Durinodic Xeric Haplargids	ashy over sandy or sandy-skeletal	glassy OVER mixed			mesic	aridic	
PEDOLI	135.0	NO	1971	NV	ID	12, 28B	Aridisols	Argids	Haplargids	Xeric Haplargids	fine-loamy	mixed	superactive		frigid	aridic	
PEEKO	488.9	NO	1986	NV		25, 28B, 29	Aridisols	Durids	Haplodurids	Xeric Haplodurids	loamy	mixed	superactive		mesic	aridic	shallow
PEEVYWELL	7.1	NO	1985	ID	MT, NV	10, 25	Mollisols	Xerolls	Durixerolls	Typic Durixerolls	fine	smectitic			frigid	xeric	
PEGLER	13.0	NO	1974	NV	CA	23	Aridisols	Cambids	Haplocambids	Vitrixerandic Haplocambids	ashy	glassy			mesic	aridic	shallow
PELIC	30.7	NO	1971	NV		27	Entisols	Aquents	Fluvaquents	Typic Fluvaquents	sandy	mixed			mesic	aquic	
PENELAS	168.9	NO	1972	NV		28B, 29	Aridisols	Argids	Haplargids	Xeric Haplargids	loamy-skeletal	mixed	superactive		mesic	aridic	shallow
PENGPONG	16.0	NO	2004	NV	UT	28A	Entisols	Orthents	Torriorthents	Xeric Torriorthents	coarse-loamy	mixed	superactive	calcareous	mesic	torric	
PENOYER	321.0	YES	1940	NV	ID, UT	28A, 29	Entisols	Orthents	Torriorthents	Typic Torriorthents	coarse-silty	mixed	superactive	calcareous	mesic	torric	
PEQUOP	52.5	NO	1986	NV		25	Mollisols	Xerolls	Argixerolls	Typic Argixerolls	loamy-skeletal	mixed	superactive		frigid	xeric	
PERAZZO	84.3	NO	1981	NV	ID	11, 26, 27	Aridisols	Argids	Haplargids	Typic Haplargids	loamy-skeletal	mixed	superactive		mesic	aridic	
PERLOR	24.8	NO	1985	NV		24, 28B	Entisols	Orthents	Torriorthents	Typic Torriorthents	loamy	mixed	superactive	calcareous	mesic	torric	shallow
PERN	61.0	NO	1990	NV		28B	Mollisols	Xerolls	Calcixerolls	Aridic Calcixerolls	fine-silty	mixed	superactive		mesic	xeric	
PERNOG	12.0	NO	1986	NV	OR	21, 25	Mollisols	Xerolls	Argixerolls	Lithic Argixerolls	loamy-skeletal	mixed	superactive		frigid	xeric	
PERNTY	157.2	NO	1974	NV	OR, UT	23, 24, 25, 26, 47	Mollisols	Xerolls	Argixerolls	Aridic Lithic Argixerolls	loamy-skeletal	mixed	superactive		frigid	xeric	
PERWASO	3.4	NO	1988	NV		24, 27	Entisols	Fluvents	Torrifluvents	Typic Torrifluvents	fine-loamy over sandy or sandy-skeletal	mixed	superactive	calcareous	mesic	torric	
PERWICK	160.6	NO	1983	NV		23, 24, 25, 28B	Entisols	Orthents	Torriorthents	Xeric Torriorthents	coarse-loamy	mixed	superactive	calcareous	mesic	torric	
PESKAH	38.0	NO	2006	NV		30	Aridisols	Argids	Petroargids	Duric Petroargids	loamy-skeletal	mixed	superactive		thermic	aridic	
PETAN	43.0	NO	1984	NV	ID	25	Mollisols	Xerolls	Durixerolls	Typic Durixerolls	clayey-skeletal	smectitic			frigid	xeric	shallow
PETSPRING	57.5	NO	1985	NV			Entisols	Orthents	Torriorthents	Xeric Torriorthents	loamy-skeletal	mixed	superactive	nonacid	mesic	torric	shallow
PHARO	107.7		1947	UT	NV	28A, 28B	Mollisols	Xerolls	Calcixerolls	Aridic Calcixerolls	loamy-skeletal	carbonatic			mesic	xeric	

(continued)

(continued)

Soil series	Area (km^2)	Bench-mark	Year Est-ablished	TL State	Other States Using	MLRAs Using	Order	Suborder	Great Group	Subgroup	Particle-size Class	Mineralogy Class	CEC Activity Class	Reaction Class	Soil temp. Class	SMR	Other Family
PHEEBS	6.4	NO	2017	NV		26	Entisols	Aquents	Fluvaquents	Aeric Fluvaquents	coarse-loamy	mixed	superactive	nonacid	mesic	aquic	
PHING	54.5	NO	1981	NV		23, 26	Aridisols	Argids	Paleargids	Vertic Paleargids	fine	smectitic			mesic	aridic	
PHLISS	125.5	NO	1988	NV		27	Aridisols	Argids	Haplargids	Lithic Xeric Haplargids	loamy-skeletal	mixed	superactive		mesic	aridic	
PIAR	50.8	NO	2004	NV		28A	Inceptisols	Cryepts	Calcicryepts	Xeric Calcicryepts	loamy-skeletal	carbonatic			cryic	xeric	
PIBLER	117.4	NO	1981	UT	NV	28A	Aridisols	Calcids	Petrocalcids	Calcic Petrocalcids	loamy-skeletal	mixed	superactive		mesic	aridic	shallow
PICKUP	610.0	NO	1988	NV	CA	23, 27	Mollisols	Xerolls	Argixerolls	Aridic Argixerolls	clayey-skeletal	smectitic			mesic	xeric	
PIE CREEK	95.8	NO	1976	NV		25	Mollisols	Xerolls	Palexerolls	Aridic Palexerolls	very-fine	smectitic			frigid	xeric	
PILINE	4.5	NO	1986	NV	ID	25	Vertisols	Aquerts	Epiaquerts	Xeric Epiaquerts	fine	smectitic			mesic	aquic	
PILTDOWN	67.7	NO	1983	NV	UT	28A, 28B	Entisols	Orthents	Torriorthents	Typic Torriorthents	coarse-loamy	mixed	superactive	calcareous	mesic	torric	
PIMOGRAN	2.1	NO	2006	CA	NV	22A	Mollisols	Xerolls	Haploxerolls	Entic Haploxerolls	sandy-skeletal	mixed			frigid	xeric	shallow
PINENUT	27.0	NO	2006	NV	CA	26	Mollisols	Xerolls	Argixerolls	Aridic Argixerolls	loamy-skeletal	mixed	superactive		frigid	xeric	shallow
PINEVAL	409.5	NO	1983	NV		24, 25, 27, 28B	Aridisols	Argids	Haplargids	Durinodic Xeric Haplargids	loamy-skeletal	mixed	superactive		mesic	aridic	
PINEZ	57.4	NO	1996	NV		30	Aridisols	Argids	Petroargids	Duric Petroargids	loamy-skeletal	mixed	superactive		thermic	aridic	
PINTWATER	715.6	YES	1972	NV		27, 29	Entisols	Orthents	Torriorthents	Lithic Torriorthents	loamy-skeletal	mixed	superactive	calcareous	mesic	torric	
PINWHEELER	14.9	NO	2004	NV		28A	Mollisols	Xerolls	Argixerolls	Typic Argixerolls	loamy-skeletal	mixed	superactive		frigid	xeric	shallow
PIOCHE	154.2	NO	1971	NV		28B, 29	Mollisols	Xerolls	Argixerolls	Lithic Argixerolls	clayey-skeletal	smectitic			mesic	xeric	
PIRAPEAK	0.8	NO	2009	NV		28A	Inceptisols	Cryepts	Haplocryepts	Xeric Haplocryepts	sandy-skeletal	mixed			cryic	xeric	
PIROUETTE	428.1	NO	1971	NV		27	Aridisols	Durids	Natridurids	Typic Natridurids	loamy-skeletal	mixed	superactive		mesic	aridic	shallow
PITTMAN	7.7	NO	1964	NV		30	Aridisols	Calcids	Petrocalcids	Typic Petrocalcids	sandy-skeletal	mixed			thermic	aridic	
PIZENE	29.5	NO	1980	NV		26, 27	Aridisols	Argids	Natrargids	Typic Natrargids	fine-loamy	mixed	superactive		mesic	aridic	
PLACERITOS	25.5	NO	1963	NV		27	Entisols	Fluvents	Torrifluvents	Oxyaquic Torrifluvents	fine-silty	mixed	superactive	calcareous	mesic	torric	
PLAYER	1.6	NO	1986	ID	NV	25	Mollisols	Xerolls	Palexerolls	Ultic Palexerolls	clayey-skeletal	smectitic			frigid	xeric	
POCAN	40.8	NO	1974	NV		24	Aridisols	Cambids	Haplocambids	Xeric Haplocambids	fine-loamy	mixed	superactive		mesic	aridic	
POCKER	16.1	NO	1974	NV	UT	24	Entisols	Fluvents	Torrifluvents	Typic Torrifluvents	fine	smectitic		calcareous	mesic	torric	
POISONCREEK	59.2	NO	1992	ID	NV	23, 25	Mollisols	Xerolls	Argixerolls	Aridic Argixerolls	loamy-skeletal	mixed	superactive		frigid	xeric	shallow
POKERGAP	196.6	NO	1988	NV		27	Aridisols	Argids	Natrargids	Durinodic Xeric Natrargids	fine-loamy	mixed	superactive		mesic	aridic	
POLUM	22.0	NO	1986	NV		24	Mollisols	Cryolls	Calcicryolls	Pachic Calcicryolls	loamy-skeletal	mixed	superactive		cryic		
PONYSPRING	57.1	NO	2006	NV		28A	Mollisols	Xerolls	Argixerolls	Vitritorrandic Argixerolls	ashy	glassy			mesic	xeric	
POOBAA	5.0	NO	2004	NV		28A	Aridisols	Argids	Calciargids	Xeric Calciargids	coarse-loamy	mixed	superactive		mesic	aridic	
POOKALOO	1483.0	YES	1990	NV		28B	Aridisols	Calcids	Haplocalcids	Lithic Xeric Haplocalcids	loamy-skeletal	carbonatic			mesic	aridic	
POORCAL	53.9	NO	1983	NV		28B	Aridisols	Calcids	Haplocalcids	Durinodic Xeric Haplocalcids	coarse-loamy	mixed	superactive		frigid	aridic	
PORRONE	24.4	NO	1986	NV		25	Aridisols	Cambids	Haplocambids	Durinodic Xeric Haplocambids	loamy-skeletal	mixed	superactive		mesic	aridic	
PORTMOUNT	99.0	NO	1974	NV		28B	Aridisols	Argids	Calciargids	Xeric Calciargids	coarse-loamy	mixed	superactive		mesic	aridic	
POTOSI	261.9	NO	2006	NV		29, 30	Entisols	Orthents	Torriorthents	Lithic Torriorthents	loamy-skeletal	carbonatic			mesic	torric	

(continued)

(continued)

Soil series	Area (km^2)	Bench-mark	Year Est-ablished	TL State	Other States Using	MLRAs Using	Order	Suborder	Great Group	Subgroup	Particle-size Class	Mineralogy Class	CEC Activity Class	Reaction Class	Soil temp. Class	SMR	Other Family
POWLOW	38.2	NO	1995	NV		23	Mollisols	Xerolls	Durixerolls	Argidic Durixerolls	clayey	smectitic			mesic	xeric	shallow
POWMENT	107.4	NO	1985	NV		26	Entisols	Orthents	Torriorthents	Xeric Torriorthents	sandy-skeletal	mixed			frigid	torric	shallow
PREBLE	165.8	NO	1974	NV		23, 24, 28B	Entisols	Orthents	Torriorthents	Oxyaquic Torriorthents	coarse-loamy	mixed	superactive	calcareous	mesic	torric	
PREY	16.3	NO	1975	NV	CA	26	Aridisols	Durids	Argidurids	Haploxeralfic Argidurids	coarse-loamy	mixed	superactive		mesic	aridic	
PRIDA	9.2	NO	1974	NV		24, 27	Entisols	Orthents	Torriorthents	Oxyaquic Torriorthents	fine-silty	mixed	superactive	calcareous	mesic	torric	
PRIDEEN	113.5	NO	1993	NV		24	Entisols	Orthents	Torriorthents	Aquic Torriorthents	fine-silty	mixed	superactive	calcareous	mesic	torric	
PRIMEAUX	39.4	NO	1979	NV		25	Mollisols	Cryolls	Argicryolls	Xeric Argicryolls	fine-loamy	mixed	superactive		cryic	xeric	
PRISONEAR	27.2	NO	2006	NV		30	Aridisols	Calcids	Petrocalcids	Calcic Petrocalcids	sandy	mixed			thermic	aridic	
PRUNIE	7.8	NO	1990	NV		26	Mollisols	Xerolls	Argixerolls	Aridic Argixerolls	fine	smectitic			frigid	xeric	
PUDDLE	3.8	NO	1974	NV	UT	28A, 29	Aridisols	Calcids	Haplocalcids	Petronodic Haplocalcids	coarse-loamy	carbonatic			mesic	aridic	
PUELZMINE	8.5	NO	2006	NV		30	Aridisols	Durids	Haplodurids	Cambidic Haplodurids	loamy-skeletal	mixed	superactive		thermic	aridic	shallow
PUETT	621.0	YES	1979	NV	UT	23, 24, 25, 26, 27, 28A, 28B	Entisols	Orthents	Torriorthents	Xeric Torriorthents	loamy	mixed	superactive	calcareous	mesic	torric	shallow
PUFFER	130.6	NO	1986	NV		24, 27	Entisols	Orthents	Torriorthents	Lithic Xeric Torriorthents	loamy-skeletal	mixed	active	calcareous	mesic	torric	
PULA	15.0	NO	1981	NV		26, 28B	Aridisols	Argids	Haplargids	Xeric Haplargids	clayey-skeletal	smectitic			mesic	aridic	
PULCAN	14.4	NO	1981	NV		26	Aridisols	Argids	Paleargids	Calcic Paleargids	clayey-skeletal	smectitic			mesic	aridic	
PULSIPHER	13.2	NO	1970	NV		30	Aridisols	Cambids	Haplocambids	Lithic Haplocambids	loamy-skeletal	mixed	superactive		thermic	aridic	
PUMEL	107.0	NO	1972	NV		27, 29	Entisols	Orthents	Torriorthents	Typic Torriorthents	loamy-skeletal	mixed	superactive	calcareous	mesic	torric	shallow
PUMPER	294.0	NO	1974	NV		24, 28B	Aridisols	Cambids	Haplocambids	Typic Haplocambids	sandy-skeletal	mixed			mesic	aridic	
PUNCHBOWL	194.2	NO	1985	NV		24, 28B	Aridisols	Argids	Haplargids	Lithic Xeric Haplargids	loamy	mixed	superactive		frigid	aridic	
PUNG	45.1	NO	1981	NV		26	Mollisols	Xerolls	Palexerolls	Vertic Palexerolls	fine	smectitic			mesic	xeric	
PUROB	496.4	NO	1996	NV		30	Aridisols	Calcids	Petrocalcids	Calcic Petrocalcids	loamy-skeletal	carbonatic			mesic	aridic	shallow
PYRAT	533.0	YES	1990	NV	ID, UT	28A, 28B	Aridisols	Calcids	Haplocalcids	Durinodic Xeric Haplocalcids	loamy-skeletal	mixed	superactive		mesic	aridic	
QUARZ	506.7	YES	1983	NV		24, 25, 28B	Mollisols	Xerolls	Argixerolls	Aridic Argixerolls	clayey-skeletal	smectitic			frigid	xeric	
QUIJINUMP	50.3	NO	2011	NV		30	Entisols	Orthents	Torriorthents	Typic Torriorthents	loamy-skeletal	carbonatic			mesic	torric	
QUIMA	26.0	NO	1972	NV		28B	Aridisols	Cambids	Haplocambids	Typic Haplocambids	coarse-loamy	mixed	superactive		mesic	aridic	
QUOMUS	34.5	NO	1993	NV		24	Mollisols	Xerolls	Haploxerolls	Duridic Haploxerolls	coarse-loamy	mixed	superactive		frigid	xeric	
QUOPANT	26.9	NO	1986	NV		25	Mollisols	Xerolls	Argixerolls	Typic Argixerolls	loamy-skeletal	mixed	superactive		frigid	xeric	shallow
QWYNN	164.8	NO	2006	NV		29	Aridisols	Argids	Haplargids	Durinodic Xeric Haplargids	fine-loamy	mixed	superactive		mesic	aridic	
RAD	213.6	NO	1974	NV	ID	11, 24, 25	Aridisols	Cambids	Haplocambids		coarse-silty	mixed	superactive		mesic	aridic	

(continued)

(continued)

Soil series	Area (km^2)	Bench-mark	Year Est-ablished	TL State	Other States Using	MLRAs Using	Order	Suborder	Great Group	Subgroup	Particle-size Class	Mineralogy Class	CEC Activity Class	Reaction Class	Soil temp. Class	SMR	Other Family
										Durinodic Xeric Haplocambids							
RADOL	141.5	NO	2006	NV		28A	Mollisols	Xerolls	Calcixerolls	Aridic Lithic Calcixerolls	loamy-skeletal	carbonatic			mesic	xeric	
RAGAMUFFIN	1.8	NO	2004	NV		28A	Entisols	Orthents	Cryorthents	Typic Cryorthents	sandy-skeletal	mixed			cryic		
RAGLAN	193.4	NO	1971	NV	CA	23, 24	Aridisols	Cambids	Haplocambids	Durinodic Haplocambids	fine-loamy	mixed	superactive		mesic	aridic	
RAGNEL	65.2	NO	2004	NV		28A, 28B	Aridisols	Cambids	Haplocambids	Xeric Haplocambids	sandy-skeletal	mixed			mesic	aridic	
RAGTOWN	504.0	NO	1971	NV	CA	23, 24, 27, 28A, 28B	Entisols	Orthents	Torriorthents	Typic Torriorthents	fine	smectitic		calcareous	mesic	torric	
RAILCITY	42.3	NO	1980	NV		22A, 26	Entisols	Orthents	Xerorthents	Typic Xerorthents	sandy-skeletal	mixed			frigid	xeric	
RAILROAD	109.6	NO	2006	NV		30	Aridisols	Calcids	Haplocalcids	Typic Haplocalcids	loamy-skeletal	mixed	superactive		thermic	aridic	
RAMIRES	178.1	NO	1974	NV		24, 25, 47	Mollisols	Xerolls	Argixerolls	Calciargidic Argixerolls	fine	smectitic			frigid	xeric	
RAMSHEAD	4.9	NO	2006	NV		30	Entisols	Orthents	Torriorthents	Typic Torriorthents	loamy-skeletal	mixed	superactive	calcareous	hyperthermic	torric	shallow
RANGERTAFT	10.8	NO	2014	NV		28B	Mollisols	Cryolls	Argicryolls	Pachic Argicryolls	loamy-skeletal	mixed	superactive		cryic		
RAPADO	74.4	NO	1992	NV		29	Aridisols	Calcids	Petrocalcids	Ustalfic Petrocalcids	loamy-skeletal	mixed	superactive		mesic	aridic	
RAPH	66.9	NO	1990	NV		28B	Aridisols	Cambids	Haplocambids	Sodic Haplocambids	fine-loamy	mixed	superactive		mesic	aridic	
RASILLE	104.1	NO	1985	NV		24, 28B	Aridisols	Cambids	Haplocambids	Durinodic Xeric Haplocambids	coarse-silty	mixed	superactive		mesic	aridic	
RASTER	2.7	NO	1993	NV		29	Mollisols	Xerolls	Haploxerolls	Pachic Haploxerolls	loamy-skeletal	mixed	superactive		mesic	xeric	
RATLEFLAT	59.6	NO	1985	NV		26, 27, 28A, 29	Aridisols	Argids	Haplargids	Xeric Haplargids	coarse-loamy	mixed	superactive		mesic	aridic	
RATSOW	19.4	NO	1986	NV		25	Aridisols	Durids	Argidurids	Xeric Argidurids	fine	smectitic			frigid	aridic	
RATTO	87.5	NO	1971	OR	NV	23, 28B	Aridisols	Durids	Argidurids	Xeric Argidurids	clayey	smectitic			frigid	aridic	shallow
RAVENDOG	108.9	NO	2004	NV		28A, 28B	Mollisols	Xerolls	Haploxerolls	Torrifluventic Haploxerolls	coarse-loamy	mixed	superactive		mesic	xeric	
RAVENELL	69.4	NO	1981	NV		26	Aridisols	Argids	Haplargids	Xeric Haplargids	loamy-skeletal	mixed	superactive		mesic	aridic	shallow
RAVENSWOOD	151.7	NO	1983	NV		27, 28B, 29	Mollisols	Xerolls	Argixerolls	Typic Argixerolls	clayey-skeletal	smectitic			frigid	xeric	
RAWE	128.3	NO	1981	NV		24, 25, 26, 27	Aridisols	Argids	Haplargids	Typic Haplargids	clayey over loamy-skeletal	smectitic OVER mixed	superactive		mesic	aridic	
REALMCOY	24.7	NO	2013	NV		28B	Mollisols	Xerolls	Argixerolls	Lithic Argixerolls	loamy-skeletal	mixed	superactive		frigid	xeric	
REBEL	344.0	NO	1974	NV		24, 25, 26, 28B	Aridisols	Cambids	Haplocambids	Xeric Haplocambids	coarse-loamy	mixed	superactive		mesic	aridic	
REDFLAME	9.3	NO	1985	NV		24	Aridisols	Argids	Haplargids	Durinodic Haplargids	loamy-skeletal	mixed	superactive		mesic	aridic	
REDHOME	7.8	NO	2006	NV	CA	23	Mollisols	Xerolls	Argixerolls	Vitritorrandic Argixerolls	fine	mixed	superactive		frigid	xeric	
REDNEEDLE	3.3	NO	2006	NV		30	Entisols	Orthents	Torriorthents	Lithic Torriorthents	loamy-skeletal	mixed	superactive	calcareous	hyperthermic	torric	
REDNIK	309.6	NO	1980	NV		24, 26, 27	Aridisols	Argids	Haplargids	Typic Haplargids	loamy-skeletal	mixed	superactive		mesic	aridic	

(continued)

(continued)

Soil series	Area (km^2)	Bench-mark	Year Est-ablished	TL State	Other States Using	MLRAs Using	Order	Suborder	Great Group	Subgroup	Particle-size Class	Mineralogy Class	CEC Activity Class	Reaction Class	Soil temp. Class	SMR	Other Family
REESE	48.9	NO	1974	NV	OR	23, 24	Inceptisols	Aquepts	Halaquepts	Duric Halaquepts	fine-loamy	mixed	superactive	calcareous	mesic	aquic	
REINA	9.1	NO	1985	NV		24, 25	Aridisols	Durids	Argidurids	Xeric Argidurids	clayey-skeletal	smectitic			mesic	aridic	shallow
RELLEY	127.2	YES	1974	NV		24, 25	Aridisols	Cambids	Haplocambids	Durinodic Haplocambids	fine-silty	mixed	superactive		mesic	aridic	
RELUCTAN	661.0	YES	1985	NV	OR	23, 24, 25	Mollisols	Xerolls	Argixerolls	Aridic Argixerolls	fine-loamy	mixed	superactive		frigid	xeric	
RENO	147.0	YES	1940	NV	CA	26	Aridisols	Durids	Argidurids	Abruptic Xeric Argidurids	fine	smectitic			mesic	aridic	
REYWAT	383.0	NO	1971	ID	CA, NV, OR, UT	10, 11, 23, 25, 26, 27, 28A, 47	Mollisols	Xerolls	Argixerolls	Aridic Lithic Argixerolls	loamy-skeletal	mixed	superactive		mesic	xeric	
REZAVE	94.2	NO	1980	NV		23, 24, 27	Aridisols	Argids	Natrargids	Lithic Natrargids	clayey	smectitic			mesic	aridic	
RICERT	463.7	NO	1983	NV		24, 27, 28B	Aridisols	Argids	Natrargids	Durinodic Natrargids	fine-loamy	mixed	superactive		mesic	aridic	
RICHINDE	391.2	NO	2006	NV		29	Aridisols	Argids	Haplargids	Lithic Xeric Haplargids	loamy-skeletal	mixed	superactive		mesic	aridic	
RIDIT	22.3	NO	1971	NV		25, 28B	Aridisols	Durids	Haplodurids	Xeric Haplodurids	loamy-skeletal	mixed	superactive		frigid	aridic	
RIO KING	64.0	NO	1992	NV	OR	24	Mollisols	Xerolls	Haploxerolls	Aridic Haploxerolls	coarse-loamy	mixed	superactive		mesic	xeric	
RIPCON	12.7	NO	2009	NV		28A, 28B	Mollisols	Xerolls	Haploxerolls	Cumulic Haploxerolls	loamy-skeletal	mixed	superactive		frigid	xeric	
RIPLEY	0.5	NO	1971	CA	AZ, NV	30, 31, 40	Entisols	Fluvents	Torrifluvents	Typic Torrifluvents	coarse-silty over sandy or sandy-skeletal	mixed	superactive	calcareous	hyperthermic	torric	
RIPPO	0.9	NO	2009	NV		28A	Entisols	Fluvents	Xerofluvents	Mollic Xerofluvents	loamy-skeletal	mixed	superactive	nonacid	frigid	xeric	
RISLEY	169.0	NO	1980	NV		10, 26, 28B	Aridisols	Argids	Haplargids	Xeric Haplargids	fine	smectitic			mesic	aridic	
RISUE	52.2	NO	1981	NV		27	Aridisols	Durids	Argidurids	Abruptic Argidurids	clayey	smectitic			mesic	aridic	shallow
RITO	10.5	NO	1971	NV		25, 28B	Mollisols	Xerolls	Haploxerolls	Calcidic Haploxerolls	loamy-skeletal	mixed	superactive		frigid	xeric	
RIVERBEND	115.3	NO	2001	AZ	CA, NV	30, 31	Aridisols	Calcids	Haplocalcids	Typic Haplocalcids	sandy-skeletal	mixed			hyperthermic	aridic	
RIXIE	43.0	NO	1974	NV		24	Mollisols	Xerolls	Haploxerolls	Aquic Duric Haploxerolls	fine-loamy	mixed	superactive		mesic	xeric	
ROBBERSFIRE	40.5	NO	2006	NV		30	Inceptisols	Ustepts	Haplustepts	Calcic Haplustepts	loamy-skeletal	carbonatic			frigid	ustic	
ROBSON	173.8	NO	1983	NV	OR	23, 24, 28B	Aridisols	Argids	Haplargids	Lithic Xeric Haplargids	clayey-skeletal	smectitic			frigid	aridic	
ROCA	770.0	YES	1971	NV	ID, OR	23, 24, 25, 28B	Aridisols	Argids	Haplargids	Xeric Haplargids	clayey-skeletal	smectitic			frigid	aridic	
ROCCONDA	469.8	NO	1992	NV		23, 24	Aridisols	Argids	Haplargids	Lithic Xeric Haplargids	clayey-skeletal	smectitic			mesic	aridic	
ROCHPAH	231.5	NO	1992	NV		29	Aridisols	Calcids	Haplocalcids	Lithic Haplocalcids	loamy-skeletal	mixed	superactive		mesic	aridic	
ROCKABIN	25.1	NO	1985	NV		26	Mollisols	Cryolls	Haplocryolls	Xeric Haplocryolls	loamy-skeletal	mixed	superactive		cryic		
RODAD	161.6	NO	1984	NV		29	Aridisols	Argids	Haplargids	Typic Haplargids	loamy-skeletal	mixed	superactive		mesic	aridic	shallow
RODELL	5.3	NO	1993	NV		23	Entisols	Orthents	Cryorthents	Lithic Cryorthents	sandy-skeletal	mixed			cryic		
RODEN	149.8	NO	1990	NV		28B	Entisols	Orthents	Torriorthents	Xeric Torriorthents	clayey-skeletal	smectitic		calcareous	mesic	torric	shallow
RODIE	76.5	NO	1986	NV		25	Mollisols	Xerolls	Haploxerolls	Duridic Haploxerolls	loamy-skeletal	mixed	superactive		frigid	xeric	

(continued)

(continued)

Soil series	Area (km^2)	Bench-mark	Year Est-ablished	TL State	Other States Using	MLRAs Using	Order	Suborder	Great Group	Subgroup	Particle-size Class	Mineralogy Class	CEC Activity Class	Reaction Class	Soil temp. Class	SMR	Other Family
RODOCK	64.7	NO	1992	NV		23, 24	Mollisols	Xerolls	Haploxerolls	Duridic Haploxerolls	loamy-skeletal	mixed	superactive		mesic	xeric	
ROIC	287.8	NO	1972	NV		27, 29	Entisols	Orthents	Torriorthents	Typic Torriorthents	loamy	mixed	superactive	calcareous	mesic	torric	shallow
ROLOC	13.6	NO	1981	NV		26	Mollisols	Xerolls	Argixerolls	Aridic Argixerolls	loamy-skeletal	mixed	superactive		mesic	xeric	shallow
ROSE CREEK	61.0	YES	1974	NV	CA	21, 24, 25, 26, 27, 28B	Mollisols	Aquolls	Endoaquolls	Fluvaquentic Endoaquolls	coarse-loamy	mixed	superactive	calcareous	mesic	aquic	
ROSITAS	30.5	YES	1918	CA	AZ, NV	30, 31, 40	Entisols	Psamments	Torripsamments	Typic Torripsamments		mixed			hyperthermic	torric	
ROSNEY	114.4	NO	1974	NV	UT	24, 28B	Entisols	Orthents	Torriorthents	Typic Torriorthents	fine-silty	mixed	superactive	calcareous	mesic	torric	
ROTINOM	27.4	NO	1985	NV		28B	Entisols	Fluvents	Torrifluvents	Vitrandic Torrifluvents	fine-silty	mixed	superactive	calcareous	mesic	torric	
ROUETTE	130.3	NO	2004	NV		28A	Aridisols	Durids	Haplodurids	Xereptic Haplodurids	loamy	mixed	superactive		mesic	aridic	shallow
ROVAL	57.7	NO	1971	NV		29	Aridisols	Durids	Argidurids	Xeric Argidurids	loamy	mixed	superactive		mesic	aridic	shallow
ROWEL	49.7	NO	1981	NV		26	Aridisols	Argids	Haplargids	Lithic Xeric Haplargids	loamy-skeletal	mixed	superactive		mesic	aridic	
ROZARA	2.7	NO	1994	NV		28B	Mollisols	Xerolls	Argixerolls	Lithic Argixerolls	loamy-skeletal	mixed	superactive		frigid	xeric	
RUBICITY	13.4	NO	1994	NV		25	Mollisols	Xerolls	Haploxerolls	Cumulic Haploxerolls	coarse-loamy	mixed	superactive		frigid	xeric	
RUBYHILL	221.7	NO	1971	NV		28B	Aridisols	Durids	Haplodurids	Xereptic Haplodurids	fine-loamy	mixed	superactive		frigid	aridic	
RUBYLAKE	16.6	NO	1994	NV		28B	Inceptisols	Aquepts	Endoaquepts	Fluvaquentic Endoaquepts	fine-silty	carbonatic			mesic	aquic	
RUGAR	16.6	NO	1986	NV		25	Mollisols	Xerolls	Argixerolls	Pachic Argixerolls	fine	smectitic			frigid	xeric	
RUHE	44.9	NO	1980	NV		27	Entisols	Psamments	Torripsamments	Typic Torripsamments		mixed			mesic	torric	shallow
RUMPAH	42.1	NO	1996	NV		30	Vertisols	Torrerts	Haplotorrerts	Sodic Haplotorrerts	fine	smectitic			thermic	torric	
RUNYON	13.6	NO	2006	NV	CA	21, 23	Mollisols	Xerolls	Argixerolls	Vitrandic Argixerolls	fine-loamy	mixed	superactive		frigid	xeric	
RUSTIGATE	266.5	NO	1984	NV		27, 28B, 29	Entisols	Orthents	Torriorthents	Oxyaquic Torriorthents	fine-loamy	mixed	superactive	calcareous	mesic	torric	
RUSTY	17.3	NO	1981	NV		27	Aridisols	Argids	Natrargids	Typic Natrargids	fine-loamy	mixed	superactive		mesic	aridic	
RUTAB	39.6	NO	1983	NV	OR	23, 28B	Aridisols	Cambids	Haplocambids	Xeric Haplocambids	loamy-skeletal	mixed	superactive		frigid	aridic	
RYEPATCH	32.4	NO	1963	NV		27	Mollisols	Xerolls	Haploxerolls	Vertic Haploxerolls	very-fine	smectitic			mesic	xeric	
SADER	26.4	NO	1971	NV		25, 28B	Aridisols	Argids	Natrargids	Aquic Natrargids	fine	smectitic			frigid	aridic	
SAGOUSPE	76.0	NO	1971	NV		27	Entisols	Fluvents	Torrifluvents	Oxyaquic Torrifluvents	sandy	mixed			mesic	torric	
SALTAIR	41.4	YES	1941	UT	NV	28A, 34B	Aridisols	Salids	Aquisalids	Typic Aquisalids	fine-silty	mixed	superactive		mesic	aridic	
SALTMOUNT	14.5	NO	2006	CA	NV	23	Aridisols	Salids	Haplosalids	Typic Haplosalids	very-fine	mixed	superactive		mesic	aridic	
SALTYDOG	28.8	NO	2006	NV		29	Aridisols	Calcids	Haplocalcids	Sodic Haplocalcids	fine-loamy	mixed	superactive		mesic	aridic	
SAMOR	45.7	NO	1986	NV		25	Aridisols	Calcids	Haplocalcids		loamy-skeletal	mixed	superactive		mesic	aridic	

(continued)

(continued)

Soil series	Area (km²)	Bench-mark	Year Est-ablished	TL State	Other States Using	MLRAs Using	Order	Suborder	Great Group	Subgroup	Particle-size Class	Mineralogy Class	CEC Activity Class	Reaction Class	Soil temp. Class	SMR	Other Family
										Lithic Xeric Haplocalcids							
SANDPAN	24.8	NO	2006	NV		30	Aridisols	Calcids	Petrocalcids	Calcic Petrocalcids	sandy-skeletal	mixed			hyperthermic	aridic	
SANWELL	170.4	NO	1996	NV		30	Entisols	Orthents	Torriorthents	Duric Torriorthents	loamy-skeletal	mixed	superactive	calcareous	thermic	torric	
SARALEGUI	72.8	NO	1971	CA	ID, NV	11, 26	Aridisols	Argids	Haplargids	Xeric Haplargids	coarse-loamy	mixed	superactive		mesic	aridic	
SARAPH	247.0	NO	1993	NV	CA	23	Aridisols	Argids	Haplargids	Vitrixerandic Haplargids	ashy	glassy			mesic	aridic	shallow
SATT	18.9	NO	1971	NV		28A	Mollisols	Xerolls	Argixerolls	Calciargidic Argixerolls	loamy-skeletal	mixed	superactive		frigid	xeric	
SAWMILLCAN	5.9	NO	2011	NV		30	Mollisols	Ustolls	Calciustolls	Pachic Calciustolls	loamy-skeletal	carbonatic			mesic	ustic	
SAY	108.3	NO	1986	NV		24	Mollisols	Xerolls	Argixerolls	Aridic Argixerolls	fine-loamy	mixed	superactive		frigid	xeric	
SCALFAR	64.0	NO	1986	NV		25	Mollisols	Xerolls	Argixerolls	Calcic Argixerolls	loamy-skeletal	mixed	superactive		frigid	xeric	
SCARINE	0.2	NO	2010	NV		29	Aridisols	Cambids	Haplocambids	Xerofluventic Haplocambids	coarse-loamy	mixed	superactive		mesic	aridic	
SCHADER	25.5	NO	2001	NV		30	Aridisols	Argids	Haplargids	Xeric Haplargids	loamy-skeletal	mixed	superactive		mesic	aridic	
SCHAMP	24.4	NO	1971	NV	CA	23	Aridisols	Argids	Haplargids	Xeric Haplargids	fine	smectitic			mesic	aridic	
SCHOER	21.2	NO	1994	NV		25	Mollisols	Xerolls	Argixerolls	Aridic Argixerolls	fine	smectitic			mesic	xeric	
SCHOOLMARM	397.7	NO	2006	NV		28A	Mollisols	Xerolls	Argixerolls	Aridic Lithic Argixerolls	ashy-skeletal	glassy			frigid	xeric	
SCHURZ	23.1	NO	2005	NV		27	Entisols	Fluvents	Torrifluvents	Vitrandic Torrifluvents	ashy	glassy		nonacid	mesic	torric	
SCHWALBE	24.9	NO	1996	NV		29	Mollisols	Xerolls	Haploxerolls	Aridic Haploxerolls	loamy-skeletal	mixed	superactive		mesic	xeric	
SCOSSA	1.9	NO	2012	NV		26	Inceptisols	Aquepts	Humaquepts	Histic Humaquepts	coarse-loamy	mixed	superactive	acid	mesic	aquic	
SCOTTCAS	40.0	NO	1984	NV		30	Aridisols	Argids	Haplargids	Durinodic Haplargids	loamy-skeletal	mixed	superactive		thermic	aridic	
SCRAPY	72.3	NO	2006	NV		30	Aridisols	Calcids	Haplocalcids	Lithic Ustic Haplocalcids	loamy-skeletal	carbonatic			mesic	aridic	
SEAMAN	5.1	NO	1965	NV	CA	29, 30	Entisols	Orthents	Torriorthents	Typic Torriorthents	coarse-loamy	mixed	superactive	calcareous	thermic	torric	
SEANNA	260.0	NO	2006	NV		30	Entisols	Orthents	Torriorthents	Typic Torriorthents	loamy-skeletal	mixed	superactive	calcareous	thermic	torric	shallow
SEARCHLIGHT	57.7	NO	1964	NV		30	Aridisols	Argids	Haplargids	Typic Haplargids	coarse-loamy	mixed	superactive		thermic	aridic	
SECREPASS	4.2	NO	1994	NV		25	Mollisols	Xerolls	Palexerolls	Typic Palexerolls	clayey-skeletal	smectitic			frigid	xeric	
SED	15.3	NO	1996	NV		30	Aridisols	Argids	Haplargids	Ustic Haplargids	loamy-skeletal	mixed	superactive		mesic	aridic	
SEDSKED	4.3	NO	2006	NV		23	Aridisols	Argids	Haplargids	Xeric Haplargids	loamy-skeletal	mixed	superactive		mesic	aridic	shallow
SEGURA	403.5	NO	1990	NV		28B	Mollisols	Xerolls	Argixerolls	Aridic Lithic Argixerolls	loamy	mixed	superactive		frigid	xeric	
SELBIT	34.9	NO	1988	NV		23	Mollisols	Xerolls	Haploxerolls	Torriorthentic Haploxerolls	sandy-skeletal	mixed			mesic	xeric	shallow
SELTI	24.3	NO	1990	NV		28B	Mollisols	Xerolls	Argixerolls	Calciargidic Argixerolls	loamy-skeletal	mixed	superactive		mesic	xeric	
SERALIN	276.2	NO	1995	NV		30	Mollisols	Ustolls	Haplustolls	Aridic Lithic Haplustolls	loamy-skeletal	mixed	superactive		mesic	ustic	
SETTLEDRAN	0.6	NO	2012	NV		26	Mollisols	Xerolls	Haploxerolls	Oxyaquic Haploxerolls	fine-loamy	mixed	superactive		mesic	xeric	

(continued)

(continued)

Soil series	Area (km²)	Bench-mark	Year Est-ablished	TL State	Other States Using	MLRAs Using	Order	Suborder	Great Group	Subgroup	Particle-size Class	Mineralogy Class	CEC Activity Class	Reaction Class	Soil temp. Class	SMR	Other Family
SETTLEMENT	96.9	NO	1984	NV		27, 28B, 29	Inceptisols	Aquepts	Halaquepts	Aeric Halaquepts	fine	smectitic		calcareous	mesic	aquic	
SETTLEMEYER	44.8	NO	1972	NV	CA	21, 24, 26, 28B, 29	Mollisols	Aquolls	Endoaquolls	Fluvaquentic Endoaquolls	fine-loamy	mixed	superactive		mesic	aquic	
SEVAL	12.1	NO	1971	NV		29	Mollisols	Xerolls	Durixerolls	Argiduridic Durixerolls	fine-loamy	mixed	superactive		mesic	xeric	
SEVENMILE	278.7	NO	1994	NV		28A, 29	Mollisols	Xerolls	Haploxerolls	Vitritorrandic Haploxerolls	coarse-loamy	mixed	superactive		mesic	xeric	
SEZNA	12.5	NO	1996	NV		30	Aridisols	Calcids	Petrocalcids	Argic Petrocalcids	loamy-skeletal	mixed	superactive		thermic	aridic	shallow
SHABLISS	1018.0	YES	1974	NV	ID, UT	11, 24, 25, 28A, 28B	Aridisols	Durids	Haplodurids	Xereptic Haplodurids	loamy	mixed	superactive		mesic	aridic	shallow
SHAFTER	22.8	NO	1986	NV		28A	Aridisols	Calcids	Petrocalcids	Calcic Petrocalcids	loamy	mixed	superactive		mesic	aridic	shallow
SHAGNASTY	186.3	NO	1983	NV		24, 28B	Mollisols	Xerolls	Argixerolls	Typic Argixerolls	fine	smectitic			frigid	xeric	
SHAKESPEARE	4.6	NO	1970	NV	CA, CO	22A	Alfisols	Cryalfs	Haplocryalfs	Xeric Haplocryalfs	loamy-skeletal	mixed	superactive		cryic	xeric	
SHALAKE	94.5	NO	1986	NV		25	Aridisols	Durids	Haplodurids	Xeric Haplodurids	coarse-loamy	mixed	superactive		mesic	aridic	
SHALCLEAV	399.2	NO	1986	NV		25	Mollisols	Xerolls	Argixerolls	Lithic Argixerolls	loamy-skeletal	mixed	superactive		frigid	xeric	
SHALPEET	1.5	NO	2012	NV		26	Mollisols	Aquolls	Endoaquolls	Fluvaquentic Endoaquolls	fine-loamy over sandy or sandy-skeletal	mixed	superactive		mesic	aquic	
SHALPER	197.0	NO	1986	NV	UT	25	Mollisols	Xerolls	Argixerolls	Aridic Lithic Argixerolls	loamy-skeletal	mixed	superactive		frigid	xeric	
SHAMOCK	74.1	NO	1996	NV		30	Aridisols	Durids	Haplodurids	Typic Haplodurids	coarse-loamy	mixed	superactive		thermic	aridic	
SHANKBA	33.9	NO	1992	NV		30	Entisols	Orthents	Torriorthents	Typic Torriorthents	loamy-skeletal	mixed	superactive	calcareous	thermic	torric	shallow
SHANTOWN	35.9	NO	1994	NV		28B	Mollisols	Xerolls	Haploxerolls	Calcidic Haploxerolls	coarse-loamy	mixed	superactive		mesic	xeric	
SHAWAVE	465.3	NO	1988	NV		23, 27	Aridisols	Argids	Haplargids	Xeric Haplargids	fine-loamy	mixed	superactive		mesic	aridic	
SHAYLA	17.7	NO	1986	NV		25	Entisols	Orthents	Torriorthents	Typic Torriorthents	loamy-skeletal	mixed	superactive	calcareous	mesic	torric	shallow
SHEEGE	55.3	NO	1969	ID	MT, NV, UT, WY	13, 28B, 43B, 47	Mollisols	Rendolls	Cryrendolls	Lithic Cryrendolls	loamy-skeletal	carbonatic			cryic		
SHEEPPASS	42.0	NO	2011	NV		30	Entisols	Orthents	Torriorthents	Lithic Torriorthents	loamy-skeletal	carbonatic			mesic	torric	
SHEEPRANGE	40.2	NO	2011	NV		30	Mollisols	Ustolls	Haplustolls	Aridic Lithic Haplustolls	loamy-skeletal	mixed	superactive		frigid	ustic	
SHEEPROCK	6.1		1969	UT			Entisols	Orthents	Torriorthents	Xeric Torriorthents	sandy-skeletal	mixed			mesic	torric	
SHEFFIT	453.2	YES	1990	NV		28A, 28B	Entisols	Orthents	Torriorthents	Xerertic Torriorthents	fine	smectitic		calcareous	mesic	torric	
SHIPLEY	78.6	NO	1971	NV		25, 28B	Entisols	Orthents	Torriorthents	Xeric Torriorthents	coarse-loamy	mixed	superactive	calcareous	frigid	torric	
SHIVELY	61.4	NO	1986	NV		24, 25, 27	Mollisols	Xerolls	Haploxerolls	Pachic Haploxerolls	coarse-loamy	mixed	superactive		frigid	xeric	
SHIVLUM	41.8	NO	1986	NV		25	Mollisols	Xerolls	Argixerolls	Aridic Argixerolls	fine-silty	mixed	superactive		frigid	xeric	
SHOKEN	18.0	NO	1974	NV		24, 26	Entisols	Orthents	Torriorthents	Xeric Torriorthents	loamy-skeletal	mixed	superactive	nonacid	mesic	torric	shallow
SHORIM	39.9	NO	1996	NV		30	Aridisols	Durids	Haplodurids	Typic Haplodurids	loamy-skeletal	mixed	superactive		thermic	aridic	
SHORT CREEK	177.8	NO	1979	NV		24, 25	Aridisols	Argids	Haplargids	Xeric Haplargids	clayey-skeletal	smectitic			frigid	aridic	
SHREE	77.3	NO	1981	NV	CA	26, 28B	Mollisols	Xerolls	Argixerolls	Aridic Argixerolls	loamy-skeletal	mixed	superactive		mesic	xeric	
SHROE	32.4	NO	1971	NV		28A, 28B, 29	Mollisols	Xerolls	Argixerolls	Aridic Argixerolls	loamy-skeletal	mixed	superactive		mesic	xeric	

(continued)

(continued)

Soil series	Area (km^2)	Bench-mark	Year Est-ablished	TL State	Other States Using	MLRAs Using	Order	Suborder	Great Group	Subgroup	Particle-size Class	Mineralogy Class	CEC Activity Class	Reaction Class	Soil temp. Class	SMR	Other Family
SHUTTLE	30.7	NO	1986	NV		28A	Aridisols	Calcids	Haplocalcids	Duric Haplocalcids	coarse-loamy	mixed	superactive		mesic	aridic	
SIBELIA	13.6	YES	1980	NV		22A	Inceptisols	Cryepts	Humicryepts	Xeric Humicryepts	loamy-skeletal	isotic			cryic	xeric	
SIEGEL	10.4	NO	2013	NV		28B	Mollisols	Xerolls	Argixerolls	Calcic Argixerolls	clayey-skeletal	smectitic			frigid	xeric	
SIEROCLIFF	10.6	NO	1971	NV		28A, 29	Aridisols	Calcids	Petrocalcids	Calcic Petrocalcids	loamy-skeletal	carbonatic			mesic	aridic	
SILENT	45.0	NO	1964	NV		29	Aridisols	Durids	Argidurids	Typic Argidurids	loamy	mixed	superactive		mesic	aridic	shallow
SILVERADO	345.2	NO	1971	NV		28B	Aridisols	Cambids	Haplocambids	Durinodic Xeric Haplocambids	coarse-loamy	mixed	superactive		frigid	aridic	
SILVERBOW	104.5	NO	1972	NV		29	Aridisols	Durids	Argidurids	Typic Argidurids	loamy-skeletal	mixed	superactive		mesic	aridic	shallow
SIMON	95.8	NO	1974	NV	OR	23, 25	Mollisols	Xerolls	Argixerolls	Aridic Argixerolls	fine-loamy	mixed	superactive		frigid	xeric	
SIMPARK	54.4	NO	1983	NV		28B	Aridisols	Durids	Argidurids	Xeric Argidurids	loamy-skeletal	mixed	superactive		frigid	aridic	shallow
SINGATSE	1022.2	YES	1980	NV		27	Entisols	Orthents	Torriorthents	Lithic Torriorthents	loamy-skeletal	mixed	superactive	calcareous	mesic	torric	
SINGLETREE	32.5	NO	1974	NV		24, 25, 28B	Mollisols	Xerolls	Argixerolls	Calciargidic Argixerolls	fine-loamy	mixed	superactive		frigid	xeric	
SIRI	29.8	NO	1971	NV		12, 25, 28B	Aridisols	Calcids	Haplocalcids	Xeric Haplocalcids	loamy-skeletal	mixed	superactive		frigid	aridic	
SISCAB	111.0	NO	1992	NV		23, 24, 25	Mollisols	Xerolls	Argixerolls	Aridic Argixerolls	loamy	mixed	superactive		mesic	xeric	shallow
SKEDADDLE	170.0	NO	1980	NV	OR	23, 26	Entisols	Orthents	Torriorthents	Lithic Xeric Torriorthents	loamy-skeletal	mixed	superactive	nonacid	mesic	torric	
SKELON	233.9	NO	1984	NV		30	Aridisols	Durids	Haplodurids	Typic Haplodurids	loamy-skeletal	mixed	superactive		thermic	aridic	
SKULL CREEK	48.4	NO	1986	NV	OR	25	Aridisols	Durids	Haplodurids	Vitrixerandic Haplodurids	coarse-loamy	mixed	superactive		mesic	aridic	
SKULLWAK	37.6	NO	1985	NV	CA	23, 24, 27	Inceptisols	Aquepts	Halaquepts	Duric Halaquepts	fine	smectitic		calcareous	mesic	aquic	
SKYHAVEN	19.6	NO	1964	NV		30	Aridisols	Calcids	Petrocalcids	Argic Petrocalcids	fine-loamy	carbonatic			thermic	aridic	
SLATERY	69.3	NO	1984	NV		29	Entisols	Orthents	Torriorthents	Typic Torriorthents	loamy	mixed	superactive	calcareous	mesic	torric	shallow
SLATTER	3.3	NO	2012	NV		26	Mollisols	Xerolls	Argixerolls	Aridic Lithic Argixerolls	loamy-skeletal	mixed	superactive		frigid	xeric	
SLAVEN	255.2	NO	1979	NV		24, 25	Mollisols	Xerolls	Argixerolls	Aridic Argixerolls	clayey-skeletal	smectitic			frigid	xeric	
SLAW	490.6	NO	1986	NV		27, 28A, 28B, 29	Entisols	Fluvents	Torrifluvents	Typic Torrifluvents	fine-silty	mixed	superactive	calcareous	mesic	torric	
SLAWHA	71.7	NO	1993	NV		24	Entisols	Fluvents	Torrifluvents	Typic Torrifluvents	fine-silty	mixed	superactive	calcareous	mesic	torric	
SLAWMASTER	14.6	NO	2005	NV		27	Entisols	Fluvents	Torrifluvents	Vitrandic Torrifluvents	fine-silty	mixed	superactive	calcareous	mesic	torric	
SLIDYMTN	144.0	NO	1992	NV		29	Mollisols	Ustolls	Argiustolls	Aridic Lithic Argiustolls	loamy-skeletal	mixed	superactive		mesic	ustic	
SLIPBACK	102.2	NO	1988	NV		27, 28B	Aridisols	Argids	Natrargids	Xeric Natrargids	fine-loamy	mixed	superactive		mesic	aridic	
SLOCAVE	62.8	NO	1988	NV		27	Entisols	Orthents	Torriorthents	Typic Torriorthents	loamy-skeletal	mixed	superactive	calcareous	mesic	torric	shallow
SLOCKEY	195.4	NO	2006	NV		28A	Mollisols	Xerolls	Argixerolls	Vitritorrandic Argixerolls	ashy-skeletal	glassy			frigid	xeric	
SMALLCONE	27.0	NO	1980	NV	CA	26	Entisols	Orthents	Torriorthents	Xeric Torriorthents	loamy-skeletal	mixed	active	nonacid	mesic	torric	shallow
SMAUG	94.5	NO	1985	NV	UT	27, 28A	Entisols	Orthents	Torriorthents	Typic Torriorthents	coarse-silty	mixed	superactive	calcareous	mesic	torric	
SMEDLEY	130.4	NO	1981	NV		26	Aridisols	Durids	Argidurids	Argidic Argidurids	clayey	smectitic			mesic	aridic	shallow

(continued)

(continued)

Soil series	Area (km^2)	Bench-mark	Year Est-ablished	TL State	Other States Using	MLRAs Using	Order	Suborder	Great Group	Subgroup	Particle-size Class	Mineralogy Class	CEC Activity Class	Reaction Class	Soil temp. Class	SMR	Other Family
SNACREEK	3.5	NO	2009	NV		28A	Mollisols	Cryolls	Haplocryolls	Pachic Haplocryolls	loamy-skeletal	mixed	superactive		cryic		
SNAG	8.9	NO	1971	NV	CA	21, 23	Mollisols	Cryolls	Argicryolls	Vitrandic Argicryolls	ashy-skeletal	glassy			cryic		
SNAPCAN	1.4	NO	2001	AZ	NV	30	Aridisols	Cambids	Haplocambids	Typic Haplocambids	loamy-skeletal	mixed	superactive		hyperthermic	aridic	
SNAPEED	4.2	NO	2004	NV		28A	Mollisols	Xerolls	Haploxerolls	Aridic Haploxerolls	loamy-skeletal	mixed	superactive		frigid	xeric	
SNAPP	310.5	NO	1986	NV		24, 25	Aridisols	Argids	Natrargids	Durinodic Xeric Natrargids	clayey over sandy or sandy-skeletal	smectitic OVER mixed			mesic	aridic	
SNOPOC	29.3	NO	1985	NV		26	Mollisols	Cryolls	Haplocryolls	Pachic Haplocryolls	loamy-skeletal	mixed	superactive		cryic		
SNOTOWN	30.5	NO	1986	NV		25	Inceptisols	Cryepts	Dystrocryepts	Xeric Dystrocryepts	loamy-skeletal	mixed	superactive		cryic	xeric	
SNOWMORE	157.0	NO	1986	NV	ID	11, 24, 25	Aridisols	Durids	Argidurids	Xeric Argidurids	fine-loamy	mixed	superactive		mesic	aridic	
SOAKPAK	18.6		1969	CA	NV		Inceptisols	Cryepts	Haplocryepts	Typic Haplocryepts	loamy-skeletal	mixed	superactive		cryic		
SOAR	141.6	NO	1988	NV		27	Aridisols	Argids	Haplargids	Xeric Haplargids	loamy-skeletal	mixed	superactive		mesic	aridic	shallow
SODA LAKE	35.3	NO	1909	NV		24, 27	Entisols	Orthents	Torriorthents	Typic Torriorthents	sandy	mixed			mesic	torric	
SODASPRING	75.3	NO	1984	NV		29	Entisols	Orthents	Torriorthents	Typic Torriorthents	coarse-loamy	mixed	superactive	calcareous	mesic	torric	
SODHOUSE	165.8	NO	1985	NV		24, 25, 28B	Aridisols	Durids	Haplodurids	Typic Haplodurids	loamy	mixed	superactive		mesic	aridic	shallow
SOFTSCRABBLE	696.0	YES	1980	NV	CA	21, 23, 24, 26, 28B	Mollisols	Xerolls	Argixerolls	Pachic Argixerolls	loamy-skeletal	mixed	superactive		frigid	xeric	
SOJUR	303.7	NO	1988	NV		24, 27	Entisols	Orthents	Torriorthents	Lithic Torriorthents	loamy-skeletal	mixed	superactive	calcareous	mesic	torric	
SOLAK	145.7	NO	1980	UT	NV	24, 25, 28A, 28B, 34, 43, 47	Entisols	Orthents	Torriorthents	Lithic Xeric Torriorthents	loamy-skeletal	mixed	superactive	calcareous	frigid	torric	
SOMBRERO	7.0	NO	1974	NV		24	Aridisols	Durids	Haplodurids	Aquicambidic Haplodurids	loamy	mixed	superactive		mesic	aridic	shallow
SONDOA	436.4	YES	1985	NV		27, 28A, 28B	Entisols	Orthents	Torriorthents	Typic Torriorthents	fine-silty	mixed	superactive	calcareous	mesic	torric	
SONOMA	406.5	YES	1963	NV		24, 25, 27, 28B	Entisols	Aquents	Fluvaquents	Aeric Fluvaquents	fine-silty	mixed	superactive	calcareous	mesic	aquic	
SOOLAKE	23.1	NO	1985	NV		24	Entisols	Orthents	Torriorthents	Typic Torriorthents	sandy	mixed			mesic	torric	
SOONAHBE	41.8	NO	1984	NV	ID	25	Alfisols	Xeralfs	Haploxeralfs	Mollic Haploxeralfs	fine-loamy	mixed	superactive		frigid	xeric	
SOONAKER	6.6	NO	1986	NV		25	Alfisols	Xeralfs	Haploxeralfs	Mollic Haploxeralfs	fine-loamy	mixed	superactive		frigid	xeric	
SOUGHE	793.9	NO	1983	NV		23, 24, 25	Aridisols	Argids	Haplargids	Lithic Xeric Haplargids	loamy-skeletal	mixed	superactive		mesic	aridic	

(continued)

Soil series	Area (km^2)	Bench-mark	Year Est-ablished	TL State	Other States Using	MLRAs Using	Order	Suborder	Great Group	Subgroup	Particle-size Class	Mineralogy Class	CEC Activity Class	Reaction Class	Soil temp. Class	SMR	Other Family
SOUTHCAMP	2.7	NO	2007	NV		22A	Alfisols	Xeralfs	Palexeralfs	Ultic Palexeralfs	loamy-skeletal	isotic			frigid	xeric	
SPAGER	43.0	NO	1981	UT	NV	28A, 28B, 29	Aridisols	Calcids	Petrocalcids	Calcic Petrocalcids	loamy-skeletal	carbonatic			mesic	aridic	shallow
SPANEL	48.2	NO	1972	NV		24, 28B	Aridisols	Durids	Argidurids	Typic Argidurids	loamy	mixed	superactive		mesic	aridic	shallow
SPASPREY	128.2	NO	1980	NV		26, 28B	Aridisols	Durids	Argidurids	Haploxeralfic Argidurids	fine-loamy	mixed	superactive		mesic	aridic	
SPECTER	2.2	NO	1965	NV		29	Aridisols	Durids	Haplodurids	Typic Haplodurids	loamy-skeletal	mixed	superactive		mesic	aridic	
SPIKE	27.2	NO	1985	NV		28B	Aridisols	Argids	Paleargids	Xeric Paleargids	loamy-skeletal	mixed	superactive		mesic	aridic	
SPILOCK	13.8	NO	1986	NV		25	Aridisols	Calcids	Petrocalcids	Xeric Petrocalcids	loamy-skeletal	mixed	active		mesic	aridic	shallow
SPINLIN	62.2	NO	1974	NV		24, 28B	Mollisols	Cryolls	Argicryolls	Xeric Argicryolls	clayey-skeletal	smectitic			cryic	xeric	
SPRING	13.1	NO	1923	NV		30	Aridisols	Gypsids	Haplogypsids	Sodic Haplogypsids	fine-silty	mixed	active		thermic	aridic	
SPRINGBAR	12.0	NO	2004	NV		28A	Aridisols	Cambids	Haplocambids	Xeric Haplocambids	sandy	mixed			mesic	aridic	
SPRINGMEADOW	8.0	NO	2010	NV		28A	Mollisols	Aquolls	Endoaquolls	Fluvaquentic Endoaquolls	fine	smectitic			mesic	aquic	
SPRINGMEYER	57.3	NO	1974	NV	CA, UT	21, 23, 26, 28A, 28B	Mollisols	Xerolls	Argixerolls	Aridic Argixerolls	fine-loamy	mixed	superactive		mesic	xeric	
SPRINGWARM	8.3	NO	1993	NV		29	Inceptisols	Aquepts	Halaquepts	Aeric Halaquepts	loamy-skeletal	mixed	superactive	calcareous	mesic	aquic	
SQUAWTIP	253.5	NO	1984	NV		29	Mollisols	Xerolls	Argixerolls	Typic Argixerolls	loamy-skeletal	mixed	superactive		frigid	xeric	
SQUAWVAL	10.8	NO	1990	NV		23	Mollisols	Cryolls	Argicryolls	Pachic Argicryolls	fine-loamy	mixed	superactive		cryic		
ST. THOMAS	875.0	YES	1970	NV	AZ	30, 40	Entisols	Orthents	Torriorthents	Lithic Torriorthents	loamy-skeletal	carbonatic			thermic	torric	
STAMPEDE	623.0	YES	1971	NV	OR	23, 25, 28B	Mollisols	Xerolls	Durixerolls	Vertic Durixerolls	fine	smectitic			frigid	xeric	
STARFLYER	167.1	NO	2006	NV		28A	Mollisols	Xerolls	Argixerolls	Lithic Argixerolls	ashy-skeletal	glassy			frigid	xeric	
STARGO	165.7	NO	1972	NV		28B, 29	Entisols	Fluvents	Torrifluvents	Duric Torrifluvents	fine-loamy over sandy or sandy-skeletal	mixed	superactive	calcareous	mesic	torric	
STEEPSHRUB	18.8	NO	2013	NV		28B	Mollisols	Cryolls	Haplocryolls	Xeric Haplocryolls	loamy-skeletal	mixed	superactive		cryic	xeric	
STEERLAKE	4.2	NO	2006	NV	CA	23	Mollisols	Xerolls	Palexerolls	Vertic Palexerolls	fine	smectitic			mesic	xeric	
STEPTOE	4.0	NO	2013	NV		28B	Mollisols	Xerolls	Haploxerolls	Cumulic Haploxerolls	loamy-skeletal	mixed	superactive		frigid	xeric	
STEWVAL	2632.5	YES	1984	NV		27, 28B, 29	Aridisols	Argids	Haplargids	Lithic Xeric Haplargids	loamy-skeletal	mixed	superactive		mesic	aridic	
STILLWATER	57.0	NO	1971	NV		27	Mollisols	Aquolls	Endoaquolls	Fluvaquentic Vertic Endoaquolls	fine	smectitic			mesic	aquic	
STINGDORN	121.9	NO	1980	NV		24, 27	Aridisols	Durids	Argidurids	Typic Argidurids	loamy-skeletal	mixed	superactive		mesic	aridic	shallow
STODICK	40.2	NO	1975	NV		26	Aridisols	Argids	Haplargids	Xeric Haplargids	loamy-skeletal	mixed	superactive		mesic	aridic	shallow
STONELL	145.8	NO	1984	NV		29	Aridisols	Argids	Haplargids	Typic Haplargids	loamy-skeletal	mixed	superactive		mesic	aridic	
STONEWALL	4.6	NO	1996	NV		29	Aridisols	Argids	Paleargids	Typic Paleargids	clayey-skeletal	smectitic			mesic	aridic	
STRAWBCREK	3.3	NO	2009	NV		28A	Inceptisols	Cryepts	Haplocryepts	Lamellic Haplocryepts	sandy-skeletal	mixed			cryic		
STRAYCOW	21.3	NO	2006	NV		30	Aridisols	Argids	Haplargids	Typic Haplargids	loamy-skeletal	mixed	superactive		thermic	aridic	shallow
STRICKLAND	13.0	NO	1984	NV	ID	25	Mollisols	Cryolls	Haplocryolls	Pachic Haplocryolls	fine-loamy	mixed	superactive		cryic		
STROZI	40.7	NO	1996	NV		30	Aridisols	Durids	Argidurids	Argidic Argidurids	fine-loamy	mixed	superactive		thermic	aridic	
STUCKY	37.6	NO	1981	NV	UT	26, 28A	Aridisols	Argids	Haplargids	Xeric Haplargids	loamy-skeletal	mixed	superactive		mesic	aridic	

(continued)

(continued)

Soil series	Area (km^2)	Bench-mark	Year Est-ablished	TL State	Other States Using	MLRAs Using	Order	Suborder	Great Group	Subgroup	Particle-size Class	Mineralogy Class	CEC Activity Class	Reaction Class	Soil temp. Class	SMR	Other Family
STUMBLE	676.6	YES	1971	NV	NM	27, 29, 35, 36	Entisols	Psamments	Torripsamments	Typic Torripsamments		mixed			mesic	torric	
SUAK	181.7	NO	1990	NV		28B, 29	Mollisols	Xerolls	Argixerolls	Typic Argixerolls	loamy-skeletal	mixed	superactive		frigid	xeric	
SUCCESSLOOP	52.0	NO	2013	NV		28B	Mollisols	Cryolls	Argicryolls	Xeric Argicryolls	loamy-skeletal	mixed	superactive		cryic	xeric	
SUMINE	1313.0	YES	1974	NV	CA, UT	23, 24, 25, 26, 27, 28A, 28B, 47	Mollisols	Xerolls	Argixerolls	Aridic Argixerolls	loamy-skeletal	mixed	superactive		frigid	xeric	
SUMMERMUTE	301.0	YES	2004	NV		28A	Aridisols	Calcids	Haplocalcids	Durinodic Haplocalcids	loamy-skeletal	carbonatic			mesic	aridic	
SUMYA	103.3	NO	1986	NV		24, 27	Entisols	Orthents	Torriorthents	Lithic Xeric Torriorthents	clayey-skeletal	smectitic		nonacid	frigid	torric	
SUNDOWN	91.4	NO	1972	NV		27, 28B, 29	Entisols	Psamments	Torripsamments	Typic Torripsamments		mixed			mesic	torric	
SUNROCK	271.0	NO	2000	AZ	CA, NV	30	Entisols	Orthents	Torriorthents	Lithic Torriorthents	loamy-skeletal	mixed	superactive	calcareous	hyperthermic	torric	
SUP	4.0	NO	1981	NV		26	Mollisols	Cryolls	Haplocryolls	Xeric Haplocryolls	loamy-skeletal	mixed	superactive		cryic	xeric	
SURGEM	11.3	NO	1980	NV		26	Aridisols	Argids	Haplargids	Xeric Haplargids	clayey-skeletal	smectitic			mesic	aridic	
SURPASS	26.0	NO	2010	NV	CA	26	Mollisols	Xerolls	Haploxerolls	Aridic Haploxerolls	coarse-loamy	mixed	superactive		mesic	xeric	
SURPRISE	12.6	NO	1931	CA	NV	21, 23, 26	Mollisols	Xerolls	Haploxerolls	Vitritorrandic Haploxerolls	ashy	glassy			mesic	xeric	
SUSIE CREEK	127.4	NO	1976	NV		25	Mollisols	Xerolls	Argixerolls	Argiduridic Argixerolls	fine	smectitic			frigid	xeric	
SUTCLIFF	33.8	NO	1980	NV		26, 27	Aridisols	Argids	Petroargids	Duric Petroargids	loamy-skeletal	mixed	superactive		mesic	aridic	
SUTRO	10.8	NO	1975	NV		24, 26	Mollisols	Xerolls	Haploxerolls	Aridic Haploxerolls	fine-loamy	mixed	superactive		mesic	xeric	
SWEETSPRING	16.3	NO	2006	NV		30	Aridisols	Argids	Calciargids	Petronodic Calciargids	sandy-skeletal	mixed			hyperthermic	aridic	
SWINGLER	243.0	NO	1971	NV	UT	27, 28A, 28B	Entisols	Orthents	Torriorthents	Typic Torriorthents	fine-silty	mixed	superactive	calcareous	mesic	torric	
SWISBOB	6.0	NO	1971	NV		28A	Mollisols	Xerolls	Argixerolls	Calcic Argixerolls	fine-loamy	mixed	superactive		frigid	xeric	
SWOPE	8.9	NO	1971	NV		27	Mollisols	Aquolls	Endoaquolls	Fluvaquentic Endoaquolls	fine-loamy over sandy or sandy-skeletal	mixed	superactive	calcareous	mesic	aquic	
SYCOMAT	340.8	NO	1991	NV		28A, 28B	Aridisols	Calcids	Haplocalcids	Durinodic Haplocalcids	coarse-loamy	mixed	active		mesic	aridic	
SYLVANIAM	15.6	NO	1984	NV		29	Mollisols	Xerolls	Calcixerolls	Typic Calcixerolls	loamy-skeletal	carbonatic			frigid	xeric	
TAGUM	6.1	NO	2012	NV		26	Mollisols	Cryolls	Argicryolls	Xeric Argicryolls	loamy-skeletal	mixed	superactive		cryic	xeric	
TAHOE	5.8	NO	2007	CA	NV	22A	Inceptisols	Aquepts	Humaquepts	Cumulic Humaquepts	coarse-loamy	mixed	superactive	acid	frigid	aquic	
TAHOMA	0.5	NO	1970	CA	NV	22A	Alfisols	Xeralfs	Haploxeralfs	Ultic Haploxeralfs	fine-loamy	isotic			frigid	xeric	
TAHQUATS	0.0	NO	1971	ID	NV	13	Mollisols	Cryolls	Argicryolls	Typic Argicryolls	loamy-skeletal	mixed	superactive		cryic		
TALLAC	19.8	YES	1970	CA	NV	22A	Inceptisols	Xerepts	Dystroxerepts	Humic Dystroxerepts	loamy-skeletal	mixed	superactive		frigid	xeric	
TANAZZA	34.5	NO	1996	NV		30	Aridisols	Gypsids	Calcigypsids	Typic Calcigypsids	fine-silty	gypsic			thermic	aridic	
TANOB	13.5	NO	1980	NV	CA	26	Mollisols	Xerolls	Argixerolls	Ultic Argixerolls	coarse-loamy	mixed	superactive		frigid	xeric	
TARLOC	7.7	NO	1975	NV		26	Aridisols	Argids	Haplargids	Xeric Haplargids	coarse-loamy	mixed	superactive		mesic	aridic	
TARNACH	340.0	NO	1985	UT	NV	28A, 28B	Aridisols	Calcids	Haplocalcids	Lithic Xeric Haplocalcids	loamy-skeletal	mixed	active		mesic	aridic	
TAYLOR CREEK	29.6	NO	1979	NV		24, 25	Mollisols	Cryolls	Argicryolls	Vertic Argicryolls	very-fine	smectitic			cryic		
TECOMAR	1166.1	NO	1986	NV		25, 28A, 28B	Aridisols	Calcids	Haplocalcids	Lithic Xeric Haplocalcids	loamy-skeletal	carbonatic			mesic	aridic	

(continued)

(continued)

Soil series	Area (km^2)	Bench-mark	Year Est-ablished	TL State	Other States Using	MLRAs Using	Order	Suborder	Great Group	Subgroup	Particle-size Class	Mineralogy Class	CEC Activity Class	Reaction Class	Soil temp. Class	SMR	Other Family
TECOPA	91.4	NO	1980	CA	NV	30	Entisols	Orthents	Torriorthents	Lithic Torriorthents	loamy-skeletal	mixed	superactive	calcareous	thermic	torric	
TEEBAR	39.7	NO	2006	NV		30	Aridisols	Calcids	Petrocalcids	Typic Petrocalcids	loamy-skeletal	carbonatic			hyperthermic	aridic	shallow
TEEBONE	41.2	NO	2006	NV		28A	Aridisols	Calcids	Haplocalcids	Vertic Haplocalcids	fine	smectitic			mesic	aridic	
TEGURO	172.0	NO	1985	NV	OR	10, 23, 24, 26, 27	Mollisols	Xerolls	Argixerolls	Lithic Argixerolls	loamy	mixed	superactive		frigid	xeric	
TEJABE	94.5	NO	1985	NV		27, 29	Entisols	Orthents	Torriorthents	Lithic Xeric Torriorthents	loamy-skeletal	mixed	superactive	nonacid	mesic	torric	
TEMAN	28.6	NO	1974	NV		24	Aridisols	Calcids	Haplocalcids	Durinodic Xeric Haplocalcids	fine-silty	mixed	superactive		mesic	aridic	
TEMO	29.6	NO	1980	NV	CA	22A	Entisols	Psamments	C ryopsamments	Typic Cryopsamments		mixed			cryic		shallow
TENABO	456.1	NO	1974	NV		24	Aridisols	Durids	Natridurids	Typic Natridurids	loamy	mixed	superactive		mesic	aridic	shallow
TENPIN	21.1	NO	1981	NV		26	Aridisols	Argids	Paleargids	Xeric Paleargids	clayey-skeletal	smectitic			mesic	aridic	
TENVORRD	18.2	NO	1983	NV		25, 28B	Aridisols	Durids	Haplodurids	Xeric Haplodurids	loamy	mixed	superactive		mesic	aridic	shallow
TENWELL	158.9	NO	2006	NV		30	Aridisols	Durids	Argidurids	Typic Argidurids	fine-loamy	mixed	superactive		thermic	aridic	
TERCA	137.7	NO	1990	NV		23, 26	Mollisols	Xerolls	Argixerolls	Aridic Lithic Argixerolls	loamy	mixed	superactive		mesic	xeric	
TERLCO	351.9	NO	1984	NV		27, 29	Aridisols	Argids	Natrargids	Typic Natrargids	fine-loamy	mixed	superactive		mesic	aridic	
TERT	41.9	NO	1985	NV		29	Entisols	Orthents	Torriorthents	Xeric Torriorthents	loamy	mixed	superactive	calcareous	mesic	torric	shallow
TESSFIVE	29.7	NO	1985	NV		24	Entisols	Orthents	Torriorthents	Lithic Xeric Torriorthents	loamy	mixed	superactive	calcareous	mesic	torric	
THEON	1462.7	YES	1981	NV		26, 27	Aridisols	Argids	Haplargids	Lithic Haplargids	loamy-skeletal	mixed	superactive		mesic	aridic	
THERIOT	294.0	NO	1940	NV	CA, UT	27, 28A, 28B, 29, 30	Entisols	Orthents	Torriorthents	Lithic Torriorthents	loamy-skeletal	carbonatic			mesic	torric	
THESISTERS	25.8	NO	2006	NV		30	Mollisols	Ustolls	Haplustolls	Aridic Lithic Haplustolls	loamy-skeletal	carbonatic			frigid	ustic	
THIEFRIDGE	6.4	NO	2006	CA	NV	22A	Mollisols	Cryolls	Argicryolls	Lithic Argicryolls	loamy-skeletal	mixed	superactive		cryic	xeric	
THIKE	20.0	NO	1984	NV		29	Aridisols	Argids	Haplargids	Lithic Xeric Haplargids	loamy-skeletal	mixed	superactive		mesic	aridic	
THREEDOGS	22.6	NO	2004	NV		28A	Aridisols	Argids	Calciargids	Typic Calciargids	fine-silty	mixed	superactive		mesic	aridic	
THREELAKES	237.3	NO	2006	NV		30	Entisols	Orthents	Torriorthents	Typic Torriorthents	loamy-skeletal	carbonatic			thermic	torric	
THREESEE	125.6	NO	1994	NV		28B	Aridisols	Calcids	Haplocalcids	Xeric Haplocalcids	loamy-skeletal over sandy or sandy-skeletal	mixed	superactive		mesic	aridic	
THULEPAH	19.2	NO	1980	NV		26	Mollisols	Cryolls	Argicryolls	Pachic Argicryolls	fine-loamy	mixed	superactive		cryic	xeric	
THUNDERBIRD	117.0	YES	1968	AZ	NM, NV	35, 39	Mollisols	Ustolls	Argiustolls	Aridic Argiustolls	fine	smectitic			mesic	ustic	
THWOOP	8.5	NO	1986	NV		25	Mollisols	Xerolls	Durixerolls	Argiduridic Durixerolls	clayey-skeletal	smectitic			frigid	xeric	
TICA	18.0		1969	ID	NV	29, 28B	Mollisols	Cryolls	Argicryolls	Lithic Argicryolls	clayey-skeletal	smectitic			cryic	xeric	
TICINO	46.4	NO	1980	NV		10, 26	Mollisols	Xerolls	Argixerolls	Typic Argixerolls	fine-loamy	mixed	superactive		frigid	xeric	
TICKAPOO	132.6	NO	1974	NV		29	Aridisols	Durids	Natridurids	Natrargidic Natridurids	clayey	smectitic			mesic	aridic	shallow
TIERNEY	0.1	NO		NV	UT	28A, 28B	Mollisols	Xerolls	Haploxerolls	Cumulic Haploxerolls	loamy-skeletal	mixed	superactive		frigid	xeric	
TIMMERCREK	15.1	NO	2009	NV		28A	Mollisols	Cryolls	Haplocryolls	Xeric Haplocryolls	loamy-skeletal	mixed	superactive		cryic	xeric	
TIMPAHUTE	5.4		1940	NV		29	Aridisols	Durids	Natridurids	Xeric Natridurids	fine	smectitic			mesic	aridic	

(continued)

(continued)

Soil series	Area (km^2)	Bench-mark	Year Est-ablished	TL State	Other States Using	MLRAs Using	Order	Suborder	Great Group	Subgroup	Particle-size Class	Mineralogy Class	CEC Activity Class	Reaction Class	Soil temp. Class	SMR	Other Family
TIMPER	97.8	NO	1972	NV		29	Aridisols	Durids	Haplodurids	Cambidic Haplodurids	loamy	mixed	superactive		mesic	aridic	shallow
TIMPIE	221.0	NO	1992	UT	NV	28A, 28B	Entisols	Orthents	Torriorthents	Typic Torriorthents	fine-silty	mixed	superactive	calcareous	mesic	torric	
TINPAN	74.6	NO	1995	NV		23	Mollisols	Xerolls	Palexerolls	Vertic Palexerolls	very-fine	smectitic			frigid	xeric	
TIPNAT	23.9	NO	2006	NV		30	Aridisols	Argids	Natrargids	Typic Natrargids	fine-loamy	mixed	superactive		thermic	aridic	
TIPPERARY	41.5	YES	1951	WY	CO, MT, NV, UT	32, 34	Entisols	Psamments	Torripsamments	Typic Torripsamments		mixed			mesic	torric	
TIPPIPAH	4.0	NO	1965	NV		29	Aridisols	Argids	Natrargids	Durinodic Natrargids	fine-loamy over sandy or sandy-skeletal	mixed	superactive		mesic	aridic	
TITIACK	36.9	NO	1991	NV		29	Entisols	Orthents	Torriorthents	Vitrandic Torriorthents	loamy-skeletal over cindery	mixed	superactive	nonacid	mesic	torric	
TOANO	176.8	NO	1986	NV		25, 28A, 28B	Entisols	Orthents	Torriorthents	Typic Torriorthents	coarse-silty	mixed	superactive	calcareous	mesic	torric	
TOBA	18.9	NO	1994	NV		28B	Aridisols	Calcids	Haplocalcids	Aquic Haplocalcids	fine-loamy over sandy or sandy-skeletal	mixed	superactive		mesic	aridic	
TOBLER	4.0	NO	1940	UT	AZ, NM, NV	30, 42, 70C	Entisols	Fluvents	Torrifluvents	Typic Torrifluvents	coarse-loamy	mixed	superactive	calcareous	thermic	torric	
TOCAN	54.6	NO	1981	NV		27	Aridisols	Argids	Haplargids	Durinodic Haplargids	fine-loamy	mixed	superactive		mesic	aridic	
TOEJA	78.2	NO	1974	NV		24, 25	Mollisols	Xerolls	Argixerolls	Aridic Argixerolls	fine-loamy	mixed	superactive		frigid	xeric	
TOEJOM	3.4	NO	2006	CA	NV	22A	Entisols	Orthents	Xerorthents	Typic Xerorthents	sandy-skeletal	mixed			mesic	xeric	shallow
TOEM	1.3	NO	1970	CA	NV	22A	Entisols	Psamments	Xeropsamments	Dystric Xeropsamments		mixed			frigid	xeric	shallow
TOGNONI	53.6	NO	1984	NV		29	Aridisols	Argids	Haplargids	Lithic Haplargids	loamy-skeletal	mixed	superactive		mesic	aridic	
TOIYABE	51.2	YES	1969	NV	CA, ID	22A, 43B	Entisols	Psamments	Xeropsamments	Typic Xeropsamments		mixed			frigid	xeric	shallow
TOKOPER	203.9	NO	1984	NV		29	Aridisols	Durids	Argidurids	Typic Argidurids	loamy-skeletal	mixed	superactive		mesic	aridic	shallow
TOLICHA	11.3	NO	1967	NV		29	Aridisols	Cambids	Haplocambids	Lithic Haplocambids	loamy-skeletal	mixed	superactive		mesic	aridic	
TOLL	16.4	NO	1974	NV	CA, OR	24, 26, 29	Entisols	Psamments	Torripsamments	Xeric Torripsamments		mixed			mesic	torric	
TOMEL	169.4	NO	1972	NV		29	Aridisols	Durids	Argidurids	Typic Argidurids	loamy-skeletal	mixed	superactive		mesic	aridic	shallow
TOMERA	157.6	NO	1980	NV		24, 25	Aridisols	Argids	Natrargids	Xeric Natrargids	fine	smectitic			mesic	aridic	
TOMSHERRY	32.0	NO	1985	ID	NV, UT	25	Mollisols	Xerolls	Durixerolls	Vitrandic Durixerolls	ashy	glassy			frigid	xeric	
TONEY	19.8	NO	1974	NV	CA, OR	23	Aridisols	Argids	Paleargids	Vertic Paleargids	fine	smectitic			frigid	aridic	
TONKIN	13.2	NO	1971	NV		25, 28B	Aridisols	Calcids	Haplocalcids	Durinodic Xeric Haplocalcids	fine-loamy	mixed	superactive		frigid	aridic	
TONOPAH	427.0	NO	1923	NV	AZ	30	Aridisols	Calcids	Haplocalcids	Typic Haplocalcids	sandy-skeletal	mixed			thermic	aridic	
TOOELE	45.0	NO	1992	UT	NV	28A, 28B	Entisols	Orthents	Torriorthents	Typic Torriorthents	coarse-loamy	mixed	superactive	calcareous	mesic	torric	
TOOPITS	43.6	NO	2004	NV		28A	Entisols	Orthents	Torriorthents	Xeric Torriorthents	fine-loamy	mixed	superactive	calcareous	mesic	torric	
TOPEKI	38.2	NO	2004	NV		28A	Mollisols	Cryolls	Haplocryolls	Lithic Haplocryolls	loamy-skeletal	mixed	superactive		cryic		
TOQUOP	52.0	NO	1965	NV	AZ	30	Entisols	Psamments	Torripsamments	Typic Torripsamments		mixed			thermic	torric	
TORNILLO	4.8		1982	TX		42	Aridisols	Cambids	Haplocambids	Ustifluventic Haplocambids	fine-loamy	mixed	superactive		hyperthermic	aridic	

(continued)

(continued)

Soil series	Area (km^2)	Bench-mark	Year Est-ablished	TL State	Other States Using	MLRAs Using	Order	Suborder	Great Group	Subgroup	Particle-size Class	Mineralogy Class	CEC Activity Class	Reaction Class	Soil temp. Class	SMR	Other Family
TORRO	134.6	NO	1974	NV		24, 25, 28B	Mollisols	Xerolls	Argixerolls	Aridic Argixerolls	loamy-skeletal	mixed	superactive		frigid	xeric	
TOSP	40.7	NO	1990	NV		23, 25	Mollisols	Cryolls	Haplocryolls	Pachic Haplocryolls	coarse-loamy	mixed	superactive		cryic		
TOSSER	122.0	NO	1985	UT	NV	28A, 28B	Aridisols	Calcids	Haplocalcids	Xeric Haplocalcids	sandy-skeletal	mixed			mesic	aridic	
TOULON	207.0	NO	1963	NV	CA	24, 27	Aridisols	Cambids	Haplocambids	Typic Haplocambids	sandy-skeletal	mixed			mesic	aridic	
TRACTUFF	20.5	NO	2004	NV		28A	Mollisols	Xerolls	Argixerolls	Aridic Lithic Argixerolls	loamy-skeletal	mixed	superactive		frigid	xeric	
TRAILAMP	42.1	NO	1984	NV		29	Mollisols	Xerolls	Argixerolls	Typic Argixerolls	loamy-skeletal	mixed	superactive		frigid	xeric	shallow
TRALEY	25.7	NO	2006	NV		29	Mollisols	Ustolls	Argiustolls	Calcidic Argiustolls	loamy-skeletal	mixed	superactive		mesic	ustic	
TREADWELL	35.5	NO	2006	NV		29	Aridisols	Durids	Haplodurids	Typic Haplodurids	loamy-skeletal	mixed	superactive		mesic	aridic	shallow
TRESED	75.7	NO	1993	NV		24, 27	Entisols	Orthents	Torriorthents	Typic Torriorthents	clayey over loamy	smectitic OVER mixed	superactive	calcareous	mesic	torric	
TRID	13.7	NO	1981	NV		26	Mollisols	Xerolls	Argixerolls	Aridic Argixerolls	loamy-skeletal	mixed	superactive		mesic	xeric	
TRINIDAD	25.8	NO	1986	NV		25	Entisols	Orthents	Torriorthents	Xeric Torriorthents	loamy	carbonatic			frigid	torric	shallow
TRISTAN	81.0	NO	1980	NV		26	Mollisols	Xerolls	Argixerolls	Aridic Argixerolls	loamy-skeletal	mixed	superactive		mesic	xeric	
TROCKEN	1236.3	YES	1980	NV		27	Entisols	Orthents	Torriorthents	Typic Torriorthents	loamy-skeletal	mixed	superactive	calcareous	mesic	torric	
TROSI	7.1	NO	1973	CA	ID, NV	11, 23, 26	Aridisols	Durids	Argidurids	Xeric Argidurids	clayey-skeletal	smectitic			mesic	aridic	shallow
TROUGHS	12.2	NO	1986	ID	NV	25	Aridisols	Durids	Argidurids	Xeric Argidurids	loamy-skeletal	mixed	superactive		mesic	aridic	shallow
TROUGHSPRING	4.7	NO	2006	NV		30	Mollisols	Ustolls	Paleustolls	Petrocalcic Paleustolls	loamy-skeletal	carbonatic			mesic	ustic	
TRUCKEE	31.6	NO	1980	NV	CA	23, 26	Mollisols	Xerolls	Haploxerolls	Fluvaquentic Haploxerolls	fine-loamy	mixed	superactive		mesic	xeric	
TRUHOY	54.1	NO	1985	NV		29	Aridisols	Durids	Haplodurids	Cambidic Haplodurids	loamy	mixed	superactive		mesic	aridic	shallow
TRUNK	343.4	NO	1974	NV		24	Aridisols	Argids	Haplargids	Xeric Haplargids	fine	smectitic			mesic	aridic	
TRUVAR	10.0	NO	1985	NV	CA	29	Aridisols	Durids	Haplodurids	Xereptic Haplodurids	loamy	mixed	superactive		mesic	aridic	shallow
TUFFMAN	34.8	NO	2009	NV		27	Aridisols	Calcids	Haplocalcids	Lithic Haplocalcids	loamy-skeletal	mixed	superactive		mesic	aridic	
TUFFO	151.0	YES	1986	NV	CA	23, 25	Entisols	Orthents	Torriorthents	Vitrandic Torriorthents	ashy	glassy		nonacid	mesic	torric	shallow
TULASE	655.0	NO	1983	NV		24, 25, 28B	Entisols	Orthents	Torriorthents	Duric Torriorthents	coarse-silty	mixed	superactive	calcareous	mesic	torric	
TULECAN	31.8	NO	1984	NV		29	Mollisols	Xerolls	Argixerolls	Aridic Argixerolls	loamy-skeletal	mixed	superactive		mesic	xeric	shallow
TULEDAD	48.3	NO	2006	NV		23	Vertisols	Xererts	Haploxererts	Lithic Haploxererts	clayey	smectitic			mesic	xeric	
TUMARION	2.5	NO	2005	AZ	NV	30	Aridisols	Durids	Haplodurids	Typic Haplodurids	loamy-skeletal	mixed	superactive		thermic	aridic	shallow
TUMTUM	143.0	NO	1991	OR	NV	24	Aridisols	Durids	Argidurids	Typic Argidurids	loamy	mixed	superactive		mesic	aridic	shallow
TUNNISON	162.0	NO	1990	CA	NV	23	Vertisols	Xererts	Haploxererts	Aridic Haploxererts	very-fine	smectitic			mesic	xeric	
TURBA	462.6	NO	1992	NV		29	Mollisols	Ustolls	Argiustolls	Vitritorrandic Argiustolls	loamy-skeletal	mixed	superactive		mesic	ustic	shallow
TURRIA	21.3	NO	1975	NV		26	Aridisols	Argids	Haplargids	Xeric Haplargids	fine-loamy	mixed	superactive		mesic	aridic	
TURUPAH	33.9	NO	2009	NV		27	Inceptisols	Aquepts	Halaquepts	Typic Halaquepts	fine-loamy	mixed	superactive	nonacid	mesic	aquic	
TUSEL	418.0	YES	1978	NV	ID	10, 23, 24, 25, 28B	Mollisols	Cryolls	Argicryolls	Vitrandic Argicryolls	loamy-skeletal	mixed	superactive		cryic		
TUSK	106.0	NO	1986	NV		24, 25	Mollisols	Xerolls	Argixerolls	Pachic Argixerolls	fine-loamy	mixed	superactive		frigid	xeric	
TUSTELL	99.2	NO	1986	NV		25	Aridisols	Argids	Haplargids	Durinodic Xeric Haplargids	fine	smectitic			mesic	aridic	

(continued)

(continued)

Soil series	Area (km^2)	Bench-mark	Year Est-ablished	TL State	Other States Using	MLRAs Using	Order	Suborder	Great Group	Subgroup	Particle-size Class	Mineralogy Class	CEC Activity Class	Reaction Class	Soil temp. Class	SMR	Other Family
TUSUNE	56.0	NO	1995	NV	CA	23	Mollisols	Cryolls	Argicryolls	Vitrandic Argicryolls	ashy-skeletal	glassy			cryic		
TWEBA	51.4	NO	1986	NV		24, 25	Entisols	Aquents	Fluvaquents	Aeric Fluvaquents	coarse-loamy	mixed	superactive	calcareous	mesic	aquic	
TWEENER	161.8	NO	1986	NV		25	Mollisols	Xerolls	Argixerolls	Lithic Argixerolls	loamy-skeletal	mixed	superactive		frigid	xeric	
TYBO	267.2	NO	1972	NV		29	Aridisols	Durids	Haplodurids	Typic Haplodurids	loamy	mixed	superactive		mesic	aridic	shallow
UANA	49.2	NO	1971	NV		29	Mollisols	Xerolls	Durixerolls	Argiduridic Durixerolls	fine	smectitic			mesic	xeric	
UAWDA	2.4	NO	2010	CA	NV	26	Mollisols	Xerolls	Argixerolls	Vitrandic Argixerolls	ashy-skeletal over clayey	glassy OVER smectitic			frigid	xeric	
UBEHEBE	134.7	NO	1984	NV		29	Mollisols	Xerolls	Argixerolls	Aridic Argixerolls	loamy-skeletal	mixed	superactive		mesic	xeric	shallow
UCOPIA	5.3	NO	1979	NV		23, 25	Aridisols	Cambids	Haplocambids	Xeric Haplocambids	coarse-loamy	mixed	superactive		frigid	aridic	
UDELOPE	28.8	NO	1993	NV		25	Mollisols	Cryolls	Haplocryolls	Lithic Haplocryolls	loamy	mixed	superactive		cryic		
UHALDI	50.6	NO	1981	NV		23, 26	Mollisols	Xerolls	Argixerolls	Aridic Argixerolls	fine-loamy	mixed	superactive		mesic	xeric	
ULTRA	7.4	NO	1981	NV		27	Aridisols	Argids	Natrargids	Typic Natrargids	fine	smectitic			mesic	aridic	
ULTRAMONT	12.9	NO	2004	NV		28A	Aridisols	Cambids	Haplocambids	Durinodic Xeric Haplocambids	coarse-loamy	mixed	superactive		mesic	aridic	
UMBERLAND	334.5	NO	1972	NV		24, 27, 28B	Inceptisols	Aquepts	Halaquepts	Aeric Halaquepts	fine	smectitic		calcareous	mesic	aquic	
UMIL	221.5	NO	1971	NV		28B, 29	Aridisols	Durids	Haplodurids	Xeric Haplodurids	loamy	mixed	superactive		frigid	aridic	shallow
UMPA	0.2	NO	1970	CA	NV	22A	Inceptisols	Xerepts	Dystroxerepts	Andic Dystroxerepts	loamy-skeletal	isotic			frigid	xeric	
UNGENE	46.2	NO	2004	NV		28A	Aridisols	Calcids	Haplocalcids	Xeric Haplocalcids	sandy-skeletal	mixed			mesic	aridic	
UNIONVILLE	32.7	NO	1963	NV		24, 27	Aridisols	Cambids	Haplocambids	Typic Haplocambids	coarse-loamy	mixed	superactive		mesic	aridic	
UNIUS	51.6	NO	1985	NV		28B	Aridisols	Durids	Haplodurids	Xereptic Haplodurids	loamy	mixed	superactive		mesic	aridic	shallow
UNIVEGA	162.6	NO	1993	NV		29	Aridisols	Durids	Haplodurids	Typic Haplodurids	loamy	mixed	superactive		mesic	aridic	shallow
UNSEL	1849.6	YES	1972	NV		28B, 29	Aridisols	Argids	Haplargids	Durinodic Haplargids	fine-loamy	mixed	superactive		mesic	aridic	
UPATAD	268.3	NO	1990	NV		28B	Mollisols	Xerolls	Argixerolls	Aridic Lithic Argixerolls	loamy-skeletal	mixed	superactive		mesic	xeric	
UPDIKE	117.8	NO	1980	NV		23, 26	Aridisols	Argids	Natrargids	Xerertic Natrargids	fine	smectitic			mesic	aridic	
UPPERLINE	90.6	NO	2006	NV		30	Aridisols	Calcids	Haplocalcids	Typic Haplocalcids	loamy-skeletal	carbonatic			thermic	aridic	
UPSEL	6.3	NO	1988	NV		23	Mollisols	Xerolls	Haploxerolls	Torripsammentic Haploxerolls		mixed			frigid	xeric	
UPSPRING	127.0	NO	1980	CA	NV	29, 30	Entisols	Orthents	Torriorthents	Lithic Torriorthents	loamy-skeletal	mixed	superactive	calcareous	thermic	torric	
UPSTEER	11.3	NO	1986	NV		25	Mollisols	Xerolls	Haploxerolls	Duridic Haploxerolls	fine-silty	mixed	superactive		frigid	xeric	
UPVILLE	47.9	NO	1986	NV		25	Mollisols	Xerolls	Haploxerolls	Aridic Haploxerolls	sandy-skeletal	mixed			frigid	xeric	
URIPNES	167.9	NO	1981	NV		27	Entisols	Orthents	Torriorthents	Typic Torriorthents	loamy-skeletal	mixed	superactive	nonacid	mesic	torric	shallow
URMAFOT	767.3	YES	1990	NV		28B	Mollisols	Xerolls	Durixerolls	Haploduridic Durixerolls	loamy	mixed	superactive		mesic	xeric	shallow
URSINE	1487.7	YES	1940	NV		28A, 29	Aridisols	Durids	Haplodurids	Xeric Haplodurids	loamy-skeletal	carbonatic			mesic	aridic	shallow
URTAH	5.7	NO	1971	NV		25	Mollisols	Xerolls	Haploxerolls	Torriorthentic Haploxerolls	loamy-skeletal	carbonatic			frigid	xeric	
USTIDUR	36.0	NO	2006	NV		30	Aridisols	Durids	Haplodurids	Cambidic Haplodurids	loamy-skeletal	mixed	superactive		thermic	aridic	shallow

(continued)

(continued)

Soil series	Area (km^2)	Bench-mark	Year Est-ablished	TL State	Other States Using	MLRAs Using	Order	Suborder	Great Group	Subgroup	Particle-size Class	Mineralogy Class	CEC Activity Class	Reaction Class	Soil temp. Class	SMR	Other Family
UVADA	75.7	NO	1942	UT	NV	24, 28A, 28B	Aridisols	Argids	Natrargids	Typic Natrargids	fine	smectitic			mesic	aridic	
UWELL	94.9	NO	1990	NV		28B	Entisols	Orthents	Torriorthents	Duric Torriorthents	fine-silty	mixed	superactive	calcareous	mesic	torric	
VACE	55.2	NO	2006	NV		30	Aridisols	Calcids	Petrocalcids	Typic Petrocalcids	loamy	mixed	superactive		thermic	aridic	shallow
VADAHO	69.0	NO	1986	NV		25	Mollisols	Xerolls	Durixerolls	Haploduridic Durixerolls	loamy	mixed	superactive		mesic	xeric	shallow
VALATIER	5.5	NO	2006	NV		30	Aridisols	Durids	Argidurids	Typic Argidurids	loamy-skeletal	mixed	superactive		mesic	aridic	
VALCREST	31.1	NO	1983	NV		24	Aridisols	Argids	Natrargids	Xeric Natrargids	clayey over loamy	smectitic OVER mixed	superactive		mesic	aridic	
VALMY	264.5	NO	1974	NV	CA	23, 24, 25, 28B	Entisols	Orthents	Torriorthents	Duric Torriorthents	coarse-loamy	mixed	superactive	calcareous	mesic	torric	
VAMP	13.8	NO	1975	NV		26	Aridisols	Durids	Haplodurids	Aquicambidic Haplodurids	coarse-loamy	mixed	superactive		mesic	aridic	
VANWYPER	589.0	NO	1985	NV	OR	23, 24, 25	Aridisols	Argids	Haplargids	Xeric Haplargids	clayey-skeletal	smectitic			mesic	aridic	
VARWASH	74.8	NO	2006	NV		30	Aridisols	Calcids	Haplocalcids	Typic Haplocalcids	sandy-skeletal	mixed			hyperthermic	aridic	
VEET	732.3	NO	1984	NV		26, 27, 28B, 29	Aridisols	Cambids	Haplocambids	Xeric Haplocambids	loamy-skeletal	mixed	superactive		mesic	aridic	
VEGASTORM	18.1	NO	2006	NV		30	Aridisols	Calcids	Haplocalcids	Petronodic Haplocalcids	coarse-loamy	carbonatic			thermic	aridic	
VERDICO	57.0	NO	1980	NV	CA	23, 26	Aridisols	Argids	Paleargids	Vertic Paleargids	fine	smectitic			mesic	aridic	
VETA	141.0	NO	1981	NV	ID	11, 24, 25, 26, 27, 28B	Aridisols	Cambids	Haplocambids	Xeric Haplocambids	loamy-skeletal	mixed	superactive		mesic	aridic	
VETAGRANDE	25.0	NO	2006	NV	CA	26	Mollisols	Xerolls	Argixerolls	Pachic Argixerolls	loamy-skeletal	mixed	superactive		frigid	xeric	
VETASH	2.1	NO	2005	CA	NV	26	Mollisols	Xerolls	Argixerolls	Vitrandic Argixerolls	ashy-skeletal	glassy			frigid	xeric	
VICEE	12.8	YES	1975	NV		22A	Mollisols	Xerolls	Haploxerolls	Entic Ultic Haploxerolls	coarse-loamy	isotic			frigid	xeric	
VIGUS	89.8	NO	1972	NV		27, 28B, 29	Aridisols	Argids	Haplargids	Durinodic Haplargids	fine-loamy	mixed	superactive		mesic	aridic	
VIL	0.1		1971	OR	NV	23	Mollisols	Xerolls	Durixerolls	Argiduridic Durixerolls	loamy	mixed	superactive		frigid	xeric	
VINDICATOR	100.1	NO	1984	NV		29	Aridisols	Argids	Haplargids	Typic Haplargids	loamy-skeletal	mixed	superactive		mesic	aridic	shallow
VININI	134.5	NO	1972	NV		28B, 29	Aridisols	Durids	Argidurids	Xeric Argidurids	loamy-skeletal	mixed	superactive		mesic	aridic	shallow
VINSAD	14.2	NO	1971	NV		25, 28B	Aridisols	Salids	Aquisalids	Typic Aquisalids	coarse-loamy	mixed	superactive		frigid	aridic	
VIRGIN PEAK	38.6	NO	1970	NV	AZ	30, 38	Mollisols	Ustolls	Haplustolls	Aridic Haplustolls	loamy-skeletal	mixed	superactive		mesic	ustic	shallow
VIRGIN RIVER	10.7	NO	1931	NV		30	Entisols	Orthents	Torriorthents	Aquic Torriorthents	clayey over loamy	smectitic OVER mixed	superactive	calcareous	thermic	torric	
VITALE	88.8	NO	1986	ID	NV, OR	10, 13, 21, 25, 47	Mollisols	Xerolls	Argixerolls	Typic Argixerolls	loamy-skeletal	mixed	superactive		frigid	xeric	
VIUM	11.9	NO	1988	NV		27	Aridisols	Argids	Haplargids	Lithic Haplargids	loamy-skeletal	mixed	superactive		mesic	aridic	
VOLTAIRE	31.0	YES	1975	NV		23, 26	Mollisols	Aquolls	Endoaquolls	Fluvaquentic Endoaquolls	fine-loamy	mixed	superactive	calcareous	mesic	aquic	
VYCKYL	25.4	NO	2004	NV		28A	Mollisols	Xerolls	Haploxerolls	Aridic Haploxerolls	loamy-skeletal	mixed	superactive		frigid	xeric	shallow
VYLACH	31.9	NO	1981	NV		27	Aridisols	Durids	Argidurids	Argidic Argidurids	loamy	mixed	superactive		mesic	aridic	shallow
WABUSKA	67.1	NO	1940	NV		26, 27	Inceptisols	Aquepts	Halaquepts	Aeric Halaquepts	coarse-loamy	mixed	superactive	calcareous	mesic	aquic	
WACA	0.5	NO	1970	CA	NV	22A	Andisols	Xerands	Vitrixerands	Humic Vitrixerands	medial-skeletal	amorphic			frigid	xeric	
WAGORE	6.1	NO	1990	NV		23	Mollisols	Xerolls	Haploxerolls	Pachic Haploxerolls	coarse-loamy	mixed	superactive		frigid	xeric	

(continued)

(continued)

Soil series	Area (km^2)	Bench-mark	Year Est-ablished	TL State	Other States Using	MLRAs Using	Order	Suborder	Great Group	Subgroup	Particle-size Class	Mineralogy Class	CEC Activity Class	Reaction Class	Soil temp. Class	SMR	Other Family
WAHGUYHE	100.5	NO	1984	NV		29, 30	Entisols	Orthents	Torriorthents	Lithic Xeric Torriorthents	loamy-skeletal	mixed	superactive	nonacid	mesic	torric	
WAKANSAPA	185.9	NO	2009	NV		29	Mollisols	Ustolls	Argiustolls	Aridic Lithic Argiustolls	ashy-skeletal	glassy			mesic	ustic	
WALA	220.3	NO	2004	NV		28A	Entisols	Orthents	Torriorthents	Lithic Xeric Torriorthents	loamy-skeletal	mixed	superactive	calcareous	mesic	torric	
WALKERIVER	12.5	NO	2005	NV		27	Entisols	Fluvents	Torrifluvents	Vitrandic Torrifluvents	sandy	mixed			mesic	torric	
WALTI	308.6	NO	1983	NV		24, 27, 28B	Mollisols	Xerolls	Argixerolls	Aridic Argixerolls	fine	smectitic			frigid	xeric	
WAMBOLT	6.5	NO	2006	NV		28A, 28B	Mollisols	Xerolls	Argixerolls	Typic Argixerolls	loamy-skeletal	mixed	superactive		frigid	xeric	
WAMP	91.8	NO	2011	NV		30	Aridisols	Calcids	Petrocalcids	Calcic Petrocalcids	loamy-skeletal	carbonatic			mesic	aridic	shallow
WANOMIE	23.3	NO	1996	NV		30	Aridisols	Durids	Haplodurids	Cambidic Haplodurids	coarse-loamy	mixed	superactive		thermic	aridic	
WARDBAY	386.8	NO	1990	NV		25, 28A, 28B	Mollisols	Cryolls	Calcicryolls	Pachic Calcicryolls	loamy-skeletal	carbonatic			cryic		
WARDCREEK	0.0	NO	2007	NV	CA	22A	Andisols	Cryands	Vitricryands	Xeric Vitricryands	ashy-skeletal	amorphic			cryic	xeric	
WARDENOT	1004.8	YES	1972	NV		27, 28B, 29	Entisols	Orthents	Torriorthents	Typic Torriorthents	sandy-skeletal	mixed			mesic	torric	
WASHOE	32.9	NO	1940	NV	NM	26, 27, 70C	Aridisols	Argids	Haplargids	Xeric Haplargids	loamy-skeletal	mixed	superactive		mesic	aridic	
WASHOVER	0.9	NO	2012	NV		28A	Mollisols	Xerolls	Calcixerolls	Pachic C alcixerolls	loamy-skeletal	mixed	superactive		frigid	xeric	
WASPO	29.0	NO	1980	NV		10, 26	Vertisols	Xererts	Haploxererts	Aridic Haploxererts	fine	smectitic			mesic	xeric	
WASSIT	172.0	NO	1985	NV		26	Aridisols	Argids	Haplargids	Lithic Xeric Haplargids	loamy-skeletal	mixed	superactive		frigid	aridic	
WATAH	1.0	NO	2007	CA	NV	22A	Inceptisols	Aquepts	Humaquepts	Histic Humaquepts	coarse-loamy	mixed	superactive	acid	frigid	aquic	
WATCHABOB	4.6	NO	1984	NV	ID	25	Mollisols	Xerolls	Argixerolls	Pachic Ultic Argixerolls	fine-loamy	mixed	superactive		frigid	xeric	
WATOOPAH	555.5	NO	1986	NV		27, 28B, 29	Aridisols	Argids	Haplargids	Durinodic Xeric Haplargids	coarse-loamy	mixed	superactive		mesic	aridic	
WATSONLAKE	1.3	NO	2007	CA	NV	22A	Alfisols	Cryalfs	Haplocryalfs	Andic Haplocryalfs	loamy-skeletal	isotic			cryic	xeric	
WAYHIGH	2.0	NO	2009	NV		28A	Inceptisols	Cryepts	Haplocryepts	Xeric Haplocryepts	loamy-skeletal	mixed	superactive		cryic	xeric	
WECHECH	670.8	NO	1994	NV		30	Aridisols	Calcids	Petrocalcids	Calcic Petrocalcids	loamy-skeletal	carbonatic			thermic	aridic	shallow
WEDEKIND	60.8	NO	1980	NV		25, 26, 27	Mollisols	Xerolls	Argixerolls	Aridic Argixerolls	loamy	mixed	superactive		mesic	xeric	shallow
WEDERTZ	106.3	NO	1980	NV		25, 26, 27, 28B	Aridisols	Argids	Haplargids	Durinodic Xeric Haplargids	fine-loamy	mixed	superactive		mesic	aridic	
WEDLAR	28.8	NO	1981	NV		26	Aridisols	Argids	Haplargids	Durinodic Xeric Haplargids	fine-loamy	mixed	superactive		mesic	aridic	
WEENA	219.9	NO	1981	NV		26, 27	Entisols	Orthents	Torriorthents	Typic Torriorthents	loamy	mixed	superactive	calcareous	mesic	torric	shallow
WEEPAH	67.7	NO	1984	NV		29	Entisols	Orthents	Torriorthents	Xeric Torriorthents	loamy-skeletal	mixed	superactive	calcareous	mesic	torric	shallow
WEEZWEED	23.2	NO	1995	NV	CA	23	Mollisols	Xerolls	Haploxerolls	Vitritorrandic Haploxerolls	ashy	glassy			mesic	xeric	
WEIGLE	19.2	NO	1983	NV		28B	Aridisols	Durids	Haplodurids	Xeric Haplodurids	loamy	mixed	superactive		frigid	aridic	shallow
WEIMER	27.8	NO	1995	NV	CA	23	Vertisols	Aquerts	Epiaquerts	Xeric Epiaquerts	very-fine	smectitic			frigid	aquic	
WEISER	1205.0	YES	1939	NV	CA, NM	30	Aridisols	Calcids	Haplocalcids	Typic Haplocalcids	loamy-skeletal	carbonatic			thermic	aridic	
WEISHAUPT	11.0	NO	1971	NV		27	Mollisols	Aquolls	Endoaquolls	Cumulic Vertic Endoaquolls	fine-loamy over clayey	mixed OVER smectitic	superactive	calcareous	mesic	aquic	
WELCH	517.0	YES	1979	NV	CA, ID, OR	10, 21, 23, 25, 26, 27, 28B	Mollisols	Aquolls	Endoaquolls	Cumulic Endoaquolls	fine-loamy	mixed	superactive		frigid	aquic	

(continued)

(continued)

Soil series	Area (km^2)	Bench-mark	Year Est-ablished	TL State	Other States Using	MLRAs Using	Order	Suborder	Great Group	Subgroup	Particle-size Class	Mineralogy Class	CEC Activity Class	Reaction Class	Soil temp. Class	SMR	Other Family
WELLINGTON	52.9	NO	1941	NV		26	Aridisols	Durids	Argidurids	Xeric Argidurids	loamy	mixed	superactive		mesic	aridic	shallow
WELLSED	39.2	NO	1981	NV		26	Aridisols	Durids	Argidurids	Xeric Argidurids	fine-loamy	mixed	superactive		mesic	aridic	
WELRING	18.2	NO	1972	UT	CO, MT, NV, WY	29, 34, 35, 49	Entisols	Orthents	Ustorthents	Lithic Ustorthents	loamy-skeletal	carbonatic			mesic	ustic	
WELSUM	39.8	NO	1986	NV		25	Mollisols	Aquolls	Endoaquolls	Cumulic Endoaquolls	fine-loamy over sandy or sandy-skeletal	mixed	superactive	calcareous	frigid	aquic	
WENDANE	793.6	YES	1974	NV		24, 27, 28B	Inceptisols	Aquepts	Halaquepts	Aquandic Halaquepts	fine-silty	mixed	superactive	calcareous	mesic	aquic	
WERELD	17.7	NO	1986	NV		24	Mollisols	Xerolls	Calcixerolls	Aridic Calcixerolls	loamy-skeletal	mixed	superactive		frigid	xeric	
WESFIL	142.8	NO	1988	NV		27, 28A	Entisols	Orthents	Torriorthents	Lithic Xeric Torriorthents	loamy-skeletal	mixed	superactive	calcareous	mesic	torric	
WESO	524.1	NO	1974	NV		11, 24	Aridisols	Cambids	Haplocambids	Durinodic Haplocambids	coarse-loamy	mixed	superactive		mesic	aridic	
WESTBUTTE	66.1	NO	1991	OR	NV	10, 23	Mollisols	Xerolls	Haploxerolls	Pachic Haploxerolls	loamy-skeletal	mixed	superactive		frigid	xeric	
WETVIT	9.7	NO	1995	NV	CA	23	Mollisols	Aquolls	Endoaquolls	Aquandic Endoaquolls	ashy	glassy			mesic	aquic	
WHEELERPASS	8.5	NO	2006	NV		30	Mollisols	Ustolls	Argiustolls	Aridic Lithic Argiustolls	loamy-skeletal	mixed	superactive		frigid	ustic	
WHEELERPEK	9.9	NO	2009	NV		28A	Entisols	Orthents	Cryorthents	Lithic Cryorthents	loamy-skeletal	mixed	active	nonacid	cryic		
WHEELERWELL	49.4	NO	2006	NV		30	Mollisols	Ustolls	Argiustolls	Aridic Argiustolls	loamy-skeletal	mixed	superactive		mesic	ustic	
WHICHMAN	19.5	NO	1981	NV		26	Mollisols	Xerolls	Haploxerolls	Aridic Haploxerolls	loamy-skeletal	mixed	superactive		mesic	xeric	
WHILPHANG	50.7	NO	1985	NV		29	Entisols	Orthents	Torriorthents	Xeric Torriorthents	loamy	mixed	superactive	calcareous	mesic	torric	shallow
WHIRLO	467.3	NO	1974	NV		24, 27, 29	Aridisols	Cambids	Haplocambids	Typic Haplocambids	loamy-skeletal	mixed	superactive		mesic	aridic	
WHITEBASIN	19.0	NO	2006	NV		30	Aridisols	Gypsids	Haplogypsids	Leptic Haplogypsids	coarse-loamy	gypsic			thermic	aridic	
WHITEPEAK	13.6	NO	1983	NV		25	Aridisols	Argids	Haplargids	Xeric Haplargids	clayey-skeletal	smectitic			frigid	aridic	
WHITMIRE	1.7	NO	2012	NV		26	Mollisols	Xerolls	Haploxerolls	Aquic Cumulic Haploxerolls	fine-loamy over sandy or sandy-skeletal	mixed	superactive		mesic	xeric	
WHITTELL	6.2		xx	CA		22A	Entisols	Orthents	Cryorthents	Typic Cryorthents	sandy-skeletal	mixed			cryic	xeric	
WHOLAN	386.0	NO	1974	NV	ID	24, 25, 27, 28B	Aridisols	Cambids	Haplocambids	Typic Haplocambids	coarse-silty	mixed	superactive		mesic	aridic	
WICKAHONEY	8.0	NO	1984	ID	NV	25	Alfisols	Xeralfs	Haploxeralfs	Lithic Mollic Haploxeralfs	clayey-skeletal	smectitic			frigid	xeric	
WICUP	27.4	NO	1986	NV		25	Mollisols	Xerolls	Argixerolls	Aridic Argixerolls	fine	smectitic			mesic	xeric	
WIELAND	1086.8	YES	1985	NV		24, 25, 28B	Aridisols	Argids	Haplargids	Durinodic Xeric Haplargids	fine	smectitic			mesic	aridic	
WIFFO	129.8	NO	1986	NV		25, 28A, 28B	Entisols	Orthents	Torriorthents	Xeric Torriorthents	loamy-skeletal	mixed	superactive	calcareous	mesic	torric	
WILE	18.1	NO	1981	NV		26	Mollisols	Xerolls	Argixerolls	Aridic Argixerolls	clayey	smectitic			mesic	xeric	shallow
WILLHILL	3.1	NO	1986	ID	NV	25	Aridisols	Argids	Calciargids	Durinodic Xeric Calciargids	loamy-skeletal	mixed	superactive		mesic	aridic	
WILLYNAT	8.7	NO	2004	NV		28A	Mollisols	Xerolls	Haploxerolls	Pachic Haploxerolls	loamy-skeletal	mixed	superactive		frigid	xeric	
WILSOR	9.8	NO	1986	NV		25	Aridisols	Argids	Haplargids	Durinodic Xeric Haplargids	fine-loamy	mixed	superactive		frigid	aridic	
WILST	20.5	NO	1996	NV		30	Entisols	Orthents	Torriorthents	Duric Torriorthents	loamy-skeletal	mixed	superactive	calcareous	thermic	torric	
WILTOP	14.0	NO	2010	NV		28A	Mollisols	Xerolls	Argixerolls	Vitrandic Argixerolls	clayey-skeletal	smectitic			frigid	xeric	
WINADA	12.9	NO	1974	NV		24	Mollisols	Cryolls	Argicryolls	Xeric Argicryolls	loamy-skeletal	mixed	superactive		cryic	xeric	
WINDWASH	10.2	NO	2009	NV		28A	Inceptisols	Cryepts	Calcicryepts	Xeric Calcicryepts	loamy-skeletal	carbonatic			cryic	xeric	

(continued)

(continued)

Soil series	Area (km^2)	Bench-mark	Year Est-ablished	TL State	Other States Using	MLRAs Using	Order	Suborder	Great Group	Subgroup	Particle-size Class	Mineralogy Class	CEC Activity Class	Reaction Class	Soil temp. Class	SMR	Other Family
WINKLO	183.8	NO	1992	NV		29	Aridisols	Argids	Haplargids	Ustic Haplargids	fine	smectitic			mesic	aridic	
WINTERMUTE	590.1	YES	1990	NV		28B	Aridisols	Calcids	Haplocalcids	Durinodic Haplocalcids	loamy-skeletal	mixed	superactive		mesic	aridic	
WINU	66.8	NO	1971	NV	ID	10, 25, 28A, 28B	Mollisols	Cryolls	Argicryolls	Pachic Argicryolls	fine-loamy	mixed	superactive		cryic		
WINZ	8.9	NO	1971	NV		28A	Alfisols	Cryalfs	Haplocryalfs	Vitrandic Haplocryalfs	ashy-skeletal	glassy			cryic	xeric	
WISKAN	199.4	NO	1986	NV		24	Aridisols	Argids	Haplargids	Xeric Haplargids	loamy-skeletal	mixed	superactive		frigid	aridic	
WISKIFLAT	50.1	NO	1985	NV		27, 29	Entisols	Orthents	Torriorthents	Xeric Torriorthents	loamy-skeletal	mixed	superactive	nonacid	mesic	torric	
WITEFELS	23.0	NO	1980	NV	CA	22A	Entisols	Psamments	Cryopsamments	Typic Cryopsamments		mixed			cryic		
WODA	17.5	NO	1996	NV		30	Aridisols	Calcids	Petrocalcids	Calcic Petrocalcids	loamy	carbonatic			thermic	aridic	shallow
WODAVAR	37.7	NO	1996	NV		30	Aridisols	Calcids	Petrocalcids	Calcic Petrocalcids	loamy-skeletal	carbonatic			thermic	aridic	shallow
WOODROW	0.8	NO	1919	UT	NV	28A, 29	Entisols	Fluvents	Torrifluvents	Xeric Torrifluvents	fine-silty	mixed	superactive	calcareous	mesic	torric	
WOODSPRING	16.1	NO	2006	NV		30	Mollisols	Ustolls	Calciustolls	Pachic Calciustolls	loamy-skeletal	mixed	superactive		mesic	ustic	
WOOFUS	97.5	NO	1986	NV		23, 25	Mollisols	Aquolls	Endoaquolls	Fluvaquentic Endoaquolls	fine-loamy over sandy or sandy-skeletal	mixed	superactive	calcareous	mesic	aquic	
WOOLSEY	8.0	NO	1957	NV		24, 27	Aridisols	Argids	Haplargids	Typic Haplargids	coarse-loamy	mixed	superactive		mesic	aridic	
WRANGO	110.0	NO	1972	NV	UT	27, 28A, 28B, 29	Entisols	Orthents	Torriorthents	Xeric Torriorthents	sandy-skeletal	mixed			mesic	torric	
WREDAH	29.9	NO	1990	NV		28B	Mollisols	Xerolls	Argixerolls	Argiduridic Argixerolls	fine-loamy	mixed	superactive		mesic	xeric	
WRENZA	60.8	NO	2016	NV		25	Inceptisols	Cryepts	Humicryepts	Xeric Humicryepts	loamy-skeletal	mixed	subactive		cryic	xeric	
WYLO	638.0	YES	1990	NV	CA	23	Mollisols	Xerolls	Argixerolls	Aridic Lithic Argixerolls	clayey	smectitic			mesic	xeric	
WYVA	215.9	NO	1992	NV		29	Aridisols	Argids	Haplargids	Lithic Ustic Haplargids	loamy-skeletal	mixed	superactive		mesic	aridic	
XERTA	7.9	NO	1975	NV		26	Mollisols	Xerolls	Durixerolls	Argiduridic Durixerolls	fine	smectitic			mesic	xeric	
XERXES	112.8	NO	1986	NV		25	Aridisols	Argids	Haplargids	Vitrixerandic Haplargids	ashy-skeletal	glassy			mesic	aridic	shallow
XICA	66.5	NO	1986	NV		25	Mollisols	Xerolls	Argixerolls	Typic Argixerolls	loamy	mixed	superactive		frigid	xeric	shallow
XINE	164.7	NO	1985	NV		24, 28B	Mollisols	Xerolls	Calcixerolls	Aridic Calcixerolls	loamy-skeletal	mixed	superactive		frigid	xeric	
XIPE	22.7	NO	1986	NV		24, 25	Mollisols	Aquolls	Endoaquolls	Fluvaquentic Endoaquolls	fine-silty over sandy or sandy-skeletal	mixed	superactive		mesic	aquic	
XMAN	113.2	NO	1980	NV		26	Aridisols	Argids	Haplargids	Xeric Haplargids	clayey	smectitic			mesic	aridic	shallow
YELBRICK	119.5	NO	2004	NV		28A, 28B	Aridisols	Calcids	Haplocalcids	Sodic Haplocalcids	fine-loamy over sandy or sandy-skeletal	mixed	superactive		mesic	aridic	
YELLOWHILLS	25.9	NO	1993	NV		23	Mollisols	Xerolls	Haploxerolls	Vitritorrandic Haploxerolls	ashy	glassy			mesic	xeric	
YERINGTON	98.4	NO	1983	NV		27	Entisols	Orthents	Torriorthents	Typic Torriorthents	sandy	mixed			mesic	torric	
YERMO	967.0	NO	1978	CA	NV	29, 30	Entisols	Orthents	Torriorthents	Typic Torriorthents	loamy-skeletal	mixed	superactive	calcareous	thermic	torric	
YIPOR	114.3	NO	1986	NV		24, 27	Entisols	Orthents	Torriorthents	Typic Torriorthents	coarse-silty	mixed	superactive	calcareous	mesic	torric	
YOBE	196.0	NO	1972	NV	CA	23, 24, 27, 28B, 29	Inceptisols	Aquepts	Halaquepts	Aeric Halaquepts	fine-silty	mixed	superactive	calcareous	mesic	aquic	
YODY	465.3	YES	1974	NV		28B	Aridisols	Durids	Argidurids	Haploxeralfic Argidurids	fine-loamy	mixed	superactive		mesic	aridic	
YOMBA	484.7	YES	1972	NV		29	Aridisols	Cambids	Haplocambids	Durinodic Haplocambids	sandy-skeletal	mixed			mesic	aridic	
YOTES	106.1	NO	2006	NV		28A	Mollisols	Xerolls	Haploxerolls	Vitritorrandic Haploxerolls	ashy	glassy			mesic	xeric	
YOUNGSTON	33.2	YES	1969	WY	CO, NM, NV, UT	29, 32, 34A, 35	Entisols	Fluvents	Torrifluvents	Typic Torrifluvents	fine-loamy	mixed	superactive	calcareous	mesic	torric	

(continued)

(continued)

Soil series	Area (km^2)	Bench-mark	Year Est-ablished	TL State	Other States Using	MLRAs Using	Order	Suborder	Great Group	Subgroup	Particle-size Class	Mineralogy Class	CEC Activity Class	Reaction Class	Soil temp. Class	SMR	Other Family
YUKO	301.8	NO	1980	NV		25, 26	Aridisols	Argids	Haplargids	Xeric Haplargids	loamy	mixed	superactive		mesic	aridic	shallow
YURM	257.4	NO	1996	NV		30	Aridisols	Calcids	Petrocalcids	Calcic Petrocalcids	loamy-skeletal	mixed	superactive		thermic	aridic	shallow
ZABA	50.8	NO	1972	NV		29	Aridisols	Argids	Haplargids	Typic Haplargids	loamy-skeletal	mixed	superactive		mesic	aridic	
ZADVAR	851.8	YES	1984	NV		29	Aridisols	Durids	Argidurids	Haploxeralfic Argidurids	loamy	mixed	superactive		mesic	aridic	shallow
ZAFOD	140.5	NO	1994	NV		28A, 28B	Aridisols	Durids	Haplodurids	Xereptic Haplodurids	loamy-skeletal	mixed	active		mesic	aridic	
ZAIDY	115.9	NO	1985	NV		28B	Aridisols	Durids	Argidurids	Haploxeralfic Argidurids	fine-loamy	mixed	superactive		mesic	aridic	
ZALDA	56.1	NO	1996	NV		30	Aridisols	Durids	Haplodurids	Typic Haplodurids	loamy	mixed	superactive		thermic	aridic	shallow
ZAPA	155.7	NO	1986	NV		25	Aridisols	Durids	Haplodurids	Xereptic Haplodurids	loamy-skeletal	mixed	superactive		mesic	aridic	
ZAQUA	274.4	NO	1992	NV		29	Aridisols	Argids	Haplargids	Ustic Haplargids	loamy-skeletal	mixed	superactive		mesic	aridic	shallow
ZARARK	157.0	NO	2009	NV		28B	Mollisols	Cryolls	Calcicryolls	Pachic Calcicryolls	loamy-skeletal	carbonatic			cryic		
ZARK	38.7	NO	1986	NV		25	Mollisols	Xerolls	Haploxerolls	Vitritorrandic Haploxerolls	ashy	glassy			mesic	xeric	
ZEHEME	795.8	NO	1992	NV		30	Aridisols	Calcids	Haplocalcids	Lithic Haplocalcids	loamy-skeletal	carbonatic			thermic	aridic	
ZEPHAN	48.6	NO	1980	NV	CA	26	Aridisols	Argids	Haplargids	Xeric Haplargids	clayey-skeletal	smectitic			mesic	aridic	
ZEPHYRCOVE	1.5	NO	2007	NV		22A	Alfisols	Xeralfs	Haploxeralfs	Ultic Haploxeralfs	coarse-loamy	isotic			frigid	xeric	
ZERK	178.2	NO	1990	NV		28B	Aridisols	Calcids	Haplocalcids	Durinodic Haplocalcids	sandy-skeletal	mixed			mesic	aridic	
ZEVADEZ	468.0	NO	1986	NV		25	Aridisols	Argids	Haplargids	Durinodic Xeric Haplargids	fine-loamy	mixed	superactive		mesic	aridic	
ZIBATE	242.6	NO	1984	NV		30	Aridisols	Argids	Haplargids	Lithic Haplargids	loamy-skeletal	mixed	superactive		thermic	aridic	
ZIMBOB	723.8	YES	1990	NV		28B	Entisols	Orthents	Torriorthents	Lithic Xeric Torriorthents	loamy-skeletal	carbonatic			mesic	torric	
ZIMWALA	53.0	NO	1990	NV		28B	Entisols	Orthents	Torriorthents	Typic Torriorthents	fine-silty	carbonatic			mesic	torric	
ZINEB	134.6	NO	1983	NV		24	Aridisols	Cambids	Haplocambids	Durinodic Xeric Haplocambids	loamy-skeletal	mixed	superactive		mesic	aridic	
ZIRAM	11.9	NO	1983	NV		25, 28B	Mollisols	Xerolls	Palexerolls	Typic Palexerolls	clayey-skeletal	smectitic			frigid	xeric	
ZOATE	20.7	NO	1971	NV		29	Mollisols	Xerolls	Durixerolls	Argiduridic Durixerolls	clayey-skeletal	smectitic			mesic	xeric	shallow
ZODA	100.7	NO	2006	NV		28A	Aridisols	Durids	Argidurids	Haploxeralfic Argidurids	ashy	glassy			mesic	aridic	
ZOESTA	80.5	NO	1985	NV		24, 28B	Aridisols	Argids	Paleargids	Vertic Paleargids	fine	smectitic			frigid	aridic	
ZORRAVISTA	133.0	NO	1968	NV	CA, OR	23, 24, 27, 28B	Entisols	Psamments	Torripsamments	Xeric Torripsamments	sandy	mixed			mesic	torric	
ZORROMOUNT	21.2	NO	2006	NV		23	Mollisols	Cryolls	Haplocryolls	Vitrandic Haplocryolls	ashy-skeletal	glassy			cryic		
ZYMANS	182.1	NO	1992	NV		23, 24	Mollisols	Xerolls	Argixerolls	Aridic Argixerolls	fine	smectitic			mesic	xeric	
ZYPLAR	13.2	NO	1996	NV		30	Aridisols	Argids	Haplargids	Lithic Haplargids	loamy	mixed	superactive		thermic	aridic	
ZYZZI	29.6	NO	1981	NV		26, 28B	Aridisols	Argids	Haplargids	Xeric Haplargids	loamy-skeletal	mixed	superactive		mesic	aridic	shallow

Appendix B
Thickness of Diagnostic Horizons of Nevada Soil Series with Areas in Excess of 78 km^2

P. W. Blackburn et al., *The Soils of Nevada*, World Soils Book Series,
https://doi.org/10.1007/978-3-030-53157-7

Series Name	Order	Suborder	Great Group	Ochric	Mollic	Umbric	Histic	Folistic	Albic	Thickness (cm)		Calcic	Duripan	Gypsic	Natric	Petrocalcic	Petrogypsic	Salic
										Cambic	Argillic							
ABGESE	Aridisols	Argids	Haplargids	13							76							
ABOTEN	Aridisols	Durids	Natridurids	18									31		20			
ACANA	Aridisols	Durids	Argidurids	8							17	18	18					
ACKETT	Aridisols	Durids	Argidurids	8							28	21	35					
ACOMA	Aridisols	Argids	Paleargids	10							66	24						
ACRELANE	Mollisols	Xerolls	Argixerolls		18						10							
ACTI	Mollisols	Ustolls	Argiustolls		46						41							
ADELAIDE	Aridisols	Durids	Haplodurids	18						15			8					
ADOBE	Mollisols	Cryolls	Calcicryolls		43							30						
ADVOKAY	Aridisols	Argids	Haplargids	8							10							
AGASSIZ	Mollisols	Xerolls	Haploxerolls		46													
AKLER	Aridisols	Argids	Haplargids	15							28							
ALHAMBRA	Entisols	Fluvents	Torrifluvents	18														
ALKO	Aridisols	Durids	Haplodurids	10									56					
ALLEY	Aridisols	Argids	Haplargids	18							28							
ALLKER	Aridisols	Argids	Haplargids	18							48							
ALLOR	Aridisols	Argids	Haplargids	30							56							
ALYAN	Mollisols	Xerolls	Argixerolls		41						58							
AMELAR	Mollisols	Xerolls	Argixerolls		38						23							
AMTOFT	Aridisols	Calcids	Haplocalcids	18								23						
ANAWALT	Aridisols	Argids	Haplargids	15							33							
ANNAW	Aridisols	Cambids	Haplocambids	18						23								
ANSPING	Mollisols	Xerolls	Calcixerolls		33							17						
APPIAN	Aridisols	Argids	Natrargids	8											20			
ARADA	Aridisols	Calcids	Haplocalcids	15								69						
ARCIA	Mollisols	Xerolls	Argixerolls		53						63							
ARCLAY	Mollisols	Xerolls	Argixerolls		28						28							
ARDIVEY	Aridisols	Argids	Haplargids	8							35							
ARGALT	Aridisols	Durids	Argidurids	8							15		5					
ARGENTA	Inceptisols	Aquepts	Halaquepts	18														
ARIZO	Entisols	Orthents	Torriorthents	18														
ARMESPAN	Aridisols	Calcids	Haplocalcids	18								56						
ARMOINE	Aridisols	Argids	Haplargids	13							17							
ARROLIME	Aridisols	Gypsids	Calcigypsids	5								79		33				
ASHTRE	Mollisols	Xerolls	Argixerolls		28						48							
ATLOW	Aridisols	Argids	Haplargids	8							30							
ATRYPA	Mollisols	Xerolls	Haploxerolls		25													
AUTOMAL	Aridisols	Calcids	Haplocalcids	18								109						
AZTEC	Aridisols	Gypsids	Calcigypsids	5								91		**30**				
BANGO	Aridisols	Argids	Natrargids	10											10			
BARD	Aridisols	Calcids	Petrocalcids	12								36				**43**		
BARTOME	Aridisols	Durids	Argidurids	18							10		35					
BATAN	Entisols	Orthents	Torriorthents	18														

(continued)

(continued)

Series Name	Order	Suborder	Great Group	Ochric	Mollic	Umbric	Histic	Folistic	Albic	Thickness (cm)		Calcic	Duripan	Gypsic	Natric	Petrocalcic	Petrogypsic	Salic
										Cambic	Argillic							
BEELEM	Entisols	Orthents	Torriorthents	10														
BELATE	Mollisols	Xerolls	Argixerolls		46						**122**							
BELLEHELEN	Mollisols	Xerolls	Argixerolls		18						15							
BELTED	Aridisols	Durids	Argidurids	20							16		10					
BENIN	Entisols	Orthents	Torriorthents	18														
BEOSKA	Aridisols	Argids	Natrargids	33								**96**			28			
BERZATIC	Entisols	Orthents	Torriorthents	18														
BESHERM	Aridisols	Calcids	Haplocalcids	5								99						
BIDDLEMAN	Aridisols	Argids	Natrargids	10											28			
BIEBER	Mollisols	Xerolls	Durixerolls		33						31		106					
BIGA	Aridisols	Argids	Natrargids	15					5						15			
BIKEN	Aridisols	Calcids	Haplocalcids	18								38						
BILBO	Aridisols	Argids	Haplargids	10							46							
BIOYA	Aridisols	Durids	Haplodurids	18									55					
BIRCHCREEK	Mollisols	Xerolls	Argixerolls		33						50							
BIRDSPRING	Entisols	Orthents	Torriorthents	10														
BLACKHAWK	Aridisols	Durids	Haplodurids	18						28			5					
BLACKTOP	Entisols	Orthents	Torriorthents	18														
BLAPPERT	Aridisols	Argids	Haplargids	8							22							
BLIMO	Entisols	Orthents	Torriorthents	18														
BLISS	Aridisols	Durids	Haplodurids	18						46			43					
BLUEGYP	Aridisols	Gypsids	Haplogypsids	5										104				
BLUEPOINT	Entisols	Psamments	Torripsamments	5														
BLUEWING	Entisols	Orthents	Torriorthents	18														
BOBS	Mollisols	Xerolls	Palexerolls		30											34		
BOJO	Aridisols	Argids	Haplargids	13							7							
BOMBADIL	Aridisols	Argids	Haplargids	5							20							
BOOFUSS	Inceptisols	Aquepts	Halaquepts	18														
BOOMSTICK	Aridisols	Argids	Haplargids	13							28							
B OREALIS	Alfisols	Xeralfs	Durixeralfs	28							30		44					
BORVANT	Mollisols	Xerolls	Palexerolls		36											49		
BOTON	Entisols	Orthents	Torriorthents	18														
BOULDER LAKE	Vertisols	Aquerts	Epiaquerts	18														
BOXSPRING	Entisols	Orthents	Torriorthents	18														
BRACKEN	Aridisols	Gypsids	Haplogypsids	5										119				
BREGAR	Aridisols	Argids	Haplargids	15							13							
BREKO	Aridisols	Argids	Haplargids	15							59							
BRICONE why not cambic?	Entisols	Orthents	Cryorthents	8														
BRIER	Mollisols	Xerolls	Argixerolls		38						28							
BROLAND	Aridisols	Durids	Argidurids	8							33		54					
BROYLES	Aridisols	Cambids	Haplocambids	18						20								
BUBUS	Entisols	Orthents	Torriorthents	18														
BUCAN	Aridisols	Argids	Haplargids	18							109							

(continued)

(continued)

Series Name	Order	Suborder	Great Group	Ochric	Mollic	Umbric	Histic	Folistic	Albic	Thickness (cm)		Calcic	Duripan	Gypsic	Natric	Petrocalcic	Petrogypsic	Salic
										Cambic	Argillic							
BUCKAROO	Aridisols	Argids	Natrargids	10											31			
BUCKLAKE	Mollisols	Xerolls	Argixerolls		30						41							
BUCKSPRING	Mollisols	Ustolls	Argiustolls		25						18							
BUDIHOL	Entisols	Orthents	Torriorthents	18														
BUFFARAN	Aridisols	Durids	Argidurids	5							36		111					
BULAKE	Alfisols	Xeralfs	Haploxeralfs	18							30							
BULLUMP	Mollisols	Xerolls	Argixerolls		58						79							
BURRITA	Aridisols	Argids	Haplargids	18							18							
CAGWIN	Entisols	Psamments	Xeropsamments	33														
CALLVILLE	Aridisols	Gypsids	Haplogypsids	5										58				
CANDELARIA	Aridisols	Calcids	Haplocalcids	18								87						
CANOTO	Entisols	Orthents	Torriorthents	18														
CARRIZO	Entisols	Orthents	Torriorthents	18														
CARSON	Vertisols	Xererts	Haploxererts		64													
CARSTUMP	Mollisols	Xerolls	Argixerolls		28						36	24						
CASAGA	Aridisols	Argids	Natrargids	2.5											51			
CASSIRO	Mollisols	Xerolls	Argixerolls		38						89							
CATH	Aridisols	Argids	Calciargids	15							36	81						
CAVANAUGH	Mollisols	Xerolls	Argixerolls		30						51							
CAVEHILL	Mollisols	Xerolls	Calcixerolls		46							28						
CEDARAN	Mollisols	Ustolls	Argiustolls		18						34							
CEEJAY	Aridisols	Argids	Haplargids	5							36							
CELETON	Entisols	Orthents	Torriorthents	18														
CHAD	Mollisols	Xerolls	Argixerolls		23						64							
CHAINLINK	Mollisols	Xerolls	Durixerolls		25							15	39					
CHALCO	Aridisols	Argids	Haplargids	8							30							
CHECKETT	Aridisols	Argids	Haplargids	15							26							
CHEME	Aridisols	Durids	Haplodurids	5								30	61					
CHEN	Mollisols	Xerolls	Argixerolls		33						13							
CHERRY SPRING	Aridisols	Durids	Argidurids	38							53		46					
CHIARA	Aridisols	Durids	Haplodurids	18									20					
CHILL	Aridisols	Argids	Haplargids	8							10							
CHILPER	Aridisols	Argids	Natrargids	33											41			
CHUBARD	Aridisols	Argids	Haplargids	10							15							
CHUCKLES	Aridisols	Cambids	Haplocambids	18						23								
CHUCKMILL	Aridisols	Durids	Argidurids	5							31		116					
CHUCKRIDGE	Aridisols	Durids	Argidurids	5							25		11					
CHUFFA	Aridisols	Cambids	Haplocambids	18						25								
CHUG	Mollisols	Xerolls	Haploxerolls		81													
CHUSKA	Aridisols	Durids	Argidurids	8							38	21	10					
CIRAC	Entisols	Fluvents	Torrifluvents	18														
CLANALPINE	Mollisols	Xerolls	Argixerolls		25						74							
CLEAVAGE	Mollisols	Xerolls	Argixerolls		23						23							

(continued)

(continued)

Series Name	Order	Suborder	Great Group	Ochric	Mollic	Umbric	Histic	Folistic	Albic	Thickness (cm)		Calcic	Duripan	Gypsic	Natric	Petrocalcic	Petrogypsic	Salic
										Cambic	Argillic							
CLEAVER	Aridisols	Durids	Argidurids	8							20		38					
CLEMENTINE	Mollisols	Aquolls	Endoaquolls		69													
CLIFFDOWN	Entisols	Orthents	Torriorthents	18														
CLOWFIN	Entisols	Orthents	Torriorthents	18														
CLURDE	Aridisols	Cambids	Haplocambids	18						15								
COBRE	Aridisols	Cambids	Haplocambids	18						28								
COILS	Aridisols	Durids	Argidurids	13							53		28					
COLBAR	Aridisols	Argids	Haplargids	23							18							
COLVAL	Aridisols	Argids	Calciargids	13							45	139						
COMMSKI	Aridisols	Calcids	Haplocalcids	12								140						
CONNEL	Aridisols	Cambids	Haplocambids	18						21								
COPPEREID	Entisols	Orthents	Torriorthents	18														
CORBILT	Aridisols	Calcids	Haplocalcids	10								101	**10**					
CORTEZ	Aridisols	Durids	Natridurids	13									66		31			
COSER	Mollisols	Xerolls	Palexerolls		41						61							
COTANT	Mollisols	Xerolls	Argixerolls		30						40							
COWGIL	Aridisols	Argids	Haplargids	8							68							
CREEMON	Aridisols	Cambids	Haplocambids	18						13								
CROOKED CREEK	Mollisols	Aquolls	Endoaquolls		84													
CROPPER	Mollisols	Xerolls	Argixerolls		18						31							
CROSGRAIN	Aridisols	Durids	Haplodurids	18									124					
DACKER	Aridisols	Durids	Argidurids	18							46		53					
DAVEY	Aridisols	Cambids	Haplocambids	18						18								
DEADYON	Aridisols	Argids	Haplargids	13							48							
DECAN	Mollisols	Xerolls	Durixerolls		20						50		30					
DEFLER	Entisols	Orthents	Torriorthents	18														
DELAMAR	Aridisols	Durids	Argidurids	13							40		76					
DELVADA	Mollisols	Aquolls	Endoaquolls		74													
DENPARK	Mollisols	Cryolls	Argicryolls		55						48							
DEPPY	Aridisols	Durids	Argidurids		15						23		15					
DESATOYA	Aridisols	Argids	Haplargids	8							28							
DEVADA	Mollisols	Xerolls	Argixerolls		33						13							
DEVEN	Mollisols	Xerolls	Argixerolls		18						23							
DEVILDOG	Aridisols	Cambids	Haplocambids	18						20								
DEVILSGAIT	Mollisols	Aquolls	Endoaquolls		71													
DEWAR	Aridisols	Durids	Argidurids	13							30		109					
DIA	Mollisols	Xerolls	Haploxerolls		30													
DIANEV	Inceptisols	Aquepts	Halaquepts	18														
DITHOD	Mollisols	Xerolls	Haploxerolls		38													
DOBEL	Aridisols	Durids	Argidurids	10							10		10					
DONNA	Mollisols	Xerolls	Durixerolls		28						23		41					
DOORKISS	Aridisols	Argids	Haplargids	10							28							
DORPER	Aridisols	Argids	Natrargids	18											25			

(continued)

(continued)

Series Name	Order	Suborder	Great Group	Ochric	Mollic	Umbric	Histic	Folistic	Albic	Thickness (cm)		Calcic	Duripan	Gypsic	Natric	Petrocalcic	Petrogypsic	Salic
										Cambic	Argillic							
DOSIE	Mollisols	Xerolls	Argixerolls		58						91							
DOUHIDE	Mollisols	Xerolls	Argixerolls		18						20							
DOWNEYVILLE	Aridisols	Argids	Haplargids	10							13							
DRYGYP	Aridisols	Gypsids	Petrogypsids	18													147	
DUCO	Mollisols	Xerolls	Argixerolls		48						35							
DUFFER	Aridisols	Calcids	Haplocalcids	18						21		142						
DUGCHIP	Aridisols	Durids	Natridurids	18									20		33			
DUN GLEN	Aridisols	Cambids	Haplocambids	18						17								
EAGLEPASS	Entisols	Orthents	Torriorthents	10														
EAGLEROCK	Mollisols	Xerolls	Argixerolls		48						66							
EAST FORK	Mollisols	Xerolls	Haploxerolls		36													
EASTGATE	Aridisols	Cambids	Haplocambids	18						30								
EASTMORE	Aridisols	Durids	Haplodurids	18								35	54					
EASTWELL	Aridisols	Durids	Haplodurids	18									5					
EBODA	Mollisols	Xerolls	Argixerolls		38						61							
EGANROC	Mollisols	Cryolls	Haplocryolls		56							63						
ENKO	Aridisols	Cambids	Haplocambids	18						36								
EQUIS	Inceptisols	Aquepts	Halaquepts	18						?								
ESCALANTE	Aridisols	Calcids	Haplocalcids	18								56						
EWELAC	Aridisols	Cambids	Haplocambids	18						58								
FANG	Entisols	Orthents	Torriorthents	18														
FERROGOLD	Aridisols	Calcids	Petrocalcids	18								15				114		
FIVEMILE	Entisols	Fluvents	Torrifluvents	18														
FLATTOP	Aridisols	Argids	Natrargids	5											56			
FLETCHERPEAK	Mollisols	Ustolls	Argiustolls		33						30							
FLUE	Aridisols	Durids	Natridurids	15									13		48			
FORTANK	Aridisols	Argids	Haplargids	23							64							
FUGAWEE	Alfisols	Xeralfs	Haploxeralfs	48							30							
FULSTONE	Aridisols	Durids	Argidurids	13							33		30					
GABBVALLY	Aridisols	Argids	Haplargids	5							15							
GANCE	Aridisols	Argids	Haplargids	10							64		99					
GARHILL	Aridisols	Durids	Haplodurids	18									35					
GEER	Entisols	Orthents	Torriorthents	18														
GENAW	Aridisols	Argids	Haplargids	15							13							
GENEGRAF	Aridisols	Argids	Natrargids	15								106			31			
GETA	Aridisols	Calcids	Haplocalcids	15						23		**114**						
GILA	Entisols	Fluvents	Torrifluvents	15														
GLEAN	Mollisols	Xerolls	Haploxerolls		74													
GLENCARB	Entisols	Fluvents	Torrifluvents	15														
GLOTRAIN	Aridisols	Argids	Haplargids	10							56							
GOCHEA	Mollisols	Xerolls	Argixerolls		30						35		97					
GOLCONDA	Aridisols	Durids	Natridurids	25									33		33			
GOLDROAD	Entisols	Orthents	Torriorthents	18														

(continued)

(continued)

Series Name	Order	Suborder	Great Group	Ochric	Mollic	Umbric	Histic	Folistic	Albic	Thickness (cm)		Calcic	Duripan	Gypsic	Natric	Petrocalcic	Petrogypsic	Salic
										Cambic	Argillic							
GOLDRUN	Entisols	Psamments	Torripsamments	18														
GOLDYKE	Entisols	Orthents	Torriorthents	15														
GOLLAHER	Entisols	Orthents	Xerorthents	15														
GOOSEL	Aridisols	Durids	Argidurids	33							20		**2**					
GOVWASH	Aridisols	Gypsids	Haplogypsids	7.5										135				
GRALEY	Mollisols	Xerolls	Argixerolls		18						18							
GRANSHAW	Aridisols	Argids	Haplargids	33					13		25							
GRAPEVINE	Aridisols	Calcids	Haplocalcids	13								163						
GRASSVAL	Aridisols	Durids	Argidurids	10							23		119					
GRAUFELS	Mollisols	Xerolls	Haploxerolls		25													
GRAVIER	Aridisols	Calcids	Haplocalcids	18								117						
GREYEAGLE	Aridisols	Durids	Haplodurids	15									41					
GRINA	Entisols	Orthents	Torriorthents	18														
GRUMBLEN	Aridisols	Argids	Haplargids	10							36							
GUARDIAN	Aridisols	Gypsids	Haplogypsids	5										43				
GYNELLE	Entisols	Orthents	Torriorthents	18														
HACKWOOD	Mollisols	Cryolls	Haplocryolls		53													
HALACAN	Mollisols	Rendolls	Cryrendolls		28													
HALEBURU	Entisols	Orthents	Torriorthents	12														
HAMTAH	Mollisols	Xerolls	Argixerolls		53						99							
HANDPAH	Aridisols	Durids	Argidurids	15							28		104					
HANDY	Aridisols	Argids	Haplargids	20							56							
HANGROCK	Aridisols	Durids	Argidurids	10							33		109					
HAPGOOD	Mollisols	Cryolls	Haplocryolls		91												28	
HARCANY	Mollisols	Cryolls	Haplocryolls		183													
HARDHAT	Aridisols	Calcids	Haplocalcids	18								99						
HARDOL	Mollisols	Cryolls	Calcicryolls		152							68						
HARDZEM	Alfisols	Cryalfs	Haplocryalfs	3							50							
HART CAMP	Mollisols	Xerolls	Argixerolls		30						21							
HARTIG	Mollisols	Xerolls	Haploxerolls		25					32								
HATPEAK	Mollisols	Xerolls	Durixerolls		25						50		75					
HAUNCHEE	Mollisols	Rendolls	Cryrendolls		34							24						
HAVINGDON	Aridisols	Argids	Haplargids	25							28							
HAWSLEY	Entisols	Psamments	Torripsamments	18														
HAYBOURNE	Aridisols	Cambids	Haplocambids	18						49								
HAYESTON	Entisols	Orthents	Torriorthents	18														
HAYMONT	Entisols	Orthents	Torriorthents	18														
HEECHEE	Mollisols	Xerolls	Argixerolls		46						56							
HEIST	Aridisols	Cambids	Haplocambids	18						29								
HESSING	Aridisols	Cambids	Haplocambids	18						14								
HIKO PEAK	Aridisols	Calcids	Haplocalcids	18								62						
HOLBROOK	Mollisols	Xerolls	Haploxerolls		38													
HOLLYWELL	Aridisols	Cambids	Haplocambids	18						28								

(continued)

(continued)

Series Name	Order	Suborder	Great Group	Ochric	Mollic	Umbric	Histic	Folistic	Albic	Thickness (cm)		Calcic	Duripan	Gypsic	Natric	Petrocalcic	Petrogypsic	Salic
										Cambic	Argillic							
HOLTVILLE	Entisols	Fluvents	Torrifluvents	43														
HOMESTAKE	Mollisols	Xerolls	Argixerolls		28						49							
HOOPLITE	Aridisols	Argids	Haplargids	10							10							
HOOT	Aridisols	Argids	Haplargids	10							26							
HOPEKA	Entisols	Orthents	Torriorthents	18														
HOPPSWELL	Aridisols	Argids	Haplargids	5							33							
HORSECAMP	Vertisols	Xererts	Haploxererts	5						64								
HOUGH	Aridisols	Argids	Haplargids	25							19							
HUEVI	Aridisols	Calcids	Haplocalcids	12								33						
HUILEPASS	Aridisols	Argids	Haplargids	13							33							
HUMBOLDT	Mollisols	Aquolls	Endoaquolls		25							25						
HUMDUN	Aridisols	Cambids	Haplocambids	18						56								
HUNDRAW	Entisols	Orthents	Torriorthents	18														
HUNEWILL	Aridisols	Argids	Haplargids	8							38							
HUNNTON	Aridisols	Durids	Argidurids	36							35		36					
HUSSA	Mollisols	Aquolls	Endoaquolls		58													
HUTCHLEY	Mollisols	Xerolls	Argixerolls		38						15							
HYLOC	Mollisols	Xerolls	Argixerolls		46						38							
HYMAS	Mollisols	Xerolls	Haploxerolls		18													
HYPOINT	Entisols	Orthents	Torriorthents	5														
HYZEN	Mollisols	Xerolls	Haploxerolls		30													
IDWAY	Aridisols	Calcids	Haplocalcids	18						20		39						
IGDELL	Mollisols	Xerolls	Durixerolls		28						49		33					
INDIAN CREEK	Aridisols	Durids	Argidurids	8					3		42		14					
INDIANO	Mollisols	Xerolls	Argixerolls		33						51							
INMO	Entisols	Orthents	Torriorthents	18														
INPENDENCE	Inceptisols	Xerepts	Humixerepts			71												
INVILLE	Alfisols	Xeralfs	Haploxeralfs			25					69							
IRONGOLD	Aridisols	Calcids	Petrocalcids	18												167		
ISOLDE	Entisols	Psamments	Torripsamments	18														
ISTER	Mollisols	Xerolls	Argixerolls		20						69							
ITCA	Mollisols	Xerolls	Argixerolls		23						20							
ITME	Entisols	Orthents	Torriorthents	18														
IVER	Mollisols	Xerolls	Haploxerolls		94					**102**								
IZAMATCH	Entisols	Orthents	Torriorthents	18														
IZAR	Entisols	Orthents	Torriorthents	18														
IZO	Entisols	Orthents	Torriorthents	18														
IZOD	Entisols	Orthents	Torriorthents	18														
JARAB	Mollisols	Xerolls	Durixerolls		25								64					
JAYBEE	Aridisols	Argids	Haplargids	20							16							
JEAN	Entisols	Orthents	Torriorthents	20														
JERICHO	Aridisols	Durids	Haplodurids	18								25	28					
JERVAL	Aridisols	Argids	Natrargids	15											41			

(continued)

(continued)

Series Name	Order	Suborder	Great Group	Ochric	Mollic	Umbric	Histic	Folistic	Albic	Thickness (cm)		Calcic	Duripan	Gypsic	Natric	Petrocalcic	Petrogypsic	Salic
										Cambic	Argillic							
JESSE CAMP	Aridisols	Cambids	Haplocambids	18						20								
JOBPEAK	Entisols	Orthents	Torriorthents	18														
JOEMAY	Aridisols	Calcids	Haplocalcids	6								142						
JUNG	Aridisols	Argids	Haplargids	20							28							
JUVA	Entisols	Fluvents	Torrifluvents	18														
KANESPRINGS	Aridisols	Durids	Argidurids	7.5							38		23					
KATELANA	Entisols	Orthents	Torriorthents	18														
KAWICH	Entisols	Psamments	Torripsamments	18														
KEEFA	Aridisols	Cambids	Haplocambids	18						15			25					
KELK	Aridisols	Cambids	Haplocambids	18						20								
KIDWELL	Aridisols	Argids	Calciargids	23							56	41						
KLECKNER	Mollisols	Xerolls	Argixerolls		30						81							
KOBEH	Aridisols	Cambids	Haplocambids	18						25								
KOLDA	Mollisols	Aquolls	Endoaquolls		56													
KOONTZ	Mollisols	Xerolls	Argixerolls		36						13							
KOYEN	Aridisols	Cambids	Haplocambids	18						35								
KRAM	Entisols	Orthents	Torriorthents	18														
KUMIVA	Entisols	Orthents	Torriorthents	18														
KUNZLER	Aridisols	Calcids	Haplocalcids	18								63						
KURSTAN	Aridisols	Calcids	Haplocalcids	23								130						
KYLER	Entisols	Orthents	Torriorthents	18														
KZIN	Entisols	Orthents	Torriorthents	18														
LABKEY	Aridisols	Cambids	Haplocambids	18						20								
LABSHAFT	Mollisols	Cryolls	Haplocryolls		28													
LAHONTAN	Entisols	Orthents	Torriorthents	25														
LANGSTON	Aridisols	Argids	Haplargids	10							41							
LANIP	Aridisols	Argids	Calciargids	15							61	43						
LAPED	Aridisols	Durids	Argidurids	15							31		12					
LAPON	Aridisols	Durids	Argidurids	5							20		26					
LARYAN	Inceptisols	Aquepts	Cryaquepts				30											
LAS VEGAS	Aridisols	Calcids	Petrocalcids	7.5								10				**91**		
LASTCHANCE	Aridisols	Calcids	Petrocalcids	5								46				102		
LATHROP	Aridisols	Argids	Haplargids	8							25							
LAXAL	Aridisols	Calcids	Haplocalcids	18								**139**						
LAYVIEW	Mollisols	Cryolls	Argicryolls		19						32							
LAZAN	Entisols	Orthents	Torriorthents	10														
LEO	Entisols	Orthents	Torriorthents	18														
LERROW	Mollisols	Xerolls	Argixerolls		38						68							
LIEN	Aridisols	Durids	Haplodurids	18									41					
LIMEWASH	Aridisols	Gypsids	Haplogypsids	15										27				
LINCO	Aridisols	Cambids	Haplocambids	18						56								
LINKUP	Aridisols	Argids	Haplargids	13							30							
LINOYER	Entisols	Orthents	Torriorthents	18														

(continued)

Series Name	Order	Suborder	Great Group	Ochric	Mollic	Umbric	Histic	Folistic	Albic	Thickness (cm)		Calcic	Duripan	Gypsic	Natric	Petrocalcic	Petrogypsic	Salic
										Cambic	Argillic							
LINROSE why not cambic?	Mollisols	Xerolls	Haploxerolls		20													
LITTLEAILIE	Aridisols	Durids	Argidurids	8							12		56					
LOCANE	Aridisols	Argids	Haplargids	13							35							
LODAR	Mollisols	Xerolls	Calcixerolls		20							18						
LOGRING	Entisols	Orthents	Torriorthents	18														
LOJET	Aridisols	Durids	Argidurids	10							79	18	15					
LOMOINE	Entisols	Orthents	Torriorthents	18														
LONCAN	Mollisols	Xerolls	Haploxerolls		36													
LONGCREEK	Mollisols	Xerolls	Argixerolls		30						28							
LONGDIS	Aridisols	Argids	Natrargids	13											101			
LONGJIM	Aridisols	Durids	Haplodurids	7.5									99					
LOOMER	Mollisols	Xerolls	Argixerolls		18						43							
LORAY	Aridisols	Calcids	Haplocalcids	18								28						
LOVELOCK	Mollisols	Aquolls	Endoaquolls		25													
LOWEMAR	Inceptisols	Cryepts	Humicryepts			47												
LUNING	Entisols	Orthents	Torriorthents	18														
LUSET	Mollisols	Cryolls	Argicryolls		50						77							
LYDA	Aridisols	Durids	Argidurids	10							20		70					
LYX	Aridisols	Cambids	Haplocambids	18						20								
MACNOT	Aridisols	Calcids	Haplocalcids	18								46						
MAJUBA	Mollisols	Xerolls	Argixerolls		28						43							
MALPAIS	Aridisols	Cambids	Haplocambids	18						56								
MANOGUE	Vertisols	Xererts	Haploxererts	18						30								
MARYJANE	Mollisols	Ustolls	Calciustolls		150							142						
MATTIER	Mollisols	Xerolls	Argixerolls		20						22							
MAU	Aridisols	Argids	Haplargids	15							56							
MAZUMA	Entisols	Orthents	Torriorthents	18														
MCCARRAN	Aridisols	Gypsids	Haplogypsids	2.5										135				
MCCONNEL	Aridisols	Cambids	Haplocambids	18						18								
MCIVEY	Mollisols	Xerolls	Argixerolls		46						119							
MCTOM	Inceptisols	Cryepts	Humicryepts			41												
MEADVIEW	Aridisols	Calcids	Haplocalcids	5								48						
MEDBURN	Entisols	Orthents	Torriorthents	20														
MEISS	Inceptisols	Cryepts	Humicryepts			33												
MEZZER	Aridisols	Cambids	Haplocambids	18						17								
MIDRAW	Aridisols	Durids	Argidurids	5							38		16					
MIJOYSEE	Entisols	Orthents	Torriorthents	15														
MINU	Aridisols	Durids	Argidurids	4							36		70					
MIRKWOOD	Aridisols	Argids	Haplargids	5							23							
MISAD	Entisols	Orthents	Torriorthents	18														
MOLION	Aridisols	Durids	Haplodurids	18									36					
MONARCH	Mollisols	Xerolls	Calcixerolls		38							18						
MORMON MESA	Aridisols	Calcids	Petrocalcids	6								15				112		
MORMOUNT	Aridisols	Calcids	Petrocalcids	7.5							41					104		

(continued)

(continued)

Series Name	Order	Suborder	Great Group	Ochric	Mollic	Umbric	Histic	Folistic	Albic	Thickness (cm)		Calcic	Duripan	Gypsic	Natric	Petrocalcic	Petrogypsic	Salic
										Cambic	Argillic							
MOSIDA	Entisols	Fluvents	Torrifluvents	18														
MOTOQUA	Mollisols	Ustolls	Argiustolls		30						11							
MOTTSVILLE	Mollisols	Xerolls	Haploxerolls		46													
MOUNTMCULL	Entisols	Orthents	Torriorthents	18														
MUIRAL	Inceptisols	Cryepts	Haplocryepts	18						61								
MUNI	Aridisols	Durids	Argidurids	20							26		68					
NADRA	Aridisols	Durids	Haplodurids	18						15			36					
NEEDLE PEAK	Entisols	Orthents	Torriorthents	18														
NEMICO	Aridisols	Durids	Natridurids	8									3		22			
NETTI	Mollisols	Xerolls	Argixerolls		30						**120**							
NEVADOR	Aridisols	Argids	Haplargids	15							46							
NEVU	Mollisols	Xerolls	Durixerolls		28						56		22					
NEWERA	Aridisols	Argids	Haplargids	5							10							
NEWVIL	Mollisols	Xerolls	Durixerolls		25						17		59					
NICKEL	Aridisols	Calcids	Haplocalcids	10								142						
NINEMILE	Mollisols	Xerolls	Argixerolls		20						31							
NIPPENO	Aridisols	Argids	Haplargids	5							10							
NIPTON	Entisols	Orthents	Torriorthents	2.5														
NOLENA	Entisols	Orthents	Torriorthents	13														
NUPART	Mollisols	Xerolls	Haploxerolls		13													
NUYOBE	Inceptisols	Aquepts	Halaquepts	18														
NYSERVA	Aridisols	Argids	Natrargids	5							13				5			
OCALA	Inceptisols	Aquepts	Halaquepts	18														
OKAN	Entisols	Orthents	Torriorthents	18														
OLA	Mollisols	Xerolls	Haploxerolls		89													
OLAC	Aridisols	Argids	Haplargids	10							81							
OLD CAMP	Aridisols	Argids	Haplargids	5							31							
OLDSPAN	Aridisols	Calcids	Haplocalcids	18						18		75						
OLEMAN	Aridisols	Durids	Argidurids	5							31		25					
ONKEYO	Mollisols	Xerolls	Calcixerolls		20							18						
OPPIO	Aridisols	Argids	Haplargids	8							45							
ORICTO	Aridisols	Argids	Calciargids	8							12	16						
ORIZABA no B	Inceptisols	Aquepts	Halaquepts	18														
OROVADA	Aridisols	Cambids	Haplocambids	18						31								
ORWASH	Entisols	Orthents	Torriorthents	18														
OSOBB	Aridisols	Durids	Haplodurids	18									3					
OUPICO	Aridisols	Durids	Haplodurids	18								54	80					
OXCOREL	Aridisols	Argids	Natrargids	20											66			
PACKER	Mollisols	Cryolls	Argicryolls		23						10							
PAHROC	Aridisols	Durids	Haplodurids	18								26	51					
PALINOR	Aridisols	Durids	Haplodurids	18								21	30					
PANLEE	Aridisols	Cambids	Petrocambids	18						49			7					
PARANAT	Mollisols	Aquolls	Endoaquolls		50													
PARISA	Aridisols	Durids	Haplodurids	18									53					

(continued)

(continued)

Series Name	Order	Suborder	Great Group	Ochric	Mollic	Umbric	Histic	Folistic	Albic	Thickness (cm)		Calcic	Duripan	Gypsic	Natric	Petrocalcic	Petrogypsic	Salic
										Cambic	Argillic							
PARRAN	Aridisols	Salids	Aquisalids	18														33
PATNA	Aridisols	Argids	Haplargids	15							46							
PAYPOINT	Aridisols	Argids	Haplargids	13							15							
PEDOLI	Aridisols	Argids	Haplargids	15							49							
PEEKO	Aridisols	Durids	Haplodurids	18									63					
PENELAS	Aridisols	Argids	Haplargids	5							8							
PENOYER	Entisols	Orthents	Torriorthents	18														
PERNTY	Mollisols	Xerolls	Argixerolls		20						28							
PERWICK	Entisols	Orthents	Torriorthents	18														
PHLISS	Aridisols	Argids	Haplargids	8							25							
PIBLER	Aridisols	Calcids	Petrocalcids	18								20				**15**		
PICKUP	Mollisols	Xerolls	Argixerolls		30						66							
PIE CREEK	Mollisols	Xerolls	Palexerolls		28						40							
PINEVAL	Aridisols	Argids	Haplargids	8							20							
PINTWATER	Entisols	Orthents	Torriorthents	18														
PIOCHE	Mollisols	Xerolls	Argixerolls		18						33							
PIROUETTE	Aridisols	Durids	Natridurids	8									2		17			
POKERGAP	Aridisols	Argids	Natrargids	15											21			
POOKALOO	Aridisols	Calcids	Haplocalcids	18								38						
PORTMOUNT	Aridisols	Argids	Calciargids	8							78	66						
POTOSI	Entisols	Orthents	Torriorthents	18														
PREBLE	Entisols	Orthents	Torriorthents	18														
PRIDEEN	Entisols	Orthents	Torriorthents	18														
PUETT	Entisols	Orthents	Torriorthents	18														
PUFFER	Entisols	Orthents	Torriorthents	18														
PUMEL	Entisols	Orthents	Torriorthents	13														
PUMPER	Aridisols	Cambids	Haplocambids	18						22								
PUNCHBOWL	Aridisols	Argids	Haplargids	15							10							
PUROB	Aridisols	Calcids	Petrocalcids	18								28				104		
PYRAT	Aridisols	Calcids	Haplocalcids	18								84						
QUARZ	Mollisols	Xerolls	Argixerolls		28						46							
QWYNN	Aridisols	Argids	Haplargids	18							61							
RAD	Aridisols	Cambids	Haplocambids	18						35			102					
RADOL	Mollisols	Xerolls	Calcixerolls		38							33						
RAGLAN	Aridisols	Cambids	Haplocambids	18						21								
RAGTOWN	Entisols	Orthents	Torriorthents	18														
RAILROAD	Aridisols	Calcids	Haplocalcids	18						20		58						
RAMIRES	Mollisols	Xerolls	Argixerolls		36						33							
RASILLE why not mollic?	Aridisols	Cambids	Haplocambids	18						23								
RATTO	Aridisols	Durids	Argidurids	15							18		10					
RAVENDOG why not cambic?	Mollisols	Xerolls	Haploxerolls		41													
RAVENSWOOD	Mollisols	Xerolls	Argixerolls		30						28							
RAWE	Aridisols	Argids	Haplargids	3							22							
REBEL	Aridisols	Cambids	Haplocambids	18						36								

(continued)

(continued)

Series Name	Order	Suborder	Great Group	Ochric	Mollic	Umbric	Histic	Folistic	Albic	Thickness (cm)		Calcic	Duripan	Gypsic	Natric	Petrocalcic	Petrogypsic	Salic
										Cambic	Argillic							
REDNIK	Aridisols	Argids	Haplargids	15							36							
REESE	Inceptisols	Aquepts	Halaquepts	18														
RELLEY	Aridisols	Cambids	Haplocambids	18						21								
RELUCTAN	Mollisols	Xerolls	Argixerolls		23						46							
RENO	Aridisols	Durids	Argidurids	20							48		19					
REYWAT	Mollisols	Xerolls	Argixerolls		18						33							
REZAVE	Aridisols	Argids	Natrargids	10											33			
RICERT	Aridisols	Argids	Natrargids	13											23			
RICHINDE	Aridisols	Argids	Haplargids	13							33							
RIVERBEND	Aridisols	Calcids	Haplocalcids	18								135						
ROBSON	Aridisols	Argids	Haplargids	25							23							
ROCA	Aridisols	Argids	Haplargids	20							41							
ROCCONDA	Aridisols	Argids	Haplargids	3							10							
ROCHPAH	Aridisols	Calcids	Haplocalcids	18								38						
RODAD	Aridisols	Argids	Haplargids	20							10							
RODEN	Entisols	Orthents	Torriorthents	18														
ROIC	Entisols	Orthents	Torriorthents	13														
ROSNEY	Entisols	Orthents	Torriorthents	18														
ROUETTE	Aridisols	Durids	Haplodurids	18						15			15					
RUBYHILL	Aridisols	Durids	Haplodurids	18						43			26					
RUSTIGATE	Entisols	Orthents	Torriorthents	18														
SALTAIR	Aridisols	Salids	Aquisalids	2.5														152
SANWELL why not cambic?	Entisols	Orthents	Torriorthents	7.5														
SARAPH	Aridisols	Argids	Haplargids	10							31							
SAY	Mollisols	Xerolls	Argixerolls		23						18							
SCHOOLMARM	Mollisols	Xerolls	Argixerolls		28						20							
SCOSSA	Inceptisols	Aquepts	Humaquepts			21	20											
SEANNA	Entisols	Orthents	Torriorthents	18														
SEGURA	Mollisols	Xerolls	Argixerolls		25						17							
SERALIN	Mollisols	Ustolls	Haplustolls		36					18								
SETTLEMENT Bk >1 m thick	Inceptisols	Aquepts	Halaquepts	18														
SETTLEMEYER	Mollisols	Aquolls	Endoaquolls		56													
SEVENMILE	Mollisols	Xerolls	Haploxerolls		43					43								
SHABLISS	Aridisols	Durids	Haplodurids	18						18			16					
SHAGNASTY	Mollisols	Xerolls	Argixerolls		30						127							
SHALAKE	Aridisols	Durids	Haplodurids	18									86					
SHALCLEAV	Mollisols	Xerolls	Argixerolls		25						17							
SHALPER	Mollisols	Xerolls	Argixerolls		23						7							
SHAWAVE	Aridisols	Argids	Haplargids	20							33							
SHEFFIT	Entisols	Orthents	Torriorthents	18														
SHORT CREEK	Aridisols	Argids	Haplargids	5							147							

(continued)

(continued)

Series Name	Order	Suborder	Great Group	Ochric	Mollic	Umbric	Histic	Folistic	Albic	Thickness (cm)		Calcic	Duripan	Gypsic	Natric	Petrocalcic	Petrogypsic	Salic
										Cambic	Argillic							
SHREE	Mollisols	Xerolls	Argixerolls		30						48							
SIBELIA	Inceptisols	Cryepts	Humicryepts			18		18		20								
SILVERADO	Aridisols	Cambids	Haplocambids	18						18								
SILVERBOW	Aridisols	Durids	Argidurids	5							28		8					
SIMON	Mollisols	Xerolls	Argixerolls		30						115							
SINGATSE	Entisols	Orthents	Torriorthents	15														
SISCAB	Mollisols	Xerolls	Argixerolls		20						12							
SKEDADDLE	Entisols	Orthents	Torriorthents	13														
SKELON	Aridisols	Durids	Haplodurids	18						23			41					
SLAVEN	Mollisols	Xerolls	Argixerolls		25						43							
SLAW	Entisols	Fluvents	Torrifluvents	18														
SLIDYMTN	Mollisols	Ustolls	Argiustolls		20						33							
SLIPBACK	Aridisols	Argids	Natrargids	23											28			
SLOCKEY	Mollisols	Xerolls	Argixerolls		23						43							
SMEDLEY	Aridisols	Durids	Argidurids	5							15		63					
SNAPP	Aridisols	Argids	Natrargids	13							40				30			
SNOWMORE	Aridisols	Durids	Argidurids	10							23		15					
SOAKPAK	Inceptisols	Cryepts	Haplocryepts	2						17								
SOAR	Aridisols	Argids	Haplargids	8							7							
SODHOUSE	Aridisols	Durids	Haplodurids	18						33			59					
SOftSCRABBLE	Mollisols	Xerolls	Argixerolls		76						175							
SOJUR	Entisols	Orthents	Torriorthents	10														
SOLAK	Entisols	Orthents	Torriorthents	12														
SONDOA	Entisols	Orthents	Torriorthents	18														
SONOMA	Entisols	Aquents	Fluvaquents	18														
SOUGHE	Aridisols	Argids	Haplargids	10							26							
SPASPREY	Aridisols	Durids	Argidurids	5							25		43					
SPRING	Aridisols	Gypsids	Haplogypsids	12								**43**		81				
SQUAWTIP	Mollisols	Xerolls	Argixerolls		28						84							
ST. THOMAS	Entisols	Orthents	Torriorthents	18														
STAMPEDE	Mollisols	Xerolls	Durixerolls		24						41		13					
STARFLYER	Mollisols	Xerolls	Argixerolls		46						38							
STARGO	Entisols	Fluvents	Torrifluvents	18														
STEWVAL	Aridisols	Argids	Haplargids	3							8							
STINGDORN	Aridisols	Durids	Argidurids	18							20		12					
STONELL	Aridisols	Argids	Haplargids	8							12							
STUCKY	Aridisols	Argids	Haplargids	15							35							
STUMBLE	Entisols	Psamments	Torripsamments	18														
SUAK	Mollisols	Xerolls	Argixerolls		33						49							
SUMINE	Mollisols	Xerolls	Argixerolls		25						56							
SUMMERMUTE	Aridisols	Calcids	Haplocalcids	28								81						

(continued)

(continued)

Series Name	Order	Suborder	Great Group	Ochric	Mollic	Umbric	Histic	Folistic	Albic	Thickness (cm)		Calcic	Duripan	Gypsic	Natric	Petrocalcic	Petrogypsic	Salic
										Cambic	Argillic							
SUMYA why not mollic?	Entisols	Orthents	Torriorthents	18														
SUNDOWN	Entisols	Psamments	Torripsamments	18														
SUNROCK	Entisols	Orthents	Torriorthents	12														
SURPRISE	Mollisols	Xerolls	Haploxerolls		23					48								
SWINGLER	Entisols	Orthents	Torriorthents	18														
SYCOMAT	Aridisols	Calcids	Haplocalcids	18								114						
TAHOE	Inceptisols	Aquepts	Humaquepts			69												
TAHOMA	Alfisols	Xeralfs	Haploxeralfs	56							124							
TANAZZA	Aridisols	Gypsids	Calcigypsids	5								74		117				
TARNACH	Aridisols	Calcids	Haplocalcids	18								28						
TECOMAR	Aridisols	Calcids	Haplocalcids	18								30						
TECOPA	Entisols	Orthents	Torriorthents	2.5														
TEGURO	Mollisols	Xerolls	Argixerolls		25						26							
TEJABE	Entisols	Orthents	Torriorthents	18														
TENABO	Aridisols	Durids	Natridurids	18									20		15			
TENWELL	Aridisols	Durids	Argidurids	10							46		**97**					
TERCA	Mollisols	Xerolls	Argixerolls		25						35							
TERLCO	Aridisols	Argids	Natrargids	5											23			
THEON	Aridisols	Argids	Haplargids	5							23							
THERIOT	Entisols	Orthents	Torriorthents	18														
THREELAKES	Entisols	Orthents	Torriorthents	18														
THREESEE	Aridisols	Calcids	Haplocalcids	18						15		114						
THUNDERBIRD	Mollisols	Ustolls	Argiustolls		48						56							
TICKAPOO	Aridisols	Durids	Natridurids	8									13		35			
TIMPER	Aridisols	Durids	Haplodurids	18						10			21					
TIMPIE	Entisols	Orthents	Torriorthents	7.5						46								
TOANO	Entisols	Orthents	Torriorthents	18														
TOIYABE	Entisols	Psamments	Xeropsamments	18														
TOKOPER	Aridisols	Durids	Argidurids	8							28		12					
TOMEL	Aridisols	Durids	Argidurids	8							15		39					
TOMERA	Aridisols	Argids	Natrargids	23							76				49			
TONOPAH	Aridisols	Calcids	Haplocalcids	18								97						
TOOELE	Entisols	Orthents	Torriorthents	23														
TORRO	Mollisols	Xerolls	Argixerolls		33						31							
TOSSER	Aridisols	Calcids	Haplocalcids	18								127						
TOULON	Aridisols	Cambids	Haplocambids	18						26								
TRISTAN	Mollisols	Xerolls	Argixerolls		43						106							
TROCKEN	Entisols	Orthents	Torriorthents	18														
TROUGHS	Aridisols	Durids	Argidurids	5							15	18	38					
TRUNK	Aridisols	Argids	Haplargids	8							68							
TUFFO	Entisols	Orthents	Torriorthents	18														

(continued)

(continued)

Series Name	Order	Suborder	Great Group	Ochric	Mollic	Umbric	Histic	Folistic	Albic	Thickness (cm)		Calcic	Duripan	Gypsic	Natric	Petrocalcic	Petrogypsic	Salic
										Cambic	Argillic							
TULASE	Entisols	Orthents	Torriorthents	18														
TUMTUM	Aridisols	Durids	Argidurids	5							25		34					
TUNNISON	Vertisols	Xererts	Haploxererts	18						76								
TURBA	Mollisols	Ustolls	Argiustolls		18						23							
TUSEL	Mollisols	Cryolls	Argicryolls		43						64							
TUSK	Mollisols	Xerolls	Argixerolls		58						**114**							
TUSTELL	Aridisols	Argids	Haplargids	13							35							
TWEENER	Mollisols	Xerolls	Argixerolls		20						7							
TYBO	Aridisols	Durids	Haplodurids	18						18			**10**					
UBEHEBE	Mollisols	Xerolls	Argixerolls		18						33							
UMBERLAND	Inceptisols	Aquepts	Halaquepts	18														
UMIL	Aridisols	Durids	Haplodurids	18									58					
UNIVEGA	Aridisols	Durids	Haplodurids	18									74					
UNSEL	Aridisols	Argids	Haplargids	10							15							
UPATAD	Mollisols	Xerolls	Argixerolls		30						30							
UPDIKE	Aridisols	Argids	Natrargids	5											46			
UPPERLINE	Aridisols	Calcids	Haplocalcids	5								94						
UPSPRING	Entisols	Orthents	Torriorthents	15														
URIPNES	Entisols	Orthents	Torriorthents	10														
URMAFOT	Mollisols	Xerolls	Durixerolls		20							16	45					
URSINE	Aridisols	Durids	Haplodurids	18								33	132			132		
UVADA	Aridisols	Argids	Natrargids	12											33			
UWELL	Entisols	Orthents	Torriorthents	18														
VALMY	Entisols	Orthents	Torriorthents	18														
VANWYPER	Aridisols	Argids	Haplargids	20							79							
VEET	Aridisols	Cambids	Haplocambids	18						44								
VETA	Aridisols	Cambids	Haplocambids	18						31								
VIGUS	Aridisols	Argids	Haplargids	18							15							
VINDICATOR	Aridisols	Argids	Haplargids	5							13							
VININI	Aridisols	Durids	Argidurids	8							25		43					
VIRGIN PEAK	Mollisols	Ustolls	Haplustolls		18													
VITALE	Mollisols	Xerolls	Argixerolls		38						43							
WACA	Andisols	Xerands	Vitrixerands			53												
WAHGUYHE	Entisols	Orthents	Torriorthents	18														
WAKANSAPA	Mollisols	Ustolls	Argiustolls		34						31							
WALA	Entisols	Orthents	Torriorthents	18														
WALTI	Mollisols	Xerolls	Argixerolls		43						43							
WAMP	Aridisols	Calcids	Petrocalcids	18								25				32		
WARDBAY	Mollisols	Cryolls	Calcicryolls		114							68						
WARDENOT	Entisols	Orthents	Torriorthents	18														
WASSIT	Aridisols	Argids	Haplargids	15							15							

(continued)

(continued)

Series Name	Order	Suborder	Great Group	Ochric	Mollic	Umbric	Histic	Folistic	Albic	Thickness (cm)		Calcic	Duripan	Gypsic	Natric	Petrocalcic	Petrogypsic	Salic
										Cambic	Argillic							
WATAH	Inceptisols	Aquepts	Humaquepts			18	20											
WATOOPAH	Aridisols	Argids	Haplargids	8							22							
WECHECH	Aridisols	Calcids	Petrocalcids	18								15				119		
WEDERTZ	Aridisols	Argids	Haplargids	23							48							
WEENA	Entisols	Orthents	Torriorthents	17														
WEISER	Aridisols	Calcids	Haplocalcids	18								127						
WELCH	Mollisols	Aquolls	Endoaquolls		71													
WENDANE	Inceptisols	Aquepts	Halaquepts	18														
WESFIL	Entisols	Orthents	Torriorthents	10														
WESO	Aridisols	Cambids	Haplocambids	18						15								
WESTBUTTE	Mollisols	Xerolls	Haploxerolls		61					31								
WHIRLO	Aridisols	Cambids	Haplocambids	18						15								
WHITEBASIN	Aridisols	Gypsids	Haplogypsids	2.5										69				
WHOLAN	Aridisols	Cambids	Haplocambids	18						41								
WICKAHONEY	Alfisols	Xeralfs	Haploxeralfs	13							35							
WIELAND	Aridisols	Argids	Haplargids	5							127							
WIFFO	Entisols	Orthents	Torriorthents	18														
WINKLO	Aridisols	Argids	Haplargids	8							50							
WINTERMUTE	Aridisols	Calcids	Haplocalcids	18								41						
WISKAN	Aridisols	Argids	Haplargids	41							30							
WOOFUS	Mollisols	Aquolls	Endoaquolls		41													
WRANGO	Entisols	Orthents	Torriorthents	18														
WRENZA	Inceptisols	Cryepts	Humicryepts			85												
WYLO	Mollisols	Xerolls	Argixerolls		18						28							
WYVA	Aridisols	Argids	Haplargids	5							33							
XERXES	Aridisols	Argids	Haplargids	5							20							
XINE	Mollisols	Xerolls	Calcixerolls		25							59						
XMAN	Aridisols	Argids	Haplargids	5							31							
YELBRICK	Aridisols	Calcids	Haplocalcids	18								51						
YERINGTON	Entisols	Orthents	Torriorthents	20														
YERMO	Entisols	Orthents	Torriorthents	25														
YIPOR	Entisols	Orthents	Torriorthents	18														
YOBE	Inceptisols	Aquepts	Halaquepts	18														
YODY	Aridisols	Durids	Argidurids	8							33	5	81					
YOMBA	Aridisols	Cambids	Haplocambids	18						15								
YOTES	Mollisols	Xerolls	Haploxerolls		30					23								
YOUNGSTON	Entisols	Fluvents	Torrifluvents	30														
YUKO	Aridisols	Argids	Haplargids	5							15							
YURM	Aridisols	Calcids	Petrocalcids	18								33				112		
ZADVAR	Aridisols	Durids	Argidurids	15							13		43					
ZAFOD	Aridisols	Durids	Haplodurids	18								79	3					

(continued)

(continued)

Series Name	Order	Suborder	Great Group	Ochric	Mollic	Umbric	Histic	Folistic	Albic	Thickness (cm)		Calcic	Duripan	Gypsic	Natric	Petrocalcic	Petrogypsic	Salic
										Cambic	Argillic							
ZAIDY	Aridisols	Durids	Argidurids	13							23		**88**					
ZAPA	Aridisols	Durids	Haplodurids	18								34	43					
ZAQUA	Aridisols	Argids	Haplargids	8							35							
ZARARK	Mollisols	Cryolls	Calcicryolls		70							60						
ZEHEME	Aridisols	Calcids	Haplocalcids	8								25						
ZERK	Aridisols	Calcids	Haplocalcids	18								134						
ZEVADEZ	Aridisols	Argids	Haplargids	13							28							
ZIBATE	Aridisols	Argids	Haplargids	15							33							
ZIMBOB	Entisols	Orthents	Torriorthents	18														
ZINEB	Aridisols	Cambids	Haplocambids	18						18								
ZODA	Aridisols	Durids	Argidurids	13							48		**91**					
ZOESTA	Aridisols	Argids	Paleargids	18							61							
ZORRAVISTA	Entisols	Psamments	Torripsamments	18														
ZYMANS	Mollisols	Xerolls	Argixerolls		30						132							
				15	38	44	23	18	7	29	39	60	44	79	33	89	88	93
				6	24	23	6		5	16	27	42	33	42	18	44	84	84
				511	169	11	3	1	3	79	279	101	115	12	33	14	2	2
				73.2	24.2	1.6	0.4	0.1	0.4	11.3	40.0	14.5	16.5	0.7	4.7	2.0	0.1	0.3
				698	698	698	698	1811	698	698	698	698	698	1811	698	698	1811	698

Appendix C
Soil-Forming Factors for Soil Series with an Area Greater than 78 km^2

P. W. Blackburn et al., *The Soils of Nevada*, World Soils Book Series,
https://doi.org/10.1007/978-3-030-53157-7

Series Name	Order	Suborder	Great Group	MAAT (°C)	MAP (mm)	Vegetation	Max. slope (%)	Med. elev. (m)	Parent material	Landform1	Landform2
ABGESE	Aridisols	Argids	Haplargids			sagebrush			alluvium, colluvium	alluvial terraces	fan remnants
ABOTEN	Aridisols	Durids	Natridurids	11	150	greasewood, bud sage, squirreltail	30	1600	alluvium	piedmont remnants	
ACANA	Aridisols	Durids	Argidurids			sagebrush, rabbitbrush			alluvium, lacustrine	fan remnants	
ACKETT	Aridisols	Durids	Argidurids			sagebrush, wheatgrass			alluvium, colluvium	alluvial terraces	fan piedmont remnants, partial ballenas
ACOMA	Aridisols	Argids	Paleargids	8	250	sagebrush, blue grama, IN ricegrass	15	2000	alluvium	fan remnants	
ACRELANE	Mollisols	Xerolls	Argixerolls	9	250	sagebrush	50	1565	residuum, colluvium	mtns granite	
ACTI	Mollisols	Ustolls	Argiustolls			Gamble oak, UT serviceberry, Moj ceanothus, manzanita			residuum, colluvium	mtns tuff	
ADELAIDE	Aridisols	Durids	Haplodurids			sagebrush, bluegrass, squirreltail			loess/alluvium	fan piedmont remnants	
ADOBE	Mollisols	Cryolls	Calcicryolls			sagebrush, wheatgrass			residuum, colluvium	mtns ls	
ADVOKAY	Aridisols	Argids	Haplargids			shadscale, In ricegrass, greasewood, bud sage, ephedra			residuum, colluvium	mtns tuff	
AGASSIZ	Mollisols	Xerolls	Haploxerolls			sagebrush, mtn mahogany			colluvium/ls	mtns ls	
AKELA	Mollisols	Cryolls	Calcicryolls	65	203	creosotebush, tarbush, black grama	70	1220	residuum, eolian	mtns ls	
AKERUE	Aridisols	Durids	Argidurids	7	250	black sage, Sandb bluegrass, Thurb needlegrass	30	2000	residuum	mtns rhyolite	
AKLER	Aridisols	Argids	Haplargids			sagebrush, rabbitbrush, needlegrass, squirreltail			residuum	mtns tuff	
ALHAMBRA	Entisols	Fluvents	Torrifluvents			sagebrush			alluvium	floodplains	alluvial fans
ALKO	Aridisols	Durids	Haplodurids			creosotebush, shadscale			alluvium	fan piedmonts	
ALLEY	Aridisols	Argids	Haplargids	9	200	sagebrush	75	1700	loess/alluvium, colluvium	fan piedmont remnants	
ALLKER	Aridisols	Argids	Haplargids			sagebrush			alluvium	fan remnants	
ALLOR	Aridisols	Argids	Haplargids			sagebrush			alluvium	fan remnants	
ALYAN	Mollisols	Xerolls	Argixerolls			sagebrush			residuum, colluvium	mtns volcanic	
AMELAR	Mollisols	Xerolls	Argixerolls			sagebrush, bitterbrush, wheatgrass			colluvium, alluvium	fan remnants	ballenas
AMTOFT	Aridisols	Calcids	Haplocalcids	8	250	wheatgrass	80	2000	residuum, colluvium	mts ls	
ANAWALT	Aridisols	Argids	Haplargids	6	250	wheatgrass, ID fescue	50	1650	colluvium, residuum	mtns volcanic	
ANNAW	Aridisols	Cambids	Haplocambids			greasewood, galleta			alluvium	alluvial fans	fan remnants
ANSPING	Mollisols	Xerolls	Calcixerolls			pinyon-juniper			alluvium, colluvium	mtn sideslopes	erosional fan piedmonts

(continued)

(continued)

Series Name	Order	Suborder	Great Group	MAAT (°C)	MAP (mm)	Vegetation	Max. slope (%)	Med. elev. (m)	Parent material	Landform1	Landform2
APPIAN	Aridisols	Argids	Natrargids	12	125	greasewood, shadscale	2	1500	alluvium/lacustrine	lake plains	alluvial flats, basin-floor remnants
ARADA	Aridisols	Calcids	Haplocalcids			IN ricegrass, galleta, w bursage, creosotebush			eolian/alluvium	fan piedmonts	sand sheets, terraces, alluvial fans
ARCIA	Mollisols	Xerolls	Argixerolls			sagebrush, bitterbrush, snowberry, wheatgrass			loess, ash/colluvium, residuum	mtns volcanic	
ARCLAY	Mollisols	Xerolls	Argixerolls			sagebrush, bluegrass, squirreltail			residuum, colluvium	mtns granite	
ARDIVEY	Aridisols	Argids	Haplargids	12	150	greasewood, shadscale	15	1600	alluvium	fan remnants	
ARGALT	Aridisols	Durids	Argidurids			bl sage, shadscale, ephedra			residuum, colluvium	plateaus basalt	
ARGENTA	Inceptisols	Aquepts	Halaquepts			wildrye, sacaton, saltgrass, greasewood			alluvium, ash	inset fans	basin-floor remnants, lake plains
ARIZO	Entisols	Orthents	Torriorthents	17	180	creosotebush	15	800	alluvium	alluvial fans	inset fans, fan skirts, floodplains
ARMESPAN	Aridisols	Calcids	Haplocalcids	11	250	sagebrush	50	1850	alluvium	fan remnants	
ARMOINE	Aridisols	Argids	Haplargids			sagebrush, ephedra, rabbitbrush			residuum, colluvium	pediments granite	
ARROLIME	Aridisols	Gypsids	Calcigypsids			creosotebush, bursage			alluvium, gypsif	alluvial fans	
ASHTRE	Mollisols	Xerolls	Argixerolls			sagebrush/ID fescue, needlegrass, wheatgrass			residuum, colluvium	plateaus, tuff	
ATLOW	Aridisols	Argids	Haplargids	8	250	sagebrush	75	1900	residuum	mtns chert	
ATRYPA	Mollisols	Xerolls	Haploxerolls			pinyon-juniper			residuum	mtns ls	
AUTOMAL	Aridisols	Calcids	Haplocalcids			sagebrush			alluvium	fan remnants	beach plains
AYMATE	Aridisols	Calcids	Petrocalcids	14	200	creosotebush, big galleta, Anders wolfberry, spiny hopsage,	8	915	alluvium ls	fan piedmont remnants	
AZTEC	Aridisols	Gypsids	Calcigypsids			creosotebush, bursage			alluvium w/ gypsum	fan remnant	
BADHAP	Mollisols	Cryolls	Haplocryolls	4	460	big sage, snowberry, bluebunch wheatgrass	75	2600	colluvium, residuum w/ loess	mtns qtzite	
BAKERPEAK	Inceptisols	Cryepts	Calcicryepts	5	965	Jeff pine, red fir, w fir, ww pine	70	2550	colluvium	mtns ls	
BANGO	Aridisols	Argids	Natrargids			greasewood, shadscale, bud sage			alluvium/lacustrine	basin-floor remnants	
BARD	Aridisols	Calcids	Petrocalcids	17	125	creosote bush	15	700	alluvium ls	valley fill terraces	alluvial fans, fan remnants
BARTOME	Aridisols	Durids	Argidurids			sagebrush			loess, ash/alluviuum	fan piedmont remnants	
BASKET	Aridisols	Argids	Haplargids	10	280	WY big sage, spiny hopsage, Indian ricegrass	30	1950	alluvium	fan remnants	
BATAN	Entisols	Orthents	Torriorthents	9	180	greasewood, shadshale	4	1550	alluvium	alluvial terraces	alluvial flat remnants

(continued)

Series Name	Order	Suborder	Great Group	MAAT (°C)	MAP (mm)	Vegetation	Max. slope (%)	Med. elev. (m)	Parent material	Landform1	Landform2
BEELEM	Entisols	Orthents	Torriorthents			pinyon-juniper			residuum, colluvium	mtns tuff	
BELATE	Mollisols	Xerolls	Argixerolls			low sage, ID fescue, pine bluegrass, prickly gilia			colluvium, residuum	mtns tuff	
BELLEHELEN	Mollisols	Xerolls	Argixerolls	8	300	sagebrush	75	2250	residuum, colluvium	mtns volcanic	
BELLENMINE	Mollisols	Xerolls	Argixerolls	5	450	pinyon pine-juniper	50	2250	colluvium, residuum	mtns qtzite	
BELTED	Aridisols	Durids	Argidurids			sagebrush			residuum, colluvium	mtns volcanic	
BENIN	Entisols	Orthents	Torriorthents			shadscale			alluvium/lacustrine	lake terraces	basin-floor remnants
BEOSKA	Aridisols	Argids	Natrargids	9	180	sagebrush	15	1450	loess, ash/alluvium	fan remnants	
BERZATIC	Entisols	Orthents	Torriorthents			shadscale, wolfberry galleta, ind ricegrass			colluvium, residuum	mtns ls	
BESHERM	Aridisols	Calcids	Haplocalcids			shadscale, spinescale saltbush, forbs			alluvium	lake plains	
BIDART	Inceptisols	Aquepts	Cryaquepts	3	1295	sedge	2	2450	alluvium	floodplains	
BIDDLEMAN	Aridisols	Argids	Natrargids			greasewood, shadscale			alluvium/lacustrine	beach terraces	
BIEBER	Mollisols	Xerolls	Durixerolls	9	380	sagebrush, grasses	15	1400	alluvium	alluvial terraces	fan remnants
BIGA	Aridisols	Argids	Natrargids			shadscale, greasewood			alluvium	fan remnants	
BIKEN	Aridisols	Calcids	Haplocalcids			sagebrush, rabbitbrush, IN ricegrass			alluvium/residuum	pediments	
BILBO	Aridisols	Argids	Haplargids			sagebrush			alluvium	fan piedmont remnants	fan remnants
BIOYA	Aridisols	Durids	Haplodurids	9	230	IN ricegrass, squirreltail, sagebrush	15	1530	loess/alluvium	fan piedmont remnants	plateaus
BIRCHCREEK	Mollisols	Xerolls	Argixerolls			sagebrush/wheatgrass, ID fescue			alluvium, colluvium, residuum	mtns mixed	
BIRDSPRING	Entisols	Orthents	Torriorthents			shadscale, ephedra			residuum, colluvium	mtns ls	
BITTER SPRING	Aridisols	Argids	Calciargids	67	100	creosotebush, white bursage	15	792	alluvium	fan remnants	
BLACKHAWK	Aridisols	Durids	Haplodurids			shadscale, sagebrush			loess/alluvium	fan piedmont remnants	fan remnants, fan aprons,
BLACKTOP	Entisols	Orthents	Torriorthents	12	175	shadscale	75	1700	residuum	mtns igneous	
BLAPPERT	Aridisols	Argids	Haplargids			shadscale, spiny menodora, galleta, rabbitbrush, bottlebrush			residuum, colluvium	sideslopes granite	
BLIMO	Entisols	Orthents	Torriorthents			sagebrush			alluvium	fan skirts, beach plains	
BLISS	Aridisols	Durids	Haplodurids			sagebrush			alluvium	fan remnants	fan piedmont remnants
BLUEGYP	Aridisols	Gypsids	Haplogypsids			shadscale, sandpaper bush, dalea, seepweed			residuum	dissected pediments, gypsif	
BLUEPOINT	Entisols	Psamments	Torripsamments	19	125	creosote bush	50	850	eolian	dunes	sand sheets
BLUEWING	Entisols	Orthents	Torriorthents	12	150	sagebrush	30	1450	alluvium	inset fans, ballenas	
BOBS	Mollisols	Xerolls	Palexerolls	7	300	wheatgrass, ricegrass	30	2000	alluvium, loess, ash	fan remnants	

(continued)

Series Name	Order	Suborder	Great Group	MAAT (°C)	MAP (mm)	Vegetation	Max. slope (%)	Med. elev. (m)	Parent material	Landform1	Landform2
BOJO	Aridisols	Argids	Haplargids			sagebrush, shadscale, rabbitbrush, needlegrass			residuum, colluvium	mtns mixed	
BOMBADIL	Aridisols	Argids	Haplargids			sagebrush			colluvium, residuum	mtns volcanic	
BOOFUSS	Inceptisols	Aquepts	Halaquepts			greasewood, inland saltgrass			alluvium	floodplains	alluvial flats, lake plains
BOOMSTICK	Aridisols	Argids	Haplargids			sagebrush, bluegrass, hopsage			residuum, colluvium	mtns metam	
BOREALIS	Alfisols	Xeralfs	Durixeralfs			pinyon/bitterbrush, ephedra			ash/residuum	plateaus basalt	
BORVANT	Mollisols	Xerolls	Palexerolls	9	350	sagebrush	80	1900	alluvium	fan remnants	
BOTON	Entisols	Orthents	Torriorthents	12	180	greasewood	2	1300	loess, alluvium, ash/lacustrine	lake plains	
BOULDER LAKE	Vertisols	Aquerts	Epiaquerts			sagebrush			lacustrine	lake plains	plateau depressions
BOXSPRING	Entisols	Orthents	Torriorthents	13	230	blackbrush	75	1600	residuum, colluvium	mtns ls	
BRACKEN	Aridisols	Gypsids	Haplogypsids			dalea, ephedra			residuum, colluvium	hills from gyp sed rocks	
BREGAR	Aridisols	Argids	Haplargids	5	330	sagebrush	75	2000	residuum, colluvium	mtns volcanic	
BREKO	Aridisols	Argids	Haplargids			sagebrush, shadscale, hopsage			alluvium	fan remnants	fan piedmonts, alluvial fans
BRICONE	Entisols	Orthents	Cryorthents			bristlecone, limber pines/gooseebrry-juniper			residuum, colluvium	mtns ls	
BRIER	Mollisols	Xerolls	Argixerolls			sagebrush			residuum, colluvium	mtns volcanic	
BROLAND	Aridisols	Durids	Argidurids			sagebrush, IN ricegrass			alluvium	fan remnants	
BROYLES	Aridisols	Cambids	Haplocambids	9	180	shadscale, sagebrush	8	1600	loess, ash/alluvium	fan remnants	fan skirts, inset fans, alluvial flats
BUBUS	Entisols	Orthents	Torriorthents	9	180	greasewood, shadshale	8	1650	alluvium	alluvial flat remnants	inset fans, fan skirts
BUCAN	Aridisols	Argids	Haplargids			sagebrush			loess, ash/residuum	mtns volcanic	
BUCKAROO	Aridisols	Argids	Natrargids			greasewood, shadscale, bud sage			alluvium	fan piedmont remnants	lake terrace remnants
BUCKLAKE	Mollisols	Xerolls	Argixerolls	8	280	sagebrush	50	1600	colluvium, residuum	mtns basalt	
BUCKSPRING	Mollisols	Ustolls	Argiustolls			pinyon-juniper			residuum, colluvium	backslopes ls	
BUDIHOL	Entisols	Orthents	Torriorthents			sagebrush, bluegrass, hopsage, ephedra			residuum, colluvium	mtns granitic	
BUFFARAN	Aridisols	Durids	Argidurids			sagebrush			alluvium	fan remnants	ballenas
BULAKE	Alfisols	Xeralfs	Haploxeralfs	7	350	sagebrush, ID fescue, wheatgrass	25	1650	loess, ash/residuum	plateaus	
BULLUMP	Mollisols	Xerolls	Argixerolls	6	400	sagebrush	75	2050	colluvium, loess	mtns mixed	
BURRITA	Aridisols	Argids	Haplargids	9	230	sagebrush	75	1750	residuum, colluvium	mtns mixed	
CAGWIN	Entisols	Psamments	Xeropsamments	5	1015	white fir-jeffrey pine-red fir	75	2025	residuum	sideslopes-granite	
CALLAT	Inceptisols	Cryepts	Humicryepts	5	1295	lodgepole, pine, w white pine	50	2400	colluvium/till	mtn slopes	

(continued)

Series Name	Order	Suborder	Great Group	MAAT (°C)	MAP (mm)	Vegetation	Max. slope (%)	Med. elev. (m)	Parent material	Landform1	Landform2
CALLVILLE	Aridisols	Gypsids	Haplogypsids			creosotebush, desert holly, bursage			residuum	pediments w/ gyp ss, sist	
CANDELARIA	Aridisols	Calcids	Haplocalcids	12	150	shadscale	30	1750	alluvium	ballenas	fan piedmonts
CANOTO	Entisols	Orthents	Torriorthents			creosotebush, w bursage			alluvium	alluvial fans	inset fans,fan skirts, fan remnant
CANYOUNG	Mollisols	Cryolls	Calcicryolls	4	600	mtn mahogany, mtn big sagebrush	75	2700	colluvium	mtns ls	
CARRIZO	Entisols	Orthents	Torriorthents	19	100	creosotebush	15	400	alluvium	floodplains	fan piedmons, bolson floors
CARSON	Vertisols	Xererts	Haploxererts			greasewood, saltgrass			alluvium	floodplains	
CARSTUMP	Mollisols	Xerolls	Argixerolls			sagebrush, rabbitbrush, cheatgrass			residuum, colluvium	mtns volcanic	
CASAGA	Aridisols	Argids	Natrargids			creosotebush, shadscale, desertholly			alluvium	erosional fan remnants	
CASSIRO	Mollisols	Xerolls	Argixerolls			sagebrush, horsebrush			alluvium, colluvium	fan remnants	inset fans, hills
CATH	Aridisols	Argids	Calciargids	11	250	sagebrush	15	1700	alluvium	fan remnants	
CAVANAUGH	Mollisols	Xerolls	Argixerolls			low sage, ID fescue, bluebunch wheatgrass, Doug rabbibrush			alluvium, colluvium	sideslopes tuff	
CAVE	Aridisols	Calcids	Petrocalcids	63	203	creosotebush, blackbrush	35	1280	alluvium	fan remnants	
CAVEHILL	Mollisols	Xerolls	Calcixerolls	7	360	pinyon-sagebrush	75	2300	loess/residuum, colluvium	mtns ls	
CEDARAN	Mollisols	Ustolls	Argiustolls			sagebrush, serviceberry, bitterbrush			residuum, colluvium	mtns tuff	
CEEJAY	Aridisols	Argids	Haplargids			sagebrush			residuum, colluvium	mtns volcanic	
CELETON	Entisols	Orthents	Torriorthents			In ricegrass, greaswood, needlegrass, ephedra			colluvium/residuum	pediments diatomite	
CHAD	Mollisols	Xerolls	Argixerolls			mtn big sage, rabbitbrush, bluebunch wheatgrass, pine bluegrass			residuum, some loess, ash	sideslopes	
CHAINLINK	Mollisols	Xerolls	Durixerolls			bl sage, bluebunch wheatgrass, bottlebrush			alluvium ls	fan remanants	ballenas
CHALCO	Aridisols	Argids	Haplargids			sagebrush, rabbitbrush			colluvium/residuum	mtns volcanic	
CHARPEAK	Inceptisols	Cryepts	Calcicryepts	4	500	muttongrass, Kern milkvetch	50	3300	colluvium, residuum	backslopes ls	
CHECKETT	Aridisols	Argids	Haplargids	9	250	sagebrush	50	1840	alluvium, colluvium, residuum	mtns igneous	
CHEME	Aridisols	Durids	Haplodurids			creosotebush, w bursage, ratany, brittlebush			alluvium/bedrock	fan remnants	
CHEN	Mollisols	Xerolls	Argixerolls	6	330	sagebrush	75	2300	loess/residuum, colluvium	mtns volcanic	
CHERRY SPRING	Aridisols	Durids	Argidurids			sagebrush			loess, ash/alluvium	fan piedmont remnants	
CHIARA	Aridisols	Durids	Haplodurids	9	230	sagebrush	30	1600	loess, ash/alluvium	fan remnants	
CHILL	Aridisols	Argids	Haplargids			sagebrush, horsebrush, ephedra, needlegrass, IN ricegrass			residuum	mtns granite	
CHILPER	Aridisols	Argids	Natrargids			shadscale, bud sage, squirreltail			alluvium	fan remnants	

(continued)

(continued)

Series Name	Order	Suborder	Great Group	MAAT (°C)	MAP (mm)	Vegetation	Max. slope (%)	Med. elev. (m)	Parent material	Landform1	Landform2
CHUBARD	Aridisols	Argids	Haplargids	9	250	needlegrass, rice grass	75	1950	residuum, colluvium, ash	mtns tuff	
CHUCKLES	Aridisols	Cambids	Haplocambids			greasewood, shadscale, bud sage			alluvium/lacustrine	lake plains	lake terraces
CHUCKMILL	Aridisols	Durids	Argidurids			IN ricegrass, needleandthread, galleta			alluvium	fan remnants	
CHUCKRIDGE	Aridisols	Durids	Argidurids			sagebrush, ephedra			alluvium	fan piedmonts	fan remnants
CHUFFA	Aridisols	Cambids	Haplocambids			sagebrush			alluvium/lacustrine	fan skirts	basin floors, lake plains
CHUG	Mollisols	Xerolls	Haploxerolls			sagebrush, wildrye			alluvium	alluvial terraces	
CHUSKA	Aridisols	Durids	Argidurids	9	250	WY big sage, basin big sage, bluebunch wheatgrass	25	1500	alluvium	fan remnants	
CIRAC	Entisols	Fluvents	Torrifluvents	12	100	greasewood, shadscale	4	1650	alluvium	alluvial flats	lake plains, lagoons, fan skirts, inset fans
CLANALPINE	Mollisols	Xerolls	Argixerolls			pinyon/ID fescue, sagebrush			residuum, colluvium	mtns volcanic	
CLEAVAGE	Mollisols	Xerolls	Argixerolls	7	350	sagebrush, ID fescue	75	2150	residuum, colluvium	mtns rhyolite	
CLEAVER	Aridisols	Durids	Argidurids			greasewood, shadscale			alluvium	fan remnants	
CLEMENTINE	Mollisols	Aquolls	Endoaquolls			hairgrass, sedges, willows			alluvium, loess	floodplains	alluvial terraces, inset fans
CLIFFDOWN	Entisols	Orthents	Torriorthents	12	175	sagebrush	15	1600	alluvium	fan remnant	fan skirts, inset fans
CLOWFIN	Entisols	Orthents	Torriorthents			winterfat, In ricegrass			alluvium mixed	inset fan piedmonts	
CLURDE	Aridisols	Cambids	Haplocambids			sagebrush, horsebrush			loess, ash/alluvium	fan remnants	inset fans, lake terraces
COBRE	Aridisols	Cambids	Haplocambids			sagebrush, rabbitbrush, IN ricegrass			residuum, colluvium	mtns tuff	
COILS	Aridisols	Durids	Argidurids			sagebrush, rabbitbrush, In ricegrass			alluvium	fan piedmonts	
COLBAR	Aridisols	Argids	Haplargids			sagebrush			residuum, colluvium	mtns volcanic	
COLOROCK	Aridisols	Calcids	Petrocalcids			creosotebush, w bursage, cacti			alluvium ls	alluvial fans	
COLVAL	Aridisols	Argids	Calciargids			saltbush, shadscale, kochia, winterfat			alluvium	basin floors	
COMMSKI	Aridisols	Calcids	Haplocalcids			desertholly, bursage			alluvium	ballenas	fan remnants, inset fans
CONNEL	Aridisols	Cambids	Haplocambids			sagebrush			loess, ash/alluvium	beach terraces	barrier bars, lagoons
COPPEREID	Entisols	Orthents	Torriorthents			sagebrush, hopsage, bluegrass, squirreltail			residuum	mtns shale	
CORBILT	Aridisols	Calcids	Haplocalcids			creosotebush, hopsage, IN ricegrass			alluvium	alluvial fans	fan skirts, fan piedmonts
CORTEZ	Aridisols	Durids	Natridurids			sagebrush, rabbitbrush			loess, ash/alluvium	fan remnants	

(continued)

(continued)

Series Name	Order	Suborder	Great Group	MAAT (°C)	MAP (mm)	Vegetation	Max. slope (%)	Med. elev. (m)	Parent material	Landform1	Landform2
COSER	Mollisols	Xerolls	Palexerolls			sagebrush, ID fescure, bluegrass			colluvium/residuum	mtns tuff	
COTANT	Mollisols	Xerolls	Argixerolls			sagebrush, rabbitbrush			residuum, colluvium	mtns volcanic	
COWGIL	Aridisols	Argids	Haplargids			sagebrush, rabbitbrush, IN ricegrass			alluvium	fan remnants	
CREEMON	Aridisols	Cambids	Haplocambids			shadscale, sagebrush			alluvium, ash, loess	fan skirts	alluvial terraces, inset fans
CROOKED CREEK	Mollisols	Aquolls	Endoaquolls	7	360	willows, sedges	4	1750	alluvium	alluvial plains	alluvial terraces, inset fans, valley floors
CROPPER	Mollisols	Xerolls	Argixerolls	6	350	pinyon-mtn mahogany	75	2200	residuum, colluvium	mtns. Volcanic	
CROSGRAIN	Aridisols	Durids	Haplodurids			creosotebush			alluvium	ballenas	fan remnants
DACKER	Aridisols	Durids	Argidurids	9	230	sagebrush	15	1650	alluvium, loess, ash	fan piedmont remnants	plateaus, partial ballenas
DAGGET	Entisols	Orthents	Cryorthents	5	965	Jeff pine, red fir, w fir, ww pine	70	2550	colluvium/residuum	sideslopes	
DAVEY	Aridisols	Cambids	Haplocambids	9	230	sagebrush	45	1500	alluvium	sand sheets	alluvial fans, basin-floor remnants, fan skirts
DEADYON	Aridisols	Argids	Haplargids			sagebrush, hopsage, IN ricegrass			alluvium	fan skirts	inset fans, fan piedmonts
DECAN	Mollisols	Xerolls	Durixerolls			sagebrush			alluvium	fan remnants	
DEFLER	Entisols	Orthents	Torriorthents			winterfat, In ricegrass, bottlebrush squirreltail			alluvium, loess, ash	inset fans	
DELAMAR	Aridisols	Durids	Argidurids			hopsage, horsebrush, rabbitbrush			alluvium	fan remnants	
DELVADA	Mollisols	Aquolls	Endoaquolls			wildrye, rushes			alluvium	floodplains	alluvial terraces
DENMARK	Aridisols	Calcids	Petrocalcids	9	250	WY big sage, yellowbrush, ind ricegrass	20	1750	alluv sedim; residuum ign	fan remnants	hillslopes
DENPARK	Mollisols	Cryolls	Argicryolls	5	550	mtn big sage, ant bitterbrush, UT serviceberry	50	2500	residuum, colluvium	mtns tuff	
DEPPY	Aridisols	Durids	Argidurids			shadscale, bud sagebrush			alluvium	fan remnants	lake terraces
DESATOYA	Aridisols	Argids	Haplargids			sagebrush, shadscale, squirreltail			alluvium	fan remnants	
DEVADA	Mollisols	Xerolls	Argixerolls	8	300	sagebrush	50	1800	loess, ash/residuum	mtns rhyolite	
DEVEN	Mollisols	Xerolls	Argixerolls	8	300	juniper-sagebrush	50	1500	residuum, colluvium	mtsn volcanic	
DEVILDOG	Aridisols	Cambids	Haplocambids			sagebrush, hopsage, IN ricegrass			ash, alluvium	inset fans	fan skirts, drainageways
DEVILSGAIT	Mollisols	Aquolls	Endoaquolls	8	250	wildrye, bluegrass, saltgrass	4	1750	alluvium	floodplains	lake plains
DEWAR	Aridisols	Durids	Argidurids	8	230	sagebrush	50	1900	loess, ash, alluvium	fan remnants	
DIA	Mollisols	Xerolls	Haploxerolls			sagebrush, wildrye			alluvium	alluvial plains	alluvial terraces
DIANEV	Inceptisols	Aquepts	Halaquepts			greasewood, sacaton, saltgrass			alluvium	floodplains	lake plains

(continued)

(continued)

Series Name	Order	Suborder	Great Group	MAAT (°C)	MAP (mm)	Vegetation	Max. slope (%)	Med. elev. (m)	Parent material	Landform1	Landform2
DITHOD	Mollisols	Xerolls	Haploxerolls			sagebrush, grasses, greasewood, saltgrass			alluvium	floodplains	alluvial terraces
DOBEL	Aridisols	Durids	Argidurids			greasewood, shadscale, ephedra			alluvium	fan remnants	
DONNA	Mollisols	Xerolls	Durixerolls	7	280	Sandberg bluegrass, needlegrass	15	1950	loess, ash/alluvium	fan piedmont remnants	
DOORKISS	Aridisols	Argids	Haplargids			sagebrush, bluegrass, squirreltail			residuum, colluvium	mtns volcanic	
DORPER	Aridisols	Argids	Natrargids	12	125	shadscale	30	1500	loess, ash/alluvium	fan remnants	
DOSIE	Mollisols	Xerolls	Argixerolls			sagebrush			colluvium, residuum	plateaus, volcanic	
DOUHIDE	Mollisols	Xerolls	Argixerolls			pinyon-juniper			residuum	mtns volcanic	
DOWNEYVILLE	Aridisols	Argids	Haplargids	11	150	sagebrush	75	1700	colluvium	mts volcanic	
DRYGYP	Aridisols	Gypsids	Petrogypsids			dalea, sandpaper, saltbush, ephedra, galleta			alluvium	fan remnants	
DUCO	Mollisols	Xerolls	Argixerolls			pinyon-juniper			colluvium, residuum	mtns volcanic	
DUFFER	Aridisols	Calcids	Haplocalcids	8	200	rabbitbrush, greasewood	2	1700	alluvium, loess, lacustrine	alluvial plains	lake plains
DUGCHIP	Aridisols	Durids	Natridurids			WY big sage, spiny hopsage, Thurber needlegrass, In ricegrass			loess, ash/alluvium	fan piedmont remnants	
DUN GLEN	Aridisols	Cambids	Haplocambids			shadscale, sagebrush			loess, ash/alluvium	fan skirts	inset fans, basin-floor remnants
EAGLEPASS	Entisols	Orthents	Torriorthents	9	250	mtn mahogany, black sagebrush	75	2150	residuum, colluvium	mtns ls	colluvium, residuum, mtns ls
EAGLEROCK	Mollisols	Xerolls	Argixerolls			sagebrush, ephedra, balsamroot			residuum, colluvium	mtns granite	
EAST FORK	Mollisols	Xerolls	Haploxerolls			sagebrush, wildrye, inl saltgrass			alluvium	floodplains	alluvial terraces
EASTGATE	Aridisols	Cambids	Haplocambids			greasewood, shadscale			alluvium	alluvial ans	fan skirts, fan piedmonts
EASTMORE	Aridisols	Durids	Haplodurids			sagebrush black, needleandthread, IN ricegrass			alluvium, loess	fan remnants	
EASTWELL	Aridisols	Durids	Haplodurids			sagebrush, squirreltail, IN ricegrass			alluvium	fan remnants	
EBODA	Mollisols	Xerolls	Argixerolls			sagebrush, rabbitbrush, wheatgrass, ID fescue, Sandberg bluegrass			loess/colluvium, residuum	plateaus tuff	
EGANROC	Mollisols	Cryolls	Haplocryolls			w fir, bristlecone pine			residuum, colluvium	mtns ls	
ENKO	Aridisols	Cambids	Haplocambids	9	230	sagebrush	30	1600	loess, ash/alluvium	fan remnant	inset fans, fan skirts
EQUIS	Inceptisols	Aquepts	Halaquepts			greasewood, rabbitbrush, saltgrass, sacaton			lacustrine, alluvium	alluvial flats	lake plains
ESCALANTE	Aridisols	Calcids	Haplocalcids	9	250	sagebrush	15	1400	alluvium	alluvial flats	lake plains, inset fans, fan skirts, fan remnants
EWELAC	Aridisols	Cambids	Haplocambids			bl greasewood, iodinebush, shadscale, inl saltgrass			lacustrine	lake plains	basin floors, alluvial flats
FANG	Entisols	Orthents	Torriorthents			saltbush, IN ricegrass, galleta			alluvium	alluvial flats	fan skirts, inset fans
FERROGOLD	Aridisols	Calcids	Petrocalcids			blackbrush, creosotebush, buckwheat, winterfat			alluvium ls	ballenas	fan remnants

(continued)

(continued)

Series Name	Order	Suborder	Great Group	MAAT (°C)	MAP (mm)	Vegetation	Max. slope (%)	Med. elev. (m)	Parent material	Landform1	Landform2
FIVEMILE	Entisols	Fluvents	Torrifluvents	7	175	Gardn saltbush, greasewood, bluestem wheatgrass, squirreltail	6	nd	alluvium mixed	inset fan piedmonts	
FLATTOP	Aridisols	Argids	Natrargids			creosotebush, w bursage, cholla, yucca			alluvium	alluvial fans	
FLETCHERPEAK	Mollisols	Ustolls	Argiustolls			pinyon-juniper, mtnmahog			colluvium, residuum	backslopes ls	
FLUE	Aridisols	Durids	Natridurids	9	230	sagebrush, hopsage, squirreltail	30	1500	loess, ash/alluvium	fan remnants	plateaus
FORTANK	Aridisols	Argids	Haplargids			Wyo bi sage, spiny hop, rabbitbrush, squirreltail			residuum andesite	sideslopes	
FUGAWEE	Alfisols	Xeralfs	Haploxeralfs	6	1000	red fir, w fir, jeffrey pine, lodgepole pine	50	2100	residuum	mtns volcanic	
FULSTONE	Aridisols	Durids	Argidurids	11	180	sagebrush	30	1700	alluvium	fan remnants	
GABBVALLY	Aridisols	Argids	Haplargids	11	250	sagebrush	75	1900	colluvium, ash	mtns volcanic	
GANCE	Aridisols	Argids	Haplargids			sagebrush			alluvium	fan piedmont remnants	
GARHILL	Aridisols	Durids	Haplodurids			galleta, menodora, shadscale, IN ricegrass			eolian, ash/residuum, colluvium	pediments	
GEER	Entisols	Orthents	Torriorthents	12	175	ricegrass, galleta, winterfat, saltbush	4	1400	alluvium	alluvial flats	inset fans, fan skirts, alluvial fans
GENAW	Aridisols	Argids	Haplargids			sagebrush, squirreltail, Sandb bluegrass, IN ricegrass			residuum	plateaus tuff	
GENEGRAF	Aridisols	Argids	Natrargids			greasewood, shadscale			alluvium	fan remnants	
GETA	Aridisols	Calcids	Haplocalcids	14	200	big galleta, Anders wolfberry, Joshua tree	4	1100	alluvium	alluvial flats	inset fans, stream terraces, fan skirts
GILA	Entisols	Fluvents	Torrifluvents	18	175	creosotebush, mesquite	5	1065	alluvium	alluvial fans	
GLEAN	Mollisols	Xerolls	Haploxerolls			sagebrush			colluvium	mtns mixed	
GLENCARB	Entisols	Fluvents	Torrifluvents			creosotebush, saltbush, sacaton, inland saltgrass			alluvium	floodplains	alluvial terraces, basin floors
GLOTRAIN	Aridisols	Argids	Haplargids			shadscale, winterfat, saltbush, In ricegrass			alluvium	fan remnants	fan skirts
GOCHEA	Mollisols	Xerolls	Argixerolls			sagebrush, IN ricegrass			alluvium	fan piedmonts	alluvial terraces
GOLCONDA	Aridisols	Durids	Natridurids	9	180	shadscale, bud sage	30	1550	loess, ash/alluvium	fan piedmonts	
GOLDROAD	Entisols	Orthents	Torriorthents	23	150	creosotebush	75	610	alluvium, residuum, colluvium/bedrock	mtns granite	
GOLDRUN	Entisols	Psamments	Torripsamments	9	200	sagebrush	30	1420	eolian, ash/lacustrine	sand dunes	sand sheets
GOLDYKE	Entisols	Orthents	Torriorthents			greasewood, galleta, menodora, bud sage			residuum, colluvium	pediments volcan	
GOLLAHER	Entisols	Orthents	Xerorthents			sagebrush			residuum, colluvium	mtns ls	
GOODSPRINGS	Aridisols	Calcids	Petrocalcids	18	125	creosotebush, w bur, globemallow	15	760	alluvium ls	fan remnants	alluvial flats
GOODWATER	Aridisols	Calcids	Petrocalcids	12	280	UT juniper, blackbrush, Stansb cliffrose, snakeweed	50	1850	alluvium ls	fan remnants	ballenas
GOOSEL	Aridisols	Durids	Argidurids			sagebrush, wheatgrass, needlegrass			residuum, loess, ash	plateaus volcanic	

(continued)

(continued)

Series Name	Order	Suborder	Great Group	MAAT (°C)	MAP (mm)	Vegetation	Max. slope (%)	Med. elev. (m)	Parent material	Landform1	Landform2
GOVWASH	Aridisols	Gypsids	Haplogypsids			creosotebush, bursage			alluvium	dissected pediments, gypsif	
GRALEY	Mollisols	Xerolls	Argixerolls			wheatgrass, ID fescue			residuum, colluvium	mtns mixed	
GRANSHAW	Aridisols	Argids	Haplargids			shadscale bud sagebrush			alluvium	fan aprons	fan skirts
GRAPEVINE	Aridisols	Calcids	Haplocalcids			creosotebush, w bursage, IN ricegrass			alluvium	fan piedmonts	alluvial flats
GRASSVAL	Aridisols	Durids	Argidurids			sagebrush			alluvium	fan piedmont remnants	
GRAUFELS	Mollisols	Xerolls	Haploxerolls			sagebrush, bitterbrush			residuum, colluvium	mtns granitic	
GRAVIER	Aridisols	Calcids	Haplocalcids			ricegrass, squirreltail, galleta			alluvium	fan skirts	fan remnants, beach plains
GREYEAGLE	Aridisols	Durids	Haplodurids			creosotebush			alluvium	fan terraces	
GRINA	Entisols	Orthents	Torriorthents			juniper, sagebrush, rabbitbrush			residuum, colluvium	mtns sedim.	
GRUBE	Mollisols	Xerolls	Calcixerolls						colluvium	mtns qtzite	
GRUMBLEN	Aridisols	Argids	Haplargids			sagebrush			residuum, colluvium	mtns volcanic	
GUARDIAN	Aridisols	Gypsids	Haplogypsids			dalea, sandpaper bush			residuum	pediments w/ gyp ss, sist	
GUMBLE	Aridisols	Argids	Haplargids	8	250	WY big sagebrush, wheatgrass	50	1800	colluium	pediments tuff	
GYNELLE	Entisols	Orthents	Torriorthents	12	100	shadscale	15	1450	alluvium	fan remnants	
HACKWOOD	Mollisols	Cryolls	Haplocryolls	5	460	aspen/brome, bluegrass	80	2600	alluvium, colluvium, loess	mtns mixed	
HALACAN	Mollisols	Rendolls	Cryrendolls			sagebrush/ID fescue, bluegrass, squirreltail			residuum, colluvium	mtns ls	
HALEBURU	Entisols	Orthents	Torriorthents	19	125	creosotebush	75	990	colluvium, residuum	mtns volcanic	
HAMTAH	Mollisols	Xerolls	Argixerolls			sage, serviceberry, bitterbrush, mtnmahogany			colluvium, residuum	mtns tuff	
HANDPAH	Aridisols	Durids	Argidurids	12	230	sagebrush	30	1700	alluvium	fan remnants	
HANDY	Aridisols	Argids	Haplargids			sagebrush, ephedra, bluegrass			alluvium, colluvium	fan remnanys	hills
HANGROCK	Aridisols	Durids	Argidurids			sagebrush, squirreltail, needlegrass			alluvium	fan remnants	
HAPGOOD	Mollisols	Cryolls	Haplocryolls	6	500	sagebrush	75	2350	loess, ash/colluvium, residuum	mtns andesite	
HARCANY	Mollisols	Cryolls	Haplocryolls	3	330	sagebrush, snowberry	75	2300	loess, ash/colluvium	mtns. Mixed	
HARDHAT	Aridisols	Calcids	Haplocalcids			shadscale, squirreltail			alluvium/lacustrine	fan skirts	inset fans, lake terraces
HARDOL	Mollisols	Cryolls	Calcicryolls			wheatgrass, sagebrush, mtnmahogany			residuum, colluvium	mtns ls	
HARDZEM	Alfisols	Cryalfs	Haplocryalfs			Subalpine w fir, limber pine			residuum, colluvium	mtns ls	
HART CAMP	Mollisols	Xerolls	Argixerolls			sagebrush			residuum	mtns tuff	
HARTIG	Mollisols	Xerolls	Haploxerolls			sagebrush/ID fescue, wheatgrass			colluvium, residuum	mtns volcanic	
HATPEAK	Mollisols	Xerolls	Durixerolls			sagebrush, wheatgrass, ID fescue, Sandberg bluegrass			alluvium	fan piedmont remnants	plateaus

(continued)

(continued)

Series Name	Order	Suborder	Great Group	MAAT (°C)	MAP (mm)	Vegetation	Max. slope (%)	Med. elev. (m)	Parent material	Landform1	Landform2
HAUNCHEE	Mollisols	Rendolls	Cryrendolls	6	500	mtn mahogany, sagebrush	75	2550	residuum, colluvium	mtns ls	
HAVINGDON	Aridisols	Argids	Haplargids			sagebrush, hopsage, horsebrush, Sandb bluegrass, squirreltail			residuum, loess, ash	mtns chert	
HAWSLEY	Entisols	Psamments	Torripsamments	11	150	greasewood	30	900	eolian, alluvium	sand sheets	
HAYBOURNE	Aridisols	Cambids	Haplocambids			sagebrush, rabbitbrush			alluvium	inset fans	alluvial fans
HAYESTON	Entisols	Orthents	Torriorthents			sagebrush, IN ricegrass			alluvium	inset fans	fan skirts, beach terraces
HAYMONT	Entisols	Orthents	Torriorthents			saltbush, shadscale			alluvium	fan skirts	alluvial flats
HEECHEE	Mollisols	Xerolls	Argixerolls			sagebrush, bitterbrush, wheatgrass			alluvium	fan piedmont remnants	
HEIST	Aridisols	Cambids	Haplocambids	8	250	sagebrush	15	1800	alluvium	alluvial fans	lake plains, inset fans, fan skirts
HESSING	Aridisols	Cambids	Haplocambids			shadscale, bud sage			alluvium	beach plains	fan skirts, inset fans, lagoons
HIKO PEAK	Aridisols	Calcids	Haplocalcids	9	250	sagebrush	60	1700	alluvium, colluvium	alluvial fan	fan remnants
HOLBROOK	Mollisols	Xerolls	Haploxerolls			WY big sage, antelope bitterbrush, Thurber's needlegrass, Sand bluegrass			alluvium mixed	alluvial fans	fan aprons, inset fans
HOLLACE	Aridisols	Calcids	Petrocalcids	12	200	blackbrush, green ephedra, desert bitterbrush, desert needlegrass, galleta	30	1500	colluvium, residuum mixed	sideslopes	
HOLLYWELL	Aridisols	Cambids	Haplocambids			saltbrush, IN ricegrass, bluegrass			alluvium	fan piedmonts	fan skirts
HOLTVILLE	Entisols	Fluvents	Torrifluvents	24	100	desert shrubs	3	200	alluvium	floodplains	
HOMESTAKE	Mollisols	Xerolls	Argixerolls			sagebrush, bitterbrush/wheatgrass, needlegrasses			alluvium/lacustrine	eroded fan remnants	
HOOPLITE	Aridisols	Argids	Haplargids			sagebrush, IN ricegrass			residuum, colluvium	mtns volcanic	
HOOT	Aridisols	Argids	Haplargids	9	180	shadscale	75	1200	residuum, colluvium	mtns volcanic	
HOPEKA	Entisols	Orthents	Torriorthents			pinyon-juniper			residuum, colluvium	mtns ls	
HOPPSWELL	Aridisols	Argids	Haplargids			blackbrush, spanish dagger, black grama, galleta, desert needlegrass			alluvium ign	fan remnants	
HORSECAMP	Vertisols	Xererts	Haploxererts			sagebrush, wheatgrass, wildrye, squirreltail			residuum	plateaus volc	
HOUGH	Aridisols	Argids	Haplargids			horsebrush, rabbitbrush, IN ricegrass, dalea, greasewood			alluvium, lacustrine	lake plains	lake terraces
HUEVI	Aridisols	Calcids	Haplocalcids	22	125	creosote bush	70	765	alluvium	fan remnants	ballenas, fan terraces
HUILEPASS	Aridisols	Argids	Haplargids			WY big sage, In ricegrass, bottlebrush squirreltail, galleta			alluvium	fan remnants	barrier beaches
HUMBOLDT	Mollisols	Aquolls	Endoaquolls	10	180	sedges, saltgrass	2	1600	alluvium, ash	alluvial plains	
HUMDUN	Aridisols	Cambids	Haplocambids			big sage, Sandb bluegrass, bluebunch wheatgrass, bottlebrush			loess, ash/alluvium	fan piedmont remnants	fan terraces
HUNDRAW	Entisols	Orthents	Torriorthents			sagebrush			residuum, colluvium	mtns tuff	

(continued)

(continued)

Series Name	Order	Suborder	Great Group	MAAT (°C)	MAP (mm)	Vegetation	Max. slope (%)	Med. elev. (m)	Parent material	Landform1	Landform2
HUNEWILL	Aridisols	Argids	Haplargids			sagebrush, needlegrass, IN ricegrass, squirreltail			alluvium	alluvial terraces, fan remnants, ballenas, inset fans	
HUNNTON	Aridisols	Durids	Argidurids	9	230	sagebrush	30	1650	loess, ash/alluvium	fan remnant	
HUSSA	Mollisols	Aquolls	Endoaquolls			rabbitbrush, wildrye, sagebrush			alluvium	floodplains	lake terraces, inset fans
HUTCHLEY	Mollisols	Xerolls	Argixerolls			sagebrush, wheatgrass, bluegrass			residuum, colluvium	mtns volcanic	
HYLOC	Mollisols	Xerolls	Argixerolls			pinyon-juniper			residuum	mtns basalt	
HYMAS	Mollisols	Xerolls	Haploxerolls	7	330	sagebrush	75	2050	residuum, colluvium	mtns ls	
HYPOINT	Entisols	Orthents	Torriorthents			creosotebush, w bursage			alluvium	fan aprons	fan skirts, alluvial fans
HYZEN	Mollisols	Xerolls	Haploxerolls	4	360	pinyon-juniper	75	2000	residuum, colluvium	mtns ls	
IDWAY	Aridisols	Calcids	Haplocalcids			sagebrush, greasewood, IN ricegrass, wildrye			alluvium	alluvial flats	basin-floor remnants
IGDELL	Mollisols	Xerolls	Durixerolls			sagebrush, rabbitbrush, ID fescue			loess/alluvium	fan piedmont remnants	plateau summits
INDIAN CREEK	Aridisols	Durids	Argidurids			sagebrush			alluvium	alluvial terraces	dissected fan remnants
INDIANO	Mollisols	Xerolls	Argixerolls			sagebrush, bitterbrush			residuum, colluvium	mtns volcanic	
INMO	Entisols	Orthents	Torriorthents			shadscale, greasewood, bud sage, IN ricegrass			alluvium	alluvial fans	inset fans, fan skirts, fan piedmonts
INPENDENCE	Inceptisols	Xerepts	Humixerepts			Subalpine aspen/snowberry			colluvium	mtns mixed	
INVILLE	Alfisols	Xeralfs	Haploxeralfs	6	635	jeffrey pine-white fir	30	1875	alluvium	alluvial fans	
IRONGOLD	Aridisols	Calcids	Petrocalcids	16	150	blackbrush	15	1200	alluvium ls	fan remnants	
ISOLDE	Entisols	Psamments	Torripsamments	11	150	horsebrush	30	1500	eolian sand	dunes	
ISTER	Mollisols	Xerolls	Argixerolls			sagebrush, ephedra			residuum	mtns ls	
ITCA	Mollisols	Xerolls	Argixerolls	6	360	pinyon-juniper	75	2050	residuum, colluvium	mtns volcanic	
ITME	Entisols	Orthents	Torriorthents			hopsage, dalea, wolfberry, saltbush			alluvium	fan piedmonts	alluvial fans, fan skirts
IVER	Mollisols	Xerolls	Haploxerolls			big sage, rabbitbrush, ID fescue, Sandb bluerass			loess/colluvium, residuum	sideslopes	
IZAMATCH	Entisols	Orthents	Torriorthents	11	150	shadscale	30	1600	loess/alluvium	beach plains	fan skirts
IZAR	Entisols	Orthents	Torriorthents			sagebrush			residuum, colluvium	mtns tuff	
IZO	Entisols	Orthents	Torriorthents	12	150	shadscale, sagebrush	15	1700	alluvium	inset fan	fan skirts
IZOD	Entisols	Orthents	Torriorthents			sagebrush, horsebrush, IN ricegrass			residuum, colluvium	mtns ls	
JARAB	Mollisols	Xerolls	Durixerolls			sagebrush, rabbitbrush, ephedra			alluvium	valley fans	falluvial fans, fan remnants
JAYBEE	Aridisols	Argids	Haplargids			sagebrush			residuum, colluvium	mtns volcanic	
JEAN	Entisols	Orthents	Torriorthents			creosotebush, w bursage, big galleta, In ricegrass			eolian/alluvium	fan remnants	inset fans, channels

(continued)

(continued)

Series Name	Order	Suborder	Great Group	MAAT (°C)	MAP (mm)	Vegetation	Max. slope (%)	Med. elev. (m)	Parent material	Landform1	Landform2
JERICHO	Aridisols	Durids	Haplodurids	9	280	sagebrush	50	1850	alluvium	fan remnants	
JERVAL	Aridisols	Argids	Natrargids	12	150	shadscale	30	1300	alluvium, loess, ash	fan remnants	lake terraces
JESSE CAMP	Aridisols	Cambids	Haplocambids			sagebrush, wildrye			alluvium	inset fans	lake terraces
JOBPEAK	Entisols	Orthents	Torriorthents			pinyon-juniper			residuum, colluvium	mtns volcanic	
JOEMAY	Aridisols	Calcids	Haplocalcids			shadscale, creosotebush, w bursage			alluvium	fan remnants	fan piedmonts
JUNG	Aridisols	Argids	Haplargids			sagebrush			residuum	mtns volcanic	
JUVA	Entisols	Fluvents	Torrifluvents			greasewood, shadscale, bud sage, spiny hopsage, In ricegrass			alluvium	fan skirts	basin-floor remnants, alluvial fans
KANACKEY	Aridisols	Argids	Haplargids			Wright eriogonum, w bur, spiny menodora, ephedra			residuum qtzite, schist	sideslopes	
KANESPRINGS	Aridisols	Durids	Argidurids			blackbrush, rabbitbrush, buckwheat, creosotebush			residuum, colluvium	mtns basalt	
KATELANA	Entisols	Orthents	Torriorthents	8	180	greasewood	4	1800	alluvium/lacustrine	lake plains	
KAWICH	Entisols	Psamments	Torripsamments			greasewood, saltbush			eolian sand	dunes	
KEEFA	Aridisols	Cambids	Haplocambids	12	150	shadscale	8	1650	alluvium	inset fans	fan skirts, lake plains
KELK	Aridisols	Cambids	Haplocambids	9	200	sagebrush	15	1700	alluvium, loess	inset fans	fan remnants, fan skirts
KEMAN	Mollisols	Cryolls	Argicryolls	6	580	mtn big sage, Idaho fescue	50	2100	colluvium, till, volcanic ash	mtn slopes	
KEYOLE	Inceptisols	Cryepts	Haplocryepts	3	750	engel spr, mtn gooseberry, mtn brome, needlegrass	75	3100	colluvium	mtns qtzite	
KIDWELL	Aridisols	Argids	Calciargids			galleta, muhly, IN ricegrass, creosotebush, ephedra			alluvium	fan remnants	
KITGRAM	Mollisols	Ustolls	Calciustolls	6	510	bristlecone pine, white fir, limber pine	75	2700	colluvium, residuum	mtns ls	
KLECKNER	Mollisols	Xerolls	Argixerolls			sagebrush, wheatgrass			alluvium, colluvium	fan remnants	
KOBEH	Aridisols	Cambids	Haplocambids			sagebrush			alluvium	inset fans	fan skirts
KOLDA	Mollisols	Aquolls	Endoaquolls			sedges, rushes			alluvium/lacustrine	lake plains	basin floors
KOONTZ	Mollisols	Xerolls	Argixerolls			pinyon-juniper/sagebrush			residuum, colluvium	mtns metavolc	
KOYEN	Aridisols	Cambids	Haplocambids	12	150	sagebrush	8	1700	loess, alluvium	mtns volcanic	
KRAM	Entisols	Orthents	Torriorthents			pinyon-juniper			residuum	mtns ls	
KUMIVA	Entisols	Orthents	Torriorthents			winterfat, bud sage, squirreltail			alluvium	inset fans	axial flood plains
KUNZLER	Aridisols	Calcids	Haplocalcids	10	250	sagebrush	8	1750	alluvium	fan remnant	
KURSTAN	Aridisols	Calcids	Haplocalcids			creosotebush, w bursage, ratany, dalea			alluvium	fan piedmont remnants	
KYLER	Entisols	Orthents	Torriorthents	12	300	sagebrush	75	2200	residuum, colluvium	mtns ls	
KZIN	Entisols	Orthents	Torriorthents			juniper-pinyon/sagebrush, IN ricegrass			residuum	pediments calcar	

(continued)

(continued)

Series Name	Order	Suborder	Great Group	MAAT (°C)	MAP (mm)	Vegetation	Max. slope (%)	Med. elev. (m)	Parent material	Landform1	Landform2
LABKEY	Aridisols	Cambids	Haplocambids			shadscale, greasewood			alluvium	alluvial terraces	fan skirts, fan insets
LABSHAFT	Mollisols	Cryolls	Haplocryolls	6	360	mtnmahogany/sagebrush	75	2650	residuum, colluvium	mtns volcanic	
LAHONTAN	Entisols	Orthents	Torriorthents			greasewood, saltgrass			lacustrine, alluvium	lake plains	
LANGSTON	Aridisols	Argids	Haplargids			WY big sage, spiny hopsage, bluebunch wheatgrass, bottlebrush squirreltail			alluvium/lacustrine	fan remnants	lake terraces
LANIP	Aridisols	Argids	Calciargids	16	150	galleta, bush muhly, ricegrass, creosotebush	8	1000	alluvium	fan remnants	
LAPED	Aridisols	Durids	Argidurids			shadscale, bud sage, squirreltail			residuum, colluvium	mtns volcanic	
LAPON	Aridisols	Durids	Argidurids			sagebrush, needlegrass, squirreltail			ash, residuum, colluvium	plateaus, ign	
LARYAN	Inceptisols	Aquepts	Cryaquepts			willow, brome, bluegrass			loess, colluvium/bedrock	mtns metamor	
LAS VEGAS	Aridisols	Calcids	Petrocalcids			creosotebush, w bursage, shadscale, ephedra			alluvium ls	basin-floor remnants	alluvial flats
LASTCHANCE	Aridisols	Calcids	Petrocalcids			w bursae, creosotebush, Sp dagger, woflberry			alluvium ls	fan remnants	
LATHROP	Aridisols	Argids	Haplargids			shadscale, bud sage			alluvium	alluvial fans	fan piedmonts
LAXAL	Aridisols	Calcids	Haplocalcids			shadscale, galleta, desert grass, bud sage			alluvium	inset fans	fan piedmonts, fan skirts
LAYVIEW	Mollisols	Cryolls	Argicryolls	6	360	bluegrass, ID fescue	75	2600	residuum, colluvium	mtns volcanic	
LAZAN	Entisols	Orthents	Torriorthents			pinyon, ephedra, needlegrass			residuum, colluvium	mtns granite	
LEO	Entisols	Orthents	Torriorthents	12	200	ephedra, hopsage, needlegrass	15	1700	alluvium	alluvial fans	fan remnants, insert fans, fan skirts
LERROW	Mollisols	Xerolls	Argixerolls			sagebrush, bitterbrush, rabbitbrush			residuum, colluvium	mtns volcanic	
LIEN	Aridisols	Durids	Haplodurids			IN ricegrass, needlegrass			alluvium	ballenas	fan remnants
LIMEWASH	Aridisols	Gypsids	Haplogypsids			creosotebush, bursage, dalea			residuum, colluvium	pediments gypsif sed	
LINCO	Aridisols	Cambids	Haplocambids			sagebrush, rabbitbrush, ephedra, IN ricegrass			alluvium	fan remnants	fan aprons, pediments
LINKUP	Aridisols	Argids	Haplargids			sagebrush, wheatgrass, needlegrass			residuum, colluvium	mtns mixed	
LINOYER	Entisols	Orthents	Torriorthents	9.5	250	sagebrush	10	1800	alluvium	fan remnants	
LINPEAK	Inceptisols	Cryepts	Calcicryepts	2	750	Engl spruce, limber pine	75	3000	colluvium	mtns ls	
LINROSE	Mollisols	Xerolls	Haploxerolls			bl sage, ID fescue, bluegrass			residuum, colluvium	mtns mixed	
LITTLEAILIE	Aridisols	Durids	Argidurids			sagebrush, IN ricegrass			alluvium	fan remnants	ballenas
LOCANE	Aridisols	Argids	Haplargids			sagebrush, ephedra			residuum, colluvium	mtns mixed	
LODAR	Mollisols	Xerolls	Calcixerolls	9	330	juniper-wheatgrass	75	2100	residuum, colluvium	mtns ls, ss	
LOGRING	Entisols	Orthents	Torriorthents			pinyon-juniper			residuum, colluvium	mtns ls	
LOJET	Aridisols	Durids	Argidurids			sagebrush			alluvium	fan remnants	
LOMOINE	Entisols	Orthents	Torriorthents			sagebrush, ephedra, horsebrush, squirreltail			residuum, colluvium	mtns granite	
LONCAN	Mollisols	Xerolls	Haploxerolls	5	360	sagebrush, needlegrass, ID fescue, bluegrass	75	2000	residuum, colluvium	mtns volcanic	

(continued)

(continued)

Series Name	Order	Suborder	Great Group	MAAT (°C)	MAP (mm)	Vegetation	Max. slope (%)	Med. elev. (m)	Parent material	Landform1	Landform2
LONGCREEK	Mollisols	Xerolls	Argixerolls			WY big sage, basin big sage, mtn big sage, bluebunch wheatgrass			residuum, colluvium	mtns volcanic	
LONGDIS	Aridisols	Argids	Natrargids			sagebrush, greasewood, wildrye			alluvium/lacustrine, ash	lake terraces	alluvial fans
LONGJIM	Aridisols	Durids	Haplodurids			blackbrush, ephedra			alluvium	fan piedmont remnants	
LOOMER	Mollisols	Xerolls	Argixerolls			sagebrush, ephedra			colluvium, residuum	mtns volcanic	
LOOMIS	Aridisols	Argids	Haplargids	8	250	bl sagebrush, rabbitbrush, wheatgrass	35	1700	colluvium, residuum	mtns volcanic	
LORAY	Aridisols	Calcids	Haplocalcids			sagebrush, ricegrass			alluvium	beach plains	offshore bars, fan skirts
LOVELOCK	Mollisols	Aquolls	Endoaquolls			inl saltgrass, black greasewood			alluvium	lake plains	alluvial terraces
LOWEMAR	Inceptisols	Cryepts	Humicryepts	5	650	subalpine forbs, grasses	75	2900	residuum, colluvium	mtns metamor	
LUNING	Entisols	Orthents	Torriorthents			greasewood, saltbush			alluvium	fan aprons	fan skirts
LUSET	Mollisols	Cryolls	Argicryolls			mtn brome, letterman's needlegrass, mtn big sage			residuum, colluvium	mtns tuff	
LYDA	Aridisols	Durids	Argidurids			greasewood, shadscale			alluvium	fan remnants	fan piedmonts
LYX	Aridisols	Cambids	Haplocambids			saltbush, rabbitbrush, horsebrush			alluvium	inset fans	alluvial fans
MACNOT	Aridisols	Calcids	Haplocalcids			sagebrush, hopsage, needlegrass			alluvium, ash	beach terraces	inset fans, alluvial fans
MAJUBA	Mollisols	Xerolls	Argixerolls			sagebrush, rabbitbrush, lupine, bluegrass, Th needlegrass, bottlebrush squirreltail			residuum, colluvium	sideslopes slate, schist	
MALPAIS	Aridisols	Cambids	Haplocambids			ephedra, greasewood, saltbrush			alluvium, colluvium	fan piedmonts	alluvial fans
MANOGUE	Vertisols	Xererts	Haploxererts			horsebrush, spiny hop, WY big sage			alluvium, colluvium, residuum	plateaus	
MARYJANE	Mollisols	Ustolls	Calciustolls			pond pine, mtnmahog, currant			alluvium, colluvium	inset fans	
MATTIER	Mollisols	Xerolls	Argixerolls			bluebunch wheatgrass, mutton grass			colluvium, residuum	mtns tuff	
MAU	Aridisols	Argids	Haplargids			big sage, low sage, Sandb bluegrass, bottlebrush			residuum, colluvium	sideslopes tuff	
MAZUMA	Entisols	Orthents	Torriorthents	11	150	greasewood	30	1550	alluvium	basin floor remnants	alluvial flats, fan skirts
MCCARRAN	Aridisols	Gypsids	Haplogypsids	18	125	creosotebush, bursage, shadscale	8	1200	alluvium	alluvial flats	
MCCONNEL	Aridisols	Cambids	Haplocambids	10	200	sagebrush	50	1600	loess, ash/alluvium	inset fans	fan aprons
MCIVEY	Mollisols	Xerolls	Argixerolls	6	350	sagebrush	75	2250	colluvium	mtns mixed	
MCTOM	Inceptisols	Cryepts	Humicryepts			Subalpine pines-whitebark, limber			residuum, colluvium	mtns granodior	
MEADVIEW	Aridisols	Calcids	Haplocalcids			blackbrush, joshua-tree, yucca			alluvium	fan remnants	
MEDBURN	Entisols	Orthents	Torriorthents	10.5	250	sagebrush	8	1700	alluvium, lacustrine	alluvial fans	fan remnants, inset fans, alluvial flats
MEISS	Inceptisols	Cryepts	Humicryepts			sagebrush			residuum	mtns andesite	
MEZZER	Aridisols	Cambids	Haplocambids			sagebrush, winterfat, rabbitbrush			alluvium	inset fans	

(continued)

(continued)

Series Name	Order	Suborder	Great Group	MAAT (°C)	MAP (mm)	Vegetation	Max. slope (%)	Med. elev. (m)	Parent material	Landform1	Landform2
MIDRAW	Aridisols	Durids	Argidurids			sagebrush			residuum, colluvium	mtns volcanic	
MIJOYSEE	Entisols	Orthents	Torriorthents			mtnmahogany, sagebrush, bluegrass			residuum, colluvium	mtns ls	
MINU	Aridisols	Durids	Argidurids			sagebrush, IN ricegrass			alluvium	fan remnants, eroded	
MIRKWOOD	Aridisols	Argids	Haplargids			greasewood, shadscale, horsebrush			residuum, colluvium	mtns volcanic	
MISAD	Entisols	Orthents	Torriorthents			shadscale, bud sage, greasewood, squirreltail			alluvium, ash	alluvial fans	fan skirts, inset fans, offshore bars
MOLION	Aridisols	Durids	Haplodurids			sagebrush, rabbitbrush, IN ricegrass			alluvium	fine remnants	
MONARCH	Mollisols	Xerolls	Calcixerolls			pinyon-juniper/sagebrush			residuum, colluvium	mtns ls	fan remnants
MORMON MESA	Aridisols	Calcids	Petrocalcids	18	125	creosote bush	15	730	loess/alluvium ls	fan remnant	
MORMOUNT	Aridisols	Calcids	Petrocalcids	14	200	blackbrush, creosotebush, yucca, ephedra	15	1200	alluvium ls	fan piedmont remnants	
MOSIDA	Entisols	Fluvents	Torrifluvents			wbeatgrass, bluegrasses, IN ricegrass, sagebrush			alluvium	alluvial fans	alluvial flats, alluvial terraces
MOTOQUA	Mollisols	Ustolls	Argiustolls	10	330	sagebrush	70	1900	residuum, colluvium	mtns volcanic	
MOTTSVILLE	Mollisols	Xerolls	Haploxerolls	9	280	sagebrush	15	1500	alluvium	alluvial fans	fan remnants
MOUNTMCULL	Entisols	Orthents	Torriorthents			blackbrush, black grama, galleta, needlegrass			residuum, colluvium	mtns metam	
MUIRAL	Inceptisols	Cryepts	Haplocryepts	5	410	spr-fir	75	2800	residuum, colluvium	mtns ls	
MULHOP	Aridisols	Calcids	Haplocalcids	7	280	juniper, black sagebrush	75	2000	residuum, colluvium	mtns ls	
MUNI	Aridisols	Durids	Argidurids			sagebrush, IN ricegrass, squirreltail			alluvium	fan remnants	
NADRA	Aridisols	Durids	Haplodurids			shadscale, rabbitbrush, winterfat, IN ricegrass			alluvium	fan piedmonts	
NAYE	Aridisols	Calcids	Petrocalcids			creosotebush, w bursage, ephedra, yucca			alluvium ls	fan remnants	
NEEDLE PEAK	Entisols	Orthents	Torriorthents			sagebrush, rabbitbrush, wildrye			alluvium	inset fans	fan skirts, alluvial terraces, floodplains
NEMICO	Aridisols	Durids	Natridurids			shadscale, bud sage, Bailey greasewoo d, In ricegrass, galleta			residuum, colluvium	mtns volcanic	
NETTI	Mollisols	Xerolls	Argixerolls			sagebrush, snowberry, serviceberry, ID fescue			alluvium	fan remnants	ballenas
NEVADOR	Aridisols	Argids	Haplargids			sagebrush, hopsage, IN ricegrass			alluvium	fan remnants	
NEVU	Mollisols	Xerolls	Durixerolls			bluebunch wheatgrass, mtn big sagebrush			alluvium tuff	fan remnants	
NEWERA	Aridisols	Argids	Haplargids			blackbrush, creosotebush, needlegrass, galleta			residuum, colluvium	mtns volcanic	
NEWLANDS	Mollisols	Cryolls	Argicryolls	7	360	big sage, ant bitterbrush, ID fescue	50	2000	residuum, colluvium	mtns volcanic	
NEWVIL	Mollisols	Xerolls	Durixerolls			squirreltail, sagebrush, rabbitbrush, ephedra			alluvium	eroded fan remnants	
NICKEL	Aridisols	Calcids	Haplocalcids	18	125	creosotebush	35	945	alluvium	fan remnant	
NINEMILE	Mollisols	Xerolls	Argixerolls	6	360	sagebrush	70	2000	residuum, colluvium	mtns volcanic	

(continued)

(continued)

Series Name	Order	Suborder	Great Group	MAAT (°C)	MAP (mm)	Vegetation	Max. slope (%)	Med. elev. (m)	Parent material	Landform1	Landform2
NIPPENO	Aridisols	Argids	Haplargids			blackbrush, black grama, needlegrass, galleta, ephedra			residuum, colluvium	mtns metam	
NIPTON	Entisols	Orthents	Torriorthents			ephedra, galleta, needlegrass			colluvium, residuum	mtns mixed	
NOLENA	Entisols	Orthents	Torriorthents			blackbrush, buckwheat, ratany, creosotebush			residuum, colluvium	mtns granite	
NUPART	Mollisols	Xerolls	Haploxerolls			pinyon/sagebrush			residuum, colluvium/granite	pediments granite	
NUYOBE	Inceptisols	Aquepts	Halaquepts			saltgrass, sacaton, iodine bush			lacustrine	alluvial flats	lake plains
NYSERVA	Aridisols	Argids	Natrargids			greasewood,seepweed			alluvium	lake terraces	alluvial flats, lake plains
OCALA	Inceptisols	Aquepts	Halaquepts	10	180	greasewood, rabbitbrush, saltgrass	2	1600	alluvium, ash	alluvial terraces	lake plains, inset fans, fan skirts
OKAN	Entisols	Orthents	Torriorthents			sagebrush			alluvium	inset fans	fan skirts, barrier beaches
OLA	Mollisols	Xerolls	Haploxerolls			ID fescue, wheatgrass, bitterbrush			colluvium, residuum	mtns ign-meta	
OLAC	Aridisols	Argids	Haplargids			sagebrush, bitterbrush, bluegrass			alluvium	alluvial terraces	alluvial fans
OLD CAMP	Aridisols	Argids	Haplargids	8	250	sagebrush	75	1700	residuum, colluvium	mtns volcanic	
OLDSPAN	Aridisols	Calcids	Haplocalcids			creosotebush, w bursage			alluvium	fan remnants	
OLEMAN	Aridisols	Durids	Argidurids			blackbrush, bitterbrush, needlegrass			alluvium	fan piedmonts	
ONKEYO	Mollisols	Xerolls	Calcixerolls	7	400	bitterbrush, snowberry, sagebrush, serviceberry	75	2300	residuum, colluvium	mtns ls	
OPPIO	Aridisols	Argids	Haplargids			sagebrush, needlegrass, bluegrass			residuum	mtns volcanic	
ORICTO	Aridisols	Argids	Calciargids	12	100	shadscale	30	1400	alluvium	fan remnants	fan piedmonts
ORIZABA	Inceptisols	Aquepts	Halaquepts			rabbitbrush, greasewood, sagebrush, saltgrass, wildrye, sacaton			alluvium	alluvial flats	lake plains
OROVADA	Aridisols	Cambids	Haplocambids	8	230	sagebrush	15	1600	loess, ash/alluvium	fan remnants	
ORR	Mollisols	Xerolls	Argixerolls			WY big sage, sand bluegrass, squirreltail			alluvium mixed	fan remnants	
ORWASH	Entisols	Orthents	Torriorthents			creosotebush, bursage			alluvium	fan aprons	fan skirts, alluvial flats
OSDITCH	Inceptisols	Cryepts	Haplocryepts	4	600	white fir, Douglas-fir, limber pine	75	2800	colluvium	mtns qtzite	
OSOBB	Aridisols	Durids	Haplodurids			shadscale, greasewood			residuum, colluvium	plateaus volcanic	
OUPICO	Aridisols	Durids	Haplodurids			WY big sage, In ricegrass, squirreltail			alluvium	fan remnants	
OXCOREL	Aridisols	Argids	Natrargids	9	150	Shadscale, sagebrush	50	1600	loess/alluvium	fan remnants	
PACKER	Mollisols	Cryolls	Argicryolls	6	380	sagebrush, buckwheat	75	2500	residuum, colluvium	mtns volcanic	
PAHROC	Aridisols	Durids	Haplodurids			blackbrush, NV ephedra, dalea			alluvium	fan remnants	
PALINOR	Aridisols	Durids	Haplodurids	8	230	sagebrush	50	1900	alluvium	fan remnants	ballena
PANLEE	Aridisols	Cambids	Petrocambids			sagebrush			residuum, colluvium, loess, ash	mtns mixed	
PAPOOSE	Aridisols	Argids	Haplargids	12	175	winterfat, bud sage, squirreltail	8	1500	alluvium	fan remnants	fan piedmonts

(continued)

(continued)

Series Name	Order	Suborder	Great Group	MAAT (°C)	MAP (mm)	Vegetation	Max. slope (%)	Med. elev. (m)	Parent material	Landform1	Landform2
PARANAT	Mollisols	Aquolls	Endoaquolls			wildrye, bluegrass, saltgrass			alluvium	alluvial flats	alluvial terraces
PARISA	Aridisols	Durids	Haplodurids			sagebrush, wheatgrass, Sandberg bluegrass			alluvium	piedmont remnants	
PARRAN	Aridisols	Salids	Aquisalids	12	125	greasewood, saltgrass	2	1400	lacustrine	lake plains	basin-floor remnants
PATNA	Aridisols	Argids	Haplargids			greasewood, hopsage			eolian/lacustrine	sand sheets	dunes
PAYPOINT	Aridisols	Argids	Haplargids			big sage, rubber rabbitbrush, Sandb bluegrass, bottlebrush			ash, alluvium/lacustrine	lagoons	
PEDOLI	Aridisols	Argids	Haplargids			sagebrush, squirreltail, ephedra			alluvium	fan remnants	beach terraces
PEEKO	Aridisols	Durids	Haplodurids	9	230	sagebrush	30	1875	loess, ash/alluvium	fan remnants	
PENELAS	Aridisols	Argids	Haplargids			sagebrush, ephedra			residuum, colluvium	mtns mixed	
PENOYER	Entisols	Orthents	Torriorthents	12	175	ricegrass, galleta, saltbush	4	1550	alluvium	inset fans	fan skirts, alluvial flats, basin floors
PERAZZO	Aridisols	Argids	Haplargids	10	125	shadscale, greasewood, bud sage, IN ricegrass	15	1725	alluvium	fan piedmonts	
PERNTY	Mollisols	Xerolls	Argixerolls	6	280	big sage, rabbitbrush, juniper, ID fescue	75	2000	residuum, colluvium	mtns volcanic	
PERWICK	Entisols	Orthents	Torriorthents			juniper, sagebrush, rabbitbrush			residuum	hills, tuff	
PHARO	Mollisols	Xerolls	Calcixerolls	9	330	wheatgrass, bluegrass, squirreltail	30	1800	alluvium	alluvial fans	ballenas
PHLISS	Aridisols	Argids	Haplargids			sagebrush, hopsage, pine bluegrass			residuum, colluvium	mtns metam	
PIAR	Inceptisols	Cryepts	Calcicryepts	2	700	bristlecone pine, limber pine, Eng spruce	75	3000	colluvium, residuum	mtns ls	
PIBLER	Aridisols	Calcids	Petrocalcids			sagebrush black, yellowbrus, Mormon-tea, IN ricegrass			alluvium sedimentary	alluvial fans	fan remnants
PICKUP	Mollisols	Xerolls	Argixerolls	8	250	sagebrush, wheatgrass	75	1800	residuum, colluvium	mtns volcanic	
PIE CREEK	Mollisols	Xerolls	Palexerolls			sagebrush, S bluegrass, squirreltail, needlegrass			residuum	mtns tuff	
PINEVAL	Aridisols	Argids	Haplargids			sagebrush			alluvium	fan remnants	fan aprons
PINTWATER	Entisols	Orthents	Torriorthents	12	180	shadscale	75	1900	residuum, colluvium	mtns tuff	
PIOCHE	Mollisols	Xerolls	Argixerolls			pinyon-juniper, wheatgrass			residuum, colluvium	mtns volcanic	
PIROUETTE	Aridisols	Durids	Natridurids	11	150	greasewood, winter fat	50	1550	residuum, colluvium	mtns volcanic	
POKERGAP	Aridisols	Argids	Natrargids			sagebrush, hopsage			alluvium	fan piedmonts	
POOKALOO	Aridisols	Calcids	Haplocalcids	8	330	pinyon-juniper	75	2150	residuum, colluvium	mtns calc sist	
PORTMOUNT	Aridisols	Argids	Calciargids			sagebrush, rabbitbrush, squirreltail			alluvium	fan piedmonts	
POTOSI	Entisols	Orthents	Torriorthents			blackbrush, needlegrass			residuum, colluvium	mtns ls	
POWMENT	Entisols	Orthents	Torriorthents	7	360	pinyon-juniper	75	2200	colluvium, residuum	mtns granite	
PREBLE	Entisols	Orthents	Torriorthents			greasewood, rabbitbrush			alluvium, ash	basin-floor remnants	alluvial flats, fan remnants

(continued)

(continued)

Series Name	Order	Suborder	Great Group	MAAT (°C)	MAP (mm)	Vegetation	Max. slope (%)	Med. elev. (m)	Parent material	Landform1	Landform2
PRIDEEN	Entisols	Orthents	Torriorthents			greasewood, seepweed, wildrye, squirreltail			alluvium	alluvial flats	alluvial terraces
PUETT	Entisols	Orthents	Torriorthents	9	230	sagebrush	75	1800	residuum, colluvium	mtns tuff	
PUFFER	Entisols	Orthents	Torriorthents			sagebrush, ephedra, hopsage, IN ricegrass			residuum	mtns sedim	
PUMEL	Entisols	Orthents	Torriorthents			shadscale, hopsage, winterfat, IN ricegrass			residuum, colluvium	mtns granite	
PUMPER	Aridisols	Cambids	Haplocambids			shadscale, sagebrush			loess, ash/alluvium	inset fans	alluvial fans
PUNCHBOWL	Aridisols	Argids	Haplargids			sagebrush, IN ricegrass			residuum	mtns tuffs	
PUROB	Aridisols	Calcids	Petrocalcids	12	200	blackbrush, mojave yucca	50	1700	alluvium ls	ballenas	fan remnants
PYRAT	Aridisols	Calcids	Haplocalcids	9	250	sagebrush	30	1800	alluvium	alluvial fans	fan remnants, fan skirts, inset fans
QUARZ	Mollisols	Xerolls	Argixerolls	7	300	sagebrush	50	2000	residuum, colluvium	mtns sedimentary	
QWYNN	Aridisols	Argids	Haplargids			sagebrush, saltbush			alluvium	inset fans	fan remnants
RAD	Aridisols	Cambids	Haplocambids			sagebrush			alluvium	fan piedmont remnants	inset fans, fan skirts
RADOL	Mollisols	Xerolls	Calcixerolls			wheatgrass, IN ricegrass, needleandthread, Sandberg bluegrass			residuum, colluvium	mtns ls	
RAGLAN	Aridisols	Cambids	Haplocambids			greasewood, wildrye, hopsage			alluvium, lacustrine	fan skirts	alluvial flat remnants
RAGTOWN	Entisols	Orthents	Torriorthents	12	125	greasewood, saltgrass	4	1600	lacustrine	lake plains	basin-floor remnants
RAILROAD	Aridisols	Calcids	Haplocalcids			big galleta, bush muhly, winterfat, ephedra, creosotebush			colluvium, residuum, ash	basalt lava flows	
RAMIRES	Mollisols	Xerolls	Argixerolls			sagebrush, rabbitbrush			residuum, colluvium	mtns volcanic	
RAPADO	Aridisols	Calcids	Petrocalcids	12	230	blackbrush, NV ephedra, banana yucca, joshua tree	30	1350	alluvium mixed	fan remnants	ballenas
RASILLE	Aridisols	Cambids	Haplocambids			sagebrush, squirreltail, IN ricegrass			alluvium	beach terraces	inset fans, fan skirts
RATTO	Aridisols	Durids	Argidurids			sagebrush			alluvium, colluvium	fan remnants	
RAVENDOG	Mollisols	Xerolls	Haploxerolls			sagebrush, rabbitbrush, IN ricegrass			alluvium	alluvial terraces	inset fans
RAVENSWOOD	Mollisols	Xerolls	Argixerolls			pinyon-juniper			colluvium, residuum	mtns volcanic	
RAWE	Aridisols	Argids	Haplargids			shadscale, greasewood, IN ricegrass			alluvium	fan remnants	
REBEL	Aridisols	Cambids	Haplocambids			sagebrush, wheatgrass, bluegrass			alluvium	inset fans	fan remnants, fan skirts
REDNIK	Aridisols	Argids	Haplargids			sagebrush, shadscale			alluvium	fan remnants	
REESE	Inceptisols	Aquepts	Halaquepts			greasewood, sacaton, bluegrass, wildrye			alluvium	alluvial flats	floodplains
RELLEY	Aridisols	Cambids	Haplocambids			shadscale, halogeton			alluvium	fan skirts	inset fans
RELUCTAN	Mollisols	Xerolls	Argixerolls	7	300	sagebrush	50	1900	residuum, colluvium	mtns volcanic	

(continued)

(continued)

Series Name	Order	Suborder	Great Group	MAAT (°C)	MAP (mm)	Vegetation	Max. slope (%)	Med. elev. (m)	Parent material	Landform1	Landform2
RENO	Aridisols	Durids	Argidurids			sagebrush, hawksbeard, squirreltail			alluvium	alluvial fans	fan remnants
REYWAT	Mollisols	Xerolls	Argixerolls	9	300	sagebrush	90	1500	residuum, colluvium	mtns volcanic	
REZAVE	Aridisols	Argids	Natrargids			greasewood, shadscale, bud sage, Doug rabbitbrush			residuum, colluvium	mtns volcanic	
RICERT	Aridisols	Argids	Natrargids	9	180	shadscale	15	1550	loess, ash, alluvium	fan remnants	
RICHINDE	Aridisols	Argids	Haplargids			IN ricegrass, needlegrass			residuum, colluvium	mtns tuff	
RISLEY	Aridisols	Argids	Haplargids	9	280	WY big sage, squirreltail	50	1600	colluvium, residuum	mtns andesite	
RIVERBEND	Aridisols	Calcids	Haplocalcids			creosotebush			alluvium	fan terraces	fan remnants
ROBBERSFIRE	Inceptisols	Ustepts	Haplustepts	7	405	white fir, bristlecone pine	75	2600	colluvium	backslopes ls	
ROBSON	Aridisols	Argids	Haplargids	6	380	sagebrush, ID fescue, wheatgrass	75	2055	residuum	mtns rhyolite	
ROCA	Aridisols	Argids	Haplargids	6	250	sagebrush	75	1800	colluvium, residuum	mtns mixed	
ROCCONDA	Aridisols	Argids	Haplargids	10	200	sagebrush	50	1700	residuum, colluvium	mtns sist	
ROCHPAH	Aridisols	Calcids	Haplocalcids			blackbrush, ephedra			residuum, colluvium	mtns tuff	
RODAD	Aridisols	Argids	Haplargids			galleta, IN ricegrass, shadscale, Joshua-trees			residuum, colluvium	pediments-sedim.	
RODELL	Entisols	Orthents	Cryorthents	3	460	mtn big sage, currant, brome	75	2500	residuum	sideslopes granite	
RODEN	Entisols	Orthents	Torriorthents			sagebrush, rabbitbrush, IN ricegrass			residuum, colluvium	mtns sedim	
ROIC	Entisols	Orthents	Torriorthents			greasewood, shadscale			residuum, colluvium	mtns tuff	
ROSITAS	Entisols	Psamments	Torripsamments	21	100	creosote bush	30	260	eolian	dunes	sand sheets
ROSNEY	Entisols	Orthents	Torriorthents			greasewood, suaeda, nuttal saltbush, shadscale, inland saltgrass			alluvium, ash/loess	alluvial flat remnants	fan skirts
ROUETTE	Aridisols	Durids	Haplodurids			sagebrush, IN ricegrass			alluvium	fan remnants	
RUBYHILL	Aridisols	Durids	Haplodurids			sagebrush			alluvium	fan piedmonts	
RUSTIGATE	Entisols	Orthents	Torriorthents			greasewood, seepweed, shadscale, saltgrass			alluvium	alluvial flats	lake plains
SALTAIR	Aridisols	Salids	Aquisalids	10	355	pickleweed, saltgrass	2	1400	lacustrine	lake plains	
SANWELL	Entisols	Orthents	Torriorthents			creosotebush, wolfberry, shadscale			lacustrine	relict alluvial flats	beach plains
SARAPH	Aridisols	Argids	Haplargids			sagebrush			residuum	tuff	
SAY	Mollisols	Xerolls	Argixerolls			sagebrush, Sandberg bluegrass, squirreltail, wildrye			residuum	mtns granite	
SCHOOLMARM	Mollisols	Xerolls	Argixerolls			wheatgrass, needlegrass			residuum, colluvium	mtns tuff	
SCOSSA	Inceptisols	Aquepts	Humaquepts			sedges, rushes			alluvium	alluvial fans	
SEANNA	Entisols	Orthents	Torriorthents			bursage, senna, ephedra			residuum	mtns gr	
SEGURA	Mollisols	Xerolls	Argixerolls			sagebrush, needlegrass			residuum, colluvium	mtns tuff	
SERALIN	Mollisols	Ustolls	Haplustolls			pinyon-juniper			colluvium, residuum	mtns ls	

(continued)

(continued)

Series Name	Order	Suborder	Great Group	MAAT (°C)	MAP (mm)	Vegetation	Max. slope (%)	Med. elev. (m)	Parent material	Landform1	Landform2
SETTLEMENT	Inceptisols	Aquepts	Halaquepts			greasewood, rabbitbrush, seepweed, inland saltgrass, sacaton			alluvium	lake plains	lake terraces
SETTLEMEYER	Mollisols	Aquolls	Endoaquolls			wildrye, sedges			alluvium	flood plains	inset fans
SEVENMILE	Mollisols	Xerolls	Haploxerolls	10	250	sagebrush	8	1900	alluvium	alluvial terraces	inset fans
SHABLISS	Aridisols	Durids	Haplodurids	9	230	sagebrush	50	1500	loess, ash/alluvium	fan remnants	
SHAGNASTY	Mollisols	Xerolls	Argixerolls			pinyon-junipeer			residuum, colluvium	mtns volcanic	
SHALAKE	Aridisols	Durids	Haplodurids			sagebrush, bottlebrush squirreltail, Sand bluegrass, In ricegrass			loess, alluvium	fan piedmont remnants	
SHALCLEAV	Mollisols	Xerolls	Argixerolls			sagebrush, ricegrass			residuum, colluvium	mtns tuff	
SHALPER	Mollisols	Xerolls	Argixerolls			sagebrush			residuum, colluvium	fan piedmons, tuff	
SHAWAVE	Aridisols	Argids	Haplargids	10	230	sagebrush	30	1450	alluvium, loess, ash	fan remnants	fan skirts, mtn valley fans
SHEFFIT	Entisols	Orthents	Torriorthents	8	230	sagebrush	2	1800	alluvium/lacustrine	lake plains	alluvial flats
SHIPLEY	Entisols	Orthents	Torriorthents			winterfat			alluvium w/ loess, ash	beach terraces	inset fans, fan skirts
SHORT CREEK	Aridisols	Argids	Haplargids			sagebrush, rabbitbrush			alluvium	ballenas	fan piedmonts
SHREE	Mollisols	Xerolls	Argixerolls			WY big sage, antelope bitterbrush Thurb needlegrass, basin wildrye			alluvium metamor	inset fans	alluvial fans, fan remnants
SIBELIA	Inceptisols	Cryepts	Humicryepts			Subalpine pines-whitebark, w white			colluvium, residuum	mtns andesite	
SILVERADO	Aridisols	Cambids	Haplocambids			sagebrush, wheatgrass, bluegrass			alluvium, ash	beach terraces	insert fans, fan skirts
SILVERBOW	Aridisols	Durids	Argidurids			wolfberry, ephedra, horsebrush, hopsage, gallea			alluvium, colluvium	mtn footslopes	alluvial fans, fan piedmonts, ballenas, pediments
SIMON	Mollisols	Xerolls	Argixerolls			sagebrush, rabbitbrush, wildrye, Sandberg bluegrass, wheatgrass			loess, ash/alluvium	fan remnants	alluvial terraces
SINGATSE	Entisols	Orthents	Torriorthents	10	150	sagebrush	75	1600	residuum, colluvium	mtns volcanic	
SISCAB	Mollisols	Xerolls	Argixerolls			sagebrush, bitterbrush, wheatgrass			residuum, colluvium	mtns granite	
SKEDADDLE	Entisols	Orthents	Torriorthents			sagebrush			residuum, colluvium	mtns volcanic	
SKELON	Aridisols	Durids	Haplodurids			w bursage, creosotebush			alluvium	alluvial fans	fan piedmonts
SLAVEN	Mollisols	Xerolls	Argixerolls			sagebrush, bluegrass, wheatgrass			residuum, colluvium	mtns mixed	
SLAW	Entisols	Fluvents	Torrifluvents	12	150	greasewood, saltbush	4	1550	alluvium/lacustrine	alluvial flats	lake plains, floodplain playas
SLIDYMTN	Mollisols	Ustolls	Argiustolls			pinyon-juniper			residuum, colluvium	mtns volcanic	
SLIPBACK	Aridisols	Argids	Natrargids			sagebrush, squirreltail, hopsage			alluvium	fan piedmonts	mtn valley fans

(continued)

(continued)

Series Name	Order	Suborder	Great Group	MAAT (°C)	MAP (mm)	Vegetation	Max. slope (%)	Med. elev. (m)	Parent material	Landform1	Landform2
SLOCKEY	Mollisols	Xerolls	Argixerolls			sagebrush, bitterbrush			residuum, colluvium	mtns tuff	
SMAUG	Entisols	Orthents	Torriorthents	10	180	shadescale, winterfat, bud sage	8	1350	lacustrine	lake plains	
SMEDLEY	Aridisols	Durids	Argidurids			shadscale, greasewood, galleta			alluvium	alluvial fans	fan piedmonts, ballenas
SNAPP	Aridisols	Argids	Natrargids			sagebrush			alluvium	fan piedmonts	
SNOWMORE	Aridisols	Durids	Argidurids	8	230	sagebrush	30	1400	loess/residuum	mtns volcanic	
SOAKPAK	Inceptisols	Cryepts	Haplocryepts			grasses, shrubs			till calc	moraine	
SOAR	Aridisols	Argids	Haplargids			sagebrush, needlegrass, ephedra, horsebrush			residuum, colluvium	mtns granite	
SODHOUSE	Aridisols	Durids	Haplodurids			shadscale, bud sage			alluvium, loess, ash	fan remnants	plateaus
SOFTSCRABBLE	Mollisols	Xerolls	Argixerolls	7	400	big sage, gold currant, mulesear wyethia, Dougl rabbitbrush	75	2000	residuum, colluvium	mtns volcanic	
SOJUR	Entisols	Orthents	Torriorthents			sagebrush			residuum	mtns metamor	
SOLAK	Entisols	Orthents	Torriorthents			wheatgrass, sagebrush			residuum, colluvium	mtns conglom	
SONDOA	Entisols	Orthents	Torriorthents	12	150	greasewood, seepweed	2	1500	alluvium, lacustrine	alluvial terraces	basin-floor remnants
SONOMA	Entisols	Aquents	Fluvaquents			saltgrass			alluvium, ash	alluvial terraces	lake plains
SOUGHE	Aridisols	Argids	Haplargids	8	230	sagebrush	75	1800	residuum, colluvium	mtns andesite	
SPAGER	Aridisols	Calcids	Petrocalcids	9	250	bluebunch wheatgrass, black sage, little rabbitbrush, ind ricegrass	30	1750	alluvium ls	alluvial plains	fan remnants
SPASPREY	Aridisols	Durids	Argidurids			sagebrush, IN ricegrass			alluvium	fan piedmont remnants	alluvial fans, lake terraces
SPRING	Aridisols	Gypsids	Haplogypsids			alkali sacaton, saltgrass, saltbush			lacustrine, gypsif	lake terraces	basin floors
SQUAWTIP	Mollisols	Xerolls	Argixerolls			pinyon-juniper			residuum, colluvium	mtns volcanic	
ST. THOMAS	Entisols	Orthents	Torriorthents	16	125	creosote bush	75	1035	residuum, colluvium	mtns ls	
STAMPEDE	Mollisols	Xerolls	Durixerolls	6	300	sagebrush	20	1650	eolian, alluvium	fan remnants	
STARFLYER	Mollisols	Xerolls	Argixerolls			wheatgrass, IN ricegrass, muttongrass			residuum, colluvium	mtns tuff	
STARGO	Entisols	Fluvents	Torrifluvents			shadscale, greasewood			alluvium	fan skirts	offshore bars, lake plains
STEWVAL	Aridisols	Argids	Haplargids	11	225	sagebrush	75	2000	residuum, colluvium	mtns volcanic	
STINGDORN	Aridisols	Durids	Argidurids			shadscale, sagebrush, horsebrush, squirreltail			residuum	mtns volcanic	
STONELL	Aridisols	Argids	Haplargids			greasewood, bud sage, ephedra, IN ricegrass			alluvium	fan piedmont remnants	
STUCKY	Aridisols	Argids	Haplargids			low sage, ind ricegrass			alluvium granitic	alluvial fans	
STUMBLE	Entisols	Psamments	Torripsamments	12	150	horsebrush	15	1450	eolian/alluvium	fan remnants	
SUAK	Mollisols	Xerolls	Argixerolls			mtn mahogany, sagebrush, IN ricegrass, bluegrass, cheatgrass			residuum, colluvium	mtns mixed	

(continued)

(continued)

Series Name	Order	Suborder	Great Group	MAAT (°C)	MAP (mm)	Vegetation	Max. slope (%)	Med. elev. (m)	Parent material	Landform1	Landform2
SUMINE	Mollisols	Xerolls	Argixerolls	6	300	sagebrush	75	2050	residuum, colluvium	mtns mixed	
SUMMERMUTE	Aridisols	Calcids	Haplocalcids			shadscale, rabbitbrush			alluvium	fan remnants	fan skirts
SUMYA	Entisols	Orthents	Torriorthents			sagebrush, squirreltail, Sandb bluegrass			residuum, colluvium	mtns volcanic	
SUNDOWN	Entisols	Psamments	Torripsamments			In ricegrass, fourwing saltbush, lettleleaf horsebrush, shadscale			eolian sand, ash/alluvium	sand sheets/fan remnants, fan skirts	
SUNROCK	Entisols	Orthents	Torriorthents	24	150	creosotebush	75	550	colluvium, residuum	mtns volcanic	
SURPRISE	Mollisols	Xerolls	Haploxerolls			big sage, ant bitterbrush, wheatgrass			alluvium	fan skirts	inset fans, fan remnants
SUSIE CREEK	Mollisols	Xerolls	Argixerolls			sagebrush, rabbitbrush, blueb wheatgrass, Sandb bluegrass			residuum	mtns tuff	
SWINGLER	Entisols	Orthents	Torriorthents			greasewood, sagebrush			alluvium/lacustrine	basin-floor remnants	fan skirts
SYCOMAT	Aridisols	Calcids	Haplocalcids			shadscale, rabbitbrush			alluvium	alluvial terraces	
TAHOE	Inceptisols	Aquepts	Humaquepts			willows, sedges, rushes			alluvium	floodplains	
TAHOMA	Alfisols	Xeralfs	Haploxeralfs			w fir, jeff pine, lodge pine			residuum	mtns volcanic	
TANAZZA	Aridisols	Gypsids	Calcigypsids			creosotebush, shadscale, mesquite			lacustrine	lake terraces	
TARNACH	Aridisols	Calcids	Haplocalcids	9	250	sagebrush	70	1900	colluvium, alluvium, residuum	mtns ls	
TECOMAR	Aridisols	Calcids	Haplocalcids	7	300	sagebrush	50	2100	residuum, colluvium	mtns ls	
TECOPA	Entisols	Orthents	Torriorthents			yucca, w bursage, mormon tea, blackbrush			residuum, colluvium	sideslopes, metamor	
TEGURO	Mollisols	Xerolls	Argixerolls			pinyon/sagebrush, ID fescue, wheatgrass			colluvium, residuum	mtns tuff	
TEJABE	Entisols	Orthents	Torriorthents			sagebrush, Sand bluegrass, green ephedra			residuum, colluvium	mtns volcanic	
TEMO	Entisols	Psamments	Cryopsamments	3	1016	whitebark pine, ww pine, jeff pine, CA red fir	75	2500	residuum, colluvium	mtns volcanic	
TENABO	Aridisols	Durids	Natridurids	8	180	shadscale	30	1500	loess, ash/alluvium	fan piedmonts	plateaus
TENWELL	Aridisols	Durids	Argidurids			galleta, bush muhly, creosotebush, hopsage, NV ephedra			alluvium	fan remnants	
TERCA	Mollisols	Xerolls	Argixerolls			sagebrush, bitterbrush, needlegrass			residuum, colluvium	mtns volcanic	
TERLCO	Aridisols	Argids	Natrargids			greasewood, galleta			alluvium	fan remnants	alluvial fans
THEON	Aridisols	Argids	Haplargids	10	150	shadscale, sagebrush	75	1600	residuum, colluvium	mtns volcanic	
THERIOT	Entisols	Orthents	Torriorthents			shadscale			residuum, colluvium	mtns ls	
THREELAKES	Entisols	Orthents	Torriorthents			shadscale, bursage			alluvium	fan aprons	
THREESEE	Aridisols	Calcids	Haplocalcids			sagebrush, IN ricegrass, needleandthread, squirreltail			alluvium	beach plains	beach bars
THUNDERBIRD	Mollisols	Ustolls	Argiustolls	11	400	juniper-grama	60	1800	alluvium	mesas	
TICKAPOO	Aridisols	Durids	Natridurids			galleta, Nev ephedra, shadscale, winterfat, IN ricegrass			alluvium	alluvial fans	fan piedmonts
TIMPER	Aridisols	Durids	Haplodurids			graymolly, shadscale			alluvium	fan skirts	fan piedmonts, alluvial flats

(continued)

(continued)

Series Name	Order	Suborder	Great Group	MAAT (°C)	MAP (mm)	Vegetation	Max. slope (%)	Med. elev. (m)	Parent material	Landform1	Landform2
TIMPIE	Entisols	Orthents	Torriorthents	9.4	175	shadscale, greasewood	4	1615	alluvium	alluvial flats	
TOANO	Entisols	Orthents	Torriorthents			saltbush, bud sage			alluvium	alluvial flats	fan skirts, inset fans, beach plains
TOIYABE	Entisols	Psamments	Xeropsamments			jeff pine-ponder pine			colluvium, residuum	mtns granite	
TOEJA	Mollisols	Xerolls	Argixerolls	7	275	WY big sage, rabbitbrush, squirreltail	30	1750	loess, ash/residuum	hillslopes	
TOKOPER	Aridisols	Durids	Argidurids			shadscale, hopsage, menodora			residuum, colluvium	plateaus, volcanic	
TOMEL	Aridisols	Durids	Argidurids			greasewood, shadscale, bud sage			alluvium	fan remnants	fan piedmonts, alluvial fans
TOMERA	Aridisols	Argids	Natrargids			sagebrush, bluegrass, cheatgrass			alluvium	dissected terraces	alluvial fans
TONOPAH	Aridisols	Calcids	Haplocalcids	18	150	creosote bush, bursage	15	745	alluvium	fan remnants	fan piedmonts
TOOELE	Entisols	Orthents	Torriorthents	9	175	shadscale	5	1550	alluvium, lacustrine, eolian	lake terrace	
TOPEKI	Mollisols	Cryolls	Haplocryolls	5	600	mtn mahogany, sagebrush, snowberry, wheatgrass	75	2700	colluvium, residuum	mtns qtzite	
TORRO	Mollisols	Xerolls	Argixerolls			sagebrush, rabbitbrush, bitterbrush, wheatgrass, bluegrass			residuum, colluvium	mtn sideslopes	
TOSSER	Aridisols	Calcids	Haplocalcids			sagebrush, shadscale			alluvium	fan terraces	fan remnants
TOULON	Aridisols	Cambids	Haplocambids			shadscale, greasewood			alluvium	longshore bars	beach terraces
TRISTAN	Mollisols	Xerolls	Argixerolls			WY big sage, rabbitbrush, squirreltail, ant bitterbrush			residuum, colluvium basalt	mtns.	
TROCKEN	Entisols	Orthents	Torriorthents	10	150	greasewood	30	1500	alluvium	alluvial fan	fan remnants, inset fans, lake terraces
TROUGHS	Aridisols	Durids	Argidurids	8	230	sagebrush	8	1350	residuum, alluvium, loess	benches	plateaus
TRUNK	Aridisols	Argids	Haplargids			sagebrush			residuum, colluvium	mtns mixed	
TUFFO	Entisols	Orthents	Torriorthents			sagebrush			residuum	pediments tuff	
TULASE	Entisols	Orthents	Torriorthents	9	250	sagebrush	8	1900	alluvium with loess, ash	fan skirts	
TUMTUM	Aridisols	Durids	Argidurids			sagebrush			alluvium	alluvial fans	fan remnants
TUNNISON	Vertisols	Xererts	Haploxererts			sagebrush, rabbitbrush			colluvium/residuum	plateaus	
TURBA	Mollisols	Ustolls	Argiustolls	9	350	pinyon-juniper	30	2000	residuum, colluvium, ash	mtns tuff	
TUSEL	Mollisols	Cryolls	Argicryolls	6	430	sagebrush	75	2300	loess, ash/residuum, colluvium	mtns. Qtzite	
TUSK	Mollisols	Xerolls	Argixerolls			sagebrush, rabbitbrush, snowberry, ID fescue			colluvium	plateaus volcanic	
TUSTELL	Aridisols	Argids	Haplargids			sagebrush, rabbitbrush, Sandb bluegrass, squirreltail			alluvium	fan piedmont remnants	ballenas
TWEENER	Mollisols	Xerolls	Argixerolls			bitterbrush, ID fescue, wheatgrass, sagebrush			residuum, colluvium	mtns tuff	
TYBO	Aridisols	Durids	Haplodurids			hopsage, ephedra, galleta			alluvium	fan remnants	
UBEHEBE	Mollisols	Xerolls	Argixerolls			juniper-pinyon, sagebrush			residuum, colluvium	mtns mixed	

(continued)

(continued)

Series Name	Order	Suborder	Great Group	MAAT (°C)	MAP (mm)	Vegetation	Max. slope (%)	Med. elev. (m)	Parent material	Landform1	Landform2
UMBERLAND	Inceptisols	Aquepts	Halaquepts	9	150	greasewood, saltgrass	2	1500	lacustrine	lake plains	
UMIL	Aridisols	Durids	Haplodurids			sagebrush			alluvium	fan remnants, eroded	
UNIVEGA	Aridisols	Durids	Haplodurids			saltbush, horsebrush, spiny hopsage			alluvium	fan piedmonts	
UNSEL	Aridisols	Argids	Haplargids	12	150	greasewood	30	1600	alluvium	fan remnants	fan skirts
UPATAD	Mollisols	Xerolls	Argixerolls			sagebrush, wheatgrass, bluegrass			residuum, colluvium	mtns volcanic	
UPDIKE	Aridisols	Argids	Natrargids			greasewood, sagebrush			alluvium/lacustrine	lake terraces	alluvial terraces
UPPERLINE	Aridisols	Calcids	Haplocalcids			w bursage, creosotebush, chaffbush, range ratany			alluvium, colluvium/ss bedrock	rock pediments	
UPSPRING	Entisols	Orthents	Torriorthents			shadscale, winterfat			ash, residuum/basalt	mtns volcanic	
URIPNES	Entisols	Orthents	Torriorthents			horsebrush, hopsage, needlegrass			residuum, colluvium	mtns granite	
URMAFOT	Mollisols	Xerolls	Durixerolls	8	300	sagebrush	50	2000	alluvium	fan remnants	ballenas
URSINE	Aridisols	Durids	Haplodurids	11	225	sagebrush	30	1800	alluvium	fan remnant	
UVADA	Aridisols	Argids	Natrargids	10	175	shadscale, greasewood, kochia, seepweed	2	1600	lacustrine	lake plains	
UWELL	Entisols	Orthents	Torriorthents			sagebrush, squirreltail, winterfat			alluvium/lacustrine	alluvial flats	inset fans, lake plains
VACE	Aridisols	Calcids	Petrocalcids	18	150	blackbrush, ephedra, bitterbrush	30	915	loess/alluvium mixed	fan remnants	ballenas
VALMY	Entisols	Orthents	Torriorthents			sagebrush			loess, ash/alluvium	inset fans	fan skirts, basin floors
VANWYPER	Aridisols	Argids	Haplargids	7	250	sagebrush	70	1800	residuum, colluvium	mtns mixed	
VEET	Aridisols	Cambids	Haplocambids	12	225	sagebrush	15	1800	alluvium	alluvial fan	fan remnants, fan skirts, inset fans
VETA	Aridisols	Cambids	Haplocambids			sagebrush, hopsage			alluvium	inset fans	alluvial fans, fan remnants
VIGUS	Aridisols	Argids	Haplargids			shadscale, bud sage, galleta, spiny horsebrush, Doug rabbitbrush			alluvium ixed	fan remnants	alluvial fans, fan skirts
VINDICATOR	Aridisols	Argids	Haplargids			hopsage, ephedra, IN ricegrass, squirreltail			residuum, colluvium	mtns volcanic	
VININI	Aridisols	Durids	Argidurids			sagebrush, galleta, IN ricegrass, NV ephedra			residuum, colluvium	mtns volcanic	
VIRGIN PEAK	Mollisols	Ustolls	Haplustolls			pinyon, big sage, snowberry, Gambels oak			residuum gneiss	mtns.	
VITALE	Mollisols	Xerolls	Argixerolls	5	355	sagebush	75	1830	residuum, colluvium	lava plateaus	
WACA	Andisols	Xerands	Vitrixerands	5	125	red fir-white fir-mtn hemlock	75	1300	residuum	mtns tuff	
WAHGUYHE	Entisols	Orthents	Torriorthents			sagebrush, ephedra, menodora			residuum, colluvium	mtns volcanic	
WAKANSAPA	Mollisols	Ustolls	Argiustolls			pinyon-juniper			residuum, colluvium	mtns volcanic	
WALA	Entisols	Orthents	Torriorthents			sagebrush, hopsage, winterfat			residuum, colluvium	mtns volcanic	
WALTI	Mollisols	Xerolls	Argixerolls			sagebrush, rabbitbrush, ID fescue			residuum, colluvium	mtns volcanic	

(continued)

(continued)

Series Name	Order	Suborder	Great Group	MAAT (°C)	MAP (mm)	Vegetation	Max. slope (%)	Med. elev. (m)	Parent material	Landform1	Landform2
WAMP	Aridisols	Calcids	Petrocalcids			blackbrush, mtn big sage, blue grama			alluvium ls	fan remnants	ballenas
WARDBAY	Mollisols	Cryolls	Calcicryolls			sagebrush			residuum, colluvium	mtns ls	
WARDENOT	Entisols	Orthents	Torriorthents	12	125	sagebrush	30	1500	alluvium	alluvial fan	fan piedmonts, fan skirts, inset fans
WASSIT	Aridisols	Argids	Haplargids			pinyon-juniper			eolian ash, residuum, colluvium	mtns volcanic	
WATAH	Inceptisols	Aquepts	Humaquepts			sedges			organic/alluvium	floodplains	
WATOOPAH	Aridisols	Argids	Haplargids	10	225	sagebrush	8	1850	alluvium, ash	fan piedmonts	fan remnants
WECHECH	Aridisols	Calcids	Petrocalcids	18	150	creosote bush	30	900	alluvium ls	fan remnant	
WEDERTZ	Aridisols	Argids	Haplargids			sagebrush, IN ricegrass			alluvium	alluvial terraces	fan piedmonts
WEENA	Entisols	Orthents	Torriorthents			IN ricegrass, shadscale, greasewood			residuum, colluvium	plateaus-ss, sist	
WEISER	Aridisols	Calcids	Haplocalcids	18	125	creosote bush	30	900	alluvium	fan remnant	
WELCH	Mollisols	Aquolls	Endoaquolls	6	360	–	15	2050	alluvium	remnant fans	
WENDANE	Inceptisols	Aquepts	Halaquepts	9	180	greasewood	4	1600	loess, ash/alluvium	alluvial flats	stream terraces
WESFIL	Entisols	Orthents	Torriorthents			sagebrush, Sandb bluegrass, squirreltail, rabbitbrush			residuum, colluvium	mtns mixed	
WESO	Aridisols	Cambids	Haplocambids	9	180	shadscale	8	1475	loess, ash/alluvium	beach plains	relic lagoons, fan skirts, inset fans
WESTBUTTE	Mollisols	Xerolls	Haploxerolls	5.6	380	ID fescue, wheatgrass	75	1750	colluvium	plateaus-basalt	
WHIRLO	Aridisols	Cambids	Haplocambids	9	180	sagebrush	15	1500	loess/alluvium	inset fans	fan aprons, fan collars
WHITEBASIN	Aridisols	Gypsids	Haplogypsids			dalea, sandpaper plant, ephedra			residuum, colluvium	pediments gypsif	
WHOLAN	Aridisols	Cambids	Haplocambids			shadscale			alluvium	inset fans	fan skirts, fan remnants
WICKAHONEY	Alfisols	Xeralfs	Haploxeralfs	6	380	sagebrush, ID fescue	45	1850	residuum, colluvium	tablelands	foothills
WIELAND	Aridisols	Argids	Haplargids	9	230	sagebrush	30	1700	alluvium, loess, ash	fan remnants	ballenas
WIFFO	Entisols	Orthents	Torriorthents			sagebrush, IN ricegrass			alluvium	fan inserts	inset fans
WINKLO	Aridisols	Argids	Haplargids			blackbrush, ephedra, yucca			colluvium	mtns tuff	
WINTERMUTE	Aridisols	Calcids	Haplocalcids	8	180	shadscale, rabbitbrush	15	1850	alluvium	fan remnants	fan skirts, beach plains
WISKAN	Aridisols	Argids	Haplargids			sagebrush			loess/residuum, colluvium	mtns mixed	
WOOFUS	Mollisols	Aquolls	Endoaquolls			wildrye, NV bluegrass, inland saltgrass, willows			alluvium	floodplains	
WRANGO	Entisols	Orthents	Torriorthents			sagebrush, bud sage, desertgrass			alluvium	alluvial fans	alluvial terraces, inset fans
WRENZA	Inceptisols	Cryepts	Humicryepts	4	660	serviceberry, snowberry, chokecherry	75	2700	loess/colluvium/metamor bedrock	mtns metamorph	
WYLO	Mollisols	Xerolls	Argixerolls	10	230	sagebrush	50	1750	colluvium	mtns basalt	
WYVA	Aridisols	Argids	Haplargids			sagebrush, ephedra			residuum, colluvium	mtns volcanic	

(continued)

(continued)

Series Name	Order	Suborder	Great Group	MAAT (°C)	MAP (mm)	Vegetation	Max. slope (%)	Med. elev. (m)	Parent material	Landform1	Landform2
XERXES	Aridisols	Argids	Haplargids			sagebrush, bitterbrush, wheatgrass			residuum, colluvium, ash	plateaus tuff	
XINE	Mollisols	Xerolls	Calcixerolls	7	360	sagebrush, snowberry	75	2100	residuum, colluvium	mtns ls	
XMAN	Aridisols	Argids	Haplargids			cheatgrass, sagebrush, squirreltail, rabbitbrush			residuum	mtns volcanic	
YELBRICK	Aridisols	Calcids	Haplocalcids			galleta, IN ricegrass, shadscale, bud sage			alluvium	barrier beaches	longshore bars
YERINGTON	Entisols	Orthents	Torriorthents			IN ricegrass, dalea, rabbitbrush, shadscale			alluvium	sand sheets	fan remnants
YERMO	Entisols	Orthents	Torriorthents	17	125	creosotebush	50	990	alluvium	alluvial fan	
YIPOR	Entisols	Orthents	Torriorthents			shadscale, greasewood, suaeda, mustard			alluvium	inset fans	fan skirts
YOBE	Inceptisols	Aquepts	Halaquepts			greasewood, saltgrass, iodine bush			lacustrine	alluvial flats	lake plains
YODY	Aridisols	Durids	Argidurids	8	230	sagebrush	15	1900	alluvium	fan remnants	
YOMBA	Aridisols	Cambids	Haplocambids	12	150	greasewood, shadshale	4	1575	alluvium	alluvial flats	alluvial plains, fan piedmonts, inset fans, fan skirts
YOTES	Mollisols	Xerolls	Haploxerolls			sagebrush, needleandthread, IN ricegrass, blue grama			alluvium	fan remnants	
YOUNGSTON	Entisols	Fluvents	Torrifluvents	7.2	175	wheatgrass, sagebrush	10	1525	alluvium	alluvial terraces	fans, fan aprons
YUKO	Aridisols	Argids	Haplargids			sagebrush			residuum, colluvium	mtns volcanic	
YURM	Aridisols	Calcids	Petrocalcids	18	100	creosotebush, w burs, yucca	8	1000	alluvium qtzite, ls	ballenas	fan piedmonts
ZADVAR	Aridisols	Durids	Argidurids	8	230	sagebrush	30	1850	alluvium	fan remnants	
ZAFOD	Aridisols	Durids	Haplodurids			sagebrush, bluegrass, IN ricegrass			alluvium	fan remnants	
ZAIDY	Aridisols	Durids	Argidurids			sagebrush, Sandberg bluegrass, rabbitbrush			alluvium	fan piedmont remnants	
ZAPA	Aridisols	Durids	Haplodurids			sagebrush, shadscale, squirreltail, cheatgrass			alluvium	fan remnants	
ZAQUA	Aridisols	Argids	Haplargids			blackbrush, bitterbrush			residuum, colluvium	mtns tuff	
ZARARK	Mollisols	Cryolls	Calcicryolls			mtnmahogany, wheatgrass, snowberry			residuum, colluvium	mtns ls	
ZEHEME	Aridisols	Calcids	Haplocalcids	16	200	blackbrush	75	1250	residuum, colluvium	mtns ls	
ZERK	Aridisols	Calcids	Haplocalcids			shadscale, bud sage, squirreltail			alluvium	fan remnants	fan skirts, inset fans, beach plains
ZEVADEZ	Aridisols	Argids	Haplargids	9	230	sagebrush	50	1600	loess, ash, alluvium	fan piedmont remnants	
ZIBATE	Aridisols	Argids	Haplargids			blackbrush, ephedra			residuum	mtns tuff	
ZIMBOB	Entisols	Orthents	Torriorthents	9	300	sagebrush	75	2200	residuum, colluvium	mtns ls	
ZINEB	Aridisols	Cambids	Haplocambids			sagebrush, horsebrush, squirreltail, Sandb bluegrass			alluvium	inset fans	fan aprons, fan skirts
ZODA	Aridisols	Durids	Argidurids			sagebrush, squirreltail, IN ricegrass			alluvium	fan remnants	
ZOESTA	Aridisols	Argids	Paleargids			low sage, squirreltail, pine bluegrass			alluvium, colluvium	sideslopes, fan remnants	
ZORRAVISTA	Entisols	Psamments	Torripsamments			sagebrush, rabbitbrush			eolian	sand dunes	sand sheets
ZYMANS	Mollisols	Xerolls	Argixerolls			sagebrush, wheatgrass, needlegrass			loess, ash, colluvium	mtns volcanic	

Appendix D
List of Benchmark Soils in Nevada

Soil series	Area (km^2)
ALLEY	467.2
ANTHONY	3.7
APPIAN	422.6
ARIZO	1204.8
ARMESPAN	1248.1
ASHONE	15.5
BELLEHELEN	877.3
BEOSKA	614.9
BLIMO	321.9
BLUEWING	953.8
BOREALIS	152.2
BORVANT	11.2
BROYLES	742.7
BUBUS	517.0
BULLVILLE	10.6
CAGWIN	4.1
CALPINE	8.6
CARSON	308.0
CAVE	341.3
CAVEHILL	1065.9
CHEN	884.3
CHIARA	2197.0
CIRAC	512.9
CLEAVAGE	1679.0
CLEAVER	416.6
COBRE	145.3
CROPPER	458.5
DACKER	508.8
DECAN	222.7
DELHEW	14.9
DENMARK	0.1
DEVILSGAIT	249.0
DIA	114.8
DONNA	503.6
DOWNEYVILLE	1253.9

(continued)

P. W. Blackburn et al., *The Soils of Nevada*, World Soils Book Series,
https://doi.org/10.1007/978-3-030-53157-7

(continued)

Soil series	Area (km^2)
DRYGYP	24.6
DUCO	307.0
DUN GLEN	287.4
EAST FORK	94.4
ENKO	778.4
GABICA	32.1
GEER	461.6
GILA	0.1
GLENDALE	21.6
GOLDRUN	340.0
GRASSVAL	350.0
GRUMBLEN	340.3
HANDPAH	657.4
HAPGOOD	432.0
HAUNCHEE	623.6
HAWSLEY	1533.0
HIKO PEAK	2.7
HOLTVILLE	5.8
HOME CAMP	5.1
HUMBOLDT	288.0
HUNNTON	945.0
HYZEN	474.8
INDIAN CREEK	164.0
ITCA	383.6
JERVAL	487.4
KATELANA	949.4
KEEFA	463.5
KIMMERLING	22.1
KUNZLER	982.0
KYLER	1341.2
LAND	39.5
LINOYER	700.0
LOGAN	49.3
LYDA	276.8
MAZUMA	1601.0
MCCARRAN	298.2
MCCONNEL	277.0
MCIVEY	714.8
MEDBURN	104.7
MORMON MESA	604.0
MOTTSVILLE	78.0
NINEMILE	1534.0
OLA	103.5
OLD CAMP	1448.0
OROVADA	983.0
OXCOREL	819.5
PACKER	173.9
PALINOR	2508.2
PARRAN	383.1
PATNA	247.8
PAYPOINT	83.6

(continued)

(continued)

Soil series	Area (km^2)
PENOYER	321.0
PINTWATER	715.6
POOKALOO	1483.0
PUETT	621.0
PYRAT	533.0
QUARZ	506.7
RELLEY	127.2
RELUCTAN	661.0
RENO	147.0
ROCA	770.0
ROSE CREEK	61.0
ROSITAS	30.5
SALTAIR	41.4
SHABLISS	1018.0
SHEFFIT	453.2
SIBELIA	13.6
SINGATSE	1022.2
SOFTSCRABBLE	696.0
SONDOA	436.4
SONOMA	406.5
ST. THOMAS	875.0
STAMPEDE	623.0
STEWVAL	2632.5
STUMBLE	676.6
SUMINE	1313.0
SUMMERMUTE	301.0
TALLAC	19.8
THEON	1462.7
THUNDERBIRD	117.0
TIPPERARY	41.5
TOIYABE	51.2
TROCKEN	1236.3
TUFFO	151.0
TUSEL	418.0
UNSEL	1849.6
URMAFOT	767.3
URSINE	1487.7
VICEE	12.8
VOLTAIRE	31.0
WARDENOT	1004.8
WEISER	1205.0
WELCH	517.0
WENDANE	793.6
WIELAND	1086.8
WINTERMUTE	590.1
WYLO	638.0
YODY	465.3
YOMBA	484.7
YOUNGSTON	33.2
ZADVAR	851.8
ZIMBOB	723.8

Appendix E
Soil Series in Nevada that are Endemic, Rare, and Endangered

Series Name	Area (km^2)	Endemic	Rare	Endangered
ABALAN	2.4		Y	
ABGESE	75.4	N	N	N
ABOTEN	149.7	N	N	N
ACANA	100.3	N	N	N
ACKETT	92.6	N	N	N
ACKLEY	17	N	Y	N
ACOMA	103.2	N	N	N
ACRELANE	598.5	N	N	N
ACTI	122.2	N	N	N
ADAMATT	2.3		Y	
ADAVEN	1.5		Y	
ADELAIDE	153.9	N	N	N
ADOBE	189.3	N	N	N
ADOS	40.4	Y	Y	Y
ADVOKAY	96.3	N	Y	N
AFFEY	7.8		Y	
AGASSIZ	10.1	N	N	N
AGON	8.9		Y	
AGORT	35.3		Y	
AGUACHIQUITA	16.8		Y	
AHCHEW	21.8		Y	
AKELA	106.7	N	N	N
AKERCAN	13.8		Y	
AKERUE	64.2	Y	Y	Y
AKLER	138.7	Y	N	N
ALADSHI	27.1		Y	
ALAMOROAD	40.3	N	Y	N
ALBURZ	37.1		Y	
ALCAN	37.8		Y	
ALDAX	6.7		Y	
ALHAMBRA	142	N	N	N
ALKO	173.3	N	N	N
ALLEY	467.2	N	N	N

(continued)

P. W. Blackburn et al., *The Soils of Nevada*, World Soils Book Series,
https://doi.org/10.1007/978-3-030-53157-7

(continued)

Series Name	Area (km^2)	Endemic	Rare	Endangered
ALLKER	201.2	N	N	N
ALLOR	177.2	N	N	N
ALPHA	11.2		Y	
ALTA	40	N	Y	N
ALVODEST	6.4	N	N	N
ALYAN	228.7	N	N	N
AMBUSH	34.7		Y	
AMELAR	121.4	N	N	N
AMENE	44.5	Y	Y	Y
AMTOFT	354.8	N	N	N
ANAUD	60.5	N	Y	N
ANAWALT	265.6	N	N	N
ANED	25.1		Y	
ANNAW	331.8	N	N	N
ANOWELL	31.4		Y	
ANSPING	108.2	N	N	N
ANTEL	26.1		Y	
ANTENNAPEAK	28.3		Y	
ANTHOLOP	15.8		Y	
ANTHONY	3.7		N	N
APMAT	11.3		N	N
APPIAN	422.6	N	N	N
AQUINAS	31.4		Y	
ARADA	180	N	N	N
ARCIA	123.5	N	N	N
ARCLAY	193.7	N	N	N
ARDIVEY	475.6	N	N	N
ARGALT	96.3	N	Y	N
ARGENTA	163.1	N	N	N
ARIZO	1204.8	N	N	N
ARKSON	2.7		Y	
ARMESPAN	1248.1	N	N	N
ARMOINE	113	N	N	N
ARMPUP	19.2		Y	
ARMYDRAIN	6.6		N	N
ARROLIME	19.9	N	Y	N
ARVA	28.8		N	N
ARZO	41.3	N	Y	N
ASH SPRINGS	7.1		Y	
ASHART	45.3	N	Y	N
ASHCAMP	26.4		Y	
ASHDOS	24.4		Y	
ASHFLAT	2.1		Y	
ASHMED	53.4	N	Y	N
ASHONE	15.5		N	N
ASHTRE	27.6	N	N	N
ATLANTA	36.3		Y	
ATLOW	479.6	N	N	N
ATRYPA	121.6	Y	N	N
ATTELLA	33.4		Y	
AURUM	3.6		Y	

(continued)

(continued)

Series Name	Area (km^2)	Endemic	Rare	Endangered
AUTOMAL	402.8	N	N	N
AYCAB	64.9	N	Y	N
AYMATE	64.7	Y	Y	Y
AYSEES	9		N	N
AZSAND	11.2		Y	
AZTEC	30	N	N	N
AZURERIDGE	33.4		Y	
BABERWIT	15.9		Y	
BACHO	40	N	Y	N
BADENA	72.5	N	Y	N
BADGERCAMP	40.7	N	N	N
BADHAP	64.2	N	Y	N
BAGARD	0.1		Y	
BAKERPEAK	52.5	N	Y	N
BAKSCRATCH	3.8		N	N
BAMOS	0.3	N	N	N
BANCY	23.3		N	N
BANGO	205.7	N	N	N
BARD	555.5	N	N	N
BARERANCH	1.5		N	N
BARFAN	34.8		Y	
BARNARD	28.1	N	N	N
BARNMOT	47	N	Y	N
BARRIER	66.2	Y	Y	Y
BARSHAAD	26.2		Y	
BARTINE	74.4	N	Y	N
BARTOME	265.3	N	N	N
BASELINE	76.8	Y	Y	Y
BASINPEAK	3.3		N	N
BASKET	168.6	N	N	N
BASTIAN	1.9		Y	
BATAN	427.2	N	N	N
BEANFLAT	73.6	Y	Y	Y
BEANO	18.7		Y	
BEARBUTTE	41.8	N	Y	N
BEARSKIN	0.1		N	N
BEDELL	22.3		Y	
BEDWYR	34.1		Y	
BEDZEE	11.7		Y	
BEELEM	359.7	N	N	N
BEEOX	30.1		Y	
BEERBO	12.8		N	N
BELATE	83.8	N	Y	N
BELCHER	52.2	N	Y	N
BELLEHELEN	877.3	N	N	N
BELLENMINE	29.7		N	N
BELMILL	67	N	Y	N
BELSAC	49.2	N	Y	N
BELTED	328.2	N	N	N
BENDASTIK	10.1		Y	
BENIN	321.5	N	N	N

(continued)

(continued)

Series Name	Area (km^2)	Endemic	Rare	Endangered
BEOSKA	614.9	N	N	N
BEOWAWE	23.5		Y	
BERIT	67.7	N	Y	N
BERNING	38		Y	
BERRYCREEK	39.1		Y	
BERZATIC	79.1	N	Y	N
BESHERM	94.5	Y	Y	Y
BETRA	50.2	Y	Y	Y
BEZO	46.4	N	Y	N
BICONDOA	23.2		Y	
BIDART	0.8		Y	
BIDDLEMAN	291.7	N	N	N
BIDRIM	22.4		Y	
BIEBER	45.9	Y	N	N
BIENFAIT	59.6	N	Y	N
BIGA	255.3	Y	N	N
BIGHAT	5.3		Y	
BIGMEADOW	10.9		Y	
BIGSPRING	12.9		N	N
BIGWASH	38.5		Y	
BIJI	61.5	Y	Y	Y
BIJORJA	9.7		Y	
BIKEN	143	Y	N	N
BILBO	256.3	N	N	N
BIMMER	12.5		Y	
BIOYA	471	N	N	N
BIRCHCREEK	109.8	N	N	N
BIRDSPRING	218.9	N	N	N
BISHOP	4.3		N	N
BITNER	11.8		Y	
BITTER SPRING	118.6	Y	N	N
BITTERRIDGE	8.7		Y	
BLACK BUTTE	6.5		Y	
BLACKA	16.6		Y	
BLACKCAN	58	N	Y	N
BLACKHAWK	245.6	N	N	N
BLACKMESA	5.2		Y	
BLACKNAT	1		Y	
BLACKTOP	865.3	N	N	N
BLACKWELL	3.9		Y	
BLAPPERT	84.8	N	Y	N
BLIMO	321.9	N	N	N
BLISS	245	N	N	N
BLITZEN	16.2		Y	
BLIZZARD	0.3		Y	
BLOOR	53.9	Y	Y	Y
BLUDIAMOND	19.3		Y	
BLUEAGLE	31.4		Y	
BLUEGYP	3.3	N	Y	N
BLUEHILL	56.5	Y	Y	Y
BLUEMASS	13.3		Y	

(continued)

(continued)

Series Name	Area (km^2)	Endemic	Rare	Endangered
BLUEPOINT	198.4	N	N	N
BLUEWING	953.8	N	N	N
BOBNBOB	35.5		Y	
BOBS	301	Y	N	N
BOBZBULZ	3.2		Y	
BODIEHILL	2.7		Y	
BOGER	61.8	Y	Y	Y
BOJO	155.8	N	N	N
BOLTZ	7		Y	
BOMBADIL	165.7	N	N	N
BOOFORD	21.4		Y	
BOOFUSS	96.1	Y	Y	Y
BOOMSTICK	123.1	N	N	N
BOOMTOWN	44.7	Y	Y	Y
BOONDOCK	54.3	N	Y	N
BORDA	4.6		Y	
BOREALIS	152.2	Y	N	N
BORVANT	11.2	Y	N	N
BOSCO	8.8		Y	
BOSO	5.9		Y	
BOTLEG	4.2		Y	
BOTON	1105.2	N	N	N
BOULDER LAKE	13.6	N	N	N
BOULFLAT	55.1	N	Y	N
BOUNCER	8		Y	
BOXSPRING	817.4	N	N	N
BRACKEN	36.2	N	Y	N
BRADSHAW	10.3		N	N
BRAMWELL	0.1		N	N
BRAWLEY	42.6	Y	Y	Y
BREGAR	287.6	N	N	N
BREKO	148.3	N	N	N
BRICONE	90.1	Y	Y	Y
BRIER	287	N	N	N
BRINKER	10.1		Y	
BRINNUM	20.2		Y	
BROCK	45.5	N	Y	N
BROCKLISS	4.2		Y	
BROE	21.8		Y	
BROKIT	15.2		Y	
BROLAND	205.5	Y	N	N
BROWNSBOWL	1.9		Y	
BROYLES	742.7	Y	N	N
BRUBECK	22.9		N	N
BRUFFY	13.2		Y	
BUBUS	517	N	N	N
BUCAN	277.4	N	N	N
BUCKAROO	180.3	N	N	N
BUCKLAKE	354.5	N	N	N
BUCKSPRING	130.5	N	N	N
BUDIHOL	99.3	N	N	N

(continued)

(continued)

Series Name	Area (km^2)	Endemic	Rare	Endangered
BUFFARAN	388	N	N	N
BULAKE	335	N	N	N
BULLFOR	23.7		Y	
BULLUMP	275.6	N	N	N
BULLVARO	16.1		Y	
BULLVILLE	10.6		Y	
BUNDORF	50.8	Y	N	N
BUNEJUG	33.6		Y	
BUNKY	30.8		Y	
BURNBOROUGH	74.8	N	Y	N
BURRITA	540.5	N	N	N
BUZZTAIL	37.3		Y	
BYLO	76	Y	Y	Y
CABINPINE	6.2		Y	
CAFETAL	23.4		Y	
CAGAS	17.7		Y	
CAGLE	45.9	N	N	N
CAGWIN	4.1	N	N	N
CALICO	13.7		Y	
CALIZA	19.5	N	N	N
CALLAT	0.1		Y	
CALLVILLE	50.6	Y	Y	Y
CALPEAK	36.6		Y	
CALPINE	8.6		N	N
CALWASH	7.4		Y	
CAMEEK	55.3	N	Y	N
CANDELARIA	582.3	N	N	N
CANFIRE	0.02	N	Y	N
CANIWE	7.8		Y	
CANOTO	177.3	N	N	N
CANUTIO	28.3		N	N
CANYONFORK	13.3		Y	
CANYOUNG	43.3	N	Y	N
CAPHOR	42.7	N	Y	N
CAPSUS	62.6	N	Y	N
CARCITY	9.6		Y	
CARIOCA	12.9		Y	
CARRIZO	174.8	N	N	N
CARRWASH	36.2		Y	
CARSON	308	Y	N	N
CARSTUMP	163.3	N	N	N
CARWALKER	15.7		Y	
CASAGA	79.6	N	Y	N
CASLO	2.9		Y	
CASSENAI	32.4		Y	
CASSIRO	67.8	N	N	N
CATH	386.5	N	N	N
CAUDLE	13.4		Y	
CAVANAUGH	16.7	N	Y	N
CAVE	341.3	N	N	N
CAVEHILL	1065.9	N	N	N

(continued)

(continued)

Series Name	Area (km^2)	Endemic	Rare	Endangered
CAVEMOUNTAIN	9.8		Y	
CAVEROCK	1.5		Y	
CAVIN	6.5	N	Y	N
CEDARAN	126.6	N	N	N
CEDARCABIN	21.2		Y	
CEEBEE	5.4		Y	
CEEJAY	315.5	N	N	N
CELETON	96.1	N	Y	N
CETREPAS	23.5		Y	
CEWAT	11.3		Y	
CHAD	79	Y	Y	Y
CHAINLINK	83.9	N	Y	N
CHALCO	115.4	N	N	N
CHANYBUCK	7.5		Y	
CHAPPUIS	7.7		Y	
CHARKILN	5.9		Y	
CHARLEBOIS	14.4		Y	
CHARNOCK	17.8		Y	
CHARPEAK	202.9	Y	N	N
CHARWELL	8		Y	
CHAYSON	49.5	N	N	N
CHECKETT	130.3	N	N	N
CHEDEHAP	17		Y	
CHEME	95.2	N	Y	N
CHEN	884.3	N	N	N
CHERRY SPRING	310.5	N	N	N
CHIARA	2197	Y	N	N
CHIEFPAN	14.6		Y	
CHIEFRANGE	12.4		Y	
CHILL	98.9	N	Y	N
CHILPER	143.6	Y	N	N
CHIME	26.1		Y	
CHINKLE	31.9		Y	
CHRISTOPHER	0.9		Y	
CHUBARD	609.3	N	N	N
CHUCKLES	54.8	N	N	N
CHUCKMILL	197.3	N	N	N
CHUCKRIDGE	205.7	N	N	N
CHUFFA	193.2	N	N	N
CHUG	101.4	Y	N	N
CHURCHILL	25.4		Y	
CHUSKA	85	N	N	N
CIRAC	512.9	N	N	N
CLANALPINE	243	N	N	N
CLEAVAGE	1679	N	N	N
CLEAVER	416.6	N	N	N
CLEAVMOR	22.9		Y	
CLEMENTINE	149.2	N	N	N
CLIFFDOWN	177	N	N	N
CLIMINE	7		Y	
CLOSKEY	8.8		Y	

(continued)

(continued)

Series Name	Area (km^2)	Endemic	Rare	Endangered
CLOWFIN	88.2	N	Y	N
CLURDE	123	N	N	N
CLURO	39.3		Y	
COBATUS	20.2		Y	
COBBLYWHEEL	17.8		Y	
COBRE	145.3	Y	N	N
COFF	11		Y	
COFFEPOT	19.3		Y	
COILS	96.3	N	Y	N
COIT	8.1		Y	
COLADO	1.7		Y	
COLBAR	265.5	N	N	N
COLDENT	49.9	N	Y	N
COLOLAG	6.2		Y	
COLOROCK	66.6	N	Y	N
COLTROOP	66	N	Y	N
COLVAL	114.8	Y	N	N
COMMSKI	390.9	N	N	N
CONNEL	243.4	N	N	N
CONTACT	7.8		Y	
COOPERWASH	1.1		Y	
COPPEREID	112.4	N	N	N
COPPERSMITH	1.7		Y	
CORBETT	18.4	N	N	N
CORBILT	116.1	Y	N	N
CORMOL	20.2		Y	
CORNCREEK	27.2		Y	
CORNFLAT	3.4		Y	
CORRAL	40.9	N	N	N
CORTEZ	108	N	N	N
COSER	130	N	N	N
COTANT	370.4	N	N	N
COUCH	40.5	Y	Y	Y
COUTIS	11.6	N	N	N
COWBELL	0.2		Y	
COWGIL	73.8	N	N	N
COZTUR	27.8	N	N	N
CRADLEBAUGH	7		Y	
CREDO	35.6		Y	
CREEMON	290.9	Y	N	N
CREN	47.7	N	Y	N
CRESAL	67.2	N	Y	N
CRESTLINE	59.1	N	N	N
CRETHERS	4.4		Y	
CREVA	42.4	Y	Y	Y
CRISPY	8.9		Y	
CROCAN	25.1		Y	
CROESUS	41.5	N	Y	N
CROOKED CREEK	312.2	Y	N	N
CROPPER	458.5	N	N	N
CROSGRAIN	176.4	N	N	N

(continued)

(continued)

Series Name	Area (km^2)	Endemic	Rare	Endangered
CRUNKER	43.4	Y	Y	Y
CRUNKVAR	4.6		Y	
CRUTCHER	4		Y	
CRUZSPRING	46.3	N	Y	N
CRYSTAL SPRINGS	45.4	Y	Y	Y
CUCAMUNGO	69.8	N	Y	N
DAB	13.5		Y	
DACKER	508.8	N	N	N
DAGGET	12		Y	
DAICK	34.8		Y	
DAKENT	16.8		Y	
DALIAN	11.4		Y	
DALZELL	16.5		Y	
DANGBERG	12.8		Y	
DAPHSUE	51.1	N	Y	N
DAVEY	305.1	N	N	N
DEADYON	153.7	N	N	N
DEANRAN	10.1		Y	
DEARBUSH	39		Y	
DECAN	222.7	N	N	N
DECATHON	37.5		Y	
DECRAM	42.4	N	Y	N
DEDAS	39.4		N	N
DEDMOUNT	30.6		Y	
DEEFAN	25.9		Y	
DEEPEEK	44.5	N	Y	N
DEERHILL	3		Y	
DEERLODGE	4.2		N	N
DEFLER	91.2	N	Y	N
DEKOOM	12		Y	
DELACIT	4.7		Y	
DELAMAR	333	Y	N	N
DELEPLAIN	3.8		Y	
DELHEW	14.9		Y	
DELMO	12.5		Y	
DELP	29		Y	
DELVADA	85.8	N	Y	N
DEMILL	2.4		Y	
DENAY	38.5		Y	
DENIHLER	55.2	N	Y	N
DENIO	30.2		Y	
DENMARK	0.1	Y	Y	Y
DENPARK	92.3	N	Y	N
DEPPY	114.9	N	N	N
DESATOYA	98.3	Y	Y	Y
DESEED	13.4		Y	
DESTAZO	27.7		Y	
DEUNAH	43.2	Y	N	N
DEVADA	1311	N	N	N
DEVEN	8.4	N	N	N
DEVILDOG	139.5	Y	N	N

(continued)

(continued)

Series Name	Area (km^2)	Endemic	Rare	Endangered
DEVILS	18.6		N	N
DEVILSGAIT	249	N	N	N
DEVILSTHUMB	5.3		Y	
DEVOY	24.3		Y	
DEWAR	1059.8	N	N	N
DEWRUST	27.7		Y	
DIA	114.8	Y	N	N
DIAMONDHIL	5.5		Y	
DIANEV	182.1	Y	N	N
DIAZ	0.2		Y	
DITHOD	157	N	N	N
DOBEL	116.3	N	N	N
DOESPRING	21.4		Y	
DOMEHILL	14.5		Y	
DOMEZ	11.5		Y	
DONNA	503.6	N	N	N
DOOH	12.3		Y	
DOORKISS	167.7	N	N	N
DOOWAK	8.9		Y	
DORPER	520.4	N	N	N
DOSIE	125.3	N	N	N
DOTEN	60	N	Y	N
DOTSOLOT	20.7		Y	
DOUHIDE	280.9	N	N	N
DOWNEYVILLE	1253.9	N	N	N
DRESSLER	9		Y	
DRESSLEWET	6.8		Y	
DREWING	30.3		Y	
DRIT	17.1		Y	
DRYGYP	24.6	Y	Y	Y
DUCKHILL	18.6		Y	
DUCO	307	N	N	N
DUFF	44.6	N	N	N
DUFFER	421.4	Y	N	N
DUGCHIP	86.2	Y	Y	N
DUGWAY	60	N	Y	N
DUN GLEN	287.4	Y	N	N
DUNPHY	88.6	N	Y	N
DUTCHJOHN	13.1		Y	
EAGLEPASS	477.1	N	N	N
EAGLEROCK	112.5	N	N	N
EARCREE	8.1		N	N
EAST FORK	94.4	N	Y	N
EASTE	2.7		Y	
EASTGATE	197.9	N	N	N
EASTLAND	0.06	N	Y	N
EASTMORE	166.7	N	N	N
EASTVAL	23		Y	
EASTWELL	65.2	N	N	N
EASYCHAIR	42.4	N	Y	N
EBIC	44.8	N	Y	N

(continued)

(continued)

Series Name	Area (km^2)	Endemic	Rare	Endangered
EBODA	104.5	N	N	N
EDNAGREY	10		Y	
EENREED	22.1		Y	
EGANROC	96.9	N	Y	N
EIGHTMILE	30.5		Y	
EKIM	34.1		Y	
ELAERO	0.1	N	Y	N
ELBOWCANYON	52.4	N	Y	N
ELHINA	7.6		Y	
ELOCIN	13.6		Y	
EMAGERT	5		Y	
EMAMOUNT	1.1		Y	
ENKO	778.4	N	N	N
ENTERO	73.6	N	Y	N
ENVOL	69.7	N	Y	N
EOJ	34.5		Y	
EPVIP	11		Y	
EQUIS	124.4	Y	N	N
ERAKATAK	21.4		N	N
ERASTRA	73.2	N	Y	N
ERBER	31.5		Y	
ESCALANTE	253.6	N	N	N
ESMOD	73	N	Y	N
ESPINT	40.2	N	Y	N
ESSAL	62.2	Y	Y	Y
EWELAC	89.5	Y	Y	Y
FADOLL	45.7	Y	Y	Y
FAIRYDELL	35.3		Y	
FALERIA	32.9		Y	
FALLON	74.3	N	Y	N
FANG	125.8	N	N	N
FANU	66	N	Y	N
FAREPEAK	45.9	N	Y	N
FAWIN	28		Y	
FAX	76.9	Y	Y	Y
FENELON	24.4		Y	
FENSTER	42.2	N	Y	N
FERA	58.9	N	Y	N
FERDELFORD	29.1		Y	
FERNLEY	37.5		Y	
FERNPOINT	26		Y	
FERROGOLD	130.8	N	N	N
FERTALINE	20.7		Y	
FERVER	40.8	Y	Y	Y
FETTIC	8.7		Y	
FEZ	39.6	N	Y	N
FIDDLER	6.7	N	N	N
FIFTEENMILE	58.6	N	Y	N
FILAREE	48.5	N	N	N
FILIRAN	33.6		Y	
FINDOUT	59.7	Y	Y	Y

(continued)

(continued)

Series Name	Area (km^2)	Endemic	Rare	Endangered
FIREBALL	56	N	Y	N
FIVEMILE	18.1	N	Y	N
FLATNOSEWASH	23.5		Y	
FLATTOP	86.9	N	Y	N
FLEISCHMANN	10.7		Y	
FLETCHERPEAK	80.6	N	Y	N
FLEX	39.4		N	N
FLOER	5.1		Y	
FLUE	139.7	N	N	N
FORTANK	79.3	N	Y	N
FORVIC	14.6		Y	
FOUR STAR	11.3		Y	
FOXCAN	28.7		Y	
FOXMOUNT	20.8		Y	
FOXVIRE	12.6		Y	
FRANKTOWN	7		N	N
FRAVAL	31.9		Y	
FRENTERA	46.5	N	Y	N
FREWA	11.2		Y	
FREZNIK	4	N	N	N
FRINES	8.3		Y	
FRODO	38		Y	
FUBBLE	26.1		Y	
FUEGOSTA	26.2		Y	
FUGAWEE	14.9	N	N	N
FULSTONE	440	N	N	N
FUSULINA	32.9		Y	
FUSUVAR	3.4		Y	
GABBS	11		Y	
GABBVALLY	1233.3	N	N	N
GABEL	3.7		Y	
GABICA	32.1		Y	
GAIA	2.5		Y	
GALEHILLS	40.1	N	Y	N
GALEPPI	36.3	N	N	N
GAMGEE	26.2		Y	
GANAFLAN	54.3	N	Y	N
GANCE	264.7	Y	N	N
GANDO	59.3	N	Y	N
GARDELLA	9.3		Y	
GARDENVALLEY	25.9		Y	
GARDNERVILLE	5.7		Y	
GARFAN	31.1		Y	
GARHILL	221.8	N	N	N
GARNEL	22.1		Y	
GEER	461.6	N	N	N
GEFO	1.5		Y	
GENAW	130.6	N	N	N
GENEGRAF	399.6	N	N	N
GENOA	5.9		Y	
GENOAPEAK	0.6		Y	

(continued)

(continued)

Series Name	Area (km^2)	Endemic	Rare	Endangered
GETA	92.5	N	Y	N
GEYSEN	31.7		Y	
GILA	0.05	N	N	N
GINEX	7.9		Y	
GITAKUP	28.5		Y	
GLASSHAWK	7.9		Y	
GLEAN	141.9	N	N	N
GLENBROOK	72.6	N	N	N
GLENCARB	96.2	Y	Y	Y
GLENDALE	21.6		N	N
GLIDESKI	5.5		Y	
GLOTRAIN	82.5	N	Y	N
GLYPHS	74.2	N	Y	N
GOCHEA	190.8	Y	N	N
GODECKE	12.2		Y	
GOL	58.2	Y	Y	Y
GOLCONDA	195.4	N	N	N
GOLDBUTTE	74.7	N	Y	N
GOLDROAD	324	N	N	N
GOLDRUN	340	N	N	N
GOLDYKE	113.5	N	N	N
GOLLAHER	325.7	Y	N	N
GOLSUM	56	N	Y	N
GOMINE	0.1	N	Y	N
GOODSKI	23.3		Y	
GOODSPRINGS	72	N	N	N
GOODWATER	68.5	N	Y	N
GOOSEL	120.2	N	N	N
GORZELL	0.5		Y	
GOSUMI	46.6	N	Y	N
GOVWASH	3.6	N	Y	N
GOWJAI	41.3	N	Y	N
GRALEY	332.7	N	N	N
GRANDEPOSIT	52.4	N	Y	N
GRANDRIDGE	3.9		Y	
GRANIPEAK	8.9		Y	
GRANMOUNT	23.1		Y	
GRANQUIN	16.1		Y	
GRANSHAW	362	N	N	N
GRANZAN	12.2		Y	
GRAPEVINE	51	N	Y	N
GRASSVAL	350	N	N	N
GRASSYCAN	69.1	N	Y	N
GRAUFELS	98	N	N	N
GRAVIER	336.6	Y	N	N
GRAYLOCK	3.1		Y	
GREATDAY	2.1		Y	
GREENBRAE	69.7	N	Y	N
GREENGROVE	18.8		Y	
GREMMERS	31.9		Y	
GREYEAGLE	196.3	N	N	N

(continued)

(continued)

Series Name	Area (km^2)	Endemic	Rare	Endangered
GRIFFY	8.9	N	N	N
GRIFLEYS	29.6		Y	
GRIMLAKE	3.8		Y	
GRINA	155.7	N	N	N
GRINK	67.7	N	Y	N
GRIVER	20		Y	
GROSSCHAT	8.2		Y	
GROWSET	8.9		Y	
GRUBE	49	N	Y	N
GRUMBLEN	340.3	N	N	N
GUARDIAN	49.6	Y	Y	Y
GUISER	61.9	N	Y	N
GUMBLE	321	N	N	N
GUND	63.1	Y	Y	Y
GURDUGEE	9.4		Y	
GWENA	55.4	Y	Y	Y
GYNELLE	697.7	N	N	N
GYPWASH	26.9		Y	
HAAR	44.6	N	Y	N
HAARVAR	20.3		Y	
HACKWOOD	345.5	N	N	N
HALACAN	149.1	N	N	N
HALEBURU	447	N	N	N
HALFASH	7.1		Y	
HALLECK	49.2	N	Y	N
HALVERT	7.7		Y	
HAMACER	22		Y	
HAMTAH	144	N	N	N
HANDPAH	657.4	N	N	N
HANDY	121.5	Y	N	N
HANGROCK	118.5	Y	N	N
HAPGOOD	432	N	N	N
HARCANY	64.6	N	N	N
HARDBASIN	5.2		Y	
HARDHAT	273.3	N	N	N
HARDNUT	16.6		Y	
HARDOL	194.4	N	N	N
HARDZEM	245.4	N	N	N
HARSKEL	2		Y	
HART CAMP	5.1	N	N	N
HARTIG	48.8	N	N	N
HARVAN	8		Y	
HASHWOODS	5.5		Y	
HASTEE	13.3		Y	
HATPEAK	85.7	N	N	N
HATUR	8.9		Y	
HAUNCHEE	623.6	N	N	N
HAVINGDON	97.1	N	Y	N
HAWSLEY	1533	N	N	N
HAYBOURNE	192.2	N	N	N
HAYESTON	181.8	N	N	N

(continued)

(continued)

Series Name	Area (km^2)	Endemic	Rare	Endangered
HAYMONT	128.9	N	N	N
HAYPEAK	16.2		Y	
HAYPRESS	49.7	N	N	N
HAYSPUR	15.6		Y	
HECKISON	14	N	N	N
HEECHEE	122.1	N	N	N
HEFED	24		Y	
HEIDTMAN	6.3		Y	
HEIST	690	N	N	N
HELEWEISER	69	N	Y	N
HELKITCHEN	16.2		Y	
HENDAP	5.7		Y	
HENNINGSEN	10.9		Y	
HESSING	153.1	N	N	N
HEUSSER	16		Y	
HIDDENFOREST	4.2		Y	
HIDDENSUN	40	N	Y	N
HIGHAMS	17.4		Y	
HIGHLAND	49.1	N	Y	N
HIGHUP	36.4		Y	
HIKO PEAK	2.7	Y	N	N
HIKO SPRINGS	6.4		N	N
HILLER	66.6	Y	Y	Y
HIRIDGE	57	N	Y	N
HIRSCHDALE	26.2		Y	
HOBOG	5.2		N	N
HOCAR	7.9		Y	
HODEDO	77.5	N	Y	N
HOGMALAT	10.1		Y	
HOGUM	4.3		Y	
HOLBORN	43.6	N	Y	N
HOLBROOK	82	Y	Y	Y
HOLLACE	56.1	N	Y	N
HOLLYWELL	219	N	N	N
HOLSINE	23.3		Y	
HOLTLE	22.9		Y	
HOLTVILLE	5.8	Y	N	N
HOME CAMP	5.1		Y	
HOMESTAKE	103.9	N	N	N
HOOPLITE	205	N	N	N
HOOT	348.7	N	N	N
HOOTEN	18		Y	
HOPEKA	413.6	N	N	N
HOPPSWELL	93.6	N	Y	N
HORSECAMP	5.1	N	N	N
HORSETRACK	18.3		Y	
HOTSPRINGS	21.9		Y	
HOUGH	116	N	N	N
HOURLAND	11.1		Y	
HUEVI	415	Y	N	N
HUILEPASS	90.7	N	Y	N

(continued)

(continued)

Series Name	Area (km^2)	Endemic	Rare	Endangered
HULDERMAN	6.4		Y	
HUMBOLDT	288	N	N	N
HUMDUN	77	N	Y	N
HUNDRAW	335	N	N	N
HUNEWILL	115.7	N	N	N
HUNNTON	945	N	N	N
HUSSA	112	N	N	N
HUSSELL	19.8		Y	
HUSSMAN	7.2		Y	
HUTCHLEY	148	N	N	N
HUXLEY	12.2		Y	
HYLOC	111.2	N	N	N
HYMAS	138.8	N	N	N
HYPOINT	49.2	N	N	N
HYZEN	474.8	N	N	N
ICEBERG	36.2	N	Y	N
ICHBOD	8.2		Y	
IDLEWILD	9.2		Y	
IDWAY	109	N	N	N
IFTEEN	12.6		Y	
IGDELL	144.7	N	N	N
ILTON	2.7		Y	
INCY	54.7	N	Y	N
INDIAN CREEK	164	N	N	N
INDIANO	112	N	N	N
INDICOVE	52.5	N	Y	N
INMO	116.7	N	N	N
INPENDENCE	19.7	Y	Y	Y
INVILLE	2.4	Y	N	N
IRETEBA	15		Y	
IRON BLOSSOM	22.4		Y	
IRONGOLD	844	N	N	N
ISOLDE	1267	N	N	N
ISTER	218.4	N	N	N
ITCA	383.6	N	N	N
ITME	185	N	N	N
IVER	81.1	N	Y	N
IXIAN	28.9		Y	
IZAMATCH	329	N	N	N
IZAR	370.1	N	N	N
IZO	838	N	N	N
IZOD	124.7	N	N	N
JABU	0.6		Y	
JACARANDA	2		Y	
JACK CREEK	26.5		Y	
JACKMORE	11.9		Y	
JACKPOT	36.5		Y	
JACKROCK	36.2		Y	
JACRATZ	30.5		Y	
JAMCANVAR	2.2		Y	
JAMES CANYON	45	N	Y	N

(continued)

(continued)

Series Name	Area (km^2)	Endemic	Rare	Endangered
JAMESCANNY	3.3		Y	
JARAB	141.1	Y	N	N
JARBOE	18.7		Y	
JAYBEE	315	N	N	N
JEAN	85.5	N	Y	N
JENNESS	61.3	N	N	N
JENOR	21.9	N	N	N
JERICHO	559.4	N	N	N
JERVAL	487.4	N	N	N
JESAYNO	38		Y	
JESSE CAMP	94	Y	Y	Y
JETCOP	64.6	Y	Y	Y
JETMINE	8.9		Y	
JEVETS	56.2	Y	Y	Y
JIVAS	50.4	N	Y	N
JOB	3.4		Y	
JOBPEAK	96.3	N	Y	N
JOBSIS	1.2		Y	
JOEMAY	118.2	N	N	N
JOLAN	27.9		Y	
JONLAKE	30.2		Y	
JONNIC	34.8		Y	
JORGE	15.7	Y	N	N
JOTAVA	23		Y	
JOWEC	19.4		Y	
JUBILEE	30.9		Y	
JUMBLE	10.2		Y	
JUMBO	17		Y	
JUMBOPEAK	19.4		Y	
JUNG	326.4	N	N	N
JUNGO	41	N	Y	N
JURADO	17.9		Y	
JUVA	94.1	N	N	N
KAFFUR	14.4		Y	
KANACKEY	78.4	N	Y	N
KANESPRINGS	127.3	Y	N	N
KARLO	21.2		Y	
KARPP	49.6	N	Y	N
KASPAL	77	Y	Y	N
KATELANA	949.4	N	N	N
KAWICH	312	N	N	N
KAYO	53.4	N	Y	N
KAZUL	3.3		Y	
KEEFA	463.5	N	N	N
KELK	618.9	Y	N	N
KEMAN	79	N	N	N
KEYOLE	63.8	Y	Y	Y
KIDWELL	98.7	N	Y	N
KIMMERLING	22.1		Y	
KINGINGHAM	77	N	Y	N
KINGSBEACH	0.6		Y	

(continued)

(continued)

Series Name	Area (km^2)	Endemic	Rare	Endangered
KINGSRIVER	21.9		Y	
KIOTE	20		Y	
KIOUS	28.6		Y	
KITGRAM	20.4		Y	
KLECK	37.1		Y	
KLECKNER	159.2	N	N	N
KLEINBUSH	24.3		Y	
KNIESLEY	3.3		Y	
KNOB HILL	40	N	Y	N
KNOSS	23.2		Y	
KNOTT	55	N	Y	N
KOBEH	218	Y	N	N
KODAK	1.8		Y	
KODRA	43.8	N	Y	N
KOLCHECK	0.9		Y	
KOLDA	177.9	Y	N	N
KOONTZ	88.7	N	Y	N
KORTTY	14.6		Y	
KOYEN	1215.2	N	N	N
KOYNIK	21.8		Y	
KRAM	274.7	N	N	N
KRENKA	9.7		Y	
KREZA	9.7		Y	
KUMIVA	111.4	N	N	N
KUNZLER	982	N	N	N
KURSTAN	127.8	Y	N	N
KYLECANYON	7.9		Y	
KYLER	1341.2	N	N	N
KZIN	114.2	N	N	N
LABKEY	203.8	N	N	N
LABOU	14.3		Y	
LABSHAFT	105.5	N	N	N
LADYOFSNOW	14.5		Y	
LAHONTAN	126.9	N	N	N
LAKASH	21.5		Y	
LAMADRE	7.6		Y	
LAND	39.5		Y	
LANDCO	29.4		Y	
LANDERMEYER	13.8		Y	
LANFAIR	53.1	Y	Y	Y
LANGSTON	88	Y	Y	Y
LANIP	247.2	N	N	N
LAPED	86.2	N	N	N
LAPON	138.2	N	N	N
LAROSS	37.5		Y	
LARYAN	4.1	Y	Y	Y
LAS VEGAS	135.5	N	N	N
LASTCHANCE	136.4	Y	N	N
LASTONE	45.2	N	Y	N
LASTSUMMER	2.4		Y	
LATHROP	187.7	Y	N	N

(continued)

(continued)

Series Name	Area (km^2)	Endemic	Rare	Endangered
LAXAL	163.7	N	N	N
LAYVIEW	171.6	N	N	N
LAZAN	182.4	Y	N	N
LEALANDIC	37.4		Y	
LEECANYON	17		Y	
LEEVAN	46	N	Y	N
LEHMANDOW	5.3		Y	
LEMCAVE	6.1		Y	
LEMM	32.1		Y	
LEO	467	N	N	N
LERROW	175	N	N	N
LEVIATHAN	64.4	N	Y	N
LEWDLAC	17.2		Y	
LIDAN	16.8		Y	
LIEN	243.7	N	N	N
LIMEWASH	6.3	Y	Y	Y
LINCO	134.8	N	N	N
LINHART	29.1		Y	
LINKUP	298.3	N	N	N
LINOYER	700	N	N	N
LINPEAK	14.7		Y	
LINROSE	83.4	N	Y	N
LITTLEAILIE	257.8	N	N	N
LITTLESPRING	9.1		Y	
LOCANE	125	N	N	N
LODAR	13.2	N	N	N
LOFFTUS	0.1	Y	Y	Y
LOGAN	49.3	N	N	N
LOGRING	406.5	N	N	N
LOJET	375.2	N	N	N
LOMOINE	159.5	N	N	N
LONCAN	224.6	N	N	N
LONE	22.9		Y	
LONGCREEK	72.1	N	N	N
LONGDAY	5.3		Y	
LONGDIS	124.6	N	N	N
LONGJIM	146.2	N	N	N
LOOMER	190.6	N	N	N
LOOMIS	88	N	N	N
LOPWASH	30.9		Y	
LORAY	183	N	N	N
LOSTLEADER	3		Y	
LOUDERBACK	65.6	Y	Y	Y
LOVAMP	28.6		Y	
LOVEBOLDT	0.8		Y	
LOVELOCK	79.9	N	Y	N
LOWEMAR	55.7	Y	Y	Y
LOX	10.8		Y	
LUAP	41.2	Y	Y	Y
LUCKYSTRIKE	7.5		Y	
LUNDER	62	N	Y	N

(continued)

(continued)

Series Name	Area (km^2)	Endemic	Rare	Endangered
LUNING	315.5	N	N	N
LUPPINO	32.5		Y	
LUSET	91.1	N	Y	N
LYDA	276.8	N	N	N
LYKAL	5.3		Y	
LYNNBOW	21.1		Y	
LYRA	8.5		Y	
LYX	220.3	N	N	N
MACAREENO	13.6		Y	
MACKERLAKE	44.7	N	Y	N
MACKEY	0.5		N	N
MACKRANCH	4.5		Y	
MACKSCANYON	58.7	N	Y	N
MACNOT	114.5	N	N	N
MACYFLET	23.1		Y	
MADELINE	63.6	N	N	N
MADERBAK	37.7		Y	
MAGGIE	20.6		Y	
MAGHILLS	11.4		Y	
MAHALA	35.5		Y	
MAHOGEE	4.2		Y	
MAJORSPLACE	63.4	N	Y	N
MAJUBA	86.7	N	Y	N
MALMESA	41.7	N	Y	N
MALPAIS	213.4	N	N	N
MANARD	10		N	N
MANOGUE	81.6	N	Y	N
MAREPAS	10.7		Y	
MARLA	3.4		Y	
MARMOTHILL	0.2		Y	
MARYJANE	81.1	N	Y	N
MASCAMP	12		N	N
MASTLY	13.6		Y	
MATTIER	130.7	N	N	N
MAU	86	Y	Y	Y
MAYGAL	3.8		Y	
MAYNARD LAKE	8.6		Y	
MAZUMA	1601	N	N	N
MCCARRAN	298.2	Y	N	N
MCCLANAHAN	30.7		Y	
MCCLEARY	17.2		Y	
MCCONNEL	277	N	N	N
MCCULLOUGH	18		Y	
MCCUTCHEN	7.6		Y	
MCIVEY	714.8	N	N	N
MCTOM	0.3	N	Y	N
MCVEGAS	34.9		Y	
MCWATT	36		Y	
MEADVIEW	8.9	Y	N	N
MEDBURN	104.7	N	N	N
MEDLAVAL	4.4		Y	

(continued)

(continued)

Series Name	Area (km^2)	Endemic	Rare	Endangered
MEDVED	3.8		Y	
MEEKS	35		Y	
MEISS	10	Y	N	N
MELLOR	13.1	N	N	N
MELODY	6.7		Y	
MENBO	46.5	N	N	N
MESABASE	22.1		Y	
MEZZER	142.8	N	N	N
MICKEY	63.6	N	Y	N
MIDAS	49.2	Y	Y	Y
MIDRAW	162	N	N	N
MIGERN	15.9		Y	
MIJAY	12.4		Y	
MIJOYSEE	157.1	N	N	N
MILKIWAY	11.3		Y	
MILLAN	4.4		Y	
MILLERLUX	10.2		Y	
MIMENTOR	20.7		Y	
MINA	5.1		Y	
MINAT	32.6		Y	
MINDLEBAUGH	38.9		Y	
MINNEHA	43.7	N	Y	N
MINNYE	36.4		Y	
MINU	187.5	N	N	N
MIRKWOOD	130.6	N	N	N
MISAD	161.8	Y	N	N
MIZEL	60.3	N	Y	N
MIZPAH	10		Y	
MOAPA	52.7	N	Y	N
MOBL	34.5		Y	
MODEM	6.8		Y	
MOENTRIA	76.4	N	Y	N
MOHOCKEN	22.3		Y	
MOLION	98.2	N	Y	N
MONARCH	94.8	N	N	N
MONGER	16.4		Y	
MONTE CRISTO	46.7	N	Y	N
MOPANA	19		Y	
MORANCH	27.7		Y	
MORBENCH	18.6		Y	
MORMON MESA	604	N	N	N
MORMONWELL	40.1	N	Y	N
MORMOUNT	144.8	N	N	N
MORWEN	2.4		Y	
MOSIDA	64	N	N	N
MOSQUET	13.8		Y	
MOTOQUA	120.6	N	N	N
MOTTSKEL	7.7		Y	
MOTTSVILLE	78	N	N	N
MOUNTMCULL	109	N	N	N
MOUNTMUMMY	43.4	N	Y	N

(continued)

(continued)

Series Name	Area (km^2)	Endemic	Rare	Endangered
MOUNTROSE	10.5		Y	
MUIRAL	114.2	N	N	N
MULHOP	24.7		Y	
MUNI	107.3	N	N	N
MYSOL	49.3	Y	Y	Y
NADRA	100.7	Y	N	N
NALL	14.8		Y	
NAYE	49.7	Y	Y	Y
NAYFAN	27.1		Y	
NAYPED	47.1	Y	Y	Y
NEEDAHOE	0.4		Y	
NEEDLE PEAK	151.7	N	N	N
NELLSPRING	15.4		Y	
NEMICO	82.6	N	Y	N
NETTI	102.4	N	N	N
NEVADANILE	4.1		Y	
NEVADASH	56.8	N	Y	N
NEVADOR	130.3	N	N	N
NEVKA	14		Y	
NEVOYER	15.4		Y	
NEVTAH	15.8		Y	
NEVU	87.5	N	Y	N
NEWERA	141.6	N	N	N
NEWLANDS	69	N	N	N
NEWPASS	76.6	N	Y	N
NEWVIL	105.3	N	N	N
NIAVI	40.8	N	Y	N
NICANOR	13.3		Y	
NICKEL	71	N	N	N
NILESVAL	5.2		Y	
NINCH	2.3		Y	
NINEMILE	1534	N	N	N
NIPPENO	132.1	Y	N	N
NIPTON	228.8	N	N	N
NIRAC	63	N	N	N
NIRE	38.4		Y	
NITPAC	53.2	Y	Y	Y
NOBUCK	9.6		Y	
NODUR	11.6		Y	
NOFET	5.9		Y	
NOLENA	175.1	N	N	N
NOMARA	17.1		Y	
NOMAZU	8.5		Y	
NONAMEWASH	9.6		Y	
NOPAH	69.1	N	Y	N
NOPEG	13.2		Y	
NORFORK	40	N	Y	N
NORTHMORE	15.8		Y	
NOSAVVY	13.3		Y	
NOSKI	38.8		Y	
NOSLO	7.7		Y	

(continued)

(continued)

Series Name	Area (km^2)	Endemic	Rare	Endangered
NOSRAC	69	N	Y	N
NOTELLUMCREEK	49.1	N	Y	N
NOTUS	6		Y	
NOVACAN	10.9		Y	
NOWOY	74.1	N	Y	N
NOYSON	75.4	N	Y	N
NUAHS	19.1		Y	
NUC	52.2	Y	Y	Y
NUHELEN	71.6	N	Y	N
NUMANA	0.01	N	Y	N
NUPART	121.1	Y	N	N
NUPPER	13.1		Y	
NUTVAL	2.2		Y	
NUTZAN	2.1	N	Y	N
NUYOBE	226.3	N	N	N
NYAK	30.1		Y	
NYALA	17.9		Y	
NYSERVA	111.4	N	N	N
OBANION	15.5		Y	
OCALA	547	Y	N	N
OCASHE	11.4		Y	
OCUD	9.7		Y	
OEST	45	N	Y	N
OKAN	268.9	N	N	N
OKAYVIEW	42.9	N	Y	N
OLA	103.5	N	N	N
OLAC	148.5	N	N	N
OLD CAMP	1448	N	N	N
OLDSPAN	107.7	N	N	N
OLEMAN	94.9	Y	Y	Y
ONEIDAS	1.2		Y	
ONKEYO	178.7	Y	N	N
OPHIR	17.7		Y	
OPPIO	118.8	N	N	N
ORENEVA	0.2		N	N
ORICTO	408.6	Y	N	N
ORIZABA	97.3	Y	Y	Y
ORMSBY	6.8		Y	
OROVADA	983	N	N	N
ORPHANT	43.9	N	Y	N
ORR	78.5	N	Y	N
ORRUBO	0.3		Y	
ORUPA	44.5	Y	Y	Y
ORWASH	121	N	N	N
OSDITCH	23.5		Y	
OSOBB	175.9	N	N	N
OSOLL	29.7		Y	
OTOMO	13.9		Y	
OUPICO	82.4	N	Y	N
OUTERKIRK	39	N	N	N
OVERLAND	21.2		Y	

(continued)

(continued)

Series Name	Area (km^2)	Endemic	Rare	Endangered
OVERTON	10.8		Y	
OXCOREL	819.5	N	N	N
OXVALLEY	26.9		Y	
PACKER	173.9	N	N	N
PAGECREEK	27.6		Y	
PAHRANAGAT	28		Y	
PAHRANGE	30.1		Y	
PAHROC	164.3	N	N	N
PAHRUMP	55.2	N	Y	N
PALINOR	2508.2	N	N	N
PAMISON	39.5		Y	
PAMSDEL	20.5		Y	
PANACKER	16.7		Y	
PANLEE	217	Y	N	N
PANOR	10.7		Y	
PAPOOSE	130.8	N	N	N
PARADISE	5.3		Y	
PARANAT	193.6	N	N	N
PARISA	346.3	Y	N	N
PARRAN	383.1	Y	N	N
PATNA	247.8	Y	N	N
PATTANI	13.2		Y	
PATTER	34.2		Y	
PAYPOINT	83.6	Y	Y	Y
PEDOLI	135	N	N	N
PEEKO	488.9	N	N	N
PEEVYWELL	7.1		Y	
PEGLER	13		Y	
PELIC	30.7		Y	
PENELAS	168.9	N	N	N
PENGPONG	16		Y	
PENOYER	321	N	N	N
PEQUOP	52.5	N	Y	N
PERAZZO	84.3	N	N	N
PERLOR	24.8		Y	
PERN	61	Y	Y	Y
PERNOG	12		Y	
PERNTY	157.2	N	N	N
PERWASO	3.4		Y	
PERWICK	160.6	N	N	N
PESKAH	38		Y	
PETAN	43	Y	Y	Y
PETSPRING	57.5	N	Y	N
PHARO	107.7	Y	N	N
PHEEBS	6.4		Y	
PHING	54.5	N	Y	N
PHLISS	125.5	N	N	N
PIAR	50.8	N	Y	N
PIBLER	117.4	N	N	N
PICKUP	610	N	N	N
PIE CREEK	95.8	Y	N	N

(continued)

(continued)

Series Name	Area (km^2)	Endemic	Rare	Endangered
PILINE	4.5		Y	
PILTDOWN	67.7	N	Y	N
PIMOGRAN	2.1		Y	
PINENUT	27		Y	
PINEVAL	409.5	N	N	N
PINEZ	57.4	N	Y	N
PINTWATER	715.6	N	N	N
PINWHEELER	14.9		Y	
PIOCHE	154.2	N	N	N
PIRAPEAK	0.8		Y	
PIROUETTE	428.1	Y	N	N
PITTMAN	7.7		Y	
PIZENE	29.5		Y	
PLACERITOS	25.5		Y	
PLAYER	1.6		Y	
POCAN	40.8	N	Y	N
POCKER	16.1		Y	
POISONCREEK	59.2	N	N	N
POKERGAP	196.6	N	N	N
POLUM	22		Y	
PONYSPRING	57.1	N	Y	N
POOBAA	5		Y	
POOKALOO	1483	N	N	N
POORCAL	53.9	N	Y	N
PORRONE	24.4		Y	
PORTMOUNT	99	N	Y	N
POTOSI	261.9	N	N	N
POWLOW	38.2		Y	
POWMENT	107.4	Y	N	N
PREBLE	165.8	N	N	N
PREY	16.3		Y	
PRIDA	9.2		Y	
PRIDEEN	113.5	N	N	N
PRIMEAUX	39.4		Y	
PRISONEAR	27.2		Y	
PRUNIE	7.8		Y	
PUDDLE	3.8		Y	
PUELZMINE	8.5		Y	
PUETT	621	N	N	N
PUFFER	130.6	N	N	N
PULA	15		Y	
PULCAN	14.4		Y	
PULSIPHER	13.2		Y	
PUMEL	107	N	N	N
PUMPER	294	N	N	N
PUNCHBOWL	194.2	N	N	N
PUNG	45.1	N	Y	N
PUROB	496.4	N	N	N
PYRAT	533	N	N	N
QUARZ	506.7	N	N	N
QUIJINUMP	50.3	Y	Y	Y

(continued)

(continued)

Series Name	Area (km^2)	Endemic	Rare	Endangered
QUIMA	26		Y	
QUOMUS	34.5		Y	
QUOPANT	26.9		Y	
QWYNN	164.8	N	N	N
RAD	213.6	N	N	N
RADOL	141.5	N	N	N
RAGAMUFFIN	1.8		Y	
RAGLAN	193.4	Y	N	N
RAGNEL	65.2	N	Y	N
RAGTOWN	504	N	N	N
RAILCITY	42.3	N	Y	N
RAILROAD	109.6	N	N	N
RAMIRES	178.1	N	N	N
RAMSHEAD	4.9		Y	
RANGERTAFT	10.8		Y	
RAPADO	74.4	Y	Y	Y
RAPH	66.9	N	Y	N
RASILLE	104.1	N	N	N
RASTER	2.7		Y	
RATLEFLAT	59.6	N	Y	N
RATSOW	19.4		Y	
RATTO	87.5	N	N	N
RAVENDOG	108.9	N	N	N
RAVENELL	69.4	N	Y	N
RAVENSWOOD	151.7	N	N	N
RAWE	128.3	Y	N	N
REALMCOY	24.7		Y	
REBEL	344	N	N	N
REDFLAME	9.3		Y	
REDHOME	7.8		Y	
REDNEEDLE	3.3		Y	
REDNIK	309.6	N	N	N
REESE	48.9	N	N	N
REINA	9.1		Y	
RELLEY	127.2	N	N	N
RELUCTAN	661	N	N	N
RENO	147	N	N	N
REYWAT	383	N	N	N
REZAVE	94.2	Y	Y	Y
RICERT	463.7	N	N	N
RICHINDE	391.2	N	N	N
RIDIT	22.3		Y	
RIO KING	64	N	N	N
RIPCON	12.7		Y	
RIPLEY	0.5		N	N
RIPPO	0.9		Y	
RISLEY	169	N	N	N
RISUE	52.2	N	Y	N
RITO	10.5		Y	
RIVERBEND	115.3	N	N	N
RIXIE	43	Y	Y	Y

(continued)

(continued)

Series Name	Area (km²)	Endemic	Rare	Endangered
ROBBERSFIRE	40.5	Y	Y	Y
ROBSON	173.8	N	N	N
ROCA	770	N	N	N
ROCCONDA	469.8	N	N	N
ROCHPAH	231.5	N	N	N
ROCKABIN	25.1		Y	
RODAD	161.6	N	N	N
RODELL	5.3		Y	
RODEN	149.8	Y	N	N
RODIE	76.5	N	Y	N
RODOCK	64.7	Y	Y	Y
ROIC	287.8	N	N	N
ROLOC	13.6		Y	
ROSE CREEK	61	N	Y	N
ROSITAS	30.5	N	N	N
ROSNEY	114.4	N	N	N
ROTINOM	27.4		Y	
ROUETTE	130.3	N	N	N
ROVAL	57.7	N	N	N
ROWEL	49.7	N	Y	N
ROZARA	2.7		Y	
RUBICITY	13.4		Y	
RUBYHILL	221.7	Y	N	N
RUBYLAKE	16.6		Y	
RUGAR	16.6		Y	
RUHE	44.9	N	Y	N
RUMPAH	42.1	Y	Y	Y
RUNYON	13.6		Y	
RUSTIGATE	266.5	N	N	N
RUSTY	17.3		Y	
RUTAB	39.6		Y	
RYEPATCH	32.4		Y	
SADER	26.4		Y	
SAGOUSPE	76	N	Y	N
SALTAIR	41.4	N	N	N
SALTMOUNT	14.5		Y	
SALTYDOG	28.8		Y	
SAMOR	45.7	N	Y	N
SANDPAN	24.8		Y	
SANWELL	170.4	N	N	N
SARALEGUI	72.8	N	N	N
SARAPH	247	N	N	N
SATT	18.9		Y	
SAWMILLCAN	5.9		Y	
SAY	108.3	N	N	N
SCALFAR	64	N	Y	N
SCARINE	0.2	?	?	?
SCHADER	25.5		Y	
SCHAMP	24.4		Y	
SCHOER	21.2		Y	
SCHOOLMARM	397.7	N	N	N

(continued)

(continued)

Series Name	Area (km^2)	Endemic	Rare	Endangered
SCHURZ	23.1		Y	
SCHWALBE	24.9		Y	
SCOSSA	1.9	Y	Y	Y
SCOTTCAS	40	N	Y	N
SCRAPY	72.3	N	Y	N
SEAMAN	5.1		Y	
SEANNA	260	N	N	N
SEARCHLIGHT	57.7	N	Y	N
SECREPASS	4.2		Y	
SED	15.3		Y	
SEDSKED	4.3		Y	
SEGURA	403.5	N	N	N
SELBIT	34.9		Y	
SELTI	24.3		Y	
SERALIN	276.2	N	N	N
SETTLEDRAN	0.6		Y	
SETTLEMENT	96.9	N	Y	N
SETTLEMEYER	44.8	N	N	N
SEVAL	12.1		Y	
SEVENMILE	278.7	N	N	N
SEZNA	12.5		Y	
SHABLISS	1018	N	N	N
SHAFTER	22.8		Y	
SHAGNASTY	186.3	N	N	N
SHAKESPEARE	4.6		Y	
SHALAKE	94.5	N	Y	N
SHALCLEAV	399.2	N	N	N
SHALPEET	1.5		Y	
SHALPER	197	N	N	N
SHAMOCK	74.1	Y	Y	Y
SHANKBA	33.9		Y	
SHANTOWN	35.9		Y	
SHAWAVE	465.3	N	N	N
SHAYLA	17.7		Y	
SHEEGE	55.3	N	N	N
SHEEPPASS	42	N	Y	N
SHEEPRANGE	40.2	N	Y	N
SHEEPROCK	6.1		N	N
SHEFFIT	453.2	Y	N	N
SHIPLEY	78.6	N	Y	N
SHIVELY	61.4	N	Y	N
SHIVLUM	41.8	Y	Y	Y
SHOKEN	18		Y	
SHORIM	39.9	N	Y	N
SHORT CREEK	177.8	N	N	N
SHREE	77.3	N	Y	N
SHROE	32.4		Y	
SHUTTLE	30.7		Y	
SIBELIA	13.6	N	Y	N
SIEGEL	10.4		Y	
SIEROCLIFF	10.6		Y	

(continued)

(continued)

Series Name	Area (km^2)	Endemic	Rare	Endangered
SILENT	45	N	Y	N
SILVERADO	345.2	N	N	N
SILVERBOW	104.5	N	N	N
SIMON	95.8	N	N	N
SIMPARK	54.4	N	Y	N
SINGATSE	1022.2	N	N	N
SINGLETREE	32.5		Y	
SIRI	29.8		Y	
SISCAB	111	N	N	N
SKEDADDLE	170	N	N	N
SKELON	233.9	N	N	N
SKULL CREEK	48.4	Y	Y	Y
SKULLWAK	37.6		Y	
SKYHAVEN	19.6		Y	
SLATERY	69.3	N	Y	N
SLATTER	3.3		Y	
SLAVEN	255.2	N	N	N
SLAW	490.6	N	N	N
SLAWHA	71.7	N	Y	N
SLAWMASTER	14.6		Y	
SLIDYMTN	144	N	N	N
SLIPBACK	102.2	N	N	N
SLOCAVE	62.8	N	Y	N
SLOCKEY	195.4	N	N	N
SMALLCONE	27		Y	
SMAUG	94.5	N	N	N
SMEDLEY	130.4	Y	N	N
SNACREEK	3.5		Y	
SNAG	8.9		Y	
SNAPCAN	1.4		Y	
SNAPEED	4.2		Y	
SNAPP	310.5	Y	N	N
SNOPOC	29.3		Y	
SNOTOWN	30.5		Y	
SNOWMORE	157	N	N	N
SOAKPAK	18.6	N	Y	N
SOAR	141.6	N	N	N
SODA LAKE	35.3		Y	
SODASPRING	75.3	N	Y	N
SODHOUSE	165.8	N	N	N
SOFTSCRABBLE	696	N	N	N
SOJUR	303.7	N	N	N
SOLAK	145.7	N	N	N
SOMBRERO	7		Y	
SONDOA	436.4	N	N	N
SONOMA	406.5	N	N	N
SOOLAKE	23.1		Y	
SOONAHBE	41.8	N	Y	N
SOONAKER	6.6		Y	
SOUGHE	793.9	N	N	N
SOUTHCAMP	2.7		Y	

(continued)

(continued)

Series Name	Area (km²)	Endemic	Rare	Endangered
SPAGER	43	N	N	N
SPANEL	48.2	N	Y	N
SPASPREY	128.2	N	N	N
SPECTER	2.2		Y	
SPIKE	27.2		Y	
SPILOCK	13.8		Y	
SPINLIN	62.2	N	Y	N
SPRING	13.1	Y	Y	Y
SPRINGBAR	12		Y	
SPRINGMEADOW	8		Y	
SPRINGMEYER	57.3	N	N	N
SPRINGWARM	8.3		Y	
SQUAWTIP	253.5	N	N	N
SQUAWVAL	10.8		Y	
ST. THOMAS	875	N	N	N
STAMPEDE	623	Y	N	N
STARFLYER	167.1	N	N	N
STARGO	165.7	Y	N	N
STEEPSHRUB	18.8		Y	
STEERLAKE	4.2		Y	
STEPTOE	4		Y	
STEWVAL	2632.5	N	N	N
STILLWATER	57	N	Y	N
STINGDORN	121.9	N	N	N
STODICK	40.2	N	Y	N
STONELL	145.8	N	N	N
STONEWALL	4.6		Y	
STRAWBCREEK	3.3		Y	
STRAYCOW	21.3		Y	
STRICKLAND	13		Y	
STROZI	40.7	N	Y	N
STUCKY	37.6	N	Y	N
STUMBLE	676.6	N	N	N
SUAK	181.7	N	N	N
SUCCESSLOOP	52	N	Y	N
SUMINE	1313	N	N	N
SUMMERMUTE	301	Y	N	N
SUMYA	103.3	Y	N	N
SUNDOWN	91.4	N	Y	N
SUNROCK	271	N	N	N
SUP	4		Y	
SURGEM	11.3		Y	
SURPASS	26		Y	
SURPRISE	12.6	N	Y	N
SUSIE CREEK	127.4	Y	N	N
SUTCLIFF	33.8		Y	
SUTRO	10.8		Y	
SWEETSPRING	16.3		Y	
SWINGLER	243	N	N	N
SWISBOB	6		Y	
SWOPE	8.9		Y	

(continued)

(continued)

Series Name	Area (km^2)	Endemic	Rare	Endangered
SYCOMAT	340.8	Y	N	N
SYLVANIAM	15.6		Y	
TAGUM	6.1		Y	
TAHOE	5.8	Y	Y	Y
TAHOMA	0.5	N	Y	N
TAHQUATS	0.01	N	Y	N
TALLAC	19.8	N	N	N
TANAZZA	34.5	Y	Y	Y
TANOB	13.5		Y	
TARLOC	7.7		Y	
TARNACH	340	Y	N	N
TAYLOR CREEK	29.6		Y	
TECOMAR	1166.1	N	N	N
TECOPA	91.4	N	Y	N
TEEBAR	39.7	Y	Y	Y
TEEBONE	41.2	Y	Y	Y
TEGURO	172	N	N	N
TEJABE	94.5	N	Y	N
TEMAN	28.6		Y	
TEMO	29.6		Y	
TENABO	456.1	Y	N	N
TENPIN	21.1		Y	
TENVORRD	18.2		Y	
TENWELL	158.9	N	N	N
TERCA	137.7	N	N	N
TERLCO	351.9	N	N	N
TERT	41.9	N	Y	N
TESSFIVE	29.7		Y	
THEON	1462.7	N	N	N
THERIOT	294	N	N	N
THESISTERS	25.8		Y	
THIEFRIDGE	6.4		Y	
THIKE	20		Y	
THREEDOGS	22.6		Y	
THREELAKES	237.3	N	N	N
THREESEE	125.6	Y	N	N
THULEPAH	19.2		Y	
THUNDERBIRD	117	N	N	N
THWOOP	8.5		Y	
TICA	18		Y	
TICINO	46.4	N	Y	N
TICKAPOO	132.6	Y	N	N
TIERNEY	0.05	N	Y	N
TIMMERCREK	15.1		Y	
TIMPAHUTE	5.4		Y	
TIMPER	97.8	N	Y	N
TIMPIE	221	N	N	N
TINPAN	74.6	Y	Y	Y
TIPNAT	23.9		Y	
TIPPERARY	41.5	N	N	N
TIPPIPAH	4		Y	

(continued)

(continued)

Series Name	Area (km²)	Endemic	Rare	Endangered
TITIACK	36.9		Y	
TOANO	176.8	N	N	N
TOBA	18.9		Y	
TOBLER	4		Y	
TOCAN	54.6	N	Y	N
TOEJA	78.2	N	Y	N
TOEJOM	3.4		Y	
TOEM	1.3	Y	N	N
TOGNONI	53.6	N	Y	N
TOIYABE	51.2	Y	N	N
TOKOPER	203.9	N	N	N
TOLICHA	11.3		Y	
TOLL	16.4		Y	
TOMEL	169.4	N	N	N
TOMERA	157.6	N	N	N
TOMSHERRY	32		Y	
TONEY	19.8		Y	
TONKIN	13.2		Y	
TONOPAH	427	N	N	N
TOOELE	45	N	N	N
TOOPITS	43.6	N	Y	N
TOPEKI	38.2		Y	
TOQUOP	52	?	Y	?
TORNILLO	4.8		N	N
TORRO	134.6	N	N	N
TOSP	40.7	N	Y	N
TOSSER	122	N	N	N
TOULON	207	N	N	N
TRACTUFF	20.5		Y	
TRAILAMP	42.1	N	Y	N
TRALEY	25.7		Y	
TREADWELL	35.5		Y	
TRESED	75.7	Y	Y	Y
TRID	13.7		Y	
TRINIDAD	25.8		Y	
TRISTAN	81	N	Y	N
TROCKEN	1236.3	N	N	N
TROSI	7.1		Y	
TROUGHS	12.2	N	N	N
TROUGHSPRING	4.7		Y	
TRUCKEE	31.6		Y	
TRUHOY	54.1	N	Y	N
TRUNK	343.4	N	N	N
TRUVAR	10		Y	
TUFFMAN	34.8		Y	
TUFFO	151	N	N	N
TULASE	655	N	N	N
TULECAN	31.8		Y	
TULEDAD	48.3	Y	Y	Y
TUMARION	2.5	N	N	N
TUMTUM	143	N	N	N

(continued)

(continued)

Series Name	Area (km^2)	Endemic	Rare	Endangered
TUNNISON	162	Y	N	N
TURBA	462.6	Y	N	N
TURRIA	21.3		Y	
TURUPAH	33.9		Y	
TUSEL	418	N	N	N
TUSK	106	N	N	N
TUSTELL	99.2	N	Y	N
TUSUNE	56	N	Y	N
TWEB A	51.4	N	Y	N
TWEENER	161.8	N	N	N
TYBO	267.2	N	N	N
UANA	49.2	N	Y	N
UAWDA	2.4		Y	
UBEHEBE	134.7	N	N	N
UCOPIA	5.3		Y	
UDELOPE	28.8		Y	
UHALDI	50.6	N	Y	N
ULTRA	7.4		Y	
ULTRAMONT	12.9		Y	
UMBERLAND	334.5	N	N	N
UMIL	221.5	N	N	N
UMPA	0.2		N	N
UNGENE	46.2	N	Y	N
UNIONVILLE	32.7		Y	
UNIUS	51.6	N	Y	N
UNIVEGA	162.6	N	N	N
UNSEL	1849.6	N	N	N
UPATAD	268.3	N	N	N
UPDIKE	117.8	N	N	N
UPPERLINE	90.6	N	Y	N
UPSEL	6.3		Y	
UPSPRING	127	N	N	N
UPSTEER	11.3		Y	
UPVILLE	47.9	N	Y	N
URIPNES	167.9	N	N	N
URMAFOT	767.3	N	N	N
URSINE	1487.7	N	N	N
URTAH	5.7		Y	
USTIDUR	36		Y	
UVADA	75.7	N	N	N
UWELL	94.9	N	Y	N
VACE	55.2	N	Y	N
VADAHO	69	N	Y	N
VALATIER	5.5		Y	
VALCREST	31.1		Y	
VALMY	264.5	N	N	N
VAMP	13.8		Y	
VANWYPER	589	N	N	N
VARWASH	74.8	N	Y	N
VEET	732.3	N	N	N
VEGASTORM	18.1		Y	
VERDICO	57	N	Y	N

(continued)

(continued)

Series Name	Area (km^2)	Endemic	Rare	Endangered
VETA	141	N	N	N
VETAGRANDE	25		Y	
VETASH	2.1		Y	
VICEE	12.8		Y	
VIGUS	89.8	N	Y	N
VIL	0.1	N	Y	N
VINDICATOR	100.1	N	Y	N
VININI	134.5	N	N	N
VINSAD	14.2		Y	
VIRGIN PEAK	38.6		Y	
VIRGIN RIVER	10.7		Y	
VITALE	88.8	N	N	N
VIUM	11.9		Y	
VOLTAIRE	31		Y	
VYCKYL	25.4		Y	
VYLACH	31.9		Y	
WABUSKA	67.1	N	Y	N
WACA	0.5	Y	N	N
WAGORE	6.1		Y	
WAHGUYHE	100.5	N	N	N
WAKANSAPA	185.9	Y	N	N
WALA	220.3	N	N	N
WALKERIVER	12.5		Y	
WALTI	308.6	N	N	N
WAMBOLT	6.5		Y	
WAMP	91.8	N	Y	N
WANOMIE	23.3		Y	
WARDBAY	386.8	N	N	N
WARDCREEK	0.01	Y	Y	Y
WARDENOT	1004.8	N	N	N
WASHOE	32.9	N	N	N
WASHOVER	0.9		Y	
WASPO	29		Y	
WASSIT	172	N	N	N
WATAH	1	N	Y	N
WATCHABOB	4.6		Y	
WATOOPAH	555.5	N	N	N
WATSONLAKE	1.3		Y	
WAYHIGH	2		Y	
WECHECH	670.8	N	N	N
WEDEKIND	60.8	N	Y	N
WEDERTZ	106.3	N	N	N
WEDLAR	28.8		Y	
WEENA	219.9	N	N	N
WEEPAH	67.7	N	Y	N
WEEZWEED	23.2		Y	
WEIGLE	19.2		Y	
WEIMER	27.8		Y	
WEISER	1205	N	N	N
WEISHAUPT	11		Y	
WELCH	517	N	N	N
WELLINGTON	52.9	N	Y	N
WELLSED	39.2		Y	

(continued)

(continued)

Series Name	Area (km^2)	Endemic	Rare	Endangered
WELRING	18.2		N	N
WELSUM	39.8	Y	Y	Y
WENDANE	793.6	Y	N	N
WERELD	17.7		Y	
WESFIL	142.8	N	N	N
WESO	524.1	N	N	N
WESTBUTTE	66.1	N	N	N
WETVIT	9.7		Y	
WHEELERPASS	8.5		Y	
WHEELERPEK	9.9		Y	
WHEELERWELL	49.4	N	Y	N
WHICHMAN	19.5		Y	
WHILPHANG	50.7	N	Y	N
WHIRLO	467.3	N	N	N
WHITEBASIN	19	N	Y	N
WHITEPEAK	13.6		Y	
WHITMIRE	1.7		Y	
WHITTELL	6.2		Y	
WHOLAN	386	Y	N	N
WICKAHONEY	8	Y	N	N
WICUP	27.4		Y	
WIELAND	1086.8	N	N	N
WIFFO	129.8	N	N	N
WILE	18.1		Y	
WILLHILL	3.1		N	N
WILLYNAT	8.7		Y	
WILSOR	9.8		Y	
WILST	20.5		Y	
WILTOP	14		Y	
WINADA	12.9		Y	
WINDWASH	10.2		Y	
WINKLO	183.8	N	N	N
WINTERMUTE	590.1	N	N	N
WINU	66.8	N	N	N
WINZ	8.9		N	N
WISKAN	199.4	N	N	N
WISKIFLAT	50.1	N	Y	N
WITEFELS	23		Y	
WODA	17.5		Y	
WODAVAR	37.7		Y	
WOODROW	0.8	N	N	N
WOODSPRING	16.1		Y	
WOOFUS	97.5	N	Y	N
WOOLSEY	8		Y	
WRANGO	110	N	N	N
WREDAH	29.9		Y	
WRENZA	60.8	N	Y	N
WYLO	638	N	N	N
WYVA	215.9	N	N	N
XERTA	7.9		Y	
XERXES	112.8	Y	N	N
XICA	66.5	Y	Y	Y

(continued)

(continued)

Series Name	Area (km^2)	Endemic	Rare	Endangered
XINE	164.7	N	N	N
XIPE	22.7		Y	
XMAN	113.2	N	N	N
YELBRICK	119.5	N	N	N
YELLOWHILLS	25.9		Y	
YERINGTON	98.4	N	Y	N
YERMO	967	N	N	N
YIPOR	114.3	N	N	N
YOBE	196	N	N	N
YODY	465.3	N	N	N
YOMBA	484.7	Y	N	N
YOTES	106.1	N	N	N
YOUNGSTON	33.2	N	N	N
YUKO	301.8	N	N	N
YURM	257.4	N	N	N
ZABA	50.8	N	Y	N
ZADVAR	851.8	N	N	N
ZAFOD	140.5	Y	N	N
ZAIDY	115.9	N	N	N
ZALDA	56.1	N	Y	N
ZAPA	155.7	N	N	N
ZAQUA	274.4	N	N	N
ZARARK	157	N	N	N
ZARK	38.7		Y	
ZEHEME	795.8	N	N	N
ZEPHAN	48.6	N	Y	N
ZEPHYRCOVE	1.5		Y	
ZERK	178.2	N	N	N
ZEVADEZ	468	N	N	N
ZIBATE	242.6	N	N	N
ZIMBOB	723.8	N	N	N
ZIMWALA	53	N	Y	N
ZINEB	134.6	N	N	N
ZIRAM	11.9		Y	
ZOATE	20.7		Y	
ZODA	100.7	N	N	N
ZOESTA	80.5	N	Y	N
ZORRAVISTA	133	N	N	N
ZORROMOUNT	21.2		Y	
ZYMANS	182.1	N	N	N
ZYPLAR	13.2		Y	
ZYZZI	29.6		Y	

Bibliography

Bockheim JG (2014) Soil Geography of the USA: a Diagnostic-Horizon Approach. Springer, NY, p 320

Buol, S.W., Hole, F.D., McCracken, R.J. 1988. Soil Genesis and Classification. 3rd edit. Iowa State Univ. Press, Ames, IA

Eardley AJ, Shuey RT, Gvosdetsky V, Nash WP, Dane Picard M, Grey DC, Kukla GJ (1973) Lake cycles in the Bonneville Basin. Utah. Geol. Soc. Am. Bull. 84:211–216

Gile LH, Grossman RB (1979) The Desert Project soil monograph: Soils and landscapes of a desert region astride the Rio Grande Valley near Las Cruces, New Mexico. U.S. Department of Agriculture, Soil Conservation Service, Lincoln, NE

Kirby, M.E., Heusser, L., Scholz, C., Ramezan, R., 8 co-authors. 2018. A late Wisconsin (32-10k cal a BP) history of pluvials, droughts and vegetation in the Pacific south-west Unites States (Lake Elsinore, CA). J. Quat. Sci. 33:238-254

Schmidlin TW, Peterson FF, Gifford RO (1983) Soil temperature regimes in Nevada. Soil Sci. Soc. Am. J. 47:977–982

Thompson RS, Benson L, Hattori EM (1986) A revised chronology for the last Pleistocene lake cycle in the central Lahontan Basin. Quat. Res. 25:1–9

P. W. Blackburn et al., *The Soils of Nevada*, World Soils Book Series,
https://doi.org/10.1007/978-3-030-53157-7

Index

A
Albic horizon, 4, 124
Alfisols, 7, 12, 37, 60, 127–129, 131–133, 159–163, 168, 170, 171, 173, 177–179, 181–184, 195, 204, 207, 208, 210, 217–219, 221, 225, 226, 228–230, 237, 239, 245, 250, 251, 253, 264, 267
Alluvial deposits, 46
Alluvial fan, flat, 46, 49–54, 56, 58, 75, 78, 81, 82, 88, 92, 135, 150
Alpine zone, 28
Andisolization, 133, 135
Anthropic epipedon, 2, 59
Aquic soil moisture class, 119
Argicryolls, 51, 53, 77, 94, 95, 105, 109, 133, 135, 150
Argidurids, 38–40, 43, 45–47, 51–55, 60, 61, 74, 77, 79, 84–86, 96, 97, 133, 149, 150
Argillic horizon, 4, 7, 11, 59, 60, 78, 81, 84, 89, 91, 94–96, 109, 128, 131, 132
Argilluviation, 98, 109, 128, 133, 135, 150
Argiustolls, 77, 80, 95, 96, 109, 133, 150
Argixerolls, 38–40, 42, 45–47, 51–55, 58–60, 74, 77–79, 81, 83, 89, 96, 105, 109, 133, 135, 149, 150
Aridic soil moisture class, 128
Aridisols, 7, 11, 12, 32, 60, 61, 74, 97, 98, 103, 133, 134, 150, 152–175, 177–180, 182, 184, 186–188, 190, 192, 194–197, 199–221, 224–240, 242–261, 263–268
Axial-stream flood plains, 49, 58

B
Ballena, 49, 50, 57, 60, 81, 83
Basin and Range, 7, 9, 22, 34, 37–39, 42–46, 49, 51, 54, 56–58, 75, 89, 90, 94, 105, 149, 150
Benchmark soil, 137
Bolson, 26, 49–51, 53, 54, 56, 58, 134, 135
Bottlebrush squirreltail, 77, 248, 252, 262
Boundary Peak, 1, 22, 39, 42, 45
Bristlecone pine, 22, 23, 28, 37, 249, 254, 259, 261

C
Calciargids, 56, 58, 80, 95, 96, 98, 134
Calcic horizon, 4, 60, 61, 81, 82, 84, 85, 87, 90, 91, 95, 98, 120
Calcicryolls, 56, 57, 77, 80, 95, 109, 224, 229, 238, 240, 242, 246, 251, 267, 268
Calcification, 133–135, 150
Calcixerolls, 54–57, 77, 79, 90–92, 96, 105, 109, 134
Cambic horizon, 4, 7, 75, 82, 83, 88, 98, 103, 105, 109, 111, 113, 119, 120, 128, 132, 134
Cambisolization, 98, 109, 133–135, 150
Carson Basin, Sink, Valley, 7, 14, 37, 39, 41, 54, 94, 105
Cation-Exchange Capacity (CEC), 87–90, 92–95, 98, 105, 111, 119
CEC activity class, 87–90, 92–95, 98, 105, 111, 119
Colluvium deposits, 25, 34, 35, 57, 149
Colorado River, 22, 32, 46, 77, 142
Columbia Plateau, 22, 33
Creosote bush, 21, 24, 75, 77, 86, 243, 244, 252, 257, 261, 263, 265, 267
Cryic soil temperature class, 56, 95, 128
Cryrendolls, 56, 57, 77, 80, 95, 96, 229, 251, 252
Cumulic subgroup, 59, 61

D
Diagnostic horizons, 7, 10, 11, 59, 98, 130, 133, 223
Durinodes, 13, 27, 61, 74, 78, 83, 133
Duripan, 7, 11, 13, 26, 58–61, 70, 74, 75, 78, 83–85, 88, 89, 91, 92, 98, 103, 120, 128, 133, 138, 150, 224, 232
Durixerolls, 39, 77, 79, 88–90, 96, 105, 109, 133

E
Early Wisconsin, 33, 78, 84, 85, 88
Endangered soil, 137, 138
Endemic soil, 137, 138, 150
Endoaquolls, 40, 42, 52, 54, 58, 77, 79, 90, 91, 96, 105
Entisols, 7, 12, 27, 35, 60, 61, 74, 111, 113, 117, 120, 125, 134, 150, 152–185, 187, 189–195, 197–207, 209–211, 213, 219, 224–234, 237, 238, 242, 244–249, 254–260, 266–268
Eolian deposits, 26, 46
Epipedons, 2, 7, 10, 59, 84, 128, 150

F
Fallon-Lovelock Area, 7, 37, 39, 42, 43, 54, 55, 92
Fan apron, 49, 58
Fan piedmont, 49–58, 78, 83, 95
Fan remnant, 49–57, 60, 78, 81–86, 91, 95, 105, 133, 135, 150
Fan skirt, 49–58, 82
Frigid soil temperature class, 90, 91, 96, 105, 109, 119, 125, 127

P. W. Blackburn et al., *The Soils of Nevada*, World Soils Book Series,
https://doi.org/10.1007/978-3-030-53157-7

G

Gleization, 109, 113, 120, 133–135, 150
Greasewood, 20, 24, 25, 60, 75, 77, 78, 85, 87, 89, 91, 92, 119, 127, 242–250, 252–268
Great Basin, 1, 21, 23–28, 31, 37, 49, 51, 54, 56, 61, 81, 83, 91–93, 105, 128, 131, 133, 149
Great Salt Lake, 7, 20, 21, 37, 39, 42–44, 56, 97
Gypsic horizon, 4, 98
Gypsification, 98, 133, 135, 150

H

Halaquepts, 38, 42, 47, 52–54, 58, 77, 79, 89, 90, 96, 119, 120, 123, 133, 135, 149, 150
Haplargids, 29, 38–40, 42, 45–47, 51–56, 58–60, 74, 77–79, 82, 96–98, 134, 149, 150
Haplocalcids, 29, 39, 43, 45–47, 54–57, 60, 74, 77, 79, 81, 82, 84, 96, 97, 133, 134, 149, 150
Haplocambids, 38–40, 42, 43, 45–47, 52–57, 60, 74, 77, 79, 82, 85, 96–98, 134, 149, 150
Haplocryolls, 39, 51, 53, 55, 77, 80, 92, 93, 96, 105, 109, 150
Haplodurids, 38, 39, 43, 45–47, 52, 54, 56, 57, 60, 74, 77, 79, 83, 84, 96, 97, 134, 149, 150
Haploxeralfs, 37, 39, 77, 80, 96, 128, 129, 132, 152
Haploxerolls, 37, 39, 40, 43, 52, 54–57, 77, 79, 88, 89, 96, 105, 109, 134, 135, 150, 152
Histic epipedon, 2, 4, 59
Histosols, 7, 135
Holocene, 27, 34, 35, 75, 78, 82, 87–90, 92, 149
Humboldt Area, 7, 37–39, 51, 53, 54, 91
Humification, 113, 133
Hyperthermic soil temperature class, 95

I

Idaho fescue, 77, 88, 94, 96, 133, 254
Inceptisols, 7, 12, 60, 75, 90, 119, 120, 123, 125, 133, 134, 154–157, 159, 160, 162, 163, 165, 168–171, 174, 179, 181, 183, 184, 186, 188, 189, 194–198, 201, 202, 204, 205, 207, 208, 210, 214, 217, 219, 220, 224, 225, 227, 228, 230–233, 235–239, 243, 245, 246, 249, 254–258, 260–262, 264, 267, 268
Indian ricegrass, 24, 77, 78, 95, 133, 243
Inset fan, 49–51, 58, 82, 243
Inter-mountain plateau, 49, 77

L

Lacustrine deposits, 38, 42, 43, 45, 56, 57, 75, 77, 149
Lake Bonneville, 20, 32, 43, 56
Lake Lahontan, 13, 20, 22, 26, 27, 32, 111, 117
Lake Mead, 46, 77, 87, 142
Lake Tahoe, 37, 38, 77, 119, 122, 131, 146
Late Wisconsin, 33, 78, 81, 83, 85, 94, 95
Lithic contact, 61, 74, 138
Lithic subgroup, 74, 138
Lodgepole pine, 37, 77, 96, 127, 250

M

Major Land Resource Areas (MLRAs), 2, 7, 11, 22, 37, 39, 46, 50, 57, 75, 78, 79, 82, 84, 85, 87–97, 103, 105, 111, 117, 119, 125, 127, 135, 137, 149, 152, 176, 208
Malheur High Plateau, 7, 37, 39, 40, 51, 53, 127, 132
Melanization, 109, 120, 133, 135, 150
Mesic soil temperature class, 78, 84, 92, 95, 111
Mesozoic, 22
Middle Wisconsin, 33, 81, 91
Mojave Desert, 7, 17, 25, 37, 39, 46, 57, 75, 78, 82, 94, 95, 135
Mollic epipedon, 4, 59, 70, 78, 81, 88–95, 105, 109, 113, 128, 150
Mollisols, 7, 12, 59–61, 74, 90, 105, 109, 133, 134, 150, 152–221, 224–239, 242–246, 248–260, 262–265, 268
Montane forest, 20, 22, 23, 27
Mormon Mesa, 30, 35, 86, 87, 97, 98, 191, 232, 257, 270, 293
Mountain big sagebrush, 22, 23, 26, 77, 93–95
Mountain mahogany, 22, 27, 77, 81, 91, 95, 97, 144
Mountain Shrubland, 9
Mountain-valley fan, 49

N

Natrargids, 27, 38, 39, 42, 45, 47, 52, 53, 55, 56, 60, 77, 79, 85, 86, 96–98, 134, 135, 149, 152, 154–158, 160, 162, 164, 166–169, 172–175, 180, 184, 185, 187, 190
Natric horizon, 4, 13, 60, 85, 86, 91, 92, 98, 113, 135, 150
Natridurids, 52, 54–56, 60, 77, 79, 91, 92, 96, 133, 135, 166, 167, 170, 172, 173, 175, 176, 184, 190–193, 198, 211, 224, 227, 228, 233, 234, 237, 242, 247, 249, 250, 257, 259, 264

O

Ochric epipedon, 4, 7, 75, 78, 82–90, 92, 95, 96, 98, 103, 111–113, 117, 119, 120, 125, 128, 131, 132
Orovada soil series, 15, 83, 85, 97, 149
Owyhee High Plateau, 7, 17, 37–39, 41, 54, 55, 82, 84

P

Pachic subgroup, 59, 93
Paleargids, 29, 59, 77, 96, 98, 152, 157, 159, 173, 174, 183, 189, 191, 198, 199, 208, 209, 211, 212, 216, 221, 224, 240, 242, 268
Paludization, 133, 135
Paralithic contact, 75, 78, 82, 84, 88, 91, 93, 94
Pastureland, 143
Pedodiversity, 59, 67, 138, 149, 150
Petrocalcic horizon, 4, 13, 27, 35, 58, 60, 70, 81, 87, 98, 134, 138, 150
Petrocalcids, 57, 60, 77, 79, 86, 87, 97, 150, 152, 155, 160, 163, 165, 167, 168, 172, 175, 179, 181, 186, 188, 191, 192, 197–200, 203, 204, 206, 217, 238, 239, 243, 246–248, 250, 252, 255, 257, 259, 260, 266–268
Petrogypsic horizon, 7, 135
Pinenut Mountains, 39, 41
Pinyon-juniper woodland, 20, 21, 33, 149
Pluvial lake, 9, 13, 20–22, 26, 27, 30, 32, 43, 56, 111, 117, 149
Ponderosa pine, 22, 23, 27, 37, 96
Protozoic, 22

Q

Quaternary, 22, 30, 34, 46, 57, 149

R

Rangeland, 15, 78, 82, 84–86, 109, 139, 140, 146, 150
Rare soil, 137
Residuum, 25, 30, 34, 35, 37–39, 42, 50–58, 64, 75, 77–81, 84, 88, 90, 94–96, 105, 109, 111, 114, 119, 127, 149
Ruby Mountains, 9, 22, 26, 38, 41

S
Sagebrush grassland, 20, 21, 25, 33, 149
Salic horizon, 7, 15, 59, 78, 97, 98, 120, 135, 150
Salinization, 98, 120, 133, 135, 150
Sand desert, 150
Sand Mountains, 22, 31
Sand sheet, 26, 32, 47, 87
Semi-bolson, 26, 49–52, 54, 55, 58, 135
Shadscale, 24, 25, 60, 75, 77–79, 82, 83, 85, 91, 92, 95, 242–261, 263–268
Shallow soil, 138
Sierra Nevada Mountains, 7, 26, 37, 39, 50, 52, 97, 103, 119, 125, 127, 132, 149
Silicification, 27, 98, 113, 133–135, 150
Soil
  definition, 1, 2, 10, 87, 97, 120, 128
  family, 59, 61, 67, 98, 109, 111, 137, 138, 152, 176, 208
  great groups, 2, 7, 12, 33, 34, 38, 39, 42–46, 59–61, 64, 65, 67, 68, 70, 74, 75, 77–79, 81–90, 92–97, 105, 111, 119, 133, 134, 149, 150
  order, 2, 7, 11, 12, 37, 39, 59–61, 65, 67, 74, 75, 97, 98, 103, 105, 109, 111, 117, 120, 125, 132–135, 149, 150, 152, 176, 208
  series, 2, 7, 10, 12, 13, 15, 16, 19, 22, 25, 28, 30, 35, 37–40, 42, 43, 45, 46, 50, 51, 54–61, 64–70, 74
  subgroup, 2, 12, 59, 61, 67, 74, 82, 93, 128, 133–135, 138, 149–152, 176, 208
  suborder, 2, 7, 11, 12, 39, 59–61, 65, 67, 74, 75, 78, 81–98, 103, 105, 109, 111–114, 117, 119, 120, 122, 125, 127, 128, 132–135, 137, 138, 149, 150, 223, 273
Solonization, 133, 135, 150
Spring Mountains, 17, 26, 27
Subalpine forest, 20, 22, 33, 97, 149

T
Thermic soil temperature class, 84, 85, 95
Torric soil moisture class, 134
Torrifluvents, 56, 77, 80, 92, 93, 96, 111, 133, 153, 159, 162, 164, 172, 174, 179, 183, 188, 193, 198, 208, 221, 224, 226, 228, 230, 231, 233, 236, 239, 242, 247, 250, 252, 254, 257, 262, 263, 268
Torriorthents, 38, 39, 42–47, 54–58, 60, 74, 75, 77, 79, 81, 96, 111, 117, 133, 134, 149, 150, 152, 154–162, 164–168, 170, 171
Torripsamments, 29, 38, 47, 56, 77, 79, 87, 88, 96, 111, 150, 173–175, 178, 202, 209

U
Umbric epipedon, 4, 59, 119–122, 125, 128

V
Vertisols, 7, 12, 60, 127–129, 131, 132, 134, 160, 162, 169, 176, 180, 183, 189, 196, 198, 203, 213, 217, 225, 226, 230, 232, 238, 245, 246, 252, 256, 265
Vertization, 133–135, 150
Volcanic ash, 25, 26, 30, 35, 38, 39, 54, 55, 60, 82, 83, 91, 92, 94, 96, 97, 127, 128, 134, 149

W
Wheeler Peak, 26, 28, 31
Wyoming big sagebrush, 21, 23, 25, 26, 77, 82, 95

X
Xeric soil moisture class, 95, 96, 105